T0122622

Smart Innovation, Systems and Technologies

Volume 50

About this Series

The Smart Innovation, Systems and Technologies book series encompasses the topics of knowledge, intelligence, innovation and sustainability. The aim of the series is to make available a platform for the publication of books on all aspects of single and multi-disciplinary research on these themes in order to make the latest results available in a readily-accessible form. Volumes on interdisciplinary research combining two or more of these areas is particularly sought.

The series covers systems and paradigms that employ knowledge and intelligence in a broad sense. Its scope is systems having embedded knowledge and intelligence, which may be applied to the solution of world problems in industry, the environment and the community. It also focusses on the knowledge-transfer methodologies and innovation strategies employed to make this happen effectively. The combination of intelligent systems tools and a broad range of applications introduces a need for a synergy of disciplines from science, technology, business and the humanities. The series will include conference proceedings, edited collections, monographs, handbooks, reference books, and other relevant types of book in areas of science and technology where smart systems and technologies can offer innovative solutions.

High quality content is an essential feature for all book proposals accepted for the series. It is expected that editors of all accepted volumes will ensure that contributions are subjected to an appropriate level of reviewing process and adhere to KES quality principles.

More information about this series at http://www.springer.com/series/8767

Suresh Chandra Satapathy · Swagatam Das
Editors

Proceedings of First International Conference on Information and Communication Technology for Intelligent Systems: Volume 1

Editors
Suresh Chandra Satapathy
Department of Computer Science
and Engineering
Anil Neerukonda Institute of Technology
and Sciences
Visakhapatnam
India

Swagatam Das
Indian Statistical Institute
Jadavpur University
Kolkata
India

ISSN 2190-3018 ISSN 2190-3026 (electronic)
Smart Innovation, Systems and Technologies
ISBN 978-3-319-80921-2 ISBN 978-3-319-30933-0 (eBook)
DOI 10.1007/978-3-319-30933-0

Softcover reprint of the hardcover 1st edition 2016

Printed on acid-free paper

This Springer imprint is published by Springer Nature
The registered company is Springer International Publishing AG Switzerland

Preface

This SIST volume contains of the papers presented at the ICTIS 2015: International Conference on Information and Communication Technology for Intelligent Systems. The conference was held during November 28–29, 2015, Ahmedabad, India and organized communally by Venus International College of Technology, Association of Computer Machinery, Ahmedabad Chapter and supported by Computer Society of India Division IV—Communication and Division V—Education and Research. It targeted state of the art as well as emerging topics pertaining to ICT and effective strategies for its implementation in engineering and intelligent applications. The objective of this international conference is to provide opportunities for the researchers, academicians, industry persons, and students to interact and exchange ideas, experience and expertise in the current trend and strategies of information and communication technologies. Besides this, participants were enlightened about vast avenues, current and emerging technological developments in the field of ICT in this era and its applications were thoroughly explored and discussed. The conference attracted a large number of high quality submissions and stimulated cutting-edge research discussions among many academic pioneering researchers, scientists, industrial engineers, and students from all around the world and provided a forum to researcher. Research submissions in various advanced technology areas were received and after a rigorous peer-review process with the help of program committee members and external reviewer, 119 (Vol-I: 59, Vol-II: 60) papers were accepted with an acceptance ratio of 0.25. The conference featured many distinguished personalities like Dr. Akshai Aggarawal, Hon'ble Vice Chancellor, Gujarat Technological University, Dr. M.N. Patel, Hon'ble Vice Chancellor, Gujarat University, Dr. Durgesh Kumar Mishra, Chairman Division IV CSI, and Dr. S.C. Satapathy, Chairman, Division V, Computer Society of India, Dr. Bhushan Triverdi, Director, GLS University, and many more. Separate invited talks were organized in industrial and academia tracks for both days. The conference also hosted few tutorials and workshops for the benefit of participants. We are indebted to ACM Ahmedabad Professional Chapter, CSI Division IV, V for their immense support to make this conference possible in such a grand scale. A total of 12 sessions were organized as a part of *ICTIS 2015* including 9 technical, 1 plenary, 1 inaugural, and 1 valedictory session. A total of 93 papers were presented in the nine technical sessions with high discussion insights. The total number of accepted submissions was 119 with a focal point on ICT and intelligent systems. Our sincere thanks to all sponsors, press, print, and electronic media for their excellent coverage of this conference.

November 2015

Suresh Chandra Satapathy
Swagatam Das

Conference Committee

Patrons

Mr. Rishabh Jain, Chairman, VICT, Ahmedabad, India
Dr. Srinivas Padmanabhuni, President, ACM India

General Chairs

Dr. A.K. Chaturvedi, Director, VICT, Ahmedabad, India
Mr. Chandrashekhar Sahasrabudhe, ACM India
Dr. Nilesh Modi, Chairman, ACM Ahmedabad Chapter

Advisory Committee

Dr. Chandana Unnithan, Victoria University, Australia
Dr. Bhushan Trivedi, Professional Member, ACM
Mustafizur Rahman, Endeavour, Agency for Science Technology and Research, Australia
Mr. Bharat Patel, Professional Member, ACM
Dr. Durgesh Kumar Mishra, Chairman, Division IV, CSI
Mr. Hardik Upadhyay, GPERI, India
Dr. Dharm Singh, PON, Namibia
Mr. Ashish Gajjar, Professional Member, ACM
Hoang Pham, Professor and Chairman, Department of ISE, Rutgers University, Piscataway, NJ, USA
Mr. Maitrik Shah, Professional Member, ACM
Ernest Chulantha Kulasekere, Ph.D., University of Moratuwa, Sri Lanka
Ms. Heena Timani, Professional Member, ACM
Prof. S.K. Sharma Director PIE, PAHER, Udaipur, India

Shashidhar Ram Joshi, Ph.D., Institute of Engineering, Pulchowk Campus, Pulchowk, Nepal
Mr. Bhaumik Shah, Professional Member, ACM
Subhadip Basu, Ph.D., Visiting Scientist, The University of Iowa, Iowa City, USA
Prof. Lata Gohil, Professional Member, ACM
Abrar A. Qureshi, Ph.D., University of Virginia's College at Wise, One College Avenue, USA
Dr. Aditya Patel, Professional Member, ACM
Dr. Muneesh Trivedi, Professor, ABES, Ghaziabad, India
Dr. Ashish Rastogi (CSI), IT Department, Higher College of Technology, Muscat-Oman
Dr. Kuntal Patel, Professional Member, ACM
Dr. Aynur Unal, Standford University, USA
Prof. Rutvi Shah, Professional Member, ACM
Dr. Malaya Kumar Nayak. Director IT Buzz Ltd., UK
Dr. Nisarg Pathak, Professional Member, ACM

Organising Committee

Chair: Prof. Jayshree Upadhyay, HOD, CE, VICT, Ahmedbad, India
Secretary: Mr. Mihir Chauhan, VICT, Ahmedabad, India

Members

Prof. Mamta Jain, VICT, Ahmedabad, India
Mr. Zishan Shaikh, VICT, Ahmedabad, India
Mr. Jalpesh Ghumaliya, VICT, Ahmedabad, India
Prof. Nehal Rajput, VICT, Ahmedabad, India
Prof. Murti Patel, VICT, Ahmedabad, India
Prof. Jalpa Shah, VICT, Ahmedabad, India
Ms. Lajja Vyash, VICT, Ahmedabad, India
Ms. Ruchita Joshi, VICT, Ahmedabad, India
Mr. Aman Barot, VICT, Ahmedabad, India
Mr. Maulik Dave, VICT, Ahmedabad, India
Ms. Priyadarshini Barot, VICT, Ahmedabad, India

Technical Program Committee

Chair: Dr. Suresh Chandra Satapathy, Chairman, Division V, CSI
Co-Chair: Prof. Vikrant Bhateja, SRMGPC, Lucknow

Members

Dr. Mukesh Sharma, SFSU, Jaipur
Prof. D.A. Parikh, Head, CE, LDCE, Ahmedabad
Dr. Savita Gandhi, Head, CE, Rolwala, GU, Ahmedabad
Dr. Jyoti Parikh, Associate Professor, CE, GU, Ahmedabad
Ms. Bijal Talati, Head, CE, SVIT, Vasad
Dr. Harshal Arolkar, Member ACM
Dr. Pushpendra Singh, JK Lakshimpath University
Dr. Sanjay M. Shah, GEC, Gandhinagar
Dr. Chirag S. Thaker, GEC, Bhavnagar, Gujarat
Mr. Jeril Kuriakose, Manipal University, Jaipur
Mr. Ajay Chaudhary, IIT Roorkee, India
Mr. R.K. Banyal, RTU, Kota, India
Mr. Amit Joshi, Professional Member, ACM
Dr. Vishal Gour, Bikaner, India
Mr. Vinod Thummar, SITG, Ahmedabad, India
Mr. Nisarg Shah, 3D Blue Print, Ahmedabad, India
Mr. Maulik Patel, Emanant TechMedia, Ahmedabad, India

CSI Apex Committee

Prof. Bipin V. Mehta, President, CSI
Dr. Anirban Basu, Vice-President, CSI
Mr. Sanjay Mohapatra, Hon-Secretary, CSI
Mr. R.K. Vyas, Hon-Treasurer, CSI
Mr. H R Mohan, Immediate Past President, CSI
Dr. Vipin Tyagi, RVP, Region III, CSI

Conference Track Managers

Track#1: Image Processing, Machine Learning and Soft Computing—Dr. Steven Lawrence Fernandes
Track#2: Software and Big Data Engineering—Dr. Kavita Choudhary
Track#3: Network Security, Wireless and Mobile Computing—Dr. Musheer Ahmad

Contents

Part I
Intelligent Information Retrieval and Business Intelligence

RC6 Based Data Security and Attack Detection

Nitin Varshney and Kavindra Raghuwanshi

Abstract Java server pages (JSP) and Hypertext Preprocessor (PHP) are the most common scripting language which is used for web designing. Both are used with Hyper Text Markup Language (HTML) and Cascading Style Sheets (CSS) to make the website better look and feel. The websites make the communication easier in the real time scenario. So the need of security comes into picture in case of data sending and receiving. In this paper, we have applied RC6 encryption technique for securing the web pages for sending and receiving. For this we are using JS, HTML and CSS combination with the Apache Tomcat Server environment. We have also detected the attacked file if there is any attack will happened and compare eavesdrop time (ET) along with the alert time (AT).

Keywords JSP · HTML · CSS · Eavesdrop time · Alert time

1 Introduction

The data communication in case of web browser is possible through hypertext transfer protocol by providing all the resources [1]. The chances of attack is possible through the cookies which are provided in between the data is received and send [2, 3]. There are several works are already carried out in this direction and still the enhancement have been done in this direction to secure it. In case of data security RSA, DES etc. are suggested in [4]. This type of technique can enhance the security vivid. There are several security precautions are taken in different are for data security as like in [5, 6, 7–9]. Data security enforcement will be applicable in the security concern in the same context is applicable in different scenario which will be applicable in the same set of data security.

There are several researches are going on with the full swing. These investigations have the capability of uncovering vulnerabilities, a large portion of them

N. Varshney (✉) · K. Raghuwanshi
SIRT, Bhopal, India
e-mail: nitin.varshney@gmail.com

S.C. Satapathy and S. Das (eds.), *Proceedings of First International Conference on Information and Communication Technology for Intelligent Systems: Volume 1*, Smart Innovation, Systems and Technologies 50, DOI 10.1007/978-3-319-30933-0_1

[10–17] can't solidly reason about the string and non-string parts of an application and numerous need way affectability, though RCE assaults oblige fulfilling interesting way conditions, including both strings and non-strings. As of late, scientists have proposed methods that can display both strings and non-strings in dynamic typical execution of web applications [18–21].

The main objective of our work is to extend the direction of previous work to provide data security as well as the eavesdrop.

2 Related Works

In 2012, Jagnere et al. [22] talks about on the issue of weakness of social locales. Creators recommend RCE, JSP Instructions, HTML labels stacking are a few samples by client's session are hacked. Creators investigate the reasons and dissects them. It can help to see the weakness regarding discovering the reasons. In 2013, Nagarjun et al. [23] control variations of RTS/CTS assaults in remote systems. We personate the assaults conduct in ns2 false show aerosphere to wrangle the fake common sense as to a great extent as expertise restricted effect of these assaults on 802.11 based systems. In 2013, Choi et al. [24] scrap go Allow for flooding fake in all actuality be normal sensical for Withdrawal of Grant-in-help (Dos) in Gift Centric Irksome (CCN) taking into account the reproduction results which can influence nature of administration. They anticipate center it adds to opportune a bolt matter noticeable all around potential dangers of DoS in CCN. In 2013, Ruse et al. [25] reject a control a two-period propose to XSS vulnerabilities and avert XSS assaults. In the tricky day, they work out the Light into b upbraid engage a burr for which illustration veteran concolic testing apparatuses are close by. In 2013, Zheng et al. [26] piddling a way and setting insightful bury procedural examination to identify RCE vulnerabilities. These investigation appearances an unusual alike of dissecting both the bond and non-string conducts of a cross section application in a way delicate design.

3 Proposed Methodology

By the use of java server pages (JSP) and Hyper Text Markup Language (HTML) are used for the user and server side environment. For this we have used Netbeans7.2 environment and Cascading Style Sheets (CSS) for the structure setup. Then we have used Tomcat server to create server client environment to process this work. Apache Tomcat is a web container which permits to run servlet and Java Server Pages (JSP) based web applications. A large portion of the advanced Java web structures are in view of servlets, e.g. Java Server Faces, Struts, spring. Apache Tomcat additionally gives as a matter of course a HTTP connector on port 8080, i.e., Tomcat can likewise be utilized as HTTP server. Anyway the execution of Tomcat is not in the same class as the execution of an assigned web server, in the same way as the Apache HTTP server.

In this framework the user is been controlled by the server and the server is one and other users works as the slave user. Customer server construction modelling (customer/server) is a system building design in which the process on the system is either a user or a server. Servers are intense or methodologies committed to overseeing circle drives (record servers), or system movement (system servers).

User's workstations on which clients run applications. Customers depend on servers for assets, for example, data gathering. The user can request the data from the server. We have implemented the support for textual data like HTML, JSP, PHP, .TXT, .DOC and .PDF files. So the user can request these data. A log is maintained for the request comes from the user. It is picked by the server based on first come first server mechanism. There is the option for the request by choose it. Then server prepares the data by applying RC6 mechanism and the data is send to the user. The user already receives the key from the server which is random means change for the same file if it is applied for the next time. Data partitioning is applied for reducing the data overload. If it is try to eavesdrop by any other user it will be traced by the program tracer design by us. So the e eavesdrop time (ET) along with the alert time (AT) is been recorded. The file process and receiving is happened simultaneously to show the mean time process for the data. The process is properly understood by the flow chart shown in Fig. 1. Unlike other algorithms like DES and RSA RC6 is fast. The main advantage of this algorithm is to same RC6 key stream is not used for two different messages.

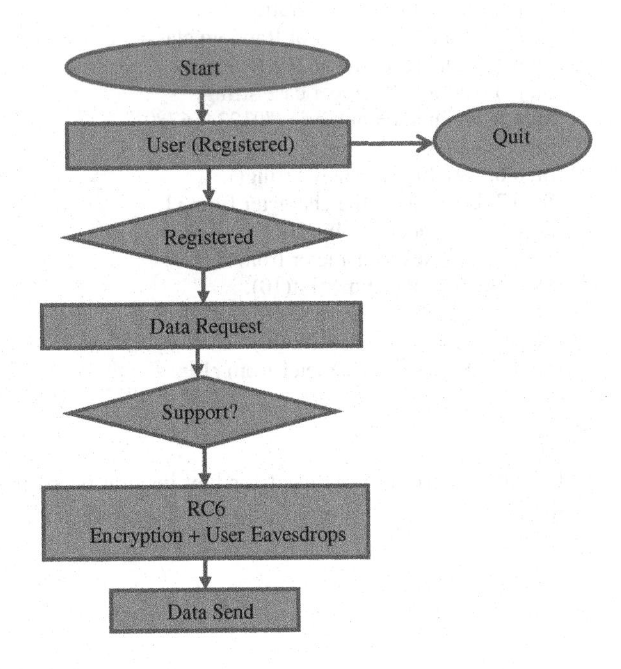

Fig. 1 Data process

Algorithm
Inputs: Textual data is in the form of Input
Output: Final Processed File and Random Password generator

```
Step 1: Pick the data object from the data pool.
Step 2: Load It in the server environment created by Apache Server Local Host.
Step 3: For every solicitation loads (FR= fr1, fr2… … frn)
Step 4: Data Encryption
Step 5: Client supplied b byte key preloaded into the c-word[29]
cluster L[0,… , c - 1]
Number r of rounds
Pw = Odd((e – 2)2w)
Qw = Odd((ø – 1)2w)
Yield:
w-bit round keys S[0,… , 2r + 3]
Strategy:
S[0] = Pw
for i = 1 to (2r + 3) do
S[i] = S[i _ 1] + Qw
A = B = i = j = 0
v = 3 x max{c, 2r + 4}
for s = 1 to v do
{
A = S[i] = (S[i] + A + B) <<< 3
B = L[j] = (L[j] + A + B) <<< (A + B)
i = (i + 1) mod (2r + 4)
j = (j + 1) mod c
}
Step 6: Random Key Generation
Step 1: initiate the object of Random class;
Step 2: variable a=lower case string;
Step 3: variable b= upper case string;
Step 4: variable c= numeric ("0123456789");
Step 5: variable r1 = object.nextInt(26);
Step 6: variable key=new String();
Step 7: key=access the character from r1.
Step 8: r1=object.nextInt(26);
Step 9: key=key+ character from r1;
Step 10: r1 = object.nextInt(10);
Step 11: key=key+ character from r1;
Step 12: r1 = object.nextInt(26);
Step 13: key=key+ character from r1;
Step 14: return (key);
```

The above procedure is used for key generation so that it is enabled for use in client side.

4 Results

Various data tables are required to be maintained at the server side. The tables are shown below. The results produced by our proposed method are shown in the tables. Our eavesdrops alerts times shows this mechanism with time calculation (in milliseconds) when server knows the information about the change of data. Our alerts times show that our mechanism is far better than previous mechanism.

Our proposed approach has an another advantage that due to automated file processing feature for the said file formats the file processing time also get reduced

Table 1 Data at the time of preparation

File	Key	Bit value	Client
MyApplet.java	0C2M3S8A5P	1	User
ab15.txt	0V5R7O6N7E3L7E	1	User
pdf1.pdf	1F2B9J1X9Y	1	User
sd.rtf	1Y3Z4K8S4U7P1 J	1	User
email.txt	2Q1H2L9T8C	1	User
wd2.html	3K5U0L4I9E9A3 V	1	User
d2.txt	3X5W2X6L7N8R7I	1	User
pdf2.pdf	4Z3C2O4K5W4O	1	User
wd17.html	5C5P0C1D7N3J8S	1	User
email.txt	5H5N5Y3Q3Z1E6Z9I	1	User

Table 2 Data after the file process

File	Key	Bit value	Sending time	Rec time	Size
ab15.txt	0V5R7O6N7E3L7E	1	9:50:44:466	9:50:44:495	29,789
email.txt	2Q1H2L9T8C	0	9:51:14:799	9:51:14:815	7213
pdf2.pdf	4Z3C2O4K5W4O	1	10:3:15:564	10:3:15:583	20,859
wd17.html	5C5P0C1D7N3J8S	1	10:25:3:10	10:25:3:25	241,889
pdf1.pdf	1F2B9J1X9Y	1	11:51:21:467	11:51:21:494	405,044

Table 3 Data status at the client

File	Bit value	Key	Client	Size	Open
MyApplet.java	0	0C2M3S8A5P	u1	180	No
ab15.txt	1	0V5R7O6N7E3L7E	u1	29789	Yes
pdf1.pdf	1	1F2B9J1X9Y	u1	405044	Yes
sd.rtf	1	1Y3Z4K8S4U7P1 J	u1	1008	Yes
wd2.html	0	3K5U0L4I9E9A3 V	u1	2348	No
d2.txt	0	3X5W2X6L7N8R7I	u1	4392	No
pdf2.pdf	1	4Z3C2O4K5W4O	u1	20859	Yes

Table 4 Data status after attack

File	Size	Server time	Attack time	Key	Client	Status
email.txt	7203	9:51:45:442	9:51:14:799	2Q1H2L9T8C	u1	Yes
anbcd.doc	26138	11:56:37:238	11:56:22:340	6X8Y7T9L7S5M7 V	u1	Yes
d2.txt	4366	11:57:19:656	11:57:1:408	3X5W2X6L7N8R7I	u1	Yes
MyApplet.java	179	11:58:11:228	11:57:54:637	0C2M3S8A5P	u1	Yes

Table 5 Time comparison

File	Traditional (ms)	Proposed (ms)
Text files	50	42
Doc files	167	137
PDF	205	160

Table 6 Other comparison

Comparison	Traditional	Proposed
Encryption standard	API based	Standard encryption
Applicability	Some specific platform	Platform independent
Data type	Specific applications	Wide application

as compared to the traditional file processing systems. We have also included data existence mechanism successfully. The security variable key length is also an enhancement from the previous approach (Tables 1, 2, 3, 4, 5 and 6).

5 Conclusions

In this paper we have proposed a server client communication system which provides better security along with timely alert in terms of data communication. In our system if data request is arrived from client to server, server automates the process if client is authorized. Then encryption and partition algorithms on requested data are applied by the server before sending the requested data. Hidden numeric adder concept is induced for preventing any unauthorized access in the communication area which also ensures that the client must beware of the content sniffed data. The proposed framework reduces the process time overhead at the server. Alongside the methodology time the reaction time is diminished to a much more noteworthy degree as contrasted with the past methodologies it additionally dynamism the presence of information document. The empowering results will outline the effect of our methodology. In future we can apply our approach on different heterogeneous files like images, video etc. We can also enhance the security mechanism (Hybrid encryption algorithm).

References

1. Gupta, S., Sharma, L., Gupta, M., Gupta, S.: Prevention of cross-site scripting vulnerabilities using dynamic hash generation technique on the server side. Int. J. Adv. Comput. Res. (IJACR), **2**(5), 49–54 (2012)
2. Barua, A., Shahriar, H., Zulkernine, M.: Server side detection of content sniffing attacks. In: 2011 22nd IEEE International Symposium on Software Reliability Engineering
3. Sharp, R., Scott, D.: Abstracting application level web security. In: Proceedings of the 11th ACM International World Wide Web Conference (WWW 2002), 7–11 May 2002
4. Dubey, A.K., Dubey, A.K., Namdev, M., Shrivastava, S.S.: Cloud-user security based on RSA and MD5 algorithm for resource attestation and sharing in java environment. In: CONSEG (2012)
5. Wurzinger, P., Platzer, C., Ludl, C., Kruegel, C.: SWAP: mitigating XSS attacks using a reverse proxy. In: Proceedings of the 2009 ICSE Workshop on Software Engineering for Secure Systems, pp. 33–39 (2009)
6. Qadri, S.I.A., Pandey, K., Tag based client side detection of content sniffing attacks with file encryption and file splitter technique. Int. J. Adv. Comput. Res. (IJACR) **2**(3)(5) (2012)
7. Saxena, P., Akhawe, D., Hanna, S., Mao, F., McCamant, S., Song, D.: A symbolic execution framework for javascript. In: 2010 IEEE Symposium on Security and Privacy (SP), pp. 513–528. IEEE (2010)
8. Chhajed, U., Kumar, A.: Detecting cross-site scripting vulnerability and performance comparison using C-Time and E-Time. Int. J. Adv. Comput. Res. (IJACR) **4**(15) 733–740 (2014)
9. Bjørner, N., Tillmann, N., Voronkov, A.: Path feasibility analysis for string-manipulating programs. In: Tools and Algorithms for the Construction and Analysis of Systems, pp. 307–321. Springer, Berlin (2009)
10. Kirda, E., Jovanovic, N., Kruegel, C., Vigna, G.: Client-side cross-site scripting protection. Sci. Direct Trans. Comput. Security 184–197 (2009
11. Ikemiya, N., Hanakawa, N.: A new web browser including a transferable function to Ajax codes. In: Proceedings of 21st IEEE/ACM International Conference on Automated Software Engineering (ASE '06), Tokyo, Japan, pp. 351–352, September 2006
12. Joshi, B., Khandelwal, A.: Rivest cipher based data encryption and clustering in wireless communication. Int. J. Adv. Technol. Eng. Explor. (IJATEE) **2**(2), 17–24 (2015)
13. Kiezun, A., Ganesh, V., Guo, P.J., Hooimeijer, P., Ernst, M.D.: HAMPI: a solver for string constraints. In: Proceedings of the Eighteenth International Symposium on Software Testing and Analysis, pp. 105–116, ACM (2009)
14. Wassermann, G., Su, Z.: Sound and precise analysis of web applications for injection vulnerabilities. In: ACM Sigplan Notices, vol. 42, no. 6, pp. 32–41. ACM (2007)
15. Shukla, N.: Data mining based result analysis of document fraud detection. Int. J. Adv. Technol. Eng. Explor. (IJATEE), **1**(1) 21–25 (2014)
16. Tateishi, T., Pistoia, M., Tripp, O.: Path-and index-sensitive string analysis based on monadic second-order logic. ACM Trans. Softw. Eng. Methodol. (TOSEM) **22**(4), 33 (2013)
17. Yu, F., Alkhalaf, M., Bultan, T.: Patching vulnerabilities with sanitization synthesis. In: Proceedings of the 33rd International Conference on Software Engineering, pp. 251–260. ACM (2011)
18. Yu, F., Bultan, T., Hardekopf, B.: String abstractions for string verification. In: Model Checking Software, pp. 20–37. Springer, Berlin (2011)
19. Zheng, Y., Zhang, X.: Static detection of resource contention problems in server-side scripts. In: Proceedings of the 34th International Conference on Software Engineering, pp. 584–594. IEEE Press (2012)
20. Halfond, W.G.J., Anand, S., Orso, A.: Precise interface identification to improve testing and analysis of web applications. In: Proceedings of the Eighteenth International Symposium on Software Testing and Analysis, pp. 285–296. ACM (2009)

21. Kaushik, M., Ojha, G.: Attack penetration system for SQL injection. Int. J. Adv. Comput. Res. (IJACR) **4**(15) 724–732 (2014)
22. Jagnere, P., Vulnerabilities in social networking sites. In: 2nd IEEE International Conference on Parallel Distributed and Grid Computing (PDGC), pp. 463, 468, 6–8 Dec 2012
23. Nagarjun, P.M.D., Kumar, V.A., Kumar, C.A., Ravi, A.: Simulation and analysis of RTS/CTS DoS attack variants in 802.11 networks. In: International Conference on Pattern Recognition, Informatics and Mobile Engineering (PRIME), vol., no., pp. 258, 263, 21–22 Feb 2013
24. Choi, S., Kim, K., Kim, S., Roh, B.-H.: Threat of DoS by interest flooding attack in content-centric networking. IEEE (2013)
25. Ruse, M.E., Basu, S.: Detecting cross-site scripting vulnerability using Concolic testing. In: Tenth International Conference on Information Technology: New Generations (ITNG), pp. 633, 638, 15–17 Apr 2013
26. Zheng, Y., Zhang, X.: Path sensitive static analysis of web applications for remote code execution vulnerability detection. In Proceedings of the 2013 International Conference on Software Engineering, pp. 652–661. IEEE Press (2013)

Application Mapping Methodology for Reconfigurable Architecture

Rahul K. Hiware and Dinesh Padole

Abstract This paper suggests mapping of application on course grain reconfigurable architecture. The Data flow graph (DFG) in form of Op Codes of an application is stored in Memory i.e. Configuration memory. The CGRA reconfigure itself according to DFG and does parallel multi-processing.

Keywords Reconfigurable architecture · CGRA · Multicore

1 Introduction

To change the shape, model or to reform is known as reconfiguration. Reconfiguration provides rearrangement of elements or settings of a system contributing to restructuring as a whole or application based. In field of research in computer architectures and software systems, Reconfigurable computing has made its important place. An application can be significantly speeded by putting the computationally concentrated parts of an application onto the reconfigurable hardware. The advantages make us realize the significance of software along with hardware walking in parallel to save computation time, complexity and cost [1]. In era we are approaching, a single chip is able to comprise more than 100 million transistors. Despite the great flexibility current general-purpose processor systems possess still they will not reach their full potential they can provide. In contrast, exceptionally high performance has been achieved by application specific integrated circuits (ASICs) by aiming every application on

R.K. Hiware (✉) · D. Padole
Department of Electronics Engineering, G.H. Raisoni College of Engineering, Nagpur, India
e-mail: rahulhiware@gmail.com

D. Padole
e-mail: dinesh.padole@raisoni.net

S.C. Satapathy and S. Das (eds.), *Proceedings of First International Conference on Information and Communication Technology for Intelligent Systems: Volume 1*, Smart Innovation, Systems and Technologies 50, DOI 10.1007/978-3-319-30933-0_2

custom circuitry. However, it is difficult to design and implement a custom chip for every application because of the vast expense.

The gap between ASICs and general-purpose computing systems can be filled by the hardware that can be reconfigured. In 1960s, the basic concept was proposed, due to which reconfigurable computing system have become easy using technological upgradation in chip design. The availability of highly dense VLSI devices that programmable switches makes possible to implement the flexible hardware architectures. Many reconfigurable systems consist of a general-purpose processor, tightly or loosely coupled with system. These systems can device definite functionality of applications on reconfigurable system afore on the general purpose processor, providing considerably better performance. The general-purpose processors in such systems are only useful for data collection and synchronization and not for the major computational power. A lot of attention has been attracted because of the high performance and high flexibility delivered by the reconfigurable computing systems as compared to an ASIC. Moreover, in recent years such system can achieve high performance for a range of applications, such as image processing, pattern recognition and encryption.

2 Methodology

Reconfigurable systems due to their blend of flexibility and efficiency have drawn increasing attention lately. Reconfigurable designs have limitation of flexibility to a specific algorithm domain. The reconfigurable systems defined in two categories one is fine-grained where the functionality of the hardware is definite at the bit level and coarse grained where functionality of the hardware is definite at the word level. System design for coarse-grained reconfigurable architectures required some High-level design tools. This paper proposes a method for mapping applications onto a coarse grained reconfigurable architecture. This is a heuristic method which tackles this complex problem in four phases: Translation, Clustering, Scheduling and Allocation [2]. In this paper a coarse-grained reconfigurable architecture shown in Fig. 1, is proposed to demonstrate the proposed mapping method.

The proposed system consists of 4 ALUs. These ALU can be either any simple ALU or it could be even advanced processor. For simplicity, here we have taken simple four bit ALU. These ALUs are connected in 2 × 2 cross bar pattern. To map DFG on ALU there is ALU selector which is an encoder named as chip selector. The memory here is used to store DFG/opcodes of an application. This memory is known as configuration memory.

User provides the interface in terms of application DFG/opcodes as input which is stored in memory [3]. Based on the user control the counter shifts the memory which selects the respective ALU to take the op code segment, and perform the operation based on the data. Op code format is as shown Fig. 2. The DGF or

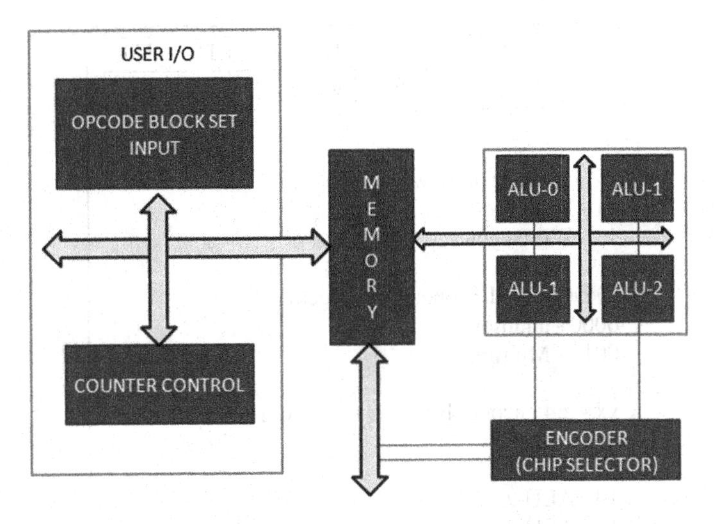

Fig. 1 Experimental CGRA architecture

opcodes for the given system are stored in configuration memory as per the format given below. Bits D7-D6 are used to select the ALU, which is nothing but input to the chip selector. Bits D1-D0 is for writing a data/result in memory or read data from memory. Rests of the bits D5-D2 are used to select operation to be performed by the selected ALU.

3 Simulation Result and Conclusion

Multi-processing is performed as shown on op code (Fig. 2) based transition phase but this reconfiguration has demerits of data transformation/flow looped with arbiter hence we suggest a programming element based alternative. This coarse grained architecture comprises of four parts: In the translation phase, an input program can be written in a high-level language and translated into a control dataflow graph; and some transformations and simplifications could be done on the control data flow graph. In the clustering phase, the data flow graph can be divided into clusters and mapped into Arithmetic Logic Units (ALUs). The structure of ALU is the main concern of this phase and the inter ALU communication is not in the consideration. In the scheduling phase the graph can be obtained from the clustering phase to know whether the scheduling taking the maximum number of ALUs into account. The scheduling algorithm should minimize the number of clock cycles used for the application under the constraints of reconfiguration. In the allocation phase, variables can be allocated to memories or registers, and data moves could be scheduled; where the main concern in this phase could arise in target architecture in terms of

D7	D6	D5	D4	D3	D2	D1	D0
X	X	@	@	@	@	Z	Z

ZZ - memory read/write cycle
01 - Read cycle
11 - Write cycle

@@@@- ALU operation selection
0000 - addition
0011 - Multiplication

xx - 2:4 encoder bits - for multi core ALU selection
00 - ALU-0
01 - ALU-1
10 - ALU-2
11 - ALU-3

Fig. 2 Op code format and operations

```
1  module top_alu(out,a,in,clk,en,reset);
2  input [7:0]in;
3  input [15:0] a;
4  output reg [7:0]out;
5  input clk,en,reset;
6  wire flag;
7  wire [5:0] count;
8  wire [15:0] dataout;
9  wire [7:0] out_temp,out_temp1,out_temp2,out_temp3;
10 wire [3:0]dout;
11 sync_reset one(count,reset,clk,en);
12 mem_ram_sync two(count,a,dataout,in[1],in[0],reset,en,clk);
13 decoder2_4 three(in[7:6],dout,clk,en);
14 ALU four_A(out_temp,dataout[15:8],dataout[7:0],in[5:2],clk,dout[3],flag);
15 ALU four_B(out_temp1,dataout[15:8],dataout[7:0],in[5:2],clk,dout[2],flag);
16 ALU four_C(out_temp2,dataout[15:8],dataout[7:0],in[5:2],clk,dout[1],flag);
17 ALU four_D(out_temp3,dataout[15:8],dataout[7:0],in[5:2],clk,dout[0],flag);
18 always@(posedge clk)
19    begin
20    if(en==1)
21    out<=out_temp;// !out_temp1 or out_temp2 or out_temp3;
22    else
23    out<=8'd0;
24    end
25 endmodule
```

Fig. 3 HDL code of proposed CGRA

area. Figure 3 shows HDL program for top level entity which is a proposed experimental architecture shown in Fig. 1.

Figure 4 shows RTL view of proposed system showing all elements of the system and Fig. 5 shows simulation result.

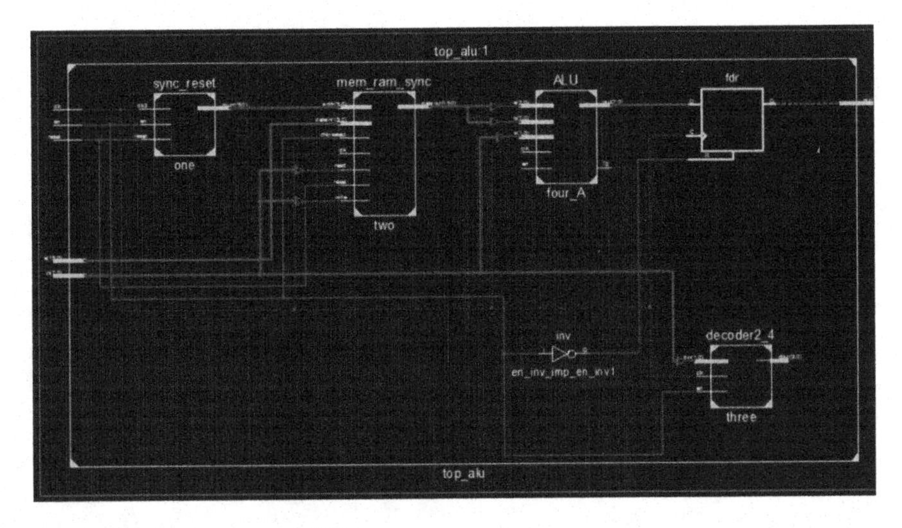

Fig. 4 RTL view of proposed CGRA

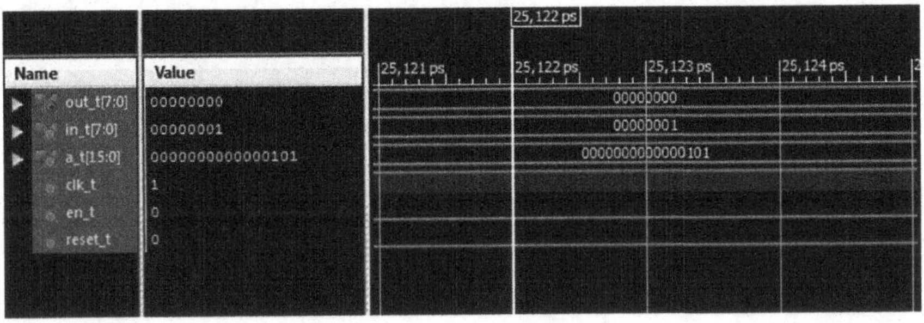

Fig. 5 Simulation result

References

1. Li, Z.: Configuration management techniques for reconfigurable computing. Northwestern University, June 2002
2. Guo, Y.: Mapping Applications to a Coarse-grained Reconfigurable Architecture. University of Twente, Enschede (2006)
3. Padole, D., Hiware, R.: Configuration memory based dynamic coarse grained reconfigurable multicore architecture. In: IEEE Region 10. Conference TENCON 2013 Xi'an, pp. 1–5, Oct 2013

Circular Quad-Octagon Bits: Stepwise Image Cubic Spread Authentication Analysis

Vijayalakshmi Kakulapati and Vijaykrishna Pentapati

Abstract In this paper, a picture analysis technique by providing authentication to photographs is planned. In our approach, invisible watermark bits were taken in circular regular sequence in exceedingly quad-octagon patterns. Quad-octagon patterns are generating by considering thirty-two bits in at random continuation eight bit planes. In every bit plane, a regular equal quad image bits were extracted and interlinked with its beside the bit plane, quad image bits circularly, forming completely thirty-two bits. These circular pattern bits were entered stepwise in an exceedingly boxlike manner in embedded image bits. Many quad-octagon patterns were addressed in our planned technique. Recovered pictures from our technique shown higher survival technique beneath numerous attacks.

Keywords Authentication · Watermark · Pattern · Embedded · Recovered

1 Introduction

Image authentication analysis is presently growing subject within the field of image diagnosis [1] and image security applications. Genuine image content and services have wide unfolded adoption in recent years. The following schemes are addressed [2–4, 5] to unravel the image authentication.

Lu et al. [2], a fragile watermark at the same time is projected, that contains a downside in resisting beneath geometric attacks. However, they didn't clearly able to specify the type of attack with the required watermark, resulted in non-uniform

V. Kakulapati (✉)
GNITC, Ibrahimpatnam, Hyderabad, India
e-mail: vldms@yahoo.com

V. Pentapati
SIEI, Ibrahimpatnam, Hyderabad, India
e-mail: vijay_p51@yahoo.com

S.C. Satapathy and S. Das (eds.), *Proceedings of First International Conference on Information and Communication Technology for Intelligent Systems: Volume 1*, Smart Innovation, Systems and Technologies 50, DOI 10.1007/978-3-319-30933-0_3

tamper detection. To differentiate the malicious manipulations from JPEG lossy compression a picture authentication technique was projected by Lin et al. [3]. However, their approach fails to differentiate replacement manipulations. Another disadvantage related to their approach is that JPEG high compression ratios aren't acceptable. A block wise freelance watermark technique was projected by Celik et al. [4]. However, their approach was restricted to the lowest level of signature verification. And another disadvantage of the projected technique is throughout block division, the payload will increase whereas concating the increasing signature blocks. a strong lossless knowledge activity formula to certify lossless compressed JPEG2000 pictures, followed by potential trans-coding was projected by Zhicheng atomic number 28, Shi et al. [5]. However, their approach uses multiple bit-embedding schemes to every image for authentication. And another downside of the projected technique is most block size is twenty to implant knowledge.

In [6] approach, to survive the cropping of a picture, the image sweetening and therefore the JPEG lossy compression a picture authentication technique is projected by embedding digital watermarks into pictures, by embedding visually recognizable patterns with the watermarks among the photographs, by selection modifying the middle-frequency components of the image. Few comparisons were shown in [7], suggesting the watermark most closely resembles communications with aspect data at the transmitter. Employing a suboptimal technique, scalar Costa theme (SCS) is projected in [8], to implant data in freelance identically distributed (IID) knowledge. Besides, a strong digital image watermark theme is proposed [9], JPEG-to-JPEG image watermark is proposed [1], a way to implant watermark victimizations designated pseudo-randomly designated macro blocks (MBs) in every video object is projected [10], a brand new methodology for the process and watermark three-dimensional (3-D) graphical objects is projected in [11], for exploits the human visual model for adapting the watermark knowledge to native properties of the host image, a brand new scaling-based image-adaptive watermark system is projected [12] associated with [13] a secret writing theme is projected to produce security by permitting watermark theme for a compressed image.

During this paper, we have a tendency to proposing a way for image authentication analysis, to enhance the image security, image classification beneath the class of image texture variants. On the image protection, we've projected an invisible watermark, which may resist completely different forms of attacks even geometric attacks by considering each uniform and non uniform attacks to relevant tamper detection. Throughout the generation of quad-octagon pattern pixels worth can modification leading to replacement manipulations creating use of acceptable high compression ratios. In our projected technique, verification of image levels vary from lower to higher and higher to lower, relying upon the image three-dimensional unfold technique. Here in image three-dimensional unfold technique; blocks are dividing into uniform size, keeping the payload constant by considering each lower and the higher level of image picture element verification. Once image three-dimensional generation, the single bit layer is made, creating the block size to extend up to the most block size of 128 to implant knowledge. Our paper concentrates on the look of invisible watermark bit victimization quad-octagon patterns and image three-dimensional unfold technique, embedding bits were generated creating the image authentication to think about beneath embedding authentication bits in the original image.

In the next section, ways image, texture classifications are explained. In Sect. 3, quad-octagon bit selection is implemented. In Sect. 4, image cubic spread is explained with its variants. In Sect. 5, image authentication analysis is described. In Sects. 6 and 7, experimental results and conclusions were stated.

2 Texture Classification

As the image texture [14, 15], work is assumed to be the classification of quantizing the coefficients of (i, j)th frame as well as the image file header. So as to classify the authentication bits, image texture coefficients are classified as reference points to embedded pattern bits in watermark bits.

In our paper, we have a tendency to propose a texture classification [2–5, 14] technique known as quantity, constant Image Textures (QCIT) as shown in Fig. 1. This technique determines the image texture coefficients by distinctive the scaled versions of rotation and site primarily based variants in a picture. The rotation primarily based variants are classified into circular, elliptical and polar, these variants depend upon the R, wherever represents the structural and cubic variations among spot textures and variants textures. The placement primarily based variants are classified into linear, quad and diagonal, these variants depend upon the L^, wherever ^ represents the classifying issue among multi-axes textures.

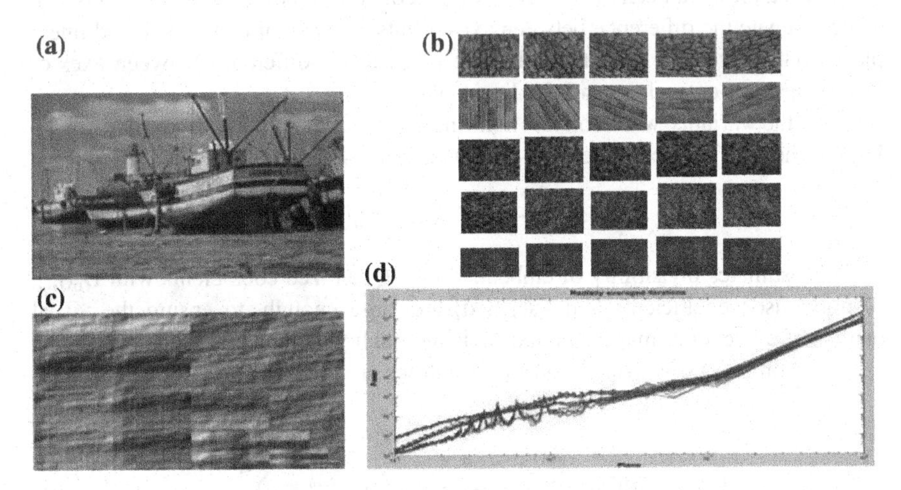

Fig. 1 Quantized coefficient image textures (QCIT) texture classification method. **a** Embedding image. **b** Rotational spatial variants. **c** Embedding £(x) constant image texture. **d** QCIT texture classification plot

In rotation based variants the QCIT coefficients of *d*th block is represented as

$$D_r(i,j,d) = Q_r(i,j,d) * D_q(i,j,d) \tag{1}$$

where Q_r(i, j, d) is the (i, j)th element of quantization table Q_t.

Suppose the invisible watermarks are just noticeable mean, the distortion magnitude is more than D_q(i, j, d), causing D_r(i, j, d) to negative. It should be noted that D_r(i, j, d) should be non negative, to make it the amount of distortion magnitude modification on D_q(i, j, d), which is limited to D_o(i, j, d), is the factor of magnitude estimation in our proposed authentication bits embedding process.

During quantizing the image texture coefficients the wavelet size of the variants are defined by

$$£(x) = (1 - \|x\|) * log - [\|x\| - \|k\|]^{-2} \tag{2}$$

where $\|x\| = (i^2 + j^2)^{1/2}$ and $\|k\| = (i^{-2} + j^{-2})+$, which should be always constant as proposed.

In our proposed QCIT method based on above, rotation based variants were considered and constant image texture coefficients were quantized. Based on first consideration the quantities are:

$$\begin{aligned} K_{mn}(x) &= L_m(x) - \sigma * L_n(x), O_m(x) = (x^{-m}.\rho(x^{-m}.x)) * I \; and \; P_n(x) \\ &= (x^{-n}.\rho(x^{-n}.x)) * I \end{aligned} \tag{3}$$

where $O_m(x)$ and $P_n(x)$ represents the rotation spatial locations of scales m and n parameters. I is the input image, σ and ρ are variants determines the estimation of rotation scaling parameters and * denotes the convolution operation. σ variants will angle estimate the difference between axes of bits along semi-quad plane and linear plane. Where are in ρ variants will angle estimate the difference between axes of bits along semi-octa plane and the linear plane?

Both the variants were rotated non-linearly to obtain rotation based variants. During the QCIT process the block non negative estimation is defined by

$$D_n(i,j,d) - D_o(i,j,d) \geq J_o(i,j,d) \tag{4}$$

This estimates invisibility of embedded bits. Quantized coefficients with D_n(i, j, d) normalised coefficients and J_o(i, j, d) are large enough, to ensure the visual quality of embedded image bits and invisible quality of embed bits.

Based on second consideration the quantities are:

$$£(x)\,positive = \sum_{rp=0}^{rp=p} P_{r1} \; and \; £(x)\,negative = \sum_{rn=0}^{rn=n} P_{r2}. \tag{5}$$

Here the constant £(x) is given by

$$£(x)constant = \sum_{I=m}^{N} (£(x)\,positive)^{i}(1 - £(x)\,negative)^{N-i} * \binom{N}{i}, \tag{6}$$

where $P_{r1} = (2)^{n}\,(n!/(n - x)!)$, $P_{r2} = (2)^{n-2}\,(n!/(n - x^{2})!)$ and N is the total number of non negative iterations.

3 Quad-Octagon Bits Selection

In order to improve the perceptual invisibility and authentication, in our paper we proposed a Quad-Octagon Bit Selection method. Embed image texture classification is taken into octa planes. In each level, quad bits were selected as shown in Fig. 2.

For each plane of image texture classification, of sizes 1 × 4, 2 × 2, 1 × 2, 1 × 3, 1 × 1 variances are considered for extracting quad bits and sorted for invisibility of embedded bits. Then, each embed bits were shuffled in special position accordingly the sorting order of embedded bits, represented as

$$\begin{aligned} E_b = \{E_O(r \times (i_1 \times 4/j) + ij, r \times (i_2 \times 2/j) + ij, r \times (i_1 \times 2/j) + ij, \\ r \times (i_1 \times 3/j) + ij, r \times (i_1 \times 1/j) + ij, r \times (i_K \times L/j) + ij) \end{aligned} \tag{7}$$

Here K and L represent the sizes and E_O represents embed classification bits. Figure 2, shows the octa planes with Quad E_b bits, which are random in nature.

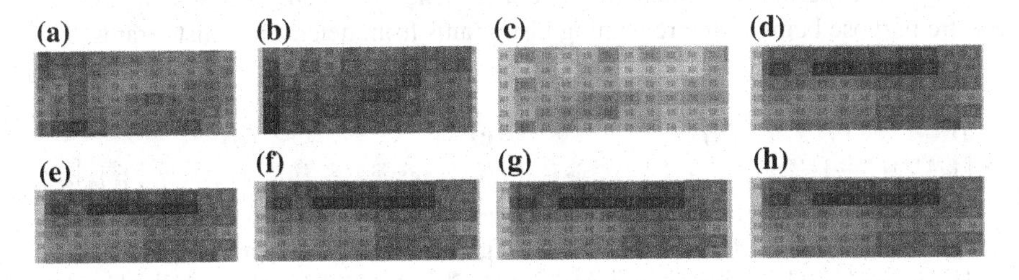

Fig. 2 The octa planes with quad E_b bits patterns. **a** 1 × 1 variance. **b** 1 × 2 variance. **c** 2 × 2 variance. **d** 1 × 4 variance. **e** 1 × 2 variance. **f** 1 × 3 variance. **g** 2 × 2 variance. **h** 1 × 2 variance

4 The Effect of Image Cubic Spread

In every bit plane, a regularized equal quad image bits were extracted and interlinked with its beside bit plane quad image bits circularly, forming all thirty-two bits.

The octal planes with Quad Eb bits pattern were circularly patterned in middle frequency vary of image. For every variance, solely four coefficients square measure designated in every plane out of sixty-four QCIT coefficients listed in Fig. 3a–h. In our new image authentication bits generation method, consists of 3 unfold levels.

At the primary level for unfold; we have a tendency to mix all plane bits into one plane bits. At the second level of unfold, image cuboidal the unfold filter has been designed. Let S * b * k colour area is employed to represent colour options. The S* element in our planned model denotes brightness level, whereas b* and k* along represent the chromatic data. Here we have a tendency to thought of 2 categories, equivalent to most often determined histograms Hfo, that square measure Quad-Octagon chose bits listed in Fig. 3i and the remaining histograms hour square measure accomplished mistreatment Image cuboidal unfold methodology listed in Fig. 3j. From the distribution of determined histograms, we will simply estimate the remaining histograms by the conditional estimators E(Ix,y | Hfo) and E(Ix,y | Hr).

The expression E(Ix,y Hfo) is that the conditional estimation of getting the centre worth Ix,y provided that it's computed from the Hfo. Similarly, E(Ix,y | Hr) is that the conditional estimation of getting the centre worth Ix,y provided that it's computed from the hour. If W is that the total variety of pixels enclosed within the cuboidal filter and Nr and Nfo of them belong to hour and Hfo severally, then E(Ix, y | Hfo) = Nr/E and E(Ix,y | Hr) = Nfo/E. we have a tendency to selected the previous estimation E(Ix,y | Hfo) as zero.7 and E(Ix,y | Hr) as zero.3, that made throughout experiments. Then, we have a tendency to calculate the chance of the centre purpose being in the remaining histograms from determined histograms. The

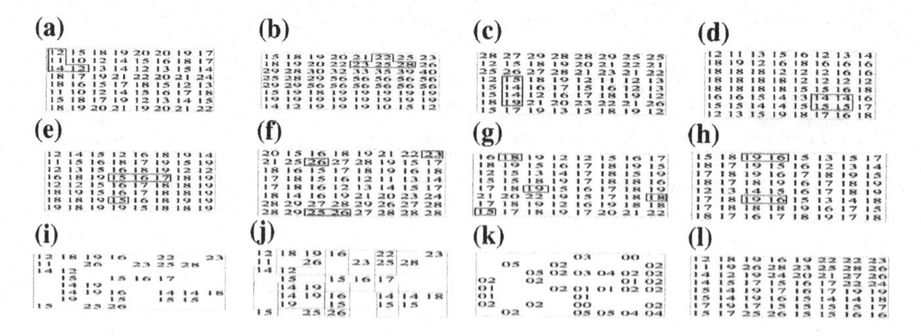

Fig. 3 Generation of authentication bits using image cubic spread method. **a** Plane 1: quad. **b** Plane 2: quad. **c** Plane 3: quad. **d** Plane 4: quad. **e** Plane 5: quad. **f** Plane 6: quad. **g** Plane 7: quad. **h** Plane 8: quad. **i** Quad-octagon bits selection. **j** Image cubic spread. **k** Effect of image cubic spread. **l** Authentication bits

remaining histograms square measure listed in figure (k). Truncating the image cuboidal unfold bits with the remaining bits, ends up in authentication bits as listed in Fig. 31 for embedding to watermark bits.

5 Image Authentication Analysis

We extract the average (As, Ab, Ak) and variance (Vs, Vb, Vk) of S * b * k color space associated with each region to improve the authentication bits recovery. The distances between observed histograms and the remaining histograms are measured by the block distance:

$$g^{H}{}_{fo,r} = \left(\sum_{H=As,Ab,Ak} |fo_H - r_H| + \sum_{H=Vs,Vb,Vk} |fo_H - r_H| \right) / 2^{-i}. \tag{8}$$

The distance in texture between the subsequent two regions fo and r is computed by the following block distance:

$$g^{H}{}_{fo,r} = \left| Y_{gfo}/X_{gfo} - Y_{gr}/X_{gr} \right|. \tag{9}$$

The featured area of distance between two regions is computed by the following block distance:

$$g^{H,area}{}_{fo,r} = \left| A^{H}_{fo} - A^{H}_{r} \right|. \tag{10}$$

Feature localisation can be computed by the following block ratio:

$$FL = \left| Max^{H}{}_{fo}/Min^{H}{}_{fo} - Max^{H}{}_{r}/Min^{H}{}_{r} \right| \tag{11}$$

The Centroid (C_{fo}, C_r) of authentication bits between two histograms is measured by the distance equation:

$$gDH fo, r = \sqrt{\left[\left(C_{xfo} - C_{xr} \right)^2 + \left(C_{yfo} - C_{yr} \right)^2 \right)}. \tag{12}$$

To estimate the remaining histograms above equations are used. After detecting the 32 remaining histograms and embedding those with 32 observed histograms as shown in Fig. 31, the shape of 64 bits is represented by authentication function:

$$A^{E}_{g} = H_{dfo} * \sum_{i=1}^{fo} H_{fo}/d_{fo} + H_{dr} * \sum_{i=1}^{r} H_r/d_r \tag{13}$$

where d_{fo} and d_r represents the cyclic rotation in the clockwise and anti clock wise movement of the remaining histograms.

5.1 Embedding Algorithm

In our proposed embedding scheme, three analysis constraints were drawn. They are: histogram texture authentication, binary realised variable modulation technique and improving invisibility of embed bits by the use of multi-image spread technique. H_{fo} and H_r represents two histogram texture classifications have formed an authentication bits as shown in Fig. 31. The binary realised variable modulation techniques (BMT) defined as follows: BMT (i, j) = +q for integer positive modulation, −q for integer negative modulation and 0 for integer zero modulation. From the sign of BMT (:) we can know where embed bits are located. In third analysis, by calculating the order of embedding bits in authentication bits, invisibility can be improved.

Then the embedding is done by bit by bit as given in $E_m = E(I, A)$, where I is original image bits, A is authentication bits, E are estimators and E_m is embed signal bits. The embedding process is done based on E_m I A values; real values are based on the integer modulation threshold units (IMUs). Each IMU is represented by a integer value E^i_m, which is given as:

$$E^i_m(|I(i,j)| * A) = |W_i(x,y)|. \tag{14}$$

For integer, positive modulation $|Wp,o^{\,i}(x,y)| = E^{ip}_m(|I(i,j)| * A_p) * B_o(I(i,j)| * A_p)$, if $E^{ip}_m(|I(i,j)| * A) \geq E^i_m(|I(i,j)| * A)$.

For integer, negative modulation $|Wn,o^{\,i}(x,y)| = E^{in}_m(|I(i,j)| * A_n) * B_o(I(i,j)| * A)$, if $E^{in}_m(|I(i,j)| * A) < E^i_m(|I(i,j)| * A)$.

For integer, zero modulation $|W_{z,o}{}^{i}(x,y)| = E^{iz}_m(|I(i,j)| * A_z) * B_o(I(i,j)| * A)$, if $E^{iz}_m(|I(i,j)| * A_o) = -E^i_m(|I(i,j)| * A)$.

The modulated wavelet coefficient, either fall into $E^{ip}_m(|I(i,j)| * A_p)$ after positive modulation is applied, $E^{in}_m(|I(i,j)| * A_n)$ after negative modulation is applied and $E^{iz}_m(|I(i,j)| * A_z)$ after zero modulation applied.

5.2 Watermark Detection

Let $E^i_m(|I(i, j)| * A)$ represent the modulated wavelet coefficients which has attacks, the positive/negative/zero integer modulated watermarks value can be extracted using the following process $R^i_m(|I(i, j)| * A) = |W_i(x, y)/J_r(x, y)|$, which depends on the sign of (x, y). By comparing E^i_m and R^i_m the purpose of robust watermark can be achieved. $W_i(x, y) = (R_m(p(x, y) + 1) * J_r(x, y) + 1)$ and $Q_r(p(x, y) = (Q_r(p(x, y) + 1) * J_r(x, y) + 1)$, for positive modulation, $W_i(x, y) = (R_m(n(x, y) - 1) * J_r(x, y) - 1)$ and $Q_r(n(x, y) = (Q_r(n(x, y) - 1) * J_r(x, y) - 1)$, for negative modulation, and $W_i(x, y) = (R_m(z(x, y)) * J_r(x, y))$ and $Q_r(z(x, y) = (Q_r(z(x, y)) * J_r(x, y))$, for zero modulation.

On the other hand, by comparing the embed watermark, the extracting embed bits and the original image information the goal of robust watermark can be achieved using the following block functions: In robust if any tampering is detected, can be analysed by the relations

T(i) =+1 if $R^i_m > |R^{ip}_m|$ for positive tampering, = −1 if $R^i_m < |R^{in}_m|$ for negative tampering and = 0 if

$$R^i_m = |R^{iz}_m| \; for\ zero\ tampering. \tag{15}$$

Detection of robust embed bits are given by relations TR_{robust} = MAX ($|R^{ip}_m|$,$|R^{in}_m|$)/MIN($|R^{ip}_m|$,$|R^{in}_m|$).

During the watermark detection if any synchronization errors between the encoder and decoder, cause of geometric attacks, can be reduced by increasing received image energy levels. Equation 15, can be written as

$$\begin{aligned} \mathrm{T(i)} &= +1 * g \text{ if } \mathrm{R^i_m} > |\min(\mathrm{R^{ip}_m})| \text{ for positive tampering,} \\ &= -1 * g \text{ if } \mathrm{R^i_m} < |\min(\mathrm{R^{in}_m})| \text{ for negative tampering and} \\ &= 0 \text{ if } \mathrm{R^i_m} = |\min(\mathrm{R^{iz}_m})| \text{ for zero tampering.} \end{aligned} \tag{16}$$

Detection of robust embed bits after resisted the geometric attack are given by relations

$$\mathrm{TR_{robust}} = \mathrm{MAX}\left(\left|g_{max}, \mathrm{R^{ip}_m}\right|, \left|g_{max}, \mathrm{R^{in}_m}\right|\right) / \mathrm{MIN}\left(\left|g_{min}, \mathrm{R^{ip}_m}\right|\left|g_{min}, \mathrm{R^{in}_m}\right|\right).$$

6 Experimental Results and Discussions

Figure 1, shows twenty-five movement special variants introduce the image leading to £(x) constant the image texture with the QCIT texture classification plot, wherever boat image is employed because the take a look at the image.

Movement special Variants offer twenty-five variable texture pictures, these texture pictures area unit classified by embedding the image to extract the £(x) image texture. These values are units clustered into 1 cluster, forming embedding the constant image texture. These constant image textures are classed into octal planes with Quad Eb bit patterns, shown in Fig. 2.

Figure 3 shows methodology of Image cube-shaped unfold method, octa plane area units taken into equal four bits. Every plane four bits taken indiscriminately supported the Eb bit pattern elite. Every pattern is generated from its quad partitions. These Eb bit quad partitions were combined as quad-octagon bits. Unknown component positions and value area units big by the image cube-shaped unfold technique. Figure 3 shows the bit pattern generation mistreatment image cube-shaped unfold technique.

Embedding quad-octagon bits and image cube-shaped unfold bits; we tend to generate authentication bits. These authentication bits of 8 × 8 area unit embedded into a 256 × 256 the original image, the resultant watermark image is shown in Fig. 4a–c. Authentication bits were extracted from the watermark image with attacks is shown in Fig. 4d. The extracted original image texture beneath positive, negative and nil modulation is shown in Fig. 4e–h. Figure 4i–k shows the comparative results of [2] and our planned technique. Here, the tampered image is analysed to sight the comparison between the original and watermark attacked image. Figure 4k shows the watermark attack change of state visual estimation. Figure 4l–n shows the comparative results of [3] and our planned technique. Here, the authentication of 3 areas manipulated the image is reconstructing by our

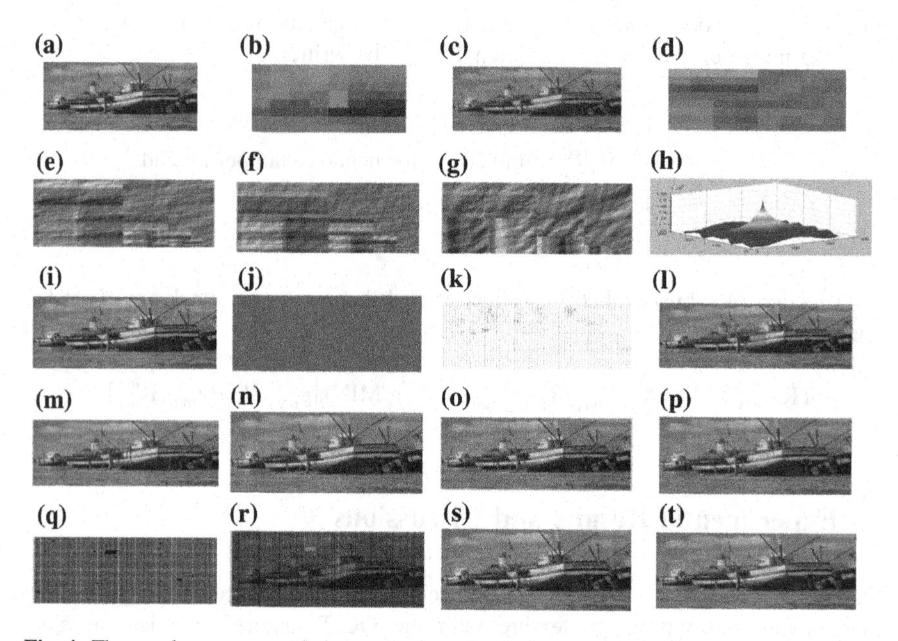

Fig. 4 The resultant watermark image. **a** Embedding image. **b** Image cubic spread method authentication bits. **c** Embed image. **d** Authentication bits detection with watermark attacks. **e** Extracted image texture with TR_{robust} is positive. **f** Extracted image texture TR_{robust} is negative. **g** Extracted image texture TR_{robust} is zero. **h** Autocorrelation map of normalized TR_{robust} in positive modulation. **i** Image after tampering [2]. **j** The tampering detection results at the $2^2 \sim 2^4$ scales with respect to t = 1 [2]. **k** Tampering estimation visualization [2]. **l** Authentication result of (**a**) [3]. **m** Ship tower, pole and ship manipulated [3]. **n** Authentication result of (**e**) [3]. **o** Watermark detection output: upper level signature verified [4]. **p** Watermark detection output: lower level signature verified [4]. **q** 128 block mask [5]. **r** Embed image 128 block mask [5] of (**l**). **s** Tamper detection of (**r**). **t** Recovered image

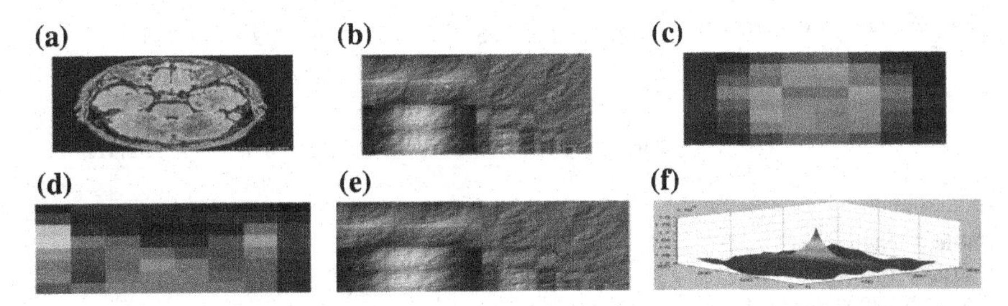

Fig. 5 Another example of the proposed approach. **a** Embedding image. **b** Embedding £(x) constant image texture. **c** Image cubic spread method authentication bits. **d** Authentication bits detection with watermark attacks. **e** Extracted image texture with TR_{robust} is positive. **f** Autocorrelation map of normalized TR_{robust} in positive modulation

proposed technique is shown. Figure 4o, p shows the comparative results of [4] and our planned technique. Here, the higher level and the lower level signature verification area units thought of. Figure 4q, r shows the comparative results of [5] and our planned technique. Here, for constant payload, most sixty-four introduce knowledge blocks area unit classified. Figure 4s, shows the tamper detection of (r) is shown, here the tampers are removed fully. Figure 4t, shows the recovered image, as a duplicate of the original embedding image. Figure 5 shows Associate in nursing another example of the planned approach on the tomogram image.

Similarity quality measuring is the analysed mistreatment normalized correlation outlined because the similarity measuring among the reference watermark and extracted watermark.

Table 1 shows the Block based PSNR and Normalized Comparisons, Tables 2 and 3 show the number of correctly read R_b bits within blocks in boat and tomogram image. Figure 6 shows the confidence measure of Boat and Tomogram images.

Tables 4 and 5 show the effectiveness of the proposed method with [2, 5].

Table 1 Block based PSNR and normalized comparisons

Image/block size	8 × 8			32 × 32			128 × 128		
	CR	PSNR	NC	CR	PSNR	NC	CR	PSNR	NC
Boat	1.25	48.52	0.99	2.35	49.56	0.98	3.12	50.22	0.98
Tomogram	1.89	47.25	0.98	2.68	48.22	0.97	3.58	51.69	0.96

Table 2 Number of correctly read R_b bits in blocks in boat image

Code length/bitrate	8	32	128
8	56	248	1016
64	448	2016	8128
256	1792	8160	32,512

Table 3 Number of Correctly read R_b bits in blocks in tomogram image

Code length/bit rate	8	32	128
8	48	240	1008
64	440	2008	8120
256	1784	8154	32,504

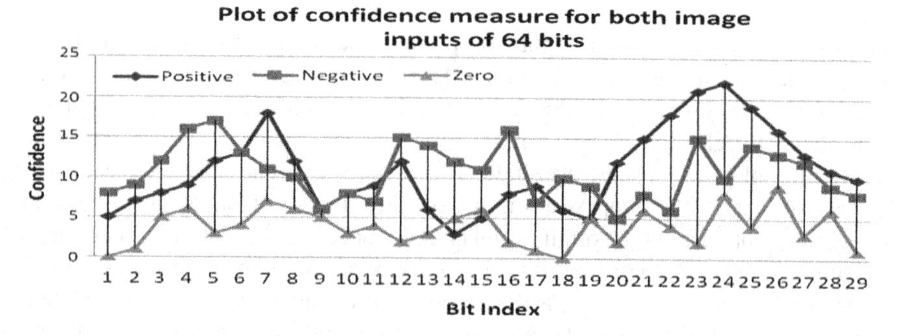

Fig. 6 The plot of the confidence measure of boat and tomogram images compare with [2]

Table 4 Tampering degree evaluation under SPIHT compression comparative with [2]

compression ratio	Degree of tampering						
	t = 1		t = 2		t = 3		INV
	NM	PM	NM	PM	NM	PM	
4	0.001	0.020	0.000	0.012	0.000	0.011	0.014
8	0.010	0.075	0.009	0.071	0.008	0.072	0.020
16	0.052	0.325	0.039	0.258	0.035	0.310	0.023
32	0.120	0.620	0.052	0.526	0.032	0.623	0.021
64	0.223	0.830	0.120	0.752	0.078	0.812	0.029
128	0.256	0.851	0.140	0.812	0.081	0.830	0.035

Table 5 Comparative test results with [5] in the proposed database

Images (512 × 512)	PSNR of marked image (db)			Data embedding capacity (bits)	Robustness (bpp)			Capacity (bits)
	Max	Min	Avg		Max	Min	Avg	
Boat	50.22	48.52	49.37	532	2.0	0.2	1.21	516
Tomogram	51.69	47.25	49.47	548	4.0	0.2	2.10	524

7 Conclusions

We introduce a brand new strong image authentication analysis methodology as Circular Quad-Octagon Bits embedding mistreatment Stepwise Image cubic unfold analysis. The options of planned methodology are, it will find the tamper beneath any physical attacks, manipulated image is recovered with genuine bits, and signature verification is done at each higher and lower level and may be able to insert 128 blocks at a time.

In our planned methodology, inserting image text variants were taken for authentication bits and are embedded with quad-octal bits to get embeds bits. Insert bits beneath 3 modulation schemes were conferred, results clearly shows the positive modulation theme extract the first image text variants. Here the octal planes with Quad Eb bits, patterns are at random generated with their equal and unequal variances. These variances are varied from one embedding image bits to a different that is a very important advantage of our planned methodology. Quad-Octagon Bits choice methodology generated eight Quads at a time, by reducing the issue in watermark insert bits generation and image cubic unfold methodology can perform the transcription of quad-octagon elite bits to get an even authentication bits to insert in original. These ways, robustly generated watermark bits from octal planes to demonstrate the first image, by overcoming the bit choice and insertion attacks. Our planned methodology is applied to the serious payload image codes, settles for serious attacks manipulation, and prevents malicious manipulations, strong to JPEG compression and information embedding capability ranges up to 65536 bits.

The planned methodology eliminates the attacks of image bit manipulations, because the attack effort is raised to 128 blocks, tamper localization accuracy is improved. Planned theme conjointly permits the detection of cropping attacks and block artifacts the attacks with the high level of confidence by still authenticating insert bits. Our planned methodology found to realize a unique methodology for a generation of authentication bits for secure image watermark bits.

References

1. Wong, Peter H.W., Au, Oscar C.: A capacity estimation technique for JPEG-to-JPEG image watermarking. IEEE Trans. Circuits Syst. Video Technol. **13**(8), 746–752 (2003)
2. Lu, C.-S., Mark, H.-Y.: Multipurpose watermarking for image authentication and protection. IEEE Trans. Image Process. **10**(10), 1579–1592 (2001)

3. Lin, C.-Y., Chang, S.-F.: A robust image authentication method distinguishing JPEG compression from malicious manipulation. IEEE Trans. Circuits Syst. Video Technol. **11**(2), 153–168 (2001)
4. Celik, M.U., Sharma, G., Saber, E., Tekal, A.M.: A hierarchical image authentication watermark with improved localization and security. In: Proceedings IEEE International Conference on Image Processing, vol. II, pp. 502–505, ©2001 IEEE, pp. 502–505, Oct 2001
5. Ni, Zhicheng, Shi, Yun Q., Ansari, Nirwan, Wei, Su, Sun, Qibin, Lin, Xiao: Robust lossless image data hiding designed for semi-fragile image authentication. IEEE Trans. Circuits Syst. Video Technol. **18**(4), 497–509 (2008)
6. Hsu, Chiou-Ting, Ja-Ling, Wu: Hidden digital watermarks in images. IEEE Trans. Image Process. **8**(1), 58–68 (1999)
7. Cox, J., Miller, M.L., Mc Kellips, A.L.: Watermarking as communications with side information. In: Proceedings of IEEE, Special Issue on Identification and Protection of Multimedia Information, vol. 87, pp. 1127–1141, July 1999
8. Eggers, Joachim J., Bäuml, Robert, Tzschoppe, Roman, Girod, Bernd: Scalar costa scheme for information embedding. IEEE Trans. Signal Process. **51**(4), 1003–1019 (2003)
9. Tang, Chih-Wei, Hang, Hsueh-Ming: A feature-based robust digital image watermarking scheme. IEEE Trans. Signal Process. **51**(4), 950–959 (2003)
10. Barni, Mauro, Bartolini, Franco, Checcacci, Nicola: Watermarking of MPEG-4 video objects. IEEE Trans. Multimedia **7**(1), 23–32 (2005)
11. Bors, Adrian G.: Watermarking mesh-based representations of 3-D objects using local moments. IEEE Trans. Image Process. **15**(3), 687–701 (2006)
12. Akhaee, Mohammad Ali, Mohammad Ebrahim Sahraeian, S., Sankur, Bulent, Marvasti, Farokh: Robust scaling-based image watermarking using maximum-likelihood decoder with optimum strength factor. IEEE Trans. Multimedia **11**(5), 822–833 (2009)
13. Subramanyam, A.V., Emmanuel, Sabu, Kankanhalli, Mohan S.: Robust watermarking of compressed and encrypted JPEG2000 images. IEEE Trans. Multimedia **14**(3), 703–716 (2012)
14. Chang, T., Kuo, C.-C.J.: Texture analysis and classification with tree-structure wavelet transform. IEEE Trans. Image Processing **2**(4), 429–441 (1993)
15. Unser, M.: Texture classification and segmentation using wavelet frames. IEEE Trans. Image Processing **4**, 1549–1560 (1995)

Towards the Next Generation of Web of Things: A Survey on Semantic Web of Things' Framework

Farhat Jahan, Pranav Fruitwala and Tarjni Vyas

Abstract The concept behind Semantic Web of Things (SWoT) is to extend the IoT (Internet of Things) architectures and to provide advanced resource management and discovery. It also uses integration of knowledge representation and reasoning techniques which are basically devised from the Semantic Web. The combining of two technologies aim towards the association of semantic annotations to real world objects. This paper discusses such three frameworks which are SWoT framework based on Ubiquitous Knowledge Base, CoAP based framework and smart gateway framework that have been developed and proposed for the extended IoT (SWoT) along with the basic challenges faced in the IoT.

Keywords Internet of things · CoAP · Semantic web · Knowledge base · Resource discovery · Ontology · RDF

1 Introduction

The next big thing that will entirely change the environment of the services around the users is Internet of Things (IoT). A world-wide connectivity between the objects will be enabled using IoT. Three major technologies which provides first layer of heterogeneity on IoT are: (I) Attached Devices (II) Sensing and Actuating Devices (III) Embedded Devices. IoT can be considered as a set of heterogeneous devices and various heterogeneous communications strategies. IoT would evolve a heterogeneous system where devices or things can be uniformly discoverable and enabled to communicate with other devices or things. When Semantic Web technology is

F. Jahan (✉) · P. Fruitwala · T. Vyas
Institue of Technology, Nirma University, Ahmedabad 382481, India
e-mail: 14mcen08@nirmauni.ac.in

P. Fruitwala
e-mail: 14mcen06@nirmauni.ac.in

T. Vyas
e-mail: tarjni.vyas@nirmauni.ac.in

S.C. Satapathy and S. Das (eds.), *Proceedings of First International Conference on Information and Communication Technology for Intelligent Systems: Volume 1*, Smart Innovation, Systems and Technologies 50, DOI 10.1007/978-3-319-30933-0_4

used to extend IoT, it is referred as SWoT (Semantic Web of Things). The goal of SWoT is to provide easy accessible information to the real world objects as well as locations and events and also associate semantically rich content. Resource discovery is a very important feature for SWoT. Many technologies like 6LOWPAN [1] and CoAP [2, 3] (Constrained Application Protocol) are gaining acceptance due to the optimization. Real-world entities are not that much of use when considered in isolation; the ability to put multiple entities into a common semantic context is needed [4].

The details of the paper are as given: Sect. 2 discusses the open issues in IoT, Sect. 3 discuss the SWoT framework based on the Ubiquitous Knowledge Base. Section 4 gives a brief introduction about the CoAP based framework. Section 5 discuses about the smart gateway framework.

2 Issues and Challenges in Internet of Things (IoT)

There are many issues which are still needed to be tackled to explore the full potential of Internet of Things (IoT).

2.1 *Heterogeneity and Scalability [5]*

In general, IoT handles the heterogeneity problem but still there are many minor problems which still needs to be resolved. IoT will require a huge number of heterogeneous devices to be integrated to the current web. Also the devil level heterogeneity will be a serious issue for the IoT. IoT will just be a castle in the air if there won't be interoperation support from lower levels [5]. Even the customers may have heterogeneous requirements as many may need real time information and others may need archived data. The performance may degrade drastically if the IoT is not designed using an efficient management design as the environment is completely distributed.

2.2 *Security and Privacy [5]*

There has always been a contradiction between sharing and security. Enabling IoT in the existing web may lead to problems like ensuring security to the shared resources along with protecting the privacy of the users and provide reliability. Even though there are lot of security approaches which have become mature but not every approach can be applied to IoT environment as it is distributed and heterogeneous. A universal boot strapping method needs to be implemented whenever two nodes want to join.

2.3 *Search and Discovery [5]*

There are basically two approaches which are used to develop a search engine for the IoT. The people and the things should be able to discover the existence, functionality and information of their desired web services [5]. The two approaches are: (1) Push approach where the outputs of the sensors are proactively pushed to the search engine but this method lacks scalability and only works with limited number of devices. (2) Pull Approach: A user query is generated and then only the search engine forwards the query to the sensors to pull relevant data and also this method is scalable.

3 SWoT Framework Based on Ubiquitous Knowledge Base

This is a general framework for the SWoT evolved over the basic knowledge base model and can be called ubiquitous Knowledge Base. A ubiquitous knowledge base consists of individuals who are physically tied to micro devices which belongs to a big distributed knowledge and no central coordination is required [6]. There are basically two kinds of knowledge in the semantic web. One is the conceptual knowledge which is to cover the problem domain and other is factual knowledge which deals to the specific instances. General properties of the relationships can be described using the domain conceptualization which is an *ontology*. Knowledge Base can be created from an ontology along with set of asserted facts. In this framework, the Knowledge Base is intended as a fixed and centralized component. There are several object classes defined which can exist in a physical environment and system infrastructure can be shared between them (Fig. 1).

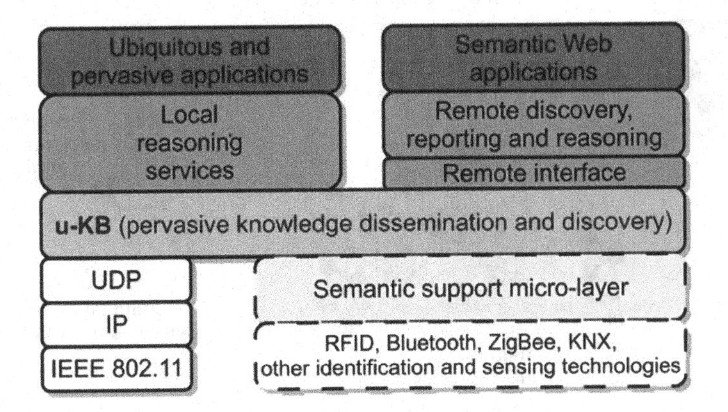

Fig. 1 Semantic web of things framework [6]

To mark the ontologies unambiguously, they adopt Ontology Universally Unique Identifier (OUUID) codes. The framework OOUID's are mostly of fixed length and are basically much shorter than URL's. OUUID-to-URL mapping mechanism is granted whenever the internet connection is available. Each resource in the framework can be characterized as: (I) 96-bit globally unique ID. (II) 64-bit OUUID (III) Set of data oriented attributes (IV) semantic annotation. Data oriented attributes can allow to integrate and extend logic based inferences whereas semantic annotation is used to stored RDF/OWL [7] in a compressed form [8]. The u-KB needs no centralized supervision and the basic tell and ask paradigm is inherited from the KBs but implemented with extensions.

Figure 2 depicts the two level infrastructure. The field layer exploits pervasive sensing and identification technologies. Inter-host communication which is basically required for knowledge dissemination and retrieval is dealt by the discovery layer. Using various available interfaces, every network host becomes cluster head for the devices present in the field. For interaction among hosts, a cooperative IP protocol is used which will also help in information dissemination and resource retrieval. The following protocol has four interaction stages: (1) Resource Parameters extraction (2) Information dissemination of resource (3) Peer to peer resource discovery (Collaborative) (4) Selected resource annotations are extracted. A smart pervasive environment is populated by the framework by providing access

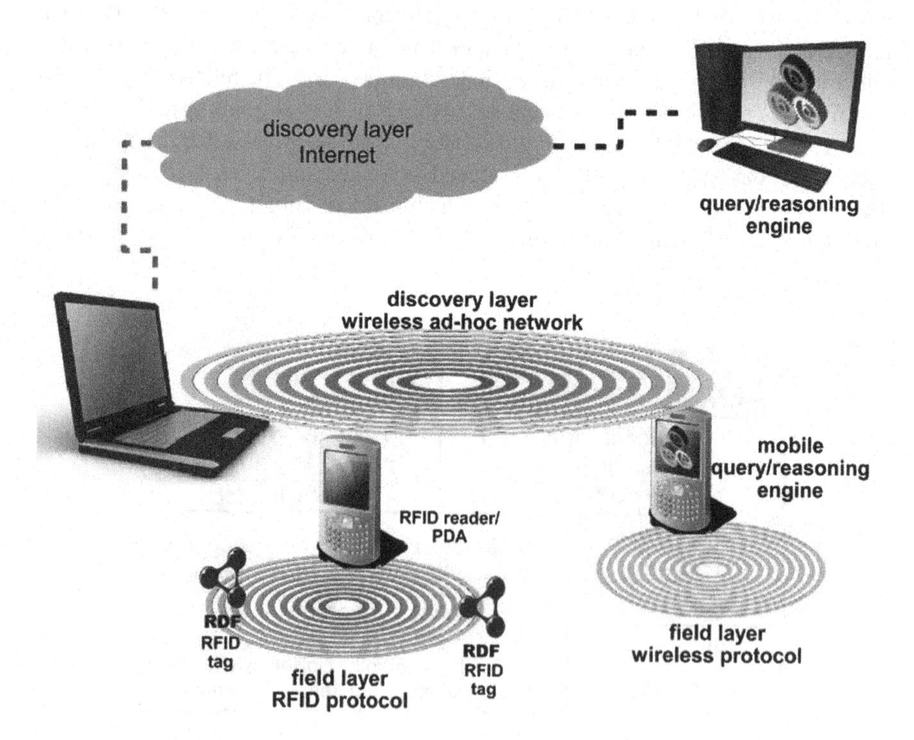

Fig. 2 Proposed semantic web of things framework architecture [6]

to common information embedded into semantic enhanced micro devices. Local hosts can perform information processing and reasoning tasks or the same can be performed by remote entity through a gateway exposing a high level interface. To adapt to the framework, a semantic micro layer is required by communication, identification and sensing technology.

4 A CoAP Based Framework for Resource Annotation, Dissemination and Discovery

The following SWoT framework was proposed the backward compatible extension of the CoAP. The Constrained Application Protocol [2, 3] supported non-standard inference services and allowed retrieval and logic based ranking of the annotated resources. The framework introduces the following: (I) CoRE Link Format, a resource discovery protocol and modified CoAP i.e. backward compatible extension. (II) To detect and annotate high level events from raw data, efficient data mining procedures are introduced. (III) Retrieve and rank resources by semantic based match making.

4.1 Semantic Enhanced CoAP and CoRE Link Format

URI is used as identification for each resource which is a server controlled abstraction and unambiguous. In general, CoAP message is formed of: (I) 32 bit header (Request method or response status). (II) Optional token value (Associate replies to requests). (III) Options fields (URI and payload media type). (IV) Payload data. Here, each sensor can be considered as a server and hence exposes both reading and internal information as resources. CoAP also supports proxies, cluster-head or sink nodes and hence a reply can be generated on behalf of more constrained sensor nodes, decreasing the load at the edge of network. Standard URI-query options has been added to the CoAP protocol for its improvement. The other three attributes that have been improved are: (I) Reference Ontology (Contains the URI). (II) Semantic Description (Annotated Request). (III) Annotation Type (Compression format). Geographical location is achieved by the longitude and latitude attributes.

4.2 Sensory Data Annotation

Figure 3 shows an explicative architecture of the framework. Several sensors deployed in a given area and communicating using a sink node who acts as a cluster head and a gateway to interface the network to the rest of world composes a CoAP based SSN. Sink nodes or Cluster Heads will allow sensors to register as a CoAP

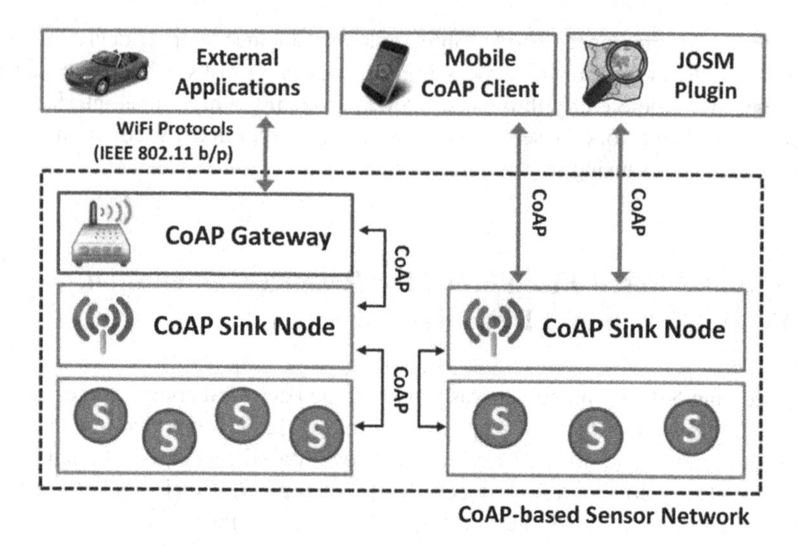

Fig. 3 CoAP based framework architecture [9]

resource and embed a lightweight matchmaker. Two types of access are possible to access the devices within the SSN: (1) CoAP Clients (Exploit semantic based discovery). (2) Remote Applications (Based on Wireless Protocols). Several stages for identification of sensory events are as follows:

A. Standard CoAP GET requests are used to read data from the sensors. A list of elements is built consisting of ID, identifier of sensor and the value of the data containing the *timestamp*. Time slots of application defined period T are created for the group measurements.
B. For current time slot, average, variance and standard deviation values are computed for each data set to know the variability.
C. To know the significant event changes in the area, an incremental ratio is computed from the statistical indexes of the elapsed periods.
D. A binary or multiple classifier is defined by the application for every data collection to reveal a situation when some condition occurs.
E. Every classifier generates an output which is a logic based expression and it is mapped according to the knowledge in the reference ontology. Logical conjunction of all derived expressions is done to obtain the final semantic description.

4.3 Resource Discovery via Concept Covering

The basic CoAP resource discovery protocol only allows syntactic string matching which lacks resource semantics and hence advanced discovery services should be adapted which should the following capabilities: (1) Ranking Resources,

(2) Identify partial correspondences. The following gives the detailed description how resource discovery is carried in the refined version of CoAP. Aim is to select minimum resources to best cover a request. Given a request R and resources $P = \{P_1, P_2, \ldots, P_k\}$ where request R and resources P are satisfiable with respect to ontology O. Concept Covering Problem aims to find a pair (P_c, H) where P_c includes concepts in D covering R with respect to T and H is the part of D not covered by concept in P_c.

$$rank(D, H_i) = 100 * \left[1 - s\_match(D, H_i) * \left(1 + \frac{distance(P, s_i)}{md}\right)\right] \quad (1)$$

where s_match measures the semantic distance between the D and a description H_i; it can be considered as the geographical distance of Sensor S_i from reference point P [9]. The sensor with the highest rank (S_{max}) is selected.

5 The Semantic Smart Gateway Framework

Integrating 'things' seamlessly with the current web infrastructure is an attempt by IoT (Internet of Things). To extend WoT (Web of Things), semantic web [10] can be used and it is referred as SWoT. This can be achieved by involving computing and storing alignments of ontologies in such a manner that they can be utilized together. A new level of entities is introduced here referred as 'Smart Entities'. The 'smart entities' can either cross domain or domain specific. The basic requirements that should be satisfied by a Semantic Smart Gateway are: (I) Registering smart entities by a semantic method. (II) On the fly ontology for the support of semantic description of the entities. (III) On the fly ontology alignment is used for the similarities between the entities during the run time. (IV) For allowing agents to place ontology driven queries, a semantic retrieval component should be provided. Also, automatically registering smart entities facility has to be provided to agents.

Smart entities which are already registered must coordinate in a semantic manner towards providing: (I) Retrieve data which is a requirement of intelligent applications. (II) Cross domain entities or domain specific should be automatically shaped to clusters. (III) For efficiency reasons, it should be able to merge similar smart entities.

5.1 *IoT Ontology*

As there is limit of space, IoT ontology can't be discussed but the following URL http://purl.org/IoT/iot [11] gives us the latest populated version of ontology where vocabularies are already imported. Task of computing alignments and ontology learning task between the ontological definitions is performed using the semantic registry or IoT Ontology.

5.2 On the Fly Ontology Learning

Lightweight domain ontology is to be created automatically or semi-automatically using the information given at run time which is performed by the following component [12]. The following learning strategy can be outlined as: (I) Mapping of OWL ontology elements to RDF schema by following some specific rules and automatically computes Relation to OWL mappings. (II) Inspection of the classes and modifying their names should be allowed to the agents or Designers.

5.3 On the Fly Ontology Alignment

For semi-automated discovery of similarities between smart entities, this component is to be designed. Given two Ontologies: $O_1 = (S_1, A_1)$ and $O_2 = (S_2, A_2)$ where S_i denotes the signature and A_i are the set of axioms and locating corresponding element E_i in the signature of S_1 and E_j in the signature of S_2 such that a relation (E_i, E_j, r) holds between them where r can be any relation such as equivalence [11]. A mapping method may relate a value γ that represents the preference to relating E_i with E_j via r.

6 Conclusion

The paper discussed on the various proposed and developed frameworks for the Semantic Web of Things. Infrastructural protocols are used for information management along with knowledge storage and processing. Backward compatible CoAP extension is used for supporting flexible resource description, management and discovery. Smart gateway framework is proposed for the support of on the fly semi-automated translation process at the run time and minimizes the human factor. Thus, the future devices will be smart as the Semantic Web technology is combined with the Internet of Things providing semantic based communication between the devices as well as resource discovery.

References

1. Kushalnagar, N., Montenegro, G., Schumacher, C.: IPv6 over low-power wireless personal area networks (6LoWPANs): overview, assumptions, problem statement, and goals. No. RFC 4919 (2007)
2. Bormann, C., Castellani, A.P., Shelby, Z.: Coap: an application protocol for billions of tiny internet nodes. IEEE Internet Comput. **2**, 62–67 (2012)
3. Shelby, Z., Hartke, K., Bormann, C.: The constrained application protocol (CoAP) (2014)

4. Pfisterer, D., et al.: SPITFIRE: toward a semantic web of things. Communications Magazine, IEEE **49**(11), 40–48 (2011)
5. Zeng, D., Guo, S., Cheng, Z.: The web of things: a survey. J. Commun. **6**(6), 424–438 (2011)
6. Ruta, M., Scioscia, F., Sciascio, E.D.: Enabling the semantic web of things: framework and architecture. In: 2012 IEEE Sixth International Conference on Semantic Computing. IEEE (2012)
7. Chenzhou, Yu, et al.: Active linked data for human centric semantic web of things. In: 2012 IEEE International Conference on Green Computing and Communications (GreenCom). IEEE (2012)
8. Scioscia, F., Ruta, M.: Building a semantic web of things: issues and perspectives in information compression. In: 2009 IEEE International Conference on Semantic Computing. IEEE (2009)
9. Ruta, M., et al.: Resource annotation, dissemination and discovery in the semantic web of things: a CoAP-based framework. In: IEEE International Conference on Green Computing and Communications (GreenCom), 2013 IEEE and Internet of Things (iThings/CPSCom), and IEEE Cyber, Physical and Social Computing. IEEE (2013)
10. Berners-Lee, T., Hendler, J., Lassila, O.: The semantic web. Sci. Am. **284**(5), 28–37 (2001)
11. Kotis, K., Katasonov, A.: Semantic interoperability on the web of things: the semantic smart gateway framework. In: Sixth International Conference on Complex, Intelligent and Software Intensive Systems (CISIS). IEEE (2012)
12. Smirnov, A., et al.: On-the-fly ontology matching in smart spaces: a multi-model approach. In: Smart Spaces and Next Generation Wired/Wireless Networking, pp. 72–83. Springer, Berlin (2010)

Performance Analysis of Dynamic Addressing Scheme with DSR, DSDV and ZRP Routing Protocols in Wireless Ad Hoc Networks

Nagendla Ramakrishnaiah, Pakanati Chenna Reddy and Kuncha Sahadevaiah

Abstract Mobile ad hoc networks (MANETs) are wireless, infrastructure-less and multi-hop networks consisting of mobile nodes. All the aspects of network initialization, operation and maintenance are performed by the host nodes. Host nodes also act as routers to send the information to other nodes in the network. Most of the research in ad hoc networks focused on routing, but another important issue in network layer is addressing. All the nodes participating in a communication, needs unique address. In literature, several addressing schemes have been proposed and each one has its own strengths and weaknesses. In this paper, performance analysis of variable length addressing scheme in terms of communication cost and latency is done by considering DSR, DSDV and ZRP routing protocols. The simulation results show that our addressing scheme works effectively and gives consistent results with the routing protocols considered.

Keywords Mobile ad hoc network · Addressing schemes · Routing protocols · DSR · DSDV · ZRP

N. Ramakrishnaiah (✉) · K. Sahadevaiah
Department of Computer Science and Engineering, University College of Engineering, Jawaharlal Nehru Technological University, Kakinada, Andhra Pradesh 533003, India
e-mail: nrkrishna27@gmail.com

K. Sahadevaiah
e-mail: ksd1868@gmail.com

P.C. Reddy
Department of Computer Science and Engineering, JNTUA College of Engineering, Pulivendula, Andhra Pradesh 516390, India
e-mail: pcdreddy1@rediffmail.com

S.C. Satapathy and S. Das (eds.), *Proceedings of First International Conference on Information and Communication Technology for Intelligent Systems: Volume 1*, Smart Innovation, Systems and Technologies 50, DOI 10.1007/978-3-319-30933-0_5

1 Introduction

In MANETs address autoconfiguration is a tough job due to lack of infrastructure and node mobility. An addressing scheme not only assigns unique addresses to mobile nodes but also maintains the address pool efficiently. An addressing protocol must cope with dynamics of the ad hoc network. When a set of nodes leave the network, their addresses must be recollected for future use and also when a group of nodes join the network, new addresses are assigned or conflicts are resolved to the newly joined nodes.

In this paper we analyze the performance of distributed variable length addressing scheme by executing with various categories of routing protocol. This addressing scheme also considers the node departures and network merging.

Routing protocols in MANETs are organized as proactive, reactive and hybrid protocols. Proactive routing protocols maintains the routing information in tables and the consistency of information in tables is maintained with the periodic exchange of topological updates. These protocols suffer from high overhead in maintaining routing tables, but delay is less in constructing a route. In reactive routing protocols a route is established when there is a need. The delay in establishing a route is high and overhead is less in these protocols. Hybrid routing protocols mixes the characteristics of both proactive and reactive routing protocols. In ideal conditions, the performance of hybrid protocols is good when compared to other routing protocols.

To study the performance of our variable length addressing scheme, we choose Destination Sequenced Distance Vector (DSDV) protocol to represent proactive routing protocol, Dynamic Source Routing (DSR) protocol to represent reactive routing protocol and Zone Routing Protocol (ZRP) to represent hybrid routing protocol.

The remaining part of this paper is organized as follows: Sect. 2 gives the literature review of addressing protocols, Sect. 3 presents the description of the addressing protocol and routing protocols considered, Sect. 4 presents the simulation results and Sect. 5 concludes the paper.

2 Related Work

Address autoconfiguration protocols in MANETS are classified as stateful, stateless and hybrid protocols.

In stateful protocols, every node maintains the addresses of other nodes, so that uniqueness is guaranteed and delay is low, but the communication overhead is high to maintain address information.

In MANETConf [1], every node maintains two tables: Allocated table contains the addresses of other nodes and Allocate_Pending table contains the address whose allocation is initiated but not completed. A new node wishes to join the network sends an address request to one of its neighbor nodes. The chosen node selects an address neither in Allocated table nor in Allocate_Pending table and gets the confirmation from other nodes and assigns this address to new node. This protocol suffers from high overhead in maintaining the address information.

In Prophet protocol [2], every node uses a common function f(n) and a different seed value to generate a disjoint series of numbers to be used as addresses. The function f(n) is carefully planned such that the occurrence of same sequence of numbers is very low. This scheme works better for small size networks, but the size of network increases the probability of occurrence of same sequence of integers is also increases.

In Wang et al. [3], a distributed IPv6 address configuration scheme is combined with clustering mechanism. The addressing mechanism is distributed across cluster heads. The cluster head gets an address from root node. The cluster heads maintains and assigns the addresses to its members. This scheme also considers the cases of address reclamation and network merging. The cost and delay increases with the number of nodes in the network.

Fernandes et al. [4] suggested a filter based addressing scheme for MANETs. A filter is a small database used to store the addresses of configured nodes. This scheme works better for small size networks. The performance of this scheme degrades with the size of the network, because filters can't accommodate large number of addresses.

In stateless schemes, nodes do not maintain the information about addresses. Instead, a new node chooses an address for itself and confirms its distinctiveness by executing a Duplicate Address Detection (DAD) process. These protocols using DAD schemes suffer from high delay.

Perkins et al. [5] suggested an addressing scheme called as query based DAD. In this scheme, a new node floods an address request having chosen address. A configured node using the same address sends reply message indicating address clash. The new node discards previous address and selects another address and repeat the same process to obtain an address.

Weniger [6] suggested another stateless scheme called as passive DAD. This protocol depends on routing protocol. A new node selects an address and assigns itself. Every node observes the incoming routing packets and detects address conflicts. In this scheme address uniqueness is not guaranteed.

Munjal et al. [7] presented a stateless protocol by using IPv4 private address block. The network is organized as hierarchical clusters and cluster headers assign the addresses to new nodes. The new node sends an address request and cluster heads send reply messages. The new node chooses the cluster head whose hop distance is less. Address allocation latency depends on the hop distance, so, as the size of the network increases latency also increases.

Hybrid protocols merge the characteristics of both stateless and stateful schemes. These protocols give better performance than other schemes, but suffer from high overhead and high complexity.

Sun et al. [8] suggested a hybrid addressing scheme called as Hybrid Centralized Query based Auto-configuration (HCQA). It uses query-based DAD with centralized allocation table. A node randomly chooses an address and verifies its duplication using QDAD in centralized allocation table available at central agent. In this scheme the control overhead is high in maintaining the central agent.

Weniger [9] suggested another hybrid addressing scheme called as Passive Auto-configuration for Mobile Ad hoc Networks (PACMAN). It uses passive DAD with a common allocation table distributed all over the network. This protocol also suffers from high overhead to synchronize common allocation table distributed across the network.

Wang and Qian [10] suggested a IPv6 addressing scheme, which combines the features of both centralized and distributed addressing schemes. The network is formed as clusters and centralized scheme is used to assign addresses to cluster heads. The cluster headers uses distributed scheme to assign addresses to its members. This scheme works better, but results in high complexity.

From the above discussion, still there is no acceptable protocol which satisfies all the addressing issues. An Addressing protocol has to assign unique addresses to mobile nodes and maintain address pool with less cost and latency, scalable and also work efficiently with all routing protocols. Some addressing schemes depend on routing protocols and these schemes give good results with those routing protocols and worst results with other routing protocols. The addressing scheme used in this paper does not depend on routings protocols, has less cost and latency, scalable and also works efficiently with all routing protocols.

3 Description of Protocols

This section briefly describes the addressing scheme and routing protocols used in this paper.

3.1 Tree Based Variable Length Addressing Scheme

We used the variable length address denoted in binary codes starting with some initial length, say n and assign addresses from 0 to $2^n - 1$ and increments address length when its address range is exhausted. The network is formed as a tree and its nodes are classified as header, coordinator and normal nodes. The first coordinator node is called as header and it is responsible for defining network identifier, incrementing address length and address reclamation. The header node maintains the records of address allocation, list of other coordinator nodes and leak address list. The coordinator node is in charge for address assignment and has the records of address allocation and other coordinator nodes list. The normal node just act as relaying node between coordinator node and new node. The tree is established with header as first node and subsequently added nodes are denoted as normal node if the hop distance is less than four, else denoted as coordinator node. If the address range is exhausted then address increment by one bit takes place and all nodes in the network update their address by adding 0 to the left of its address.

When a node gets a data packet with dissimilar network id, it identifies network merging and passes the merge notification to the headers of both networks. The headers exchange their records and higher network header will be the header of whole network, new address length is higher network length + 1 and new network id is generated. The new records are defined by combining the records of both networks and address update takes place by adding 1 s to the address of higher network nodes and adding 0 s to the left of lower network nodes.

3.2 Destination Sequenced Distance Vector Routing Protocol (DSDV) [11]

This protocol maintains the information of all routes in the form of table. This table contains destination node, first node in the path, path distance and update message id. The consistency of the table is maintained by continuously advertising the topological changes of the network. The updates are either incremental or full dumps. The incremental update takes place with single network data packet unit (NDPU), when there is a single change. Full dumps take place with multiple NDPUs, when there is more than one change.

3.3 Dynamic Source Routing Protocol (DSR) [12]

The route is found by flooding a Route_Request message with destination node. The target node or any middle node having the path to destination in its route cache sends a Route_Reply message to the source node in reverse direction. Upon getting the reply message source node starts sending data to destination node. It uses route cache that saves the information dig out from source route available in a data packet. Route caches help to construct the routes in less time.

3.4 Zone Routing Protocol (ZRP) [13]

This protocol defines various zones in the network with radius = 1 hop, 2 hops, etc. Within a zone it uses proactive routing mechanism to find a path and the protocol used is called as intra-zone routing protocol. The reactive routing mechanism is used across multiple zones to find a path and the protocol used is called as inter-zone routing protocol.

4 Simulation

The variable length addressing mechanism is simulated using NS—2 [14] with DSDV, DSR and ZRP routing protocols. The performance is analyzed with metrics cost and latency of the addressing scheme. The simulation parameters used are: Simulation area—500 m X 500 m, Max. Speed—15 m/s, Mobility model—Random waypoint mobility, Communication range—100 m, Pause time—15 s, No. of nodes—50 to 100 and Medium access protocol—IEEE 802.11. The nodes of the network are distributed across the simulation area in a random fashion.

4.1 Addressing Performance

4.1.1 Cost

It is the amount of control messages required to allocate an address to a single node. The network is built by joining the nodes one after another with an interval of 0.2 s. All the unicast, broadcast and flooding messages are added to find the cost of addressing scheme. The average cost of a single node to obtain an address is shown in Fig. 1.

The cost of the network is defined as total quantity of control messages used to allocate addresses to all nodes in the network. Network is built by making all the nodes enter into the network at the same time. The cost of all nodes to obtain an address is shown in Fig. 2.

From Figs. 1 and 2, it is observed that the cost of a single node or of all nodes in the network to acquire an address varies very little from one routing protocol to another. For example, in a network of 50 nodes the cost of all nodes to obtain an address in DSDV and ZRP is 507 and 519 respectively. Hence, the variable length

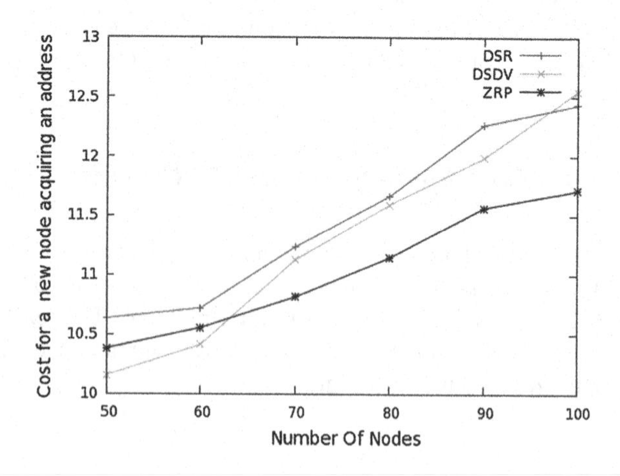

Fig. 1 Average cost of a single node obtaining an address

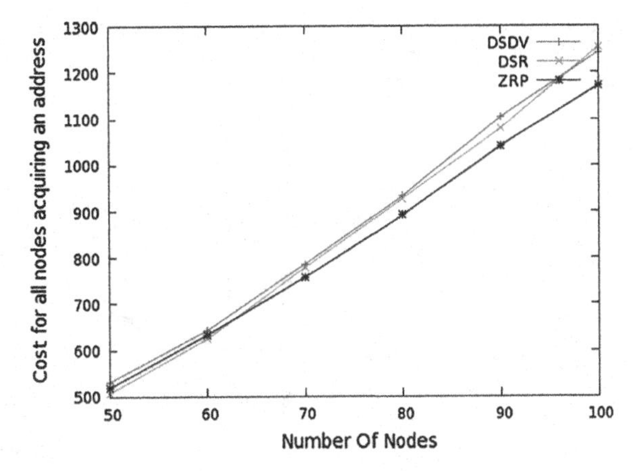

Fig. 2 Cost of all nodes obtaining an address in a network

addressing scheme used in this paper gives consistent results with all types of routing protocols.

4.1.2 Latency

It is the interval between initiation of addressing task and acquiring an address. All the values of timers, unicast, broadcast and flooding delays are added to find the latency of the addressing task. The network is built by joining the nodes one after another with an interval 0.2 s to find the average delay of the node to get an address. The average latency of a single node obtaining an address is shown in Fig. 3.

The latency for all nodes of the network is defined as total time taken to assign addresses to all nodes in the network. Network is built by making all the nodes enter into the network at the same time. The latency of all nodes acquiring an address is shown in Fig. 4.

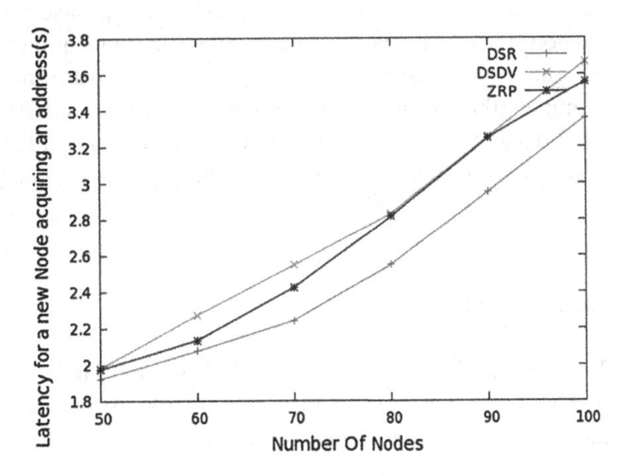

Fig. 3 Average latency of a single node obtaining an address

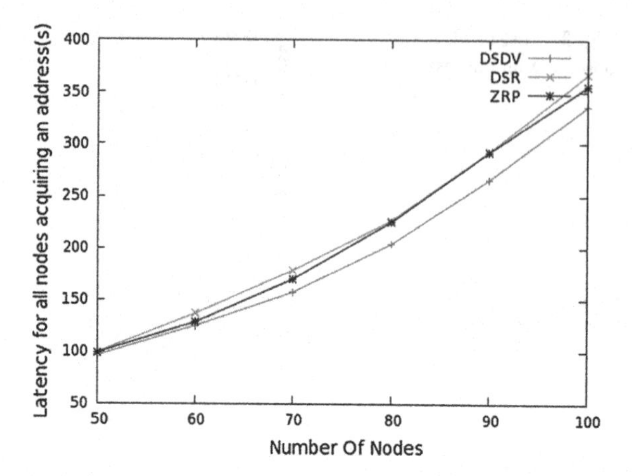

Fig. 4 Latency of all nodes obtaining an address in a network

From Figs. 3 and 4, it is observed that the latency of a single node or of all nodes in the network to acquire an address varies very little from one routing protocol to another. For example, in a network of 60 nodes the average latency of a single node to obtain an address in DSR and ZRP is 2.07 and 2.13 s respectively. So that, the variable length addressing scheme used in this paper gives consistent results with all types of routing protocols.

4.2 Merging Performance

4.2.1 Cost

The merging cost is calculated as the amount of control messages required to assign the addresses to all merged network nodes. In our simulation, two networks are randomly deployed in the simulation area such that the nodes of one network not in the communication range of other network and also configured independently. If the nodes of one network come into the communication range of other network then merging process is initiated. To find merging cost of a network with total size, for example 100, we run all the combinations of 10 and 90, 20 and 80, 30 and 70, 40 and 60 and 50 and 50 and finally took the average of these costs. The cost of other network sizes is also calculated in the same way. The merging cost is shown in Fig. 5.

4.2.2 Latency

The merging latency is calculated as the time taken to assign the addresses from first node to last node in merged network. In our simulation, two networks are randomly deployed in the simulation area such that the nodes of one network not in

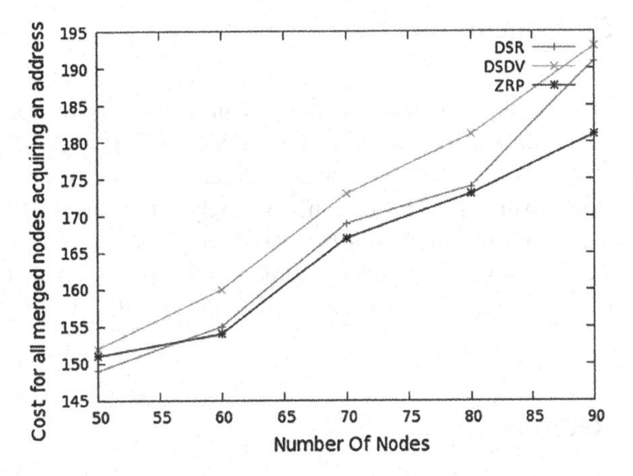

Fig. 5 Cost of network merging

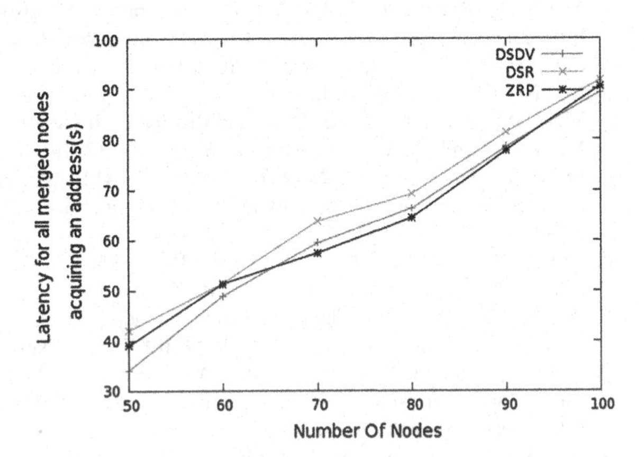

Fig. 6 Latency of network merging

the communication range of other network and also configured independently. If the nodes of one network come into the communication range of other network then merging process is initiated. To find merging latency of a network with total size, for example 100, we run all the combinations of 10 and 90, 20 and 80, 30 and 70, 40 and 60 and 50 and 50 and finally took the average of these latencies. The latency of other network sizes is also calculated in the same way. The merging latency is shown in Fig. 6.

From Figs. 5 and 6, it is seen that the merging cost and latencies are varying very low from one routing protocol to another. For example, in a network of 80 nodes the latency of network merging in DSDV and ZRP is 66.30 and 64.41 s respectively. Hence, our addressing scheme gives consistent results with all kinds of routing protocols.

5 Conclusion

This paper presented the performance analysis of variable length address autoconfiguration by executing with DSR, DSDV and ZRP routing protocols. These protocols are used to represent reactive, proactive and hybrid routing protocols respectively. The performance is analyzed in terms of cost and latency. The simulation results show that the variation of the metrics from one routing protocol to another is very less. The addressing scheme used in this paper gives consistent results with all three routing protocols considered.

References

1. Nesargi, S., Prakash, R.: MANETconf: Configuration of hosts in a mobile ad hoc network. In: Proceedings of IEEE INFOCOM 2002, vol. 2, pp. 1059–1068. IEEE Press, New York (2002)
2. Zhou, H., Ni, L.M., Mutka, M.W.: Prophet address allocation for large scale MANETs. Ad Hoc Netw. **1**, 423–434 (2003)
3. Wang, X., Qian, H.: Cluster-based and distributed IPv6 address configuration scheme for a MANET. Wireless Pers. Commun. **71**, 3131–3156 (2013)
4. Fernandes, N.C., Moreira, M.D.D., Duarte, O.C.M.B.: An efficient and robust addressing protocol for node autoconfiguration in ad hoc networks. IEEE Trans. Netw. **21**, 845–856 (2013)
5. Perkins, C., Malinen, J.T., Wakikawa, R., Belding-Royer, E.M., Sun, Y.: IP address autoconfiguration for ad hoc networks. IETF draft (2001)
6. Weniger, K.: Passive duplicate detection in mobile ad hoc networks. In: Proceedings of IEEE WCNC 2003, vol. 3, pp. 1504–1509, IEEE Press, New Orleans, LA, USA (2003)
7. Munjal, A., Singh, Y.N., Phaneendra, A.K., Roy, A.: IPv4 based hierarchical distributive autoconfiguration protocol for manets. In: Proceedings of IEEE TENCON 2014, pp. 1–6, IEEE Press, Bangkok (2014)
8. Sun, Y., Belding-Royer, E.M.: Dynamic address configuration in mobile ad hoc networks. UCSB technical report 2003–11, Santa Barbara, CA (2003)
9. Weniger, K.: PACMAN: Passive autoconfiguration for mobile ad hoc network. IEEE J. Sel. Areas Commun. **23**, 507–519 (2005)
10. Wang, X., Qian, H.: Dynamic and hierarchical IPv6 address configuration for mobile ad hoc networks. Int. J. Commun Syst **28**, 127–146 (2015)
11. Perkins, C.E., Bhagvat, P.: Highly dynamic destination-sequenced distance-vector routing for mobile computers. In: Proceedings of ACM SIGCOMM 1994, pp. 197–211, ACM Press, New York, USA (1994)
12. Johnson, D.B., Maltz, D.A.: Dynamic source routing in ad hoc wireless networks. Mobile Comput. **353**, 153–181 (1996) (The Kluwer international series in Engineering and Computer Science)
13. Haas, Z.J., Pearlman, M.R., Samar, P.: The zone routing protocol for ad hoc networks. IETF draft (2002)
14. Fall, K., Varadhan, K.: The ns manual. www.isi.edu/nsnam/nsdocumentation.html

EncryptPost: A Framework for User Privacy on Social Networking Sites

Shilpi Sharma and J.S. Sodhi

Abstract Social networking sites are gaining popularity among Internet users. As users are enjoying this new style of networking, the privacy concerns are also attracting public attention due to privacy breaches in social networking sites. We propose a framework that protects user privacy on a social networking site by shielding a user's personal post or messages, other service providers or third parties who are not explicitly authorized by the user to view the content. The architecture maintains the usability of sites services and stores sensitive information in encrypted form on a separate server. Our result shows that the proposed framework successfully conceals a user's personal information, while allowing the user and his friends to explore Social networking site services as usual.

Keywords Encryption · Web 3.0 · Privacy · Public key · Social networking site

1 Introduction

Web 3.0 is a new concept which focuses on giving user's the complete control of their data and persons with whom it is shared with. It follows certain strict procedures to make sure that user data is not compromised and is only shared with the users having sufficient privileges. Data is the most important asset within the internet and the ways to implement and display it may vary but that should not affect the data itself and this is the concept where encrypt post picks up and opens the doors for all new possibilities of a more secure and controlled data sharing in social networking sites.

S. Sharma (✉)
ASET, Amity University, Noida, Uttar Pradesh, India
e-mail: ssharma22@amity.edu

J.S. Sodhi
AKC Data Systems Pvt. Ltd, Amity University, Noida, Uttar Pradesh, India
e-mail: jssodhi@amity.edu

S.C. Satapathy and S. Das (eds.), *Proceedings of First International Conference on Information and Communication Technology for Intelligent Systems: Volume 1*, Smart Innovation, Systems and Technologies 50, DOI 10.1007/978-3-319-30933-0_6

Considering today's scenario where the user is free to share his personal information in social networking site facing whole world, the focus upon privacy and security of user-generated data has become a major concern. The main focus is upon the security of data and the owner's control over it. In present day scenario, user data is being harnessed, mined, sold, compromised and what users can do about it, is almost comparable to nothing. The main concept behind it is to encrypt, package, disintegrate and send the data to the user. Now whenever a query is generated, only authorized users would be able to get the required packets and integrate them in correct sequence. After the integration has taken place, the user decrypts the data and only then, he/she may be able to view it.

Social networking sites generally allow a user to post private, sensitive information, although the community is regularly warned about the risk allied with social networking sites. According to the survey conducted at Carnegie Mellon University, the users reveal their phone number, residence address etc. [1]. The two biggest rulers in online social networking, Facebook and MySpace, were both found to cross site scripting attacks enabling attackers to steal user credentials [2]. Also even Facebook employees modify user's personal information [3].

After having studied various existing solutions, we believe that a privacy protection technology that can effectively circumvent the threats raised by user's unawareness and server side vulnerabilities in client side architecture will automate the process of privacy protection. Our proposed framework provides a solution to address this problem, with a tool to control user's own post by means of encryption. It enforces user privacy on social networking sites by shielding a user's personal post from other users that were not explicitly authorized by the user. At the same time, the services and the user interface provided by the site continue to function as before. More specifically, users are given the option to express what information they intend to put guard on. We evaluate the design behind EncryptPost by applying it to an online social networking site, whereas our design is applicable to other social networking sites as well.

The rest of the paper is organized as follows: In Sect. 2, we survey related works that address privacy protection on social networking sites. We explain our design and the architecture of EncryptPost in Sect. 3, which is followed by our experimental results and analysis in Sect. 4. We conclude by discussing problems and propose future work for a more usable solution.

2 Related Work

In this section we are discussing substantial amount of limitation addressing the problem of privacy protection in social networks.

In Access Control Model, the relationship certificates are granted which are encrypted using symmetric cryptographic algorithms. These algorithms are difficult to use and manage. Also, incompatible with high implementation costs [4]. Whereas Public Key Protocol requires multiple users for each new access [5]. With the expansion of hash functions and chosen cipher text attack, key management scheme works on single user environment [6].

In other framework, while performing encryption the cipher texts are labeled with set of attributes and user is able to decrypt, ABE [7]. The state-of-the-art is decentralized architecture for social network privacy. It hides user data with attribute-based encryption (ABE), allowing users to apply fine-grained policies to view their data.

The well known architecture, Persona tends to have limited computation power and limited battery life, so the operations themselves should be reasonably interactive [8].

In case of EASiER, every time a new key is generated so the proxy needs to be updated and a new revocation takes place which discards the old one. It supports access control policies and dynamic group membership by using attribute based encryption [9].

Whereas, flyByNight have many calculations that are not suitable for non technical person who does not understand the concepts of double keys for encryptions. Users encrypt their own sensitive messages using JavaScript on the client side, and not on servers [10]. In FaceCloak, the encrypted data is on a third-party server and is suitable for a single personal computer at a time as its plug-in remembers the login credentials for single users [11]. A cipher text is transformed into an encrypted specific policy to other encrypted scheme. The proxy and the delegate together can recover the delegator's private key. Revocation in CP-ABE does not consider the compromise of proxy server [12, 13]. NOYB (short for"None Of your Business") encrypts personal information using a pseudo random substitution cipher, which divides user personal information into atoms and replicates each atom with the corresponding atom from a randomly selected user. Hence, difficult to implement 3-tier systems and maintain dictionary that requires extra space [14].

Thus, EncryptPost framework combining the various advantages of above mentioned techniques or models, along with new features such as no requirement of third party servers, can be implemented on clouds systems or personal computers, limited processing time and single key for encryption that makes it easy for non technical user too.

3 Methodology

In SNS, every user's profile page has an area in which other users can post public messages. These messages remain on their profile until either deleted by the user who posted the messages or until the profile user deactivates his SNS account. Message posted by a user may contain text, video, pictures or links that are not restricted by the providers. Most of the social networking websites provide their users with the possibility to send messages, photos and attachments to their listed friends or users who are not listed as their friends.

The question arises whether the users of social networking sites want everyone to know everything that the user have posted and loaded because as a concern this pool of information is also available to people who want to abuse it. Since the social networking sites makes their money by selling users data to third party advertisers. So, it is required to be aware of privacy and security issues.

3.1 Proposed Algorithm

After studying many common encryption techniques like RSA, DES, 3DES we conclude that in order to protect the post or messages of user in social networking sites, we must develop a powerful and light encryption algorithm using existing encryption technique, ROT13 with few modification.

In ROT13 algorithm, a single letter is rotated by 13th value corresponding to the text from the point of start that is A = 1, B = 2, C = 3…Z = 26.

So, A would be changed to N, B to O and so on while Z will be changed to M starting from A again. Using this concept a novel algorithm of random rotation is implemented.

Also, ASCII-Unicode technique is merged with random rotation so that back-tracking would be difficult [15]. It was observed that even same letter could yield different forms of cipher text; also the same word could yield entirely different set of ciphers every time when encryption takes place.

Each user (here, Alice and Bob) has their own digital signature, private keys and public keys into their respective browsers or miners which will later be used to decrypt the data.

The primary goal is to protect intentional information leak in social networking sites through encryption specifically ASCII rotation. The software used for analysing the result and test cases is EverCrack [16]. Here, the encrypted data was posted on a private network and white hat penetration testers were asked to retrieve the data with given amount of information.

The miner has SSL certificates and encrypt algorithm so, whenever a query is generated, it displays the packets having relevant meta-description. Then the user picks the packets and arranges them in the right order as per the digital signature of the owner and tags provided. Thus only authorized people with correct keys would be able to do so. The service providers of social networking site offers users the complete access to their data and makes sure who is able to access that post within the social network. It is to be noted that a user 'owns' the data he shares and he needs to be given proper authority to protect it the way required. So, in the proposed framework, only the person with authorized access would be able to get the packets and even if the user manages to attain them, he will not be able to arrange them in the correct sequence unless authenticate user has the required keys (Fig. 1).

And if any authorized user has certain complaints about the data, or he needs to report it then he can easily report the data from the site itself, in this case the miner will send the unencrypted data 'without' any modification to it and the key to

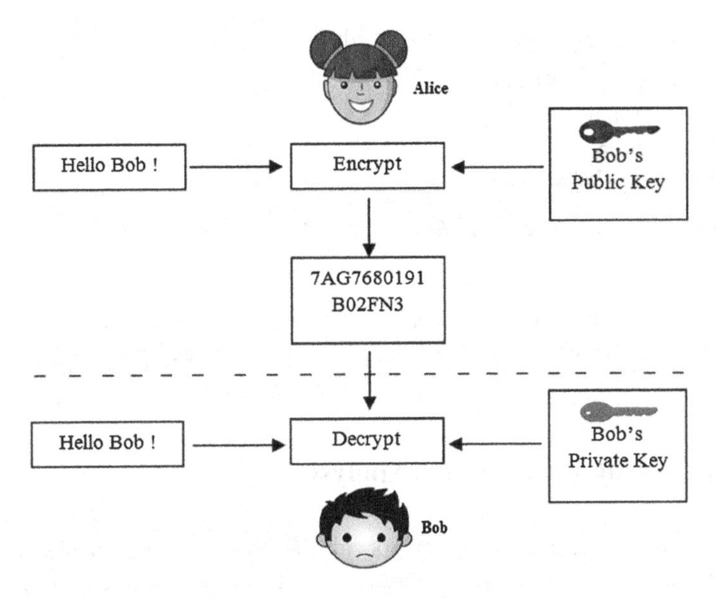

Fig. 1 Miner encrypts post using private key

decrypt, that would still be secure. For the protection of post, messages being typed by the user to be shared by end user's needs to be encrypted. If the user clicks on the check button labeled "Encrypt my text" than the data or the information being typed will be encrypted. The algorithm is as follows:

1. Input encrypted data and save it to encrypt[n].
2. Change each element of array to its ASCII value.
3. For each length of encrypt, generate a random no. and rotate element of array by that no. starting from first element.
4. Generate a random no. m. Again generate m random nos. and convert them into their ASCII value.
5. Check the length of original data.

    i. If length is even, divide length by 2.
    ii. Add random data generate in step 4 after the quotient obtained. Data added should be equal to the quotient as well.
    iii. Any extra random data should be added at last.
    iv. If length is odd, divide length by 2.
    v. Add random data generate in step 4 after the quotient obtained. Data added should be equal to quotient as well.
    vi. Add any random data before the last element of rotated data.

6. The key generated is of form length: data rotation: length function (odd, even).
7. Raise an exception if data rotation crosses 127(max ASCII value)

An example to encrypt the text of length 5, i.e. odd.

1. Original Data = "HELLO"
2. Data after Rotation = ggx{T
3. Random Data = a78"",/\tqN+
4. Final Data:gga7xT8"",/\tqN + T

Suppose we encrypt data HELO and HELLO
Then,

$$\begin{array}{ll}\text{helo : after rotation} = \text{xyZ\{} & \text{hello : after rotation} = \text{w\{\}} \\ \text{Zletter `l' rotated to Z} & \text{letter `l' rotated to \{}\end{array}$$

4 Experimental Result and Analysis

In the proposed EncryptPost framework, average results were obtained that are quite satisfactory. The proposed algorithm is implemented in a Testing machine with the following specifications:

(1) AMD A10 4600 M 2.3 GHz
(2) 4 GB RAM 2 X 4 Clocked at 1600 MHz & 1 TB Hard disk 5400 RPM
(3) GPU Cross fired hd8650 g + hd8670 m
(4) 2 Standard Cooling fans & 2 External Cooling fans (Cooler master)

Here we have implemented 4 cases to perform various test where encrypted data was posted on a private network and white hat penetration testers were asked to retrieve the data with given amount of information.

4.1 Case 1

1. Actual text data = "hellowor".
2. Crackers/Hackers provided with
   - Decryption Algorithm.
   - More than 3 letters in length.
   - Personal Text Message (Fig. 2)

Analysis for Case 1: white hat penetration testers took more than 30 min to retrieve a single letter.

Data (Char)	Time (m)
0	10
0	20
0	30
0	35
0	37
2	40
6	40
12	40

Fig. 2 White hat penetration testing with decryption algorithm and basic information showing data versus time for case 1

4.2 *Case 2*

1. Actual text data = "hellowor".
2. Crackers/Hackers provided with
   - Decryption Algorithm (Fig. 3).

Analysis for Case 2: white hat penetration testers have retrieved only 4 letters in 1–8 min of time.

4.3 *Case 3*

1. Actual text data = "hellowor".
2. Crackers/Hackers provided with
   - Decryption Algorithm.
   - Dictionary containing 60 k relevant combinations (Fig. 4)

Data (Char)	Time (m)
4	1
4	2
4	3
4	4
4	5
4	6
4	7
4	8

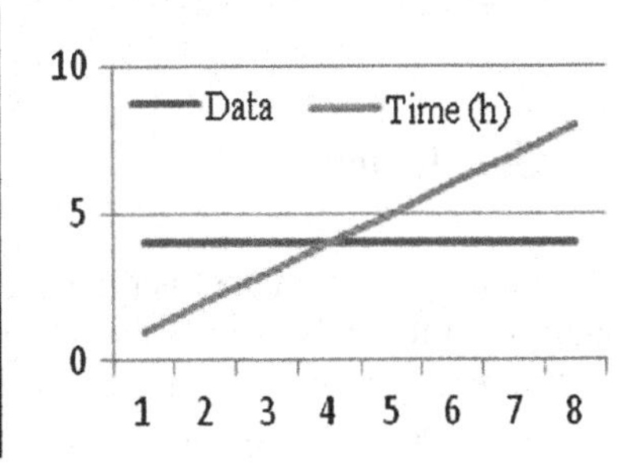

Fig. 3 White hat penetration testing with decryption algorithm showing data versus time for case 2

Data (Char)	Time (m)
0	0
2	50
4	100
5	190
5	200
5	240
6	250
8	260

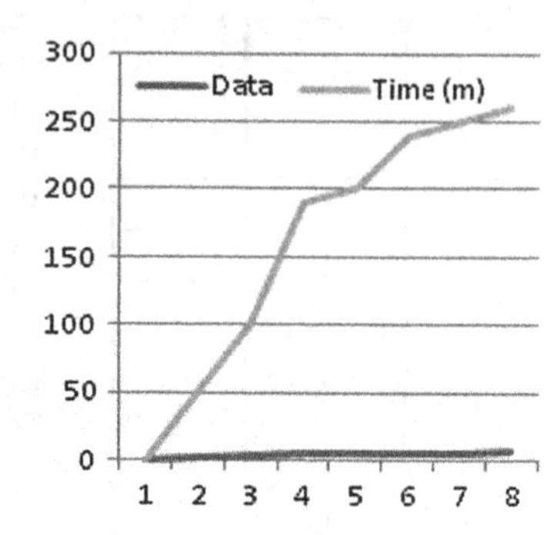

Fig. 4 White hat penetration testing with decryption algorithm for case 3 of chosen plaintext attack to show data versus time chart

Data (Char)	Time (m)
0	0
2	2
4	3
6	3
8	3

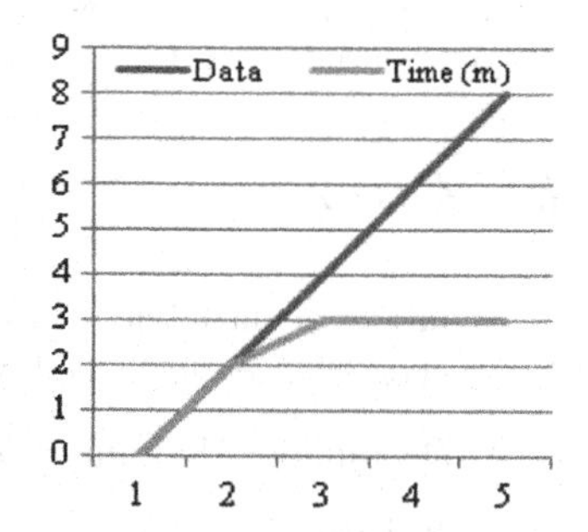

Fig. 5 White hat penetration testing with decryption algorithm showing data versus time for case 4 and all data was provided

Analysis for Case 3: White hat penetration testers were able to retrieve text easily if provided with dictionary with relevant combinations.

4.4 *Case 4*

1. Actual text data = "hellowor".
2. Crackers/Hackers provided with
   - Both Decryption Algorithm and the key was provided to the tester (Fig. 5)

Analysis for Case 4: white hat penetration testers were able to retrieve text easily if provided decryption algorithm and the key.

5 Conclusion

The main idea behind the EncryptPost framework is to encrypt the data at user level and send it to any website such as Facebook which would then encrypt it further.

The result confirms that the proposed framework consumes less encryption time and could be robust for any real time application. Hence we conclude that the algorithm can be useful and benefited in terms of security and prevention on social networking sites as (1) Same letter text data may or may not generate different encrypted data. (2) Different data may or may not generate same data. (3) Easy to implement and port in different programming languages. (4) Data length does not match with encrypted data. (5) Implemented on cloud systems. (6) Use of single key for encryption. (7) Minimum calculation with less processing power. (7) No requirement of third party servers.

We have considered all user controls in their privacy protection mechanisms to make the personalized system more usable and trustworthy. To fetch actual data, an unauthorized person would first get the encrypted data and would incorporate different technique to decrypt it. The decrypting time would depend on the skill of the person and the level of encryption.

References

1. Gross, R., Acquisti, A.: Information revelation and privacy in online social networks. In: Proceedings of the 2005 ACM Workshop on Privacy in the Electronic Society, pp. 71–80 (2005, November)
2. http://www.darknet.org.uk/2007/02/another-0-day-myspace-xss-exploit/
3. http://www.darknet.org.uk/2007/02/another-0-day-myspace-xss-exploit/
4. Carminati, B., Ferrari, E., Perego, A.: Private relationships in social networks. In: Data Engineering Workshop, 2007 IEEE 23rd International Conference on, pp. 163–171 (2007, April)
5. Domingo-Ferrer, J.: A public-key protocol for social networks with private relationships. In: Modeling Decisions for Artificial Intelligence. pp. 373–379. Springer, Berlin Heidelberg (2007)
6. Atallah, M.J., Blanton, M., Fazio, N., Frikken, K.B.: Dynamic and efficient key management for access hierarchies. ACM Trans. Inf. Syst. Secur. (TISSEC) **12**(3), 18 (2009)
7. Goyal, V., Pandey, O., Sahai, A., Waters, B.: Attribute-based encryption for fine-grained access control of encrypted data. In: Proceedings of the 13th ACM Conference on Computer and Communications Security, pp. 89–98 (2006, October)
8. Baden, R., Bender, A., Spring, N., Bhattacharjee, B., Starin, D.: Persona: an online social network with user-defined privacy. In: ACM SIGCOMM Computer Communication Review, vol. 39, No. 4, pp. 135–146 (2009, August)
9. Jahid, S., Mittal, P., Borisov, N.: EASiER: Encryption-based access control in social networks with efficient revocation. In: Proceedings of the 6th ACM Symposium on Information, Computer and Communications Security, pp. 411–415 (2011, March)
10. Lucas, M. M., Borisov, N.: Flybynight: mitigating the privacy risks of social networking. In: Proceedings of the 7th ACM Workshop on Privacy in the Electronic Society, pp. 1–8 (2008, October)
11. Luo, W., Xie, Q., Hengartner, U.: Facecloak: an architecture for user privacy on social networking sites. In: Computational Science and Engineering, 2009 CSE'09, International Conference on, Vol. 3, pp. 26–33 (2009, August)
12. Liang, X., Cao, Z., Lin, H., Shao, J.: Attribute based proxy re-encryption with delegating capabilities. In: Proceedings of the 4th International Symposium on Information, Computer, and Communications Security, pp. 276–286 (2009, March)

13. Yu, S., Wang, C., Ren, K., Lou, W.: Attribute based data sharing with attribute revocation. In: Proceedings of the 5th ACM Symposium on Information, Computer and Communications Security, pp. 261–270 (2010, April)
14. Guha, S., Tang, K., Francis, P.: NOYB: privacy in online social networks. In: Proceedings of the First Workshop on Online Social Networks, pp. 49–54 (2008, August)
15. Yergeau, F.: UTF-8, a transformation format of unicode and ISO 10646 (1996)
16. http://evercrack.sourceforge.net/

Green Routing Algorithm for Wireless Networks

Akshay Atul Deshmukh, Mintu Jothish and K. Chandrasekaran

Abstract With wireless devices gaining greater prevalence, there is a growing need for energy conservation for these devices. We propose a routing algorithm that reduces energy consumption at these mobile devices by modifying Optimized Link State Routing Protocol (OLSR). The protocol we propose is energy aware and reduces traffic to those nodes in the network that are low on battery life by using a modified Dijkstra's algorithm.

Keywords Energy saving · OLSR · Dijkstra's algorithm · Green routing

1 Introduction

The term routing refers to the selection of the best path between a source and a destination for efficient transport and delivery of packets. Routing algorithms are implemented in the network layer according to the five layer model. Routing algorithms maintain a routing table which remembers the routes to various destination nodes. Most routing protocols today give emphasis to choosing a shorter path (less hops) for quicker transmission. But this happens at the cost of high energy consumption, which results in diminished lifetime of the device. In such situations, the shortest route is not the most optimal route with respect to energy consumption.

Presently there is a massive increase in the usage of mobile devices in the world. In 2013, 90 % of people on Earth own a mobile phone and half of the majority uses

---

A.A. Deshmukh (✉) · M. Jothish · K. Chandrasekaran
Department of Computer Science and Engineering, National Institute of Technology Karnataka, Surathkal, India
e-mail: akshayd31@yahoo.co.in

M. Jothish
e-mail: mintujot@ieee.org

K. Chandrasekaran
e-mail: kchnitk@ieee.org

S.C. Satapathy and S. Das (eds.), *Proceedings of First International Conference on Information and Communication Technology for Intelligent Systems: Volume 1*, Smart Innovation, Systems and Technologies 50, DOI 10.1007/978-3-319-30933-0_7

these as their primary device to connect to the internet [1]. A major issue facing mobile device users and manufacturers is battery life.

A link-state routing protocol (LS) is one of the two main types of routing protocols used in computer communications. In this algorithm, each node in the domain has the complete topology of the domain in the form of a graph and information regarding how they are connected including the cost (metric), type and condition (up or down) of the links. Each node, then determines the best route to each other node using this data. A mobile implementation of LS protocol is Optimized Link State Routing Protocol (OLSR).

We are proposing a heuristic routing algorithm which modifies the standard link state routing protocol to be more energy efficient. In this algorithm, the traffic is routed to devices with longer battery life remaining as they will be able to sustain more data passing through them. To achieve this, the metric sent by nodes will be modified to include information regarding battery life remaining and energy consumption.

2 Related Work

First, there are studies from the upper layer point of view on energy conservation of the Internet. An example proposal, Energy Efficient TCP [2] implements congestion control with automatic bandwidth adjustment.

Second, there are studies on efficient and minimal energy utilisation at the routers. Some studies include developing a better forwarding behaviour to save energy at the Ternary CAMs in a router [3].

In the third category, there are studies that attempt to conserve energy from the level of network routing. There are studies in which change the transfer rate in networks called rate switching [4]. Reducing the transfer rate is not very attractive for many applications, especially with the increase in network traffic and proliferation of multimedia services.

There are also many studies in this category that puts network components to sleep mode. REsPoNse [5] proposes to identify energy-critical paths by analysing traffic which is a centralized schemes. Energy Aware Algorithm (EAR) [6] proposes to cluster networks into autonomous systems and switch network elements into sleep mode. A centralized algorithm [7] is still required to earmark links to sleep mode, to achieve good performance. A fully distributed approach is proposed by Bianzino et al. [8] which collects global traffic information without the need for a central controller and clusters traffic to switch suitable network links into sleep mode.

Hop-by-Hop routing solution by Yang et al. [9] is a distributed Dijkstra's based algorithm which seeks to conserve energy by using an algebraic model and the Dijkstra's algorithm with concentration on line cards. Extensive calculations at routing points can increase overheads.

The above proposals are meant for wired networks which set network links or components into sleep mode. These networks are in contrast to wireless networks as wireless networks are dynamic and unreliable compared to wired. For most wireless networks, there is an absence of such network elements and are usually adhoc. Moreover, there are no defined 'links' or interfaces in wireless networks.

There are many proposed works on green communications in general wireless and wireless adhoc networks. E^2R is an algorithm proposed by Zhu et al. [10] in which an opportunistic algorithm is used to broadcast packets to the destination. Antonio Junior et al. proposed an algorithm for energy aware multihop routing [11].

Our approach is synonymous to those in the third category, but it differs from the above proposed protocols in the following aspects. Our design considers that different energy consumption levels are a direct consequence of varying traffic levels. Our routing algorithm builds upon the existing OLSR algorithm which is prevalent in today's communication system, hence easier to implement.

3 Green Routing Algorithm

In the Green Routing (GR) algorithm that we are proposing, we add information regarding battery life remaining and battery discharge rate of the device into the Linked State Packets (LSP). We use this information to create a routing table which takes into consideration the energy consumption of the devices. This algorithm routes packets through nodes that have a longer battery life left after considering its discharge rate. However, our algorithm does not forsake performance for the purpose of optimizing energy conservation.

3.1 Parameters of the Algorithm

Let $G(V, E)$ be an undirected graph with set of V vertices and E edges of the network being simulated. The set of nodes V are the devices in the network. Edges E are the wireless links between the nodes with weights representing their costs in terms of distance and bandwidth.

Let N be the cardinality of vertices V and L be the cardinality of edges E.

$$N = |V| \tag{1}$$

$$L = |E| \tag{2}$$

For every node n_i in V, each link to n_j, is recalculated using our algorithm to take care of battery life and load.

$$C_{ij} = \frac{\sum (BL_k - \delta)}{N_p \times H \times num_{nodes}} \qquad (3)$$

where BL_k is the battery life of the mobile node, in Joules, through which the packet is routed. δ is a constant which gives the energy lost after processing a packet. N_p the average number of packets being processed at node n_j which can be considered as unity in case all the nodes are processing the same number of packets. H is the number of hops to the destination node. num_{nodes} gives the number of nodes traversed in the network from the source to the destination, excluding the source node. This variable is used to maintain a tolerable balance between performance and energy conservation.

C_j can be used to rank different routes to a destination node relatively based on the parameters mentioned above. The aim of our algorithm is to maximise the value of the summation of battery lives and, minimise the number of hops taken. Hence, it is evident from the equation that a higher value of C_j is desirable.

The resulting routing table is the GR routing table which incorporates energy information and distance constraints of the links in the various paths from one node to another.

3.2 GR Algorithm

Let *Rank*[], *SumOf Lives*[], *Hops*[] and *NodeNumber*[] be four arrays needed to formulate the source node's routing table.

Rank[] stores the ranks of the routes to destination nodes with respect to the source node. The values of *Rank*[] are obtained from C_{ij} where *i* indicates the source node and *j* represents the destination node. Since C_{ij} is highly dependent on the battery life of nodes and the number of hops in the route, its values are variable and should be considered relatively. Visited nodes are not taken into consideration for subsequent iterations. All values are initialized to zero except the source node, $\mathrm{n_i}$, which is initialised to infinity.

Hops[] stores the distance from the source node to the destination node. In the original Dijkstra's algorithm, this parameter forms the basis of all calculations. It is used to find the shortest route by comparing various options provided by the intermediate nodes. In our algorithm, the array is used to update the routing table. The values in *Rank*[] are used to select the next node to be considered. All values are initialized to zero.

SumOf Lives[] stores the sum of residual energy in the batteries of the nodes encountered in a route from the source node to the destination node. On computation of a lower rank of a node n_k, the algorithm stores this sum, expressed in Joules, of all the intermediate nodes in the selected route, including that of n_k in *SumOf Lives*[*k*]. All values are initialized to zero.

NodeNumber[] stores the number of nodes traversed from the source node to an intermediate or destination node. The values of this array do not take the source node into consideration. All values are initialized to zero.

S is used to store visited nodes. *Q* stores all the vertices or nodes in the network. *BLRank* is a temporary variable which stores the rank of a possible route from the source node to the destination node. *Sum* stores the sum of battery lives of all nodes from the source to the destination (excluding source battery life). *Distance* stores the distance between the source node and the node under consideration.

Neighbours(n_k) returns the neighbouring nodes of node n_k. *BatteryLife*(n_k) returns the battery life of node n_k. *length*(n_u, n_v) gives the distance between n_u and n_v.

maxRank(*Q*, *Rank*[]) in used to return the node such that the route from the source node to that node bears the highest rank in relation to the ranks of routes to other nodes.

Using these parameters the algorithm is given below:

Algorithm 1 Pseudo code for GR algorithm

To find energy-efficient paths to all nodes in a network from source node n_s
Given a set of nodes $V(n_1, n_2, ...n_n)$ and its edges or links in a network, a list of arrays $Rank[\,]$, $Hops[\,]$ and $SumOfLives[\,]$;

Let n_s signify the source node;

$Rank[s] \leftarrow \infty$
$Hops[s] \leftarrow 0$
$SumOfLives[s] \leftarrow 0$
$NodeNumber[s] \leftarrow 0$
for all $v \in V - \{s\}$ **do**
$Rank[v] \leftarrow 0$
$Hops[v] \leftarrow 0$
$SumOfLives[v] \leftarrow 0$
$NodeNumber[v] \leftarrow 0$
end for
$S \leftarrow \emptyset$
$Q \leftarrow V$
while $Q \neq \emptyset$ **do**
  $u \leftarrow maxRank(Q, Rank)$
  $S \leftarrow S \cup \{u\}$
  $Q \leftarrow Q - \{u\}$
  for all $v \in neighbours[u]$ **do**
    $Sum \leftarrow SumOfLives[u] + BatteryLife(v) - \delta$
    $Distance \leftarrow Hops[u] + length(u, v)$
    $BLRank \leftarrow \frac{Sum}{N_p \times Distance \times (NodeNumber[u]+1)}$
    if $BLRank > Rank[v]$ **then**
      $Rank[v] \leftarrow BLRank$
      $SumOfLives[v] \leftarrow Sum$
      $Hops[v] \leftarrow Distance$
      $NodeNumber[v] \leftarrow NodeNumber[u] + 1$
    end if
  end for
end while

The algorithm mentioned above does not change the complexity of the itsroot, Dijkstra's Algorithm, and remains $O(|E| + |V| \log|V|)$.

3.3 Example

The GR algorithm will be demonstrated with a network given below. The cost on links shows the distance and the battery lives of the nodes have been selected for the purpose of demonstration.

All values of the *SumOf Lives*[] array are given as multiples of 10^3. All values of C_{ij} obtained after calculation are rounded off to the nearest integer.

In the given network, let node 1 represent the source node. Assume that the average number of packets, N_p, is a constant. Also consider the distance between two consecutive nodes to be unit length.

Now the GR algorithm is applied. Since node n_1 has the largest rank of ∞ among all the other nodes, n_1 is chosen for the first iteration. The neighbours of this node are n_2 and n_3. The distances and battery lives of these nodes are given in the network diagram. Note that the terms δ and Np are treated as constants.

Formula (3) is applied to nodes n_2, n_3 to give the values 500 for C_{12} and 1500 for C_{13}. Hence, after one pass of execution, the tables are updated to include new information.

C_{12} and C_{13} are compared with its existing values in *Rank* [2] and *Rank* [3]. Since the existing values are 0, these values are replaced with the new values. After the first pass, all calculations for node n_1 have been completed. This node is not considered for subsequent passes and is marked as visited. Its corresponding values in all arrays are left unchanged. At this stage, node n_3 has the highest rank of 1500 among all the remaining nodes. Hence n_3 is chosen for the next phase of execution of the algorithm (Fig. 1).

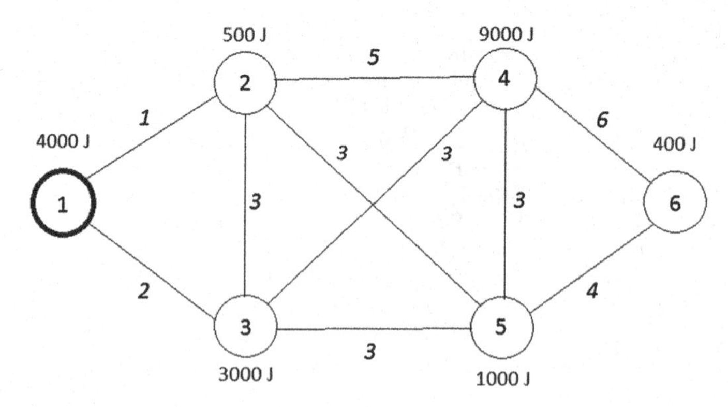

Fig. 1 A network for GR algorithm demonstration

Table 1 Final states of arrays

Array	n_1	n_2	n_3	n_4	n_5	n_6
Rank[]	∞	500	1500	1200	542	376
Hops[] (*GR*)	0	1	2	5	8	11
Sum of lives[] (*GR*)	0	0.5	3	12	13	12.4
Nodenumber[] (*GR*)	0	1	1	2	3	3
Hops[] (*Dijkstra's*)	0	1	2	5	4	8
Sumof lives[] (*Dijkstra's*)	0	0.5	3	12	1.5	1.9
Nodenumber[] (*Dijkstra's*)	0	1	1	2	2	3

The neighbours of n_3 are nodes n_2, n_4 and n_5. Formula (3) is applied to n_2, n_4 and n_5 to give values 350 for C_{12}, 1200 for C_{14} and 400 for C_{15}.

C_2, C_4 and C_5 are compared with its existing values in *Rank* [2], *Rank* [4] and *Rank* [5]. Since the existing values of *Rank* [4] and *Rank* [5] are equal to 0, these values are replaced with its corresponding new values. The value of *Rank* [2] is 500 which is greater than $C_{12} = 350$. So, no changes are made to *Rank* [3]. Hence, all the remaining arrays are changed in their respective positions to account for the new routes. The algorithm considers node n_3 to be a visited node and it is not considered for future computations. Thus, node n_4 has the highest rank of 1200 among the remaining nodes. The process is continued and the final state of tables are obtained (Table 1).

If this algorithm is compared with Dijkstra's algorithm, it can be seen that some routes take a longer path for the purpose of reducing energy usage at mobile nodes. But the net result produces a wireless network which accomodates fairly equal shares of performance and energy parameters.

3.4 Analysis of the GR Algorithm

A sample network consisting of six nodes was used to demonstrate the GR algorithm. The network given in the previous section was used for this purpose. The battery lives and the distances between every pair of nodes was selected arbitrarily.

The figure given below shows the lifespan of nodes with low battery levels. A node is considered to be low on battery life if its current energy state is lesser than or equal to 1000 J. For the sake of simplicity, it is assumed that it takes 1 min of a node's life in idle state and 5 min in processing state. The lifespan of all nodes was given in proportion to its current energy state (Fig. 2).

Advantages of GR Algorithm The advantages of the GR algorithm are similar to that of Dijkstra's algorithm. The following advantages are:

- The graph has shown an increase in the overall lifespan of mobile nodes.

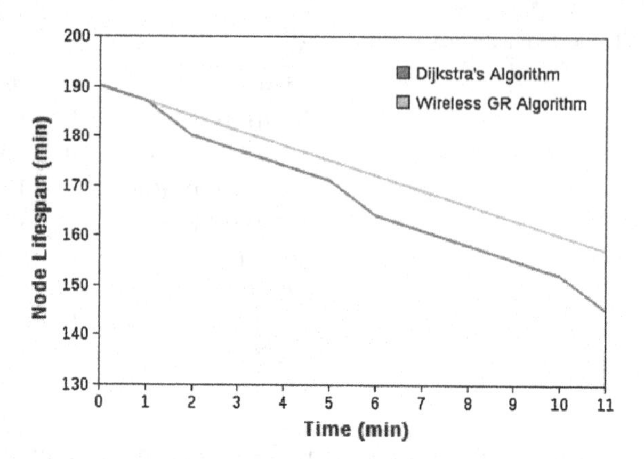

Fig. 2 Residual energy of a node

These results can be equated to longer lives of low powered nodes.

- Although the algorithm focuses on reducing battery consumption, it does not neglect the need to provide credible performance along with energy conservation. This has been taken into account through the inclusion of the number of nodes traversed and the distances between pairs of nodes in the GR equation.
- The GR algorithm is a variation of Dijkstra's algorithm which is used in OLSR algorithm. Hence the deployment of this algorithm is easy.

4 Conclusion

Our algorithm shows that energy conservation and performance can be accommodated together to provide credible results. The GR algorithm can be experimented with even further to improve its capabilities and to encompass as much performance and energy parameters as possible. A concerted effort is required to improve the energy conservation paradigm of this algorithm. Further studies can be done by deploying the algorithm over massive networks subject to other requirements.

References

1. Super Monitoring: State of mobile 2013. http://www.supermonitoring.com/blog/state-of-mobile-2013/(2013). Accessed 15 April 2015
2. Gan, L., Walid, A., Low, S.H.: Energy-efcient congestion control. In: Proceedings of ACM SIGMETRICS, pp. 89–100 (2012)
3. Lu, W., Sahni, S.: Low-power TCAMs for very large forwarding tables. IEEE/ACM Trans. Networking **18**(6), 948–959 (June 2010)

4. Gunaratne, C., Christensen, K., Nordman, B.: Managing energy consumption costs in desktop PCs and LAN switches with proxying, split TCP connections, and scaling of link speed. Int. J. Network Mgmt. **15**(5), 297310 (2005)
5. Vasic, N., Bhurat, P., Novakovic, D., Canini, M., Shekhar, S., Kostic, D.: Identifying and using energy-critical paths. In: Proceedings of CoNext11 (2011)
6. Cianfrani, A., Eramo, V., Listanti, M., Marazza, M., Vittorini, E.: An energy saving routing algorithm for a green OSPF protocol. In: INFOCOM IEEE Conference on Computer Communications Workshops, p. 15, March 2010
7. Cianfrani, A., Eramo, V., Listanti, M., Polverini, M.: An OSPF enhancement for energy saving in IP networks. In: IEEE INFOCOM Workshop on Green Communications and Networking, pp. 325–330, April 2011
8. Bianzino, A.P., Chiaraviglio, L., Mellia, M.: Distributed algorithms for green IP networks. In: IEEE INFOCOM Workshop on Green Networking and Smart Grid, pp. 121–126, 25–30 March 2012
9. Yang, Y., Wang, D., Xu, M., Li, S.: Hop-by-hop computing for green Internet routing. In: 2013 21st IEEE International Conference on Network Protocols (ICNP), p. 1, 7–10 October 2013
10. Zhu, T., Towsley, D.: E2R: Energy efcient routing for multi-hop green wireless networks. In: IEEE INFOCOM 2011 Workshop on Green Communications and Networking. doi:10.1109/INFCOMW.2011.5928821 (2011)
11. Junior, A., Soa, R.: Energy-efcient routing. In: 19th IEEE International Conference on Network Protocols, pp. 295–297 (2011)

Decision Theoretic Rough Intuitionistic Fuzzy C-Means Algorithm

Sresht Agrawal and B.K. Tripathy

Abstract The RCM algorithm may lead to undesirable solutions in practice because the points close to data points being assigned are neglected which led to the development of the decision theoretic rough set (DTRS) model and the decision theoretic rough C-means algorithm (DTRCM). It was recently improved further with the introduction of decision theoretic rough fuzzy C-means (DTRFCM). Here, we present a further improved algorithm called the decision theoretic intuitionistic fuzzy rough C-means (DTRIFCM) and provide a comparative performance analysis of DTRCM, DTRFCM and DTRIFCM through experiments and efficiency measuring indices DB, D and Acc. According to DB and D indexes DTRIFCM is better than the other two, where as far as the accuracy is concerned DTRFCM is better. We have chosen the data sets Iris, Wine, WDBC and Glass from UCI repository as input for the experimental purpose.

Keywords Rough set · Fuzzy set · Intuitionistic fuzzy set · DTRS · Clustering

1 Introduction

Cluster analysis has been studied by several researchers from different point of view. As the data sets are mostly uncertainty based, several uncertainty based clustering algorithms are found in literature to handle them. For this purpose the uncertainty models fuzzy set (FS) [1], IFS [2], rough set (RS) [3] and their hybrid models are considered. Some such algorithms are FCM, RCM [4] and the RFCM algorithm [5]. IFSs are generalizations of FSs. So, the hybrid model of rough intuitionistic fuzzy set (RIFS) is more general than the RFS model [6].

S. Agrawal · B.K. Tripathy (✉)
School of Computer Science and Engineering, VIT University, Vellore 632014, Tamil Nadu, India
e-mail: tripathybk@vit.ac.in

S. Agrawal
e-mail: agrawal.sresht@gmail.com

S.C. Satapathy and S. Das (eds.), *Proceedings of First International Conference on Information and Communication Technology for Intelligent Systems: Volume 1*, Smart Innovation, Systems and Technologies 50, DOI 10.1007/978-3-319-30933-0_8

Table 1 Data Sets

Dataset	Dimension	Sample	Class
Iris	4	150	3
Wine	13	178	3
WDBC	30	569	2
Glass	9	214	6

The DTRS was developed to generalise rough approximations [7]. Introducing proper loss functions [8, 9] several important rough set models can be derived from it. It has been introduced to the clustering arena recently [10]. This can be done by assigning a data point to a certain cluster or clusters according to the expected loss (risk) of this action. A DTRS model based clustering algorithm was introduced and studied in [10].

As mentioned above, the hybrid model RIFS is better than the individual models of RS and IFS as well as the hybrid model RFS. So, here we propose a decision theoretic rough intuitionistic fuzzy set based clustering algorithm (DTRIFCM) based upon RIFCM [11], which we establish to be more efficient than the corresponding algorithm DTRCM and DTRFCM which are based upon DTRS and RCM or the RFCM [5, 4, 3]. For the purpose of comparison we take several datasets and use three measures of efficiency; namely the DB-index [12], the D-index [13] and the accuracy measure [10].

We now describe the further organization of our work. Section 2 takes care of the various concepts to be used by us. The next section produces the algorithms proposed and those which are relevant to our work. The experiments and their

Table 2 Iris

Algorithms	DB_a	D_a	ACC_a
DTRIFCM	0.4132	0.5466	0.8891
DTRFCM	0.6112	0.5319	0.8901
DTRCM	0.7001	0.4662	0.8515
RFPCM	0.8422	0.3808	0.7945
$RFCM^P$	0.7029	0.4075	0.7963
RCM	2.6592	0.4312	0.8219
$RFCM^S$	0.7352	0.4097	0.7983

Table 3 Iris-b

Algorithms	DB_b	D_b	ACC_b
DTRIFCM	0.3627	0.4933	0.9208
DTRFCM	0.5707	0.4933	0.9217
DTRCM	0.6435	0.4933	0.9000
RFPCM	0.6594	0.4933	0.9000
$RFCM^P$	0.6438	0.4933	0.8933
RCM	0.7639	0.4933	0.8867
$RFCM^S$	0.7212	0.4933	0.8867

Table 4 Wine

Algorithms	DB_a	D_a	ACC_a
DTRIFCM	1.3734	0.1922	0.9688
DTRFCM	1.3958	0.1733	0.9746
DTRCM	1.3054	0.2163	0.9652
RFPCM	1.2465	0.2359	0.9281
$RFCM^P$	1.2857	0.2133	0.8854
RCM	2.0772	0.2329	0.9355
$RFCM^S$	1.2288	0.2199	0.9561

results along with the graphs and the values of the measures are narrated in Sect. 4. Also, we provide a comparative analysis of these results. The next section includes the concluding remarks followed by the list of works referred by us.

2 Definitions and Notations

The theoretic concepts and definitions of imprecision models are required in the sequel.

2.1 *The Uncertainty Based Models*

Extending the notion of fuzzy sets [1], intuitionistic fuzzy sets were introduced by Atanassov [2] as follows:

Definition 2.1.1 Given an universal set Y, an intuitionistic fuzzy set B on Y is characterized by the functions, $m_B and n_B$, given by, $m_B, n_B : Y \rightarrow [0, 1]$. The

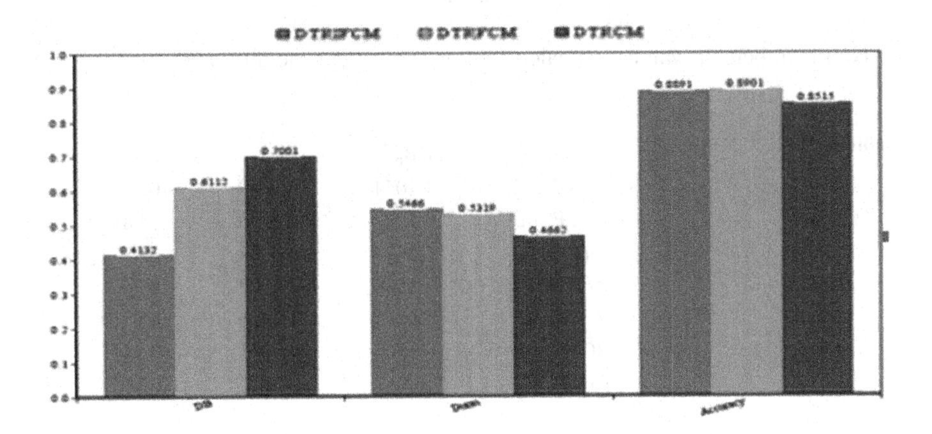

Fig. 1 Iris

hesitation function associated with B is denoted by π_B, which is the one's complement of $(m_B + n_B)$.

Let R be an equivalence relation R on U and the equivalence classes of any $x \in U$ be denoted by $[x]_R$.

Definition 2.1.2 Every $X \subseteq U$ is associated with two crisp sets $\underline{R}X$ and $\overline{R}X$ given by

$$\overline{R}X = \{x \in U | [x]_R \cap X \neq \phi\} \tag{1}$$

$$\underline{R}X = \{x \in U | [x]_R \subseteq X\} \tag{2}$$

The set X is said to be rough with respect to R if and only if $\underline{R}X \neq \overline{R}X$. Otherwise X is said to be R-definable.

The uncertainty region of X with respect to R is denoted by $BN_R(X)$ defined

$$as \quad BN_R(X) = \overline{R}X \backslash \underline{R}X \tag{3}$$

Definition 2.1.3 The points near to a data point $x_l \in X$ (except x_l) is represented as:

$$L(x_l) = \{x \in X : d(x, x_l) \leq \delta \wedge x \neq x_l\}, \tag{4}$$

where

$$\delta = \frac{\min_{1 \leq k \leq c} d(x_l, v_k)}{p} \tag{5}$$

we denote the conditional probability of x_1 being a member of C_i is given by (6) as:

$$P(C_i | x_l) = \frac{1}{\sum_{j=1}^{c} \left(\frac{d(x_l, v_i)}{d(x_l, v_j)} \right)^{\frac{2}{m-1}}} \tag{6}$$

Non-membership values are calculated, $\nu_A(x)$. 't' is generally taken as 2.

Table 5 Wine-b

Algorithms	DB_b	D_b	ACC_b
DTRIFCM	1.1044	0.3719	0.9824
DTRFCM	1.1209	0.3148	0.9851
DTRCM	1.0419	0.4465	0.9777
RFPCM	1.1380	0.2502	0.9607
$RFCM^P$	1.0215	0.3446	0.9719
RCM	1.1366	0.4465	0.9551
$RFCM^S$	1.0947	0.3446	0.9663

$$\nu_A(x) = \frac{1 - P(C_i|x_l)}{1 + t.P(C_i|x_l)}, \quad t > 0 \tag{7}$$

We derive hesitation degree as

$$\pi_i(x_l) = 1 - P(C_i|x_l) - \frac{1 - P(C_i|x_l)}{1 + t.P(C_i|x_l)}, \quad x \in X \tag{8}$$

Modifying the fuzzy membership value as $P_{new}(C_i|x_k)$, where,

$$P_{new}(C_i|x_k) = P(C_i|x_k) + \pi_i(x_l) \tag{9}$$

T_x represents the group of clusters similar to each individual data point x.

$$T_x = \left\{ C_i \in C : P_{new}(C_i|x) > \frac{1}{c} \right\} \tag{10}$$

The action set is defined as A = $\{ a_1, a_2, \ldots, a_c \}$, where a_j represents allocating a data point to C_j. The loss related with taking the action a_j for x_1 when x_1 belongs to C_i is represented by $\lambda_{x1}(a_j|C_i)$, and defined as:

$$\lambda_{x_l}(a_j|c_i) = \lambda_{C_i}^{a_j}(x_l) + \sum_{x \in L(x_l)} \beta(x)\lambda^{a_j}(x) \tag{11}$$

$$\lambda_{C_i}^{a_j}(x_l) = \begin{cases} 0, & if\ i = j \\ 1, & if\ i \neq j \end{cases} \tag{12}$$

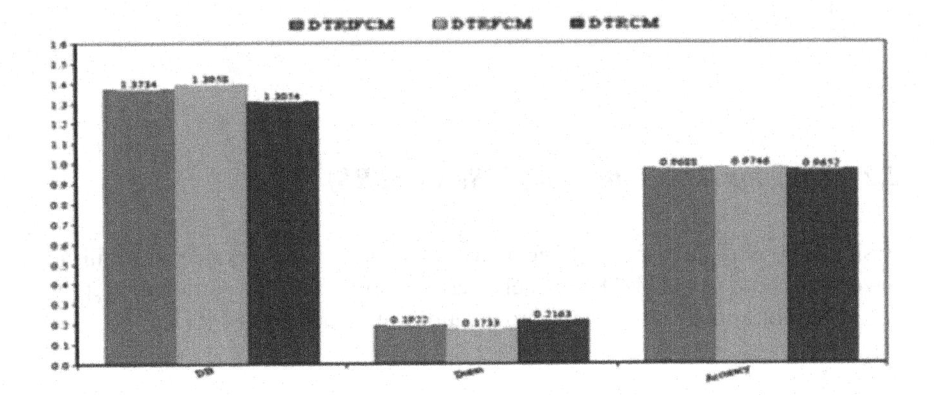

Fig. 2 Wine

$$\lambda^{a_j}(x) = \begin{cases} \dfrac{|a_j - T_x|}{a_j} = 0, & if\ a_j \in T_x \\ \dfrac{|a_j - T_x|}{a_j} = 1, & if\ a_j \notin T_x \end{cases} \tag{13}$$

$$\beta(x) = \exp\left(-\frac{d^2(x, x_l)}{2\sigma^2}\right) \tag{14}$$

The risk related when action a_j for x_1 is taken, can be represented by $R(a_j|x_l)$ and defined as:

$$R(b_j|x_l) = \sum_{i=1}^{k} \lambda_{x_l}(b_j|c_i) P_{new}(c_i|x_l) \tag{15}$$

For each data point x_1, let $a_k = \arg\min_{ai \in A}\{R(a_i|x_l)\}$. For actions closer to a_k in accordance of risk value, the index set J_D is defined as:

$$J_D = \left\{ j : \frac{R(a_j|x_l)}{R(a_k|x_l)} \leq 1 + \varepsilon \wedge j! = k \right\} \tag{16}$$

In the case when $J_D \neq \phi$, $\forall j \in J_D$, $x_l \in b_n(C_j)$.

$$v_i = \left\{ \begin{array}{l} w_{low} \dfrac{\sum_{x_k \in \underline{BU_i}} x_k}{|\underline{BU_i}|} + w_{up} \dfrac{\sum_{x_k \in BND(U_i)} (P_{new}(C_i|x_k))^m x_k}{\sum_{x_k \in BND(U_i)} (P_{new}(C_i|x_k))^m} if |\underline{BU_i}| \neq \phi \text{ and } |BND(U_i)| \neq \phi \\ \dfrac{\sum_{x_k \in BND(U_i)} (P_{new}(C_i|x_k))^m x_k}{\sum_{x_k \in BND(U_i)} (P_{new}(C_i|x_k))^m} if |\underline{BU_i}| = \phi \text{ and } |BND(U_i)| \neq \phi \\ \dfrac{\sum_{x_k \in \underline{BU_i}} x_k}{|\underline{BU_i}|} else \end{array} \right\} \tag{17}$$

2.2 Decision Theoretic Rough Sets (DTRS)

Basic rough set models are algebraic by nature. In order to incorporate probabilistic flavor into these models DTRS models were put forth. For any partition $\pi = \{A_1, A_2, \ldots, A_m\}$ of a universe U, the approximations of A_i are given by:

Table 6 WDBC

Algorithms	DB_a	D_a	ACC_a
DTRIFCM	1.0879	0.1018	0.9337
DTRFCM	1.0901	0.1005	0.9448
DTRCM	1.0769	0.0959	0.9324
RFPCM	1.0638	0.0757	0.7872
$RFCM^P$	1.0168	0.0895	0.9096
RCM	2.0935	0.0838	0.9297
$RFCM^S$	1.0772	0.0838	0.9297

$$\underline{apr}_{(\alpha,\beta)}(A_i) = POS_{(\alpha,\beta)}(A_i) = \{x \in U | P(A|[x]) \geq \alpha\},$$
$$\overline{apr}_{(\alpha,\beta)}(A_i) = POS_{(\alpha,\beta)}(A_i) \cup BND_{(\alpha,\beta)}(A_i) = \{x \in U | P(A|[x]) > \beta\}.$$

The approximations of a classification π are defined in terms of the above approximations of the elements A_i. The three approximate components are defined as:

$$POS_{(\alpha,\beta)}(\pi) = \bigcup_{1 \leq i \leq m} POS_{(\alpha,\beta)}(A_i),\ BND_{(\alpha,\beta)}(\pi) = \bigcup_{1 \leq i \leq m} BND_{(\alpha,\beta)}(A_i),$$
$$NEG_{(\alpha,\beta)}(\pi) = U - POS_{(\alpha,\beta)}(\pi) \bigcup BND_{(\alpha,\beta)}(\pi)$$

The three regions defined above may not be mutually exclusive but together they form a covering for U.

3 Algorithms

In this section we present the RFCM algorithm from [14, 5] and also the proposed algorithm DTRFCM [15], which is obtained by modifying the DTRCM algorithm proposed in [10] to incorporate the fuzzy component.

Table 7 WDBC-b

Algorithms	DB_b	D_b	ACC_b
DTRIFCM	0.7655	0.3226	0.5602
DTRFCM	0.7751	0.2701	0.5712
DTRCM	2.0054	0.2185	0.7838
RFPCM	1.7685	0.1701	0.7747
$RFCM^P$	0.7129	0.5826	0.7245
RCM	1.4083	0.1885	0.7647
$RFCM^S$	2.1686	0.1701	0.7137

3.1 DTRIFCM

The centroid for DTRIFCM is calculated by the following

$$v_i = \begin{cases} w_{low}\dfrac{\sum_{x_k\in \underline{BU}_i}\mu_{ik}^m x_k}{\sum_{x_k\in \underline{BU}_i}\mu_{ik}^m} + w_{up}\dfrac{\sum_{x_k\in(\overline{BU}_i-\underline{BU}_i)}\mu_{ik}^m x_k}{\sum_{x_k\in(\overline{BU}_i-\underline{BU}_i)}\mu_{ik}^m}, & \text{if } \underline{BU}_i \neq \phi \wedge (\overline{BU}_i - \underline{BU}_i) \neq \phi; \\ \dfrac{\sum_{x_k\in(\overline{BU}_i-\underline{BU}_i)}\mu_{ik}^m x_k}{\sum_{x_k\in(\overline{BU}_i-\underline{BU}_i)}\mu_{ik}^m}, & \text{if } \underline{BU}_i = \phi \wedge (\overline{BU}_i - \underline{BU}_i) \neq \phi; \\ \dfrac{\sum_{x_k\in \underline{BU}_i}\mu_{ik}^m x_k}{\sum_{x_k\in \underline{BU}_i}\mu_{ik}^m}, & \text{otherwise.} \end{cases} \quad (18)$$

where μ is P_{new}.

The algorithm DTRIFCM:

input: The given data set $X = \{x_1, \ldots, x_n\}$, c, w_1, ε, p, σ, m, S_{max}

output: Clustering result: $(\overline{C}_1, \underline{C}_1), \ldots, (\overline{C}_c, \underline{C}_c)$ and cluster centroids

1. randomly assign the initial centroid vi for C_i, $i = 1, 2, \ldots, c$;
2. repeat for $i \leftarrow 1$ to n do
3. for a data point $xi \in X$, calculate $P_{new}(C_j \mid x_i)$ $j = 1, \ldots, c$, by using Eq. (9);
4. determine x_i's neighboring points set $L(x_i)$ by using Eq. (4);
5. for every data point $x \in L(x_i)$, determine T_x by using Eq. (10);
6. calculate $R(a_j \mid x_i)$, $j = 1, \ldots, c$, by using Eq. (11) to Eq. (15);
7. find the action with minimal risk. $a_h = \text{argmin}_{aj\in A}\{R(a_j \mid x_i)\}$;
8. Compute u_{ik} and $u_{ik}x_i$ for the cluster h and the data point xi.
9. assign x_i to C_h, i.e. $x_i \in C_h$;
10. find the index set J_D with respect to a_h by using Eq. (16);
11. if $J_D = \emptyset$ then assign x_i to C_h, i.e. $x_i \in C_h$; else
12. assign x_i to the upper approximations of the clusters determined by J_D i.e. $x_i \in C_j, \forall j \in J_D$;
13. end
14. end
15. New centroid is calculated for each cluster by (18);
16. Repeat

4 Experimental Results

The behavior of different hybrid models, DTRCM, DTRFCM and DTRIFCM are studied by taking different data sets, like iris, wine, WDBC and glass and also we use various parameters, like DB index, Dunn index to measure the efficiency of these algorithms .

We use the following four data [16] sets as detailed below (Table 1)

4.1 *Quantitative Measures*

The following are the standard indexes available in literature for measuring the efficiency of the clustering algorithms.

1. Davis-Bouldin index (DB)

$$DB = \frac{1}{c}\sum_{i=1}^{c} \max_{j \neq i}\left(\frac{S(C_i) + S(C_j)}{d(v_i, v_j)}\right)$$

Here, we denote the average distance between data points in a cluster by $S(C_i)$. Also, $d(v_i,v_j)$ is the Euclidean distance between cluster centroids

$$S(C_i) = \frac{\sum_{x \in C_i} \|x - v_i\|}{|C_i|}$$

2. Dunn index (D)

$$D = \min_{1 \le i \le l}\left\{\min_{1 \le j \le l}\left\{\frac{dt(C_i, C_j)}{\max_{1 \le l \le c}\{\Delta(C_l)\}}\right\}\right\},$$

Table 8 Glass

Algorithms	DB_a	D_a	ACC_a
DTRIFCM	0.8921	0.2206	0.4267
DTRFCM	1.0075	0.2098	0.4338
DTRCM	1.1348	0.2014	0.4094
RFPCM	1.3874	0.0864	0.3856
$RFCM^P$	1.8937	0.1502	0.3998
RCM	4.3960	0.1950	0.3975
$RFCM^S$	1.2219	0.1920	0.3983

where

$$dt(S,T) = \min_{x \in S, y \in T} d(x,y)$$

$$\Delta(S) = \max_{x,y \in S}\{d(x,y)\}$$

3. Clustering accuracy index (ACC)

$$ACC = \frac{\sum_{i=1}^{n} \delta(t_i, map(C_i))}{n}$$

Here, we have followed the standard notations and conventions.

4.2 *Performance Comparison*

The average and the best cases of all the above mentioned parameters were obtained (average-DB, Dunn, Acc, Best-BD_b, $Dunn_b$ and Acc_b) for the data sets mentioned, i.e., iris, wine, WDBC and glass. For best results, the experiment was iterated a hundred times. Initially, the centroids are chosen randomly. We provide the DB, D and ACC values for the Iris dataset in Tables 2 and 3; Wine dataset in Tables 4 and 5; WDBC in Tables 6 and 7; Glass dataset in Tables 8 and 9 for the algorithms

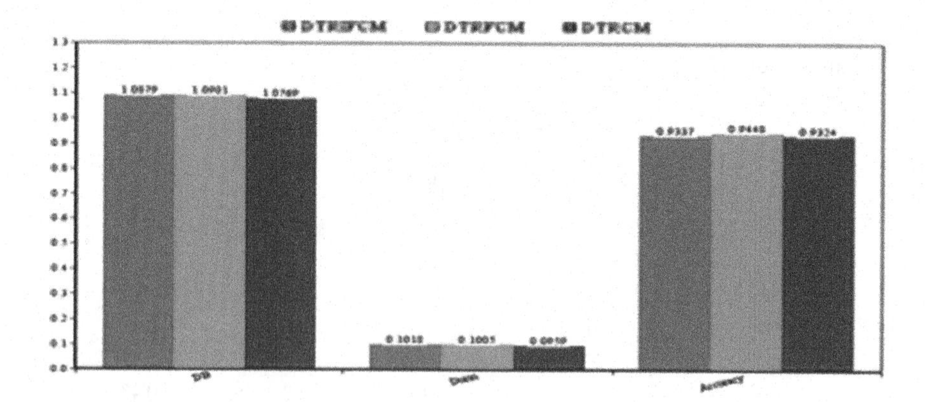

Fig. 3 WDBC

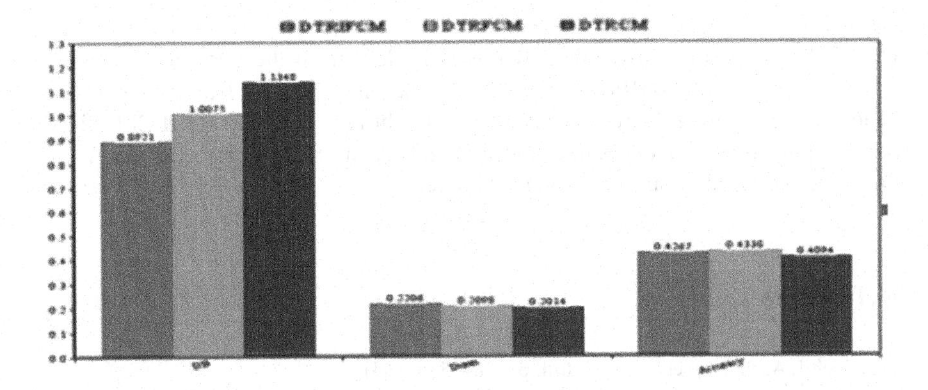

Fig. 4 Glass

Table 9 Glass-b

Algorithms	DB_b	D_b	ACC_b
DTRIFCM	1.0775	0.1197	0.9382
DTRFCM	1.0861	0.1140	0.9452
DTRCM	1.0748	0.0978	0.9332
RFPCM	0.9495	0.0758	0.9297
$RFCM^P$	0.7513	0.1508	0.9315
RCM	0.8933	0.0838	0.9297
$RFCM^S$	1.0764	0.0838	0.9315

RCM, $RFCM^P$, $RFCM^S$, RFPCM, DTRCM, DTRFCM and DTRIFCM for comparative analysis.

The following graphs (Figs. 1–4) were obtained showing the behavior of different hybrid models of decision theoretic sets and c-means algorithm. For the graphs, only the average cases are taken.

From the graphs, the following conclusions can be made:

- DTRIFCM has a better DB and Dunn index than DTRFCM in all cases.
- DTRIFCM has a better accuracy than DTRCM but DTRFCM performs best in terms of accuracy.
- In case of rough decision theoretic sets, the IFCM has lower accuracy than the FCM, which contradicts belief that the IFCM always has a better accuracy.

5 Conclusions

Here DTRIFCM, a new algorithm which extends the earlier algorithms DTRCM & DTRFCM is introduced. A comparative analysis was performed to analyze the behavior of the three algorithms by taking several well-known data sets from UCI repository. Taking different indices like DB, D and Acc it is observed that DTRIFCM outperforms the other two as is evident from the DB and D indices, whereas the Acc of DTRFCM is the best. Since the DB and Dunn indexed are better, it is possible that precision and recall might be better for DTRIFCM. Since some applications demand better precision and recall, rather than better accuracy, these factors could be an interesting look over.

References

1. Zadeh L.A.: Fuzzy sets. Inf. Control. **8**, 338–353 (1965)
2. Atanassov, K.T.: Intuitionistic fuzzy sets. Fuzzy Sets Syst. **20**(1), 87–96 (1986)
3. Pawlak, Z.: Rough Sets: Theoretical Aspects of Reasoning about Data. Kluwer Academic Publishers, Boston (1991)
4. Lingras, P., West, C.: Interval set clustering of web users with rough k-means. J. Intell. Inf. Syst. **23**(1), 5–16 (2004)
5. Maji, P., Pal, S.K.: RFCM: a hybrid clustering algorithm using rough and fuzzy sets. Fundamenta Informaticae **80**(4), 475–496 (2007)
6. Dubois, D., Prade H.: Rough fuzzy sets model. Int. J. Gen. Syst. **46**(1), 191–208 (1990)
7. Yao, Y.Y., Wong, S.K.M.: A decision theoretic framework for approximating concepts. Int. J. Man Mach. Stud. **37**(6), 793–809 (1992)
8. Peters, G., Crespo, F., Lingras, P., Weber, R.: Soft clustering—fuzzy and rough approaches and their extensions and derivatives. Int. J. Approximate Reasoning **54**(2), 307–322 (2013)
9. Yao, Y.Y.: Decision-theoretic rough set models. In: Yao J., Lingras P., Wu W.Z., et al. (eds.) Proceedings of the Second International Conference on Rough Sets and Knowledge Technology, LNCS, vol. 4481, pp.1–12 (2007)
10. Li, F., Ye, M., Chen, X.: An extension to rough c-means clustering based on decision-theoretic rough sets model. Int. J. Approximate Reasoning
11. Bhrgava, R., Tripathy, B.K., Tripathy, A.: Rough intuitionistic fuzzy c-means algorithm and a comparative analysis. 6th ACM India Computing Convention (2015)
12. Davis, D.L., Bouldin, D.W.: Clusters separation measure. IEEE Trans. Pattern Anal. Mach. Intell. Pami-**1**(2), 224–227 (1979)
13. Dunn J.C.: Fuzzy relative of the ISODATA process and its use in detecting compact well-separated clusters. J. Cuibernetic **3**(3), 32–57 (1974)
14. Mitra, S., Banka, H., Pedrycz W.: Rough–fuzzy collaborative clustering. IEEE Trans. Syst. Man Cybern. Part B Cybern. **36**(4), 795–805 (2006)
15. Agrawal, S., Tripathy, B.K: Rough fuzzy c-means clustering algorithm using decision theoretic rough set. Accepted for presentation in ICRCICN_2015
16. Blake, C.L., Merz C.J.: UCI repository of machine learning databases. http://www.ics.uci.edu/mlearn/mlrepository.html (1998)

Dual Band/Wide Band Polarization Insensitive Modified Four—Legged Element Frequency Selective Surface for 2.4 GHz Bluetooth, 2.4/5.8 GHz WLAN Applications

Vandana Jain, Sanjeev Yadav, Bhavana Peswani, Manish Jain and Ajay Dadhich

Abstract In this paper, a modified four-legged element frequency selective surface (FSS) with wide and dual band stop behaviour in two wireless local area network frequencies of 2.4 and 5.8 GHz have been proposed. The proposed design possesses 1050 and 1200 MHz bandwidths with insertion loss −41.45 and −35 dB around the centre operating frequencies 2.4 and 5.8 GHz, respectively. To obtain best characteristics, FSS unit cell has been designed using FR-4 substrate material having dielectric constant 4 and dissipation factor of 0.025. The structure exhibits a dual band from 1.8 to 2.85 GHz for GSM 1800/1900 MHz and Bluetooth and another band is from 5.25 to 6.45 GHz for WLAN for wideband application. It found applications into a number of vertical markets such as retail, warehousing, healthcare, manufacturing, retail and academic.

Keywords Frequency selective surface · Four-legged element · Resonant characteristics · Wireless local area network · Dual band

V. Jain (✉) · S. Yadav · B. Peswani
Department of ECE, Govt. Women Engineering College Ajmer (Raj).,
Ajmer, India
e-mail: j.vandana26@gmail.com

S. Yadav
e-mail: sanjeev.mnit@gmail.com

B. Peswani
e-mail: bhavnapesswani21@gmail.com

M. Jain
Rockwell Collins India Pvt. Ltd., Gachibowli, Hyderabad, India
e-mail: manishjain@rockwellcollins.com

A. Dadhich
Department of EICE, Govt. Engineering College Ajmer, Ajmer, India
e-mail: ajaydadhich13@gmail.com

S.C. Satapathy and S. Das (eds.), *Proceedings of First International Conference on Information and Communication Technology for Intelligent Systems: Volume 1*, Smart Innovation, Systems and Technologies 50, DOI 10.1007/978-3-319-30933-0_9

1 Introduction

Frequency selective surfaces (FSSs) have been studied since 1960s [1, 2]. Frequency Selective Surface (FSS) is a type of filter which is design in the form of a selective surface for some specific frequency application. When this surface is working as a filter, it is transparent to the desired frequency band and reflecting at other frequency band so we can use it as antenna reflectors, spatial filters, absorbers, radomes, artificial magnetic conductors as well as electromagnetic band gap (EBG) materials, due to their band-stop or band-pass responses.FSS characteristics include narrow Band, periodic in two dimensions and it is fully describe by reflection and transmission coefficient. FSS are usually planar periodic structures that function as spatial filters for entire electromagnetic (EM) spectrum.

In recent years, with the rapid development of wireless communication technology, the use of FSSs in wireless security, telecommunication and interference mitigation between adjacent wireless local area networks (WLAN) have been researched [3–7]. If the individual element of the designed structures are quite large, usually comparable to a half wavelength, then the element size make a barrier for FSS structure to be used at frequencies in WLAN bands application and it also leads to troublesome problem. Therefore, for most practically designed applications, it is desirable that the FSS element is as electrically small as possible. If the element size is sufficiently small, it is possible for the FSS to implement HIS at mobile bands frequency so that it may be use in handsets [8].

Frequency Selective Surfaces are basically form of passive filter which uses dielectric substrate on which conducting element are placed or set of periodic element inside the conductive sheet. These surface are excited by electromagnetic wave with different incident angle. They have band pass or band stop properties according to the selection of type of element, physical construction, element geometry, material. The band pass and band stop properties for filtering have been used as Frequency selective Horns, Surface radomes, sub reflectors and Frequency Selective surface Guides [1–7].

The purpose of research is to design new microwave, radar systems, spatial filters for communication which give accurate and improved performance and characteristics. When FSS surface comes under the field of electromagnetic wave, these surfaces generate a scattered wave which is used in many applications. The Design complexity of existing Frequency selective surface and limited size and sensitive nature of incidence angle limit their uses and functionality, but when we improve its characteristics it shows demand in many applications such as Wi-Max bands 2.3/2.5 GHz (2300–2400, 2500–2700 MHz), Bluetooth band 2.4 GHz (2400–2483.5 MHz), GSM band 1800/1900 MHz and WLANs operate in the unlicensed 2.4 GHz (2400–2484 MHz) and 5.8 GHz (5725–5825 MHz). In addition, the proposed FSS is polarization-independent and angular-insensitive under the incident angle from 0° to 75° and polarization angle of 0° to 60° [9, 10].

Wireless communications technologies have great demand in recent years due increasing demand for better and smaller wireless devices in modern market. Also, since FCC allowed potential users to make use of unlicensed frequency bands for medical, scientific and industrial application, engineers are trying constantly to achieve reduction of size so it is made possible to use that miniaturized design for multiband and broadband applications that transmit or receive information over short distances [11, 12]. As we know through the advance research in communication, wired LAN has been converted into the wireless local area network which is a wireless data communication system with a great flexibility and also provide security to data.

WLANs uses radio frequency technology which provide connectivity and user mobility to transmit and receive data over the air. As a result of all these, WLAN becomes popular in a number of vertical markets such as retail, warehousing, healthcare, manufacturing, retail and academic. This wireless network are being widely recognized as a cost effective, reliable solution for a broad range of systems and wireless high speed data connectivity that make use of hand-held manual terminal transmit real-time information to centralized hosts for further processing.

2 Wireless Technology Standards

FSS is majorly compatible with below mentioned Wireless Technology Standards.

2.1 Wi-Max

Wi-Max is a wireless communication standard used for microwave access. It is basically an alternative of DSL and cable. Wi-Max frequency bands come under IEEE std 802.16. Three WiMax bands 2.3/2.5/3.5 GHz (2300–2400, 2500–2700, 3400–3600 MHz) are widely used.

2.2 Bluetooth

Bluetooth is a wireless communication standard used for short distance communication .It is found in many electronic devices in modern world. Bluetooth devices operates in the band of 2.4 GHz (2400–2483.5 MHz).

2.3 WLAN

WLAN system is a set of computer network located in that location where installation of cable is very expensive and difficult approach. WLANs equipment provide high data rate. WLANs operate in the unlicensed 2.4 GHz (2400–2484 MHz) and the 5 GHz band −5.2 GHz (5150–5350 MHz) and 5.8 GHz (5725–5825 MHz). Some of the popular WLAN standards as follow:

IEEE 802.11a. This is a IEEE standard of physical layer which specifies eight available radio channels and operate at 5 GHz. It uses the modulation technique of orthogonal Frequency Division Multiplexing(OFDM). It has 12 non overlapping channels and 54 Mbps data rate.

IEEE 802.11b. This is a IEEE standard of physical layer which specifies three available radio channels and operate at 2.4 GHz (e.g. microwave ovens, Bluetooth devices etc.). It is able to transfer data with 11 Mbps data rate. It uses the CSMA/CA technique. It is specify to operate at basic data rate of 11 Mbps.

IEEE 802.11a. This WLAN physical layer standard specifies three available radio channels in the 5 and 2.4 GHz.

3 Design of FSS Structure

Based on Munk's FSS designing theory [1], the modified four-legged element shaped have such structure from which broad bandwidth can obtain and can be arranged to obtain resonant frequency of interest. The FSS is made by single layer of metallic structure. It is designed with array of metallic patches of square loop separated by small gaps and metallic lines which are periodic in nature. Using this type of structure, the single-layered modified four-legged element designs are expanded to an infinite array in the *x* and *y* directions. Structure is assumed to be an infinite periodic surface structure. FSS structure is excited by plane waves with different incident angles. The incident wave vector is fixed in the *x*-*z* plane and the magnetic field vector is fixed in the *y* direction.

The geometry of the proposed FSS unit cell is illustrated in Fig. 1, where the gray areas denote metallic loops and the green areas denote substrate, respectively.

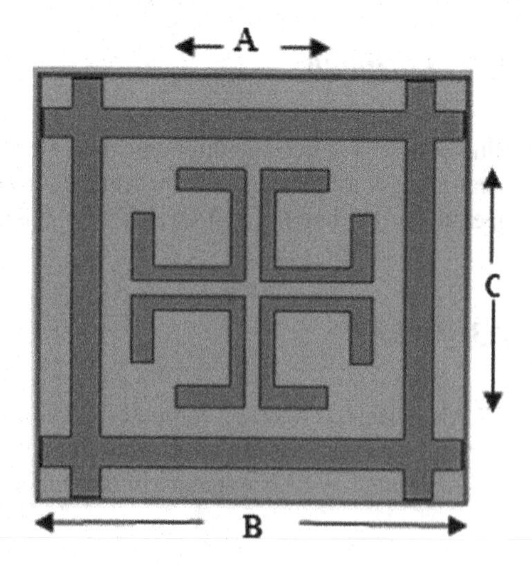

Fig. 1 Top view of the proposed unit cell

Table 1 Parameter values of the proposed design

Parameters	Values (mm)
A	10
B	28
C	16.2

A single layer FSS unit cell has been simulated in the CST Microwave Studio Software for wireless local area network standard frequency application. The FSS structure has been designed using FR-4 as the substrate material having dielectric constant of 4 and dissipation factor 0.025 and height is 1 mm. The optimized parameters have been shown in Table 1.

4 FSS Result Analysis

The parameters for the designed FSS structure have been calculated and simulated transmission coefficient results as shown in Fig. 2. The bandwidth at the 2.4 GHz band is around 1050 MHz with the corresponding value of transmission coefficient as −41.45 dB and bandwidth at the resonating frequency 5.8 GHz is 1200 MHz with the corresponding value of transmission coefficient as −35 dB. It shows demand in many applications such as Wi-Max bands 2.3/2.5 GHz (2300–2400, 2500–2700 MHz), Bluetooth band 2.4 GHz (2400–2483.5 MHz) GSM 1800/1900 MHz and WLANs operate in the unlicensed 2.4 GHz (2400–2484 MHz) and 5.8 GHz (5725–5825 MHz). In addition, the proposed FSS is polarization-independent and angular-insensitive under the incident angle from 0° to 75° and polarization angle of 0° to 60°.

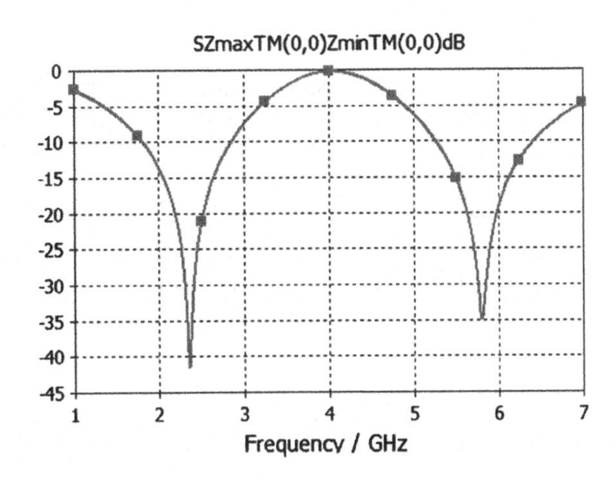

Fig. 2 Simulated transmission coefficient at normal incidence

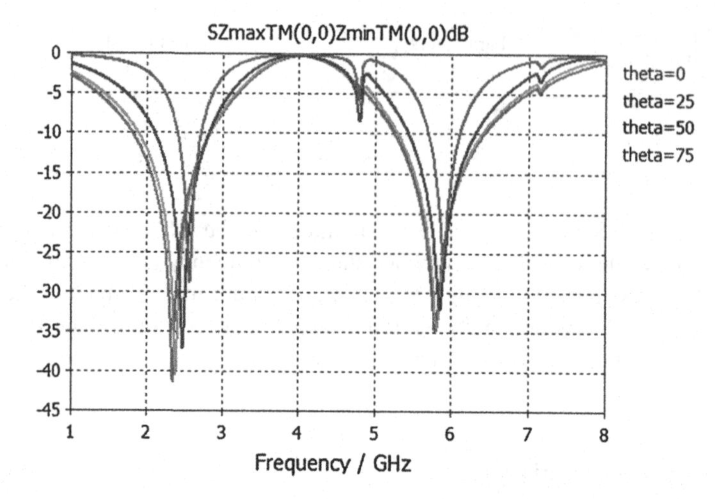

Fig. 3 Simulated transmission coefficient at different incident angle

The structure exhibits a dual band from 1.8 to 2.85 GHz for GSM 1800/1900 MHz and Bluetooth and another band is from 5.25–6.45 GHz for WLAN for wideband application.

The transmission coefficient result of FSS structure for various incidence angles and polarization angle response have been presented as shown in the figure. While increasing the incident angle, transmission response shift toward the higher frequency as shown in Fig. 3. The shifting effect of response towards higher frequency is negligible. The simulated transmission response have excellent stability toward different polarization angle as shown in Fig. 4.

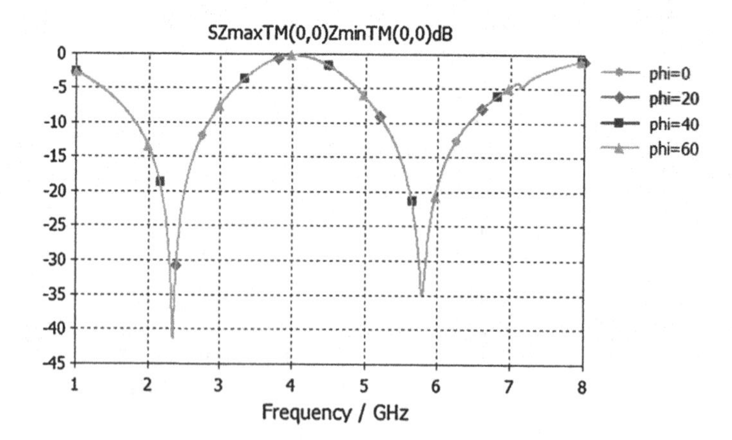

Fig. 4 CST simulated transmission coefficient under different polarization angle

5 Conclusion

In this paper, modified four—legged element frequency selective surface with dual and wide band stop response has been introduced and the corresponding simulation results have been presented. We demonstrated that the band-stop filter response could be dynamically tuned to different frequency bands over the frequency band of interest by changing the height of the substrate material. In addition, Polarization and incidence angle sensitivity of the FSS were also examined. The simulated transmission coefficient shows stable performance under different polarization angle varied from 0° to 60°. The simulated transmission coefficient shows small variation of resonant frequency under incident angle varied from 0° to 75°. We use FR-4 as substrate material (relative permittivity $\epsilon_r = 4$) to obtain the required response which support the IEEE802.11a wireless LAN technology, GSM and Wi-Fi .The FSS provides stable polarization-independent performance and exhibits good transmission/reflection performance under a wide range of incident angle from 0° to 75°.

References

1. Munk, B.A.: Frequency Selective Surfaces: Theory and Design. Wiley, New York (2000)
2. Vardaxoglou, J.C.: Frequency-Selective Surfaces: Analysis and Design. Research Studies Press Ltd., Taunton (1997)
3. Parker, E. A. Hamdy, S.M.A.: "Rings as Elements For Frequency Selective Surfaces". Electron. Lett. **17**, 612–614 (1981)
4. Sekil, Y.:"A consideration of the breakdown field strength of XLPE cable insulation". In: IEEE 5th International Conference on Conduction and Breakdown in Solid Dielectrics (1995)
5. Sarabandi, K., Behdad, N.: A frequency selective surface with miniaturized elements. IEEE Trans. Antennas Propag **55**(5), 1239–1245 (2007)
6. Zhou, H., Qu, S., Wang, J., et al.: Ultra-wideband frequency selective surface. Electron. Lett. **48**(1), 11–13 (2012)
7. Yuan, Y., Wang, X.-H., Zhou, H.: Dual-band frequency selective surface with miniaturized element in low frequencies. Prog. Electromagnet. Res. Lett. **33**, 167–175 (2012)
8. Goussetis, G., et.al.: Miniaturization of electromagnetic band gap structures for mobile applications. Radio Sci. **40**, p. RS6S04 (2005)
9. Sanz-Izquierdo, B., Parker, E.A.T., Robertson, J.-B., Batchelor, J.C.: Singly and Dual Polarized Convoluted Frequency Selective Structures. IEEE Trans. Antennas Propag. **58**, 690–696 (2010)
10. Yang, H.-Y., Gong, S.-X., Zhang, P.-F., Zha, F.-T., Ling, J.: A novel miniaturized frequency selective surface with excellent center frequency stability. Microwave Opt. Technol. Lett. **51** (10), 2513–2516 (2009)
11. Kartal, M., Pinar, S.K., Doken, B., Gungor, I.: A new narrow band frequency selective surface geometry design at the unlicensed 2.4 GHz ISM band. Microw. Opt. Technol. Lett. **55**(12), 2986–2990 (2013)
12. Natarajan, R., Kanagasabai, M., Baisakhiya, S., Sivasamy, R., Pakkathillam, J.K.: A compact frequency selective surface with stable response for WLAN applications. IEEE Antennas Wirel. Propag. Lett. **12**, 718–720 (2013)

5 Conclusion

References

Dynamic Power Allocation in Wireless Routing Protocols

Ravi Kumar Singh and M.M. Chandane

Abstract Power optimization in routing Protocol is an important factor in a mobile ad hoc network. In all of the existing routing protocols when a packet travels from source to destination, then source and all the intermediate nodes in the path assign a constant energy to the packet irrespective of the distance packet has to travel and hence there is wastage of power. In this paper, a new power control mechanism has been proposed which find the distance between two nodes, calculate the power required to cover the distance and assign only that amount of power to the packet. This mechanism works at physical layer, so this can be used with any of the existing routing protocols. After deployment of the scenario in a simulator, simulation result will show that energy consumption by using the proposed technique is less than the traditional mechanism used in the protocols.

Keywords Power optimization · Routing protocols · Wireless networks and MANET

1 Introduction

MANET (Mobile ad hoc Network) is a wireless communication network, consisting of mobile devices which communicate through wireless links [1]. As there are a lot of layers in TCP/IP, on top of Link Layer it has an additional integrated routable networking environment which helps it to achieve self-forming, self-healing and peer to peer network. Ad- hoc networks can be set up on an urgent basis and that too with very less infrastructure and at very low cost wherever needed as no setting

R.K. Singh (✉) · M.M. Chandane
Veermata Jijabai Technological Institute, Mumbai 400019, India
e-mail: ravikumar.vjti@gmail.com

M.M. Chandane
e-mail: mmchandane@vjti.org.in

S.C. Satapathy and S. Das (eds.), *Proceedings of First International Conference on Information and Communication Technology for Intelligent Systems: Volume 1*, Smart Innovation, Systems and Technologies 50, DOI 10.1007/978-3-319-30933-0_10

up of access points or base stations are required [1]. In a wireless sensor network (WSN) a sensor field is created by communication among the set of sensors.

MANETs send and receive signals to and from sensors respectively in order to achieve their functionality. A battery which acts as a power source, a computational unit small enough to fit into the sensor and able to perform basic operations when required and a small memory are among the major components of a sensor device. Controlling security needs for houses and spaces, monitoring environmental changes, monitoring military systems and many other applications in environmental, health, safety are some of the real time applications where these sensors can be used [2–4]. Most of these areas are inaccessible areas and sensors need to work in harsh and difficult conditions to keep the sensor field alive. That is why it is important that sensors should be durable and sustain power as long as possible. To deal with the problem of power consumption they should consume as low power as possible while providing all functionality of the sensor. To achieve efficiency in wireless sensor networks is much more difficult than any other compatible computer network devices. Many routing protocols use the technique of data transfer over wireless sensor nodes. Different routing algorithms prefer different parameters like some are concerned with the Quality of Service (QoS) while some are concerned with energy consumption. Of all the available routing protocols, AODV protocol is the most efficient routing protocols in terms of shortest path and power consumption. Even being such an efficient protocol, power consumption in this protocol is not optimized.

As packets are processed at every layer of OSI model so, power consumption takes place everywhere but the layers mainly considered for power consumption optimization are Data Link Layer and Physical Layer. Here the discussed technique for optimization of power is of Physical Layer. Several modes on which power consumption takes place at physical layer are: transmission mode, reception mode, ideal mode and sleep mode. Of the above described parameters maximum wastage of energy takes place in transmission mode. Power optimization can be done by making just few changes at the physical layer. This reduces not only the power consumption of reactive routing protocols but also of proactive routing protocols and can be used with any of the existing routing protocols.

2 Literature Survey

A lot of researchers have proposed different techniques to solve the problem of Power Consumption.

Author Supriya Srivastav and writing et al. proposed the method for selection of best route from source to destination on the basis of different parameters like max

Energy, max Bandwidth, min Load and min Hop Count among the entire route requests arrived [1].

Author G. Rajkumar and writing et al. have proposed a routing protocol Power-Aware Ad hoc On-demand Distance Vector (PAW-AODV) which selects its route based on a power-based cost function and thus saves power [5].

Author Hesham Abusaimeh and writing et al. suggested that most of the area will be covered by the nodes having high energy and the remaining area will be covered by the nodes having low energy, which they achieved by varying the transmission and the reception powers of the nodes and thus achieved the power optimization [6]. Furthermore, author Thrung Dung Nguyen and writing et al. in their paper suggested dynamic power allocation to the routes [7] but limited to only one routing protocol AODV.

2.1 *Cross-Layer Model*

Figure below shows a cross layer communication. When packets are received at physical layer, it calculates the transmitted power, remaining energy of the node and reports these parameters to network layer where routing protocols are deployed. Based on these parameters of physical layer, routing protocol analyses and decides routing behaviors. Routing behaviors are the best path selection, link quality control and so on. When a route is selected by network layer, data link layer will find which transmitted power is best for this path and inform the transmitted power value to physical layer. Thus, the transmitted power is assigned appropriately by physical layer [7] (Fig. 1).

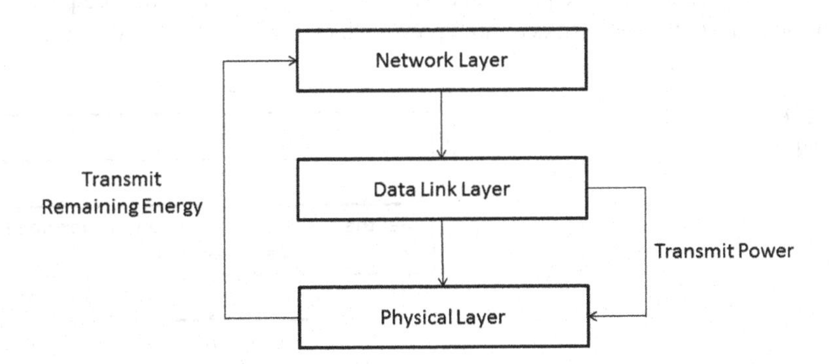

Fig. 1 Cross layer communication [7]

3 Proposed Solution

The solution proposed [7] for power optimization in routing protocol is good solution but it allocates the power to the complete route. So there might be a chance that the individual nodes will be wasting the power. So, here a new technique has been proposed which allocates power to the individual nodes.

The power control technique is described in figure below. Initially distance between the individual nodes must be calculated. This can be done by using routing tables in case of proactive routing protocols and in case of reactive of routing protocols each node keeps track of its next and previous nodes so their positions and hence the distance can be found easily. Next calculate the power required for the transmission of packets and finally assign them the power and then transmit the packets (Fig. 2).

The power optimization technique which has been used here works in energy model of physical layer, so this technique can be used with any of the existing wireless routing protocols.

$$P_{assign} = \frac{tx_{max} * V * d}{d_{max} * I_{energy} * S_{energy} * \alpha} + Rec_{load} + M_{load} \tag{1}$$

where, d_{max}, I_{energy}, S_{energy}, α can't be 0.

In the above equation, P_{assign} is the power to be assigned to the packet, tx_{max} is maximum transmitted power, V is voltage, d_{max} maximum distance for which tx_{max} is allocated, α is the routing efficiency of routing protocol being used which is decided at run-time, d is the distance between the nodes which can be gathered from the routing tables, S_{energy} Idle energy, S_{energy} is the sleep energy, Rec_{load} is the reception load and M_{load} is the marginal power.

Assigning values to these parameters is a crucial task. If big values are provided then each packet will be assigned more power which it may not use and hence there

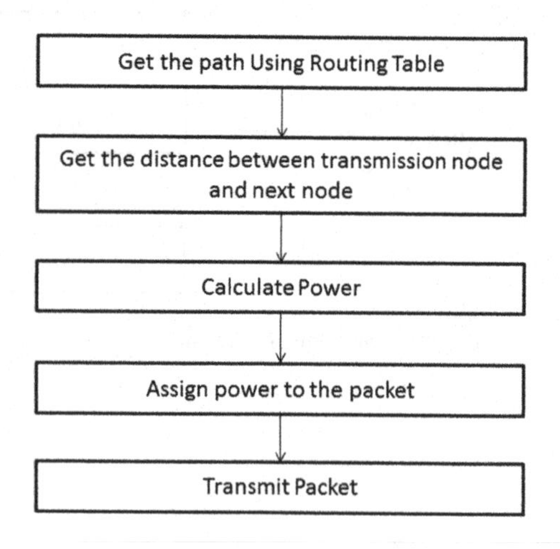

Fig. 2 Power control technique

would be a lot of wastage of energy. If less power is assigned then the packet will collapse before reaching to the destination.

To implement the above proposed formula a new energy module *Effective* has been created in exata.

4 Implementation and Results

For implementation EXATA version 5.1 has been used. The figure given below shows the topology of 10 nodes used for calculation of result (Fig. 3).

The network parameters for the simulation are given in Table 1.

The performance of Effective energy model has been evaluated and compared with Generic and Mica energy models.

Figure 4 shows power consumed by the whole topology at 10th second, 20th second, 30th second, 40th second, 50th second and 60th second.

Figure 5 shows how Effective energy module works with different routing protocols and their comparison with different energy modules.

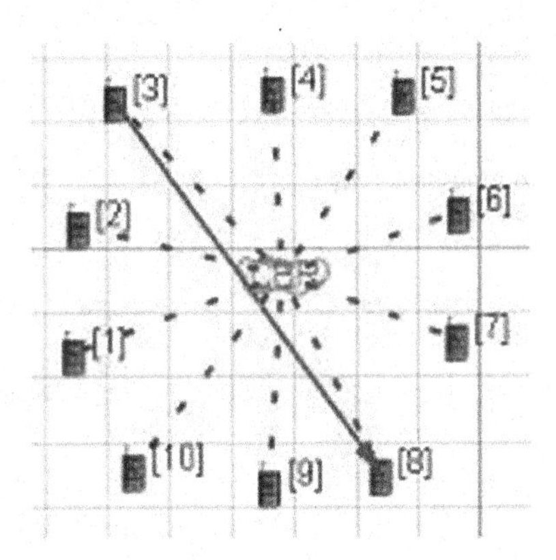

Fig. 3 Topology

Table 1 Configured parameters

Parameters	Values
PHY/MAC protocol	802.11b radio
Routing protocol	AODV
Energy module	Effective, generic, mica
Data size	70 KB (CBR)
Rate	250 Kbps
tx_{max}	280 mA
Voltage	3 V
Idle current	178 mA
Sleep current	14 mA
Recetion current	204 mA

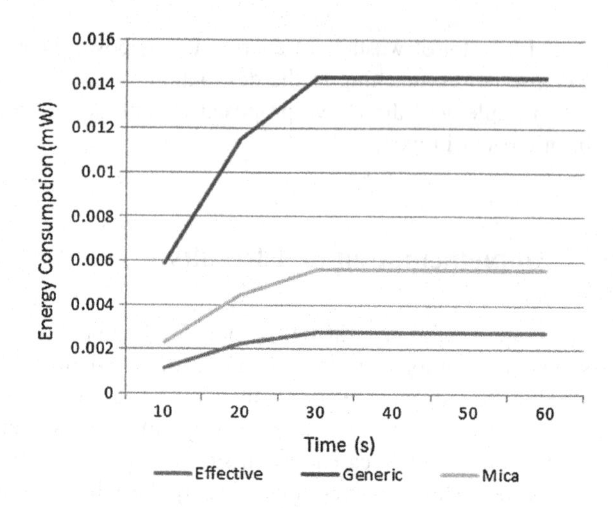

Fig. 4 Total power consumption at different time intervals

It can be seen from the above figure that the power consumed by the effective energy model which implements the proposed formula is much less than other energy models like Generic and Mica.

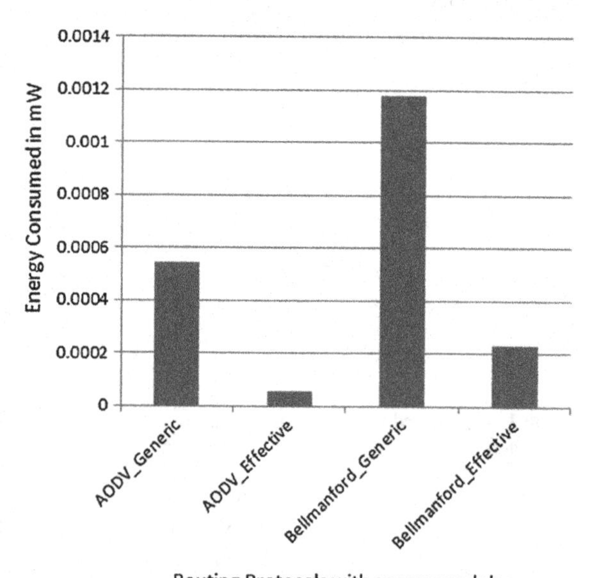

Fig. 5 Total power consumption by different routing protocols with energy modules

5 Conclusion and Future Work

Power conservation is a very important factor while working with MANETs. To address the power efficiency issue of routing a new energy module Effective which works on the proposed formula has been implemented and the results (graphs) obtained from the simulation shows that power usage has significantly reduced up to 60–80 % and hence effectively optimizes the power consumption. As this energy module works at the physical layer, hence it can work with any routing protocol.

In this paper the crucial values like α and M_{load} had been found by brute force method, which is not feasible. So there is a need to automate the process of finding the values.

References

1. Srivastava, S., Daniel, A.K., Singh, R., Saini, J.P.: Energy-Efficient Position Based Routing Protocol for Mobile Ad Hoc Networks: 978-1-4673-2758-9/12 ©2012 IEEE (2012)
2. Tan, C.-W., Bose, S.K.: Investigating Power Aware AODV for Efficient Power Routing in MANETs: 0-7803-9282-5/05/ ©2005 IEEE
3. Poongkuzhali, T., Bharathi, V., Vijayakumar, P.: An Optimized Power Reactive Routing Based on AODV Protocol for Mobile Ad-hoc Network: 978-1-4577-0590-8/11/ ©2011 IEEE
4. Romer, K., Kastin, O., Mattern, F.: Middleware challenges for wireless sensor networks. ACM SIGMOBILE Mobile Comput. Commun. Rev. **6**(4), 59–61 (2002)
5. Rajkumar G, Kasiram R, Parthiban D: Optimizing throughput with reduction in power consumption and performance comparison of DSR and AODV routing protocols: 978-1-4673-0210-4112/ ©2012 IEEE
6. Abusaimeh, H., Yang, S.-H.: Reducing the Transmission and Reception Powers in the AODV" 978-1-4244-3492-3/09/ ©2009 IEEE (2009)
7. Nguyen, T.D., Nguyen, V.D., Van, T.P., Koichiro, W., Nguen, N.T.: Power Control Combined With Routing Protocol For Wireless Sensor Network 978-1-4799-2903-0/14 ©2014 IEEE

UWB Microstrip Antenna with Inverted Pie Shaped Slot

Minakshi Sharma, Hari Shankar Mewara, Mahendra Mohan Sharma, Sanjeev Yadav and Ajay Dadhech

Abstract This paper proposed a compact planar ultra wideband (UWB) antenna with single band-notched behavior. An inverted pie-shaped slot is etched in a circular patch for getting the notch behavior. The presented antenna is successfully designed, simulated and measured showing single notch nature in UWB band as this band interfere in it. This antenna is designed on FR-4 substrate with dielectric constant 4.4 and thickness of 1.6 mm. The antenna parameters such as return loss, VSWR, gain and directivity are simulated and optimized using commercial computer simulation technology microwave studio (CST MWS). The designed antenna is showing band notch property at 7 GHz (6.4–7.5 GHz for satellite communication) as this band interfere with the UWB. The main advantage of this antenna is that the designed structure is very simple and the cost for making this antenna is also low.

Keywords Pie-shaped slots · Circular patch · Single band-notched characteristics · Ultrawideband (UWB) antennas microstrip antenna · C-band

M. Sharma (✉) · H.S. Mewara · A. Dadhech
Government Engineering College Ajmer, Ajmer, India
e-mail: minakshiiii.3991@gmail.com

H.S. Mewara
e-mail: hsmewara@gmail.com

A. Dadhech
e-mail: ajaydadhich13@gmail.com

M.M. Sharma
Department of ECE Engineering, MNIT, Jaipur, India
e-mail: mmsjpr@gmail.com

S. Yadav
Government Women Engineering College Ajmer, Ajmer, India
e-mail: Sanjeev.mnit@gmail.com

S.C. Satapathy and S. Das (eds.), *Proceedings of First International Conference on Information and Communication Technology for Intelligent Systems: Volume 1*, Smart Innovation, Systems and Technologies 50, DOI 10.1007/978-3-319-30933-0_11

1 Introduction

Since wireless communication is increasing exponentially these days, there is a need of new technology that can open new door to wireless communication. UWB provides solution to this problem. In 2002 Federal communication commission allocate a frequency band from 3.1 to 10.6 GHz for commercial use [1], researchers pay much attention on modern indoor wireless communication with high data rate. Commercial UWB systems require small low cost omnidirectional antenna and large impedance bandwidth.

There are some characteristics of UWB antenna compared with general narrow band antenna. First, UWB uses communication method using pulse, the phase must be linear over entire operating frequency range to decrease distortion of pulse during transmission and reception. Second, return loss should be maintained below −10 dB over entire Frequency range. Third, VSWR should be lower than 2 for whole bandwidth. Antenna miniaturization is also required for easy mounting and fabrication.

There are many wireless communication system exists that produce interference with ultra wide band such as 802.11a in USA (5.150–5.350 GHz and 5.725–5.825 GHz), World Interoperability for Microwave Access (WiMAX) (3.400–3.690 GHz and 5.250–5.825 GHz), HIPERLAN/2 in Europe (5.150–5.350 GHz and 5.470–5.825 GHz), c-band (4–8 GHz) etc. These pre-existing band interfere UWB band. So for proper operation in UWB band, we need to reject these bands.

Filter can be used to suppress these unwanted bands but they increase complexity and cost of antenna. By introducing several design methods and inserting U shaped slots [2, 3], L-shaped slots [4] and, H-shaped slots and other shapes of slots [5] on antenna patches, we can reject unwanted bands.

This paper proposes a microstrip antenna for UWB antenna. This antenna operates for whole bandwidth of ultra wideband and rejects a frequency band from 6.4 to −7.5 GHz which includes super extended C-band (6.425–6.725 GHz) and INSAT (6.725–7.025 GHz). This C-band used for satellite communication. Owing of simple structure this design can be easily designed and fabricated so it reduces cost and time of manufacturing. This antenna uses FR-4 substrate that is easily available material.

2 Antenna Design

The dimensions of proposed antenna are mentioned in Table 1. Design of antenna and ground plane with slots is shown in Fig. 1. In this antenna FR-4 substrate with the dimension $26 \times 30\ \text{mm}^2$ and the thickness of 1.6 mm is used dielectric

Table 1 Dimensions of proposed Antenna

Antenna parameter	Value (mm)	Antenna parameter	Value (mm)
L	26	W	30
L1	2	W1	11.6
L2	8.4	W2	3
L3	2.6	W3	1
L4	3	W4	2.5
L5	26	W5	11

constant of this substrate is 4.4 and loss tangent is 0.025. Feeding provided to this antenna with 50 Ω microstrip feed line with a width of 2.6 mm and length of 11 mm.

In this antenna a circular patch covers UWB bandwidth [6, 7] and then an inverted pie shaped slot is inserted on circular patch that provides rejection in required c-band at 7 GHz. There are various cuts and slots on partial ground plane those are useful in increasing impedance bandwidth of antenna.

3 Results and Discussion

The proposed antenna is simulated using CST Microwave Studio. Simulated results fulfill requirements of UWB antenna. Fig. 2. Shows return loss and VSWR of antenna. As per the requirements return loss is below −15 dB for whole impedance bandwidth but above −10 dB for rejected band.

VSWR is below 1.5 for whole frequency range that should be below 2 for UWB application. But for rejected band VSWR is above 2.

Figure 3a, b shows the farfield radiation patterns of this circular patch antenna at frequency 7.5 GHz and for broadband. The antenna is representing about omni-directional radiation pattern.

Figure 4. Shows gain of antenna that suddenly falls for rejected band and it is between 2 and 4 for entire UWB bandwidth.

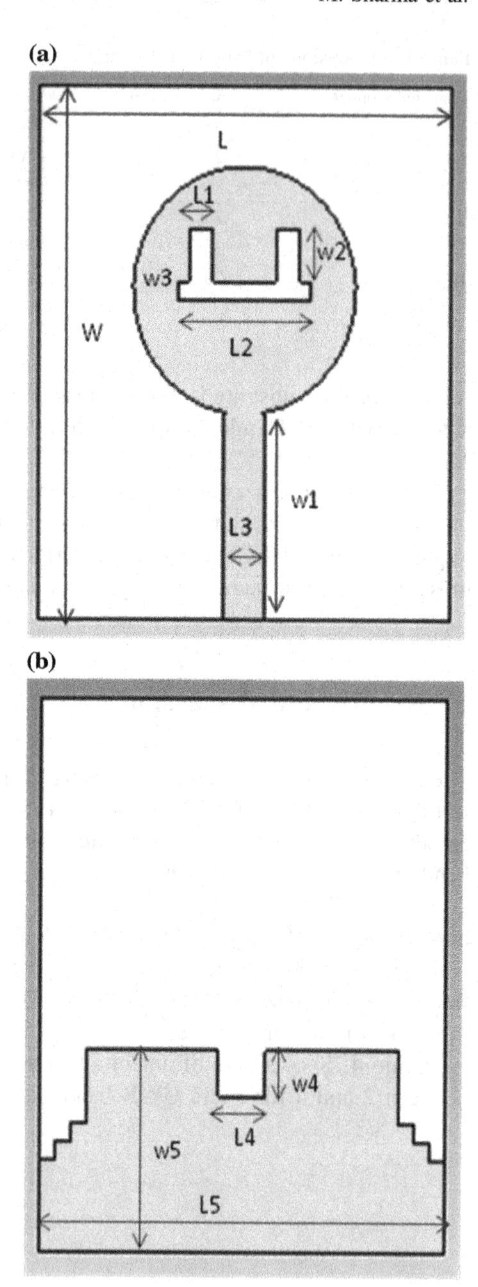

Fig. 1 Proposed antenna. **a** Antenna patch. **b** Ground plane

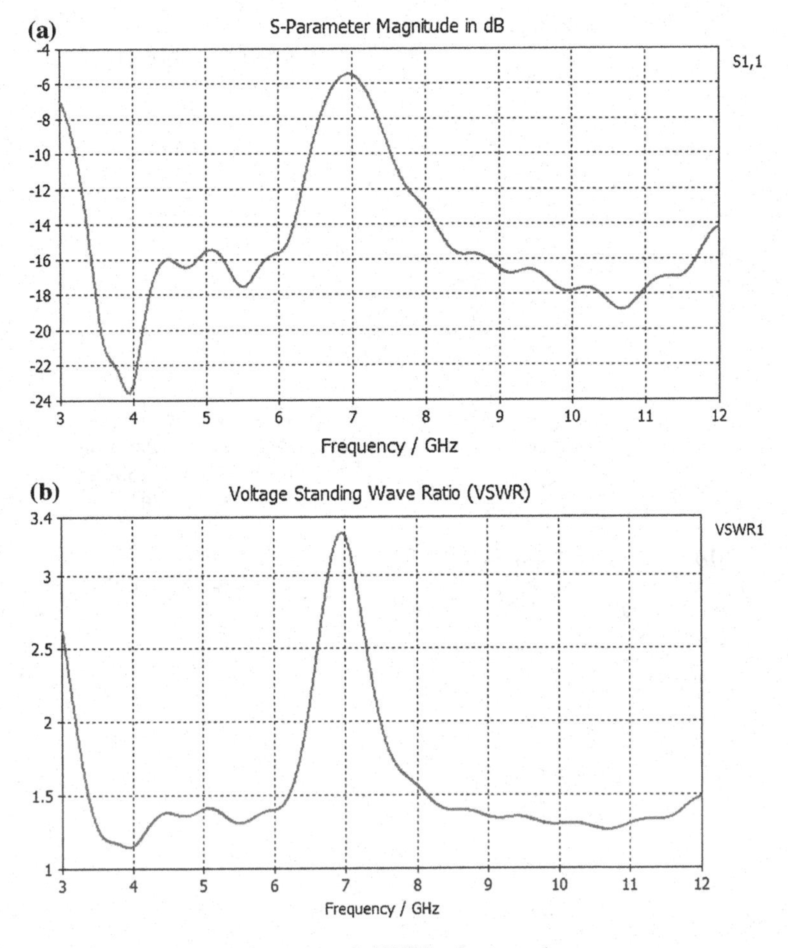

Fig. 2 Simulated results for **a** return loss, **b** VSWR of proposed antenna

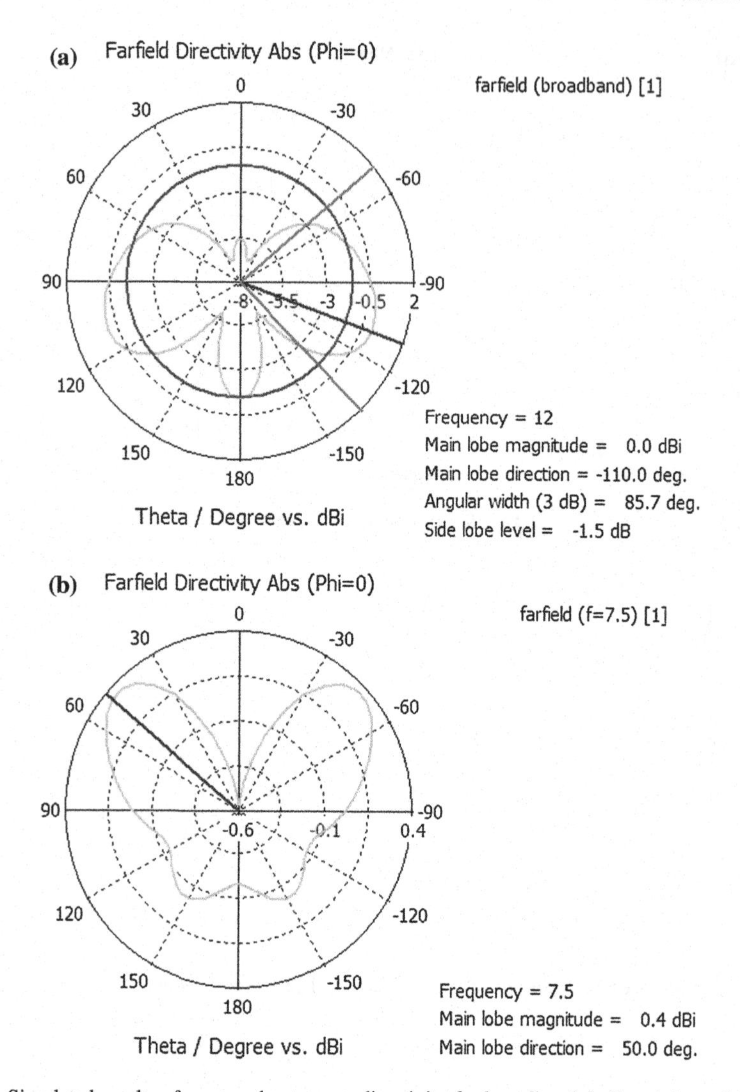

Fig. 3 Simulated results of proposed antenna **a** directivity for broadband, **b** directivity at 7.5 GHz

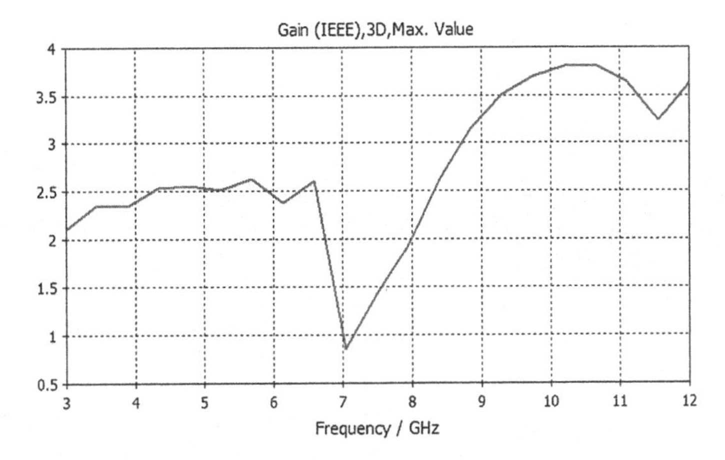

Fig. 4 Gain of proposed antenna

4 Conclusions

A small size and low cost planar microstrip antenna is proposed in this paper. FR-4 substrate with dielectric constant 4.4 is used in it. Return loss and VSWR are according to requirement. Farfield patterns of antenna are also in desired direction. Antenna gain is in range of 2–4 for whole frequency range. Thus all simulated results are good.

References

1. Federal Communications Commission, First report and order, revision of part 15 of Commission's rule regarding UWB transmission system FCC02-48 (2002)
2. Cho, Y.J., Kim, K.H., Choi, D.H., Lee, S.S., Park, S.O.: A miniature UWB planar monopole antenna with 5 GHz band rejection filter and time domain characteristics. IEEE Trans. Antenna Propag. **54**(5), 1453–1460 (2006)
3. Lee, W.S., Kim, D.Z., Yu, J.W.: Wideband planar monopole antenna with dual band-notched characteristics. IEEE Trans. Microw. Theory Tech. **54**(6), 2800–2806 (2006)
4. Zhao, Y.H., Xu, J.P., Yin, K.: Dual band-notched ultra-wideband microstrip antenna using asymmetrical spurlines. Electron Lett. **44**(18), 1051–1052 (2008)
5. Su, S.W., Wong, K.L., chang, F.S.: Compact printed ultrawide band slot antenna with a band notched operation. Microw. Opt. Technol. Lett. **15**(9), 576–578 (2005)
6. Wong, K.L., Chi, Y.W., Su, C.M., Chang, F.S.: Band-notched ultrawide band circular disc monopole antenna with arc-shaped slot. Microw. Opt. Technol. Lett. **45**(3), 188–191 (2005)
7. Sim, C.Y.D., Chung, W.T., Lee, C.H.: A circular disc monopole antenna with band-rejection function for ultrawide band application (2009)

New Architecture and Routing Protocol for VANETs

V.B. Vaghela and D.J. Shah

Abstract Vehicular Adhoc Networks (VANETs) are now famous for increasing safety on the road as well as providing value added services (i.e. entertainment and internet access). The intention to write this paper is to provide information for avoiding accidents using vehicle-to-vehicle (V2V) communication. We have provided a pioneering architecture for broadcasting the messages which contains information such as position of the vehicle with the help of latitude and longitude and the information regarding road conditioning as well as parking space. We have also defined some new parameters to avoid injuries, provide convenience to park vehicles and also provide infotainment applications for the drivers and passengers. With this paper, we propose a solution for ultimate usage of bandwidth and provide minimum delay in various situations in VANETs. We have developed a new protocol to take care of the quick and reliable delivery of information among all the users.

Keywords VANET · Service · V2V · Internet access · Entertainment · Scalable

1 Introduction

In this decade, the research communities of automotive industries and wireless networking have given more attention Vehicular Adhoc Networks (VANETs). VANETs have some special characteristics which derives them different then Mobile Adhoc Networks (MANETs). Specifically said nodes in VANET follow the lane directions; nodes also move with high speed; the number of nodes (vehicles) are more comparatively so VANETs are required to be scalable; the nodes need not to worry about energy because they are always fitted with dynamo [1].

V.B. Vaghela (✉)
J. P. Institute, Sidhpur, Gujarat, India
e-mail: vanaraj79548@yahoo.co.in

D.J. Shah
Sruj LED Ltd., Ahmedabad, Gujarat, India
e-mail: djshah99@yahoo.co.in

S.C. Satapathy and S. Das (eds.), *Proceedings of First International Conference on Information and Communication Technology for Intelligent Systems: Volume 1*,
Smart Innovation, Systems and Technologies 50, DOI 10.1007/978-3-319-30933-0_12

To achieve stable and effective communication in VANETs, following three crucial challenges play the vital role [2]:

(1) Efficient utilization of limited bandwidth
(2) Maintaining fragmented topology for communication and
(3) Achieving low-latency in delivering any kind of information.

A major challenge mentioned above is how can we use available bandwidth efficiently with number of applications and large number of network nodes in VANET. Also VANETs are sensitive to delays and jitter. In order to achieve faster propagation of broadcasted message generated by the event we need multihop transmission with minimum delays. In the proposed protocol, we have modified MAC layer, Network layer and transport layer with the use of Network simulator for the specific logic which can estimate the transmission range of nodes. We have used VanetMobiSim to generate the mobility for the urban scenario. This model can give facility to model each node with its start point and end point of journey with collision free movement.

This technique is to be used in vehicles to estimate the transmission range for effective propagation of broadcast messages with as few hops and transmissions as possible. Analysis of the data that we have received for the State of Gujarat reveals that there were over 111,431 people injured in the year 2012, in 2013 it was 105,273 and the same trend continued for 2014, giving nearby one lac injured in Gujarat state. Out of these more than 4000 people got injured in accidents every year in Gandhinagar city only. Considering the above facts, using the proposed research we are positive to provide a solution that can provide the information about number of vehicles on road in particular area, traffic jam or congestion in particular area, vehicle speed, road conditions, parking space etc.

The remaining part of the paper is organized with following sections: Sect. 2 gives a brief idea about the work related to the VANETs. Then after our proposed mechanism is explained in Sect. 3 along with the basic details of the proposed solution with best practical example. The results are shown in Sect. 4 with the proper justification while final conclusions are derived in the last section.

2 Related Work

The work proposed in [3] shows back-off mechanism which has ultimately tried to reduces message retransmissions due congestions and collisions. While in [4], the procedure to detect the repeated transmission is mentioned. It is found that, none of these techniques have given importance to the parameters which determines the end to end delay. One of these parameters is the number of hops essential to cover the targeted area.

Last two schemes explained in [5, 6] get the parameter of a minimum number of hops during message broadcast. The cars communicating with V2V logic, randomly select a waiting time for a CW before message forwarding. In all way, the mentioned schemes are somewhat unrealistic. So as an alternative, we propose a new scheme which is aware with vehicle positioning and can reduce the basic parameter i.e. hop count, based on the proposed algorithms.

3 Proposed Solution (REMBP)

We have proposed a new architecture in which the messages generated by generic application will be broadcasted via multi hop transmission. It is assumed that all the vehicles are fitted with OBUs and Global Positioning System (GPS). The packets received by OBU will be processed by our modified code of network simulator. Ultimately, the proposed solution is used to provide V2V communication between vehicles in a targeted area with the following features.

- It provides rapid propagation of messages generated by any event or on timer reset.
- It also provides a competent system to select the next-hop vehicle for forwarding the message.
- It estimates transmission range. This is the fundamental feature of the solution for a highly dynamic vehicular ad-hoc networks. It causes small overhead.

This solution works intelligently otherwise all the nodes in the area will once again broadcast the received message. The overall effect will lead to congestion, increase in collisions, higher delays, and will tend to transmission problem in the vehicular network. Such phenomenon is known as "broadcast storm problem" in networking literature [7].

In our proposed solution, as mentioned, vehicles are fitted with OBUs for vehicle-to-vehicle (V2V) communication, monitoring, entertainment and location awareness (GPS). Also we assume that a specific number of cars move at various speeds ranging from 40 to 140 km/h on a highway road with multiple lanes. As per the characteristics of VANETs, V2V communication may have different transmission range and available bandwidth. This we can assume because of the availability of the DSRC, which promises to provide vehicles with communication capabilities. Also in our proposed architecture we have considered two practical case studies (i) safety applications and (ii) non safety applications. Safety applications include accident alert communication, road conditioning, congestion etc. and non safety applications include real time or non real time multimedia streaming and interactive communications such as video conferencing and interactive games [8].

The developed mechanism is used to broadcast quickly, even through multi-hops, send alert messages from a vehicle behaving abnormally to all following vehicles in a range or video triggering messages to activate a video stream sent in a certain geographical location back to a requesting vehicle (e.g., first responders travelling toward an emergency area) [9]. With the use of hello messages each vehicle can compute the transmission range and utilize the same to reduce the number of hops, the total transmitted messages, and hence the delay to cover the whole area-of-interest till destination [10]. The vehicles, engaged in the online session are having this common protocol and one of the most important feature of this protocol is transmission range estimation in both the directions i.e. forward and backward. This can be achieved via a mechanism called transmission range estimation, explained below.

3.1 Range Estimation

In this stage, by the use of hello messages, each car estimates its transmission range. For updating the value of transmission range, time is divided into slots. There are two main parameters (i) Current Front Maximum Transmission Range (CFMTR) and (ii) Current Back Maximum Transmission Range (CBMTR). The information

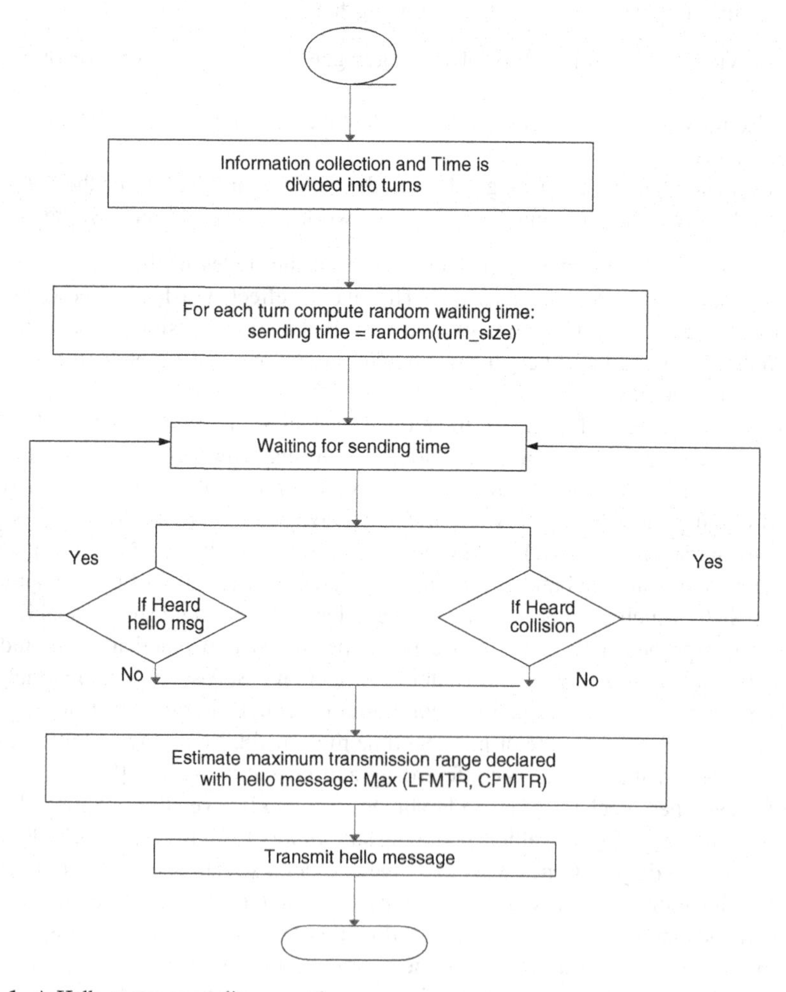

Fig. 1 A Hello message sending procedure

collected from the packet received for the current slot is called Current Front Maximum Transmission Range (CFMTR). LFMTR represents the estimated maximum forward distance from which the considered car can hear another car along the road and similarly CBMTR represents maximum backward distance. We have provided the mechanism to keep both the variables updated continuously via an event called hello messages receiving procedure, at this stage, their values are stored as LFMTR (Latest Front Maximum Transmission Range) and LBMTR (Latest Back Maximum Transmission Range), respectively.

The range estimation stage is best explained with the help of Figs. 1 and 2 as flow diagram. The logic of our scheme is best explained as hello messages sending (Fig. 1) and receiving hello messages (Fig. 2) procedures, respectively. During the procedure of hello message sending, a random waiting time is selected by each car determines in every slot. After the waiting time, if no other event happens (i.e. new transmission or collision), it ensues with transmitting a hello message, having estimation of maximum forward transmission range.

The flow diagram of Fig. 2, best explains hello message receiving procedure. After extracting the information of sender's position, its own position and the included estimation of the maximum transmission range, it determines the distance between itself and the sender. If the hello message is received from front the value of CFMTR is updated, otherwise CBMTR is changed.

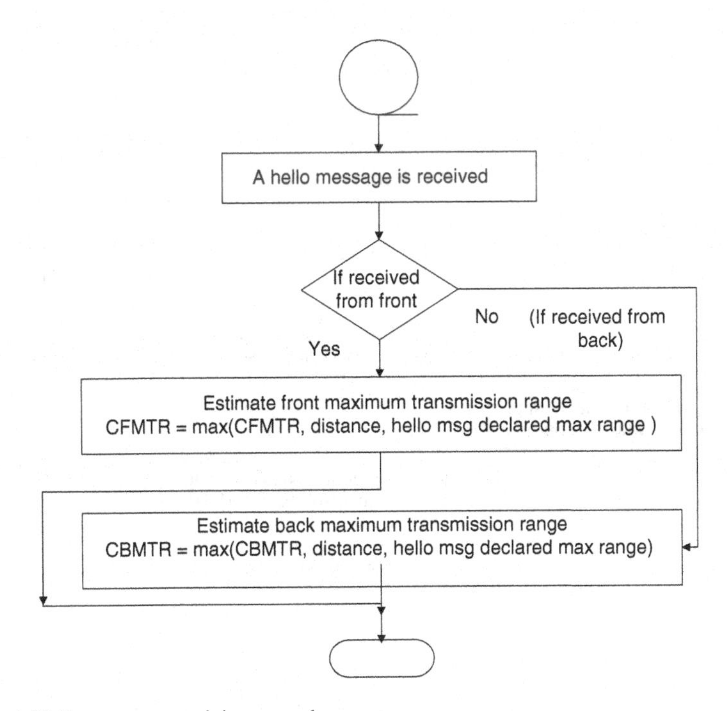

Fig. 2 A Hello message receiving procedure

3.2 Message Broadcast

As depicted from Fig. 3, message broadcast stage is used to reduce transmission time and get a quick receipt of message. Priority of forwarder is decided from this stage. The logic applied in this case is higher the relative distance, higher the priority. Max_Range() is a typical parameter that corresponds to how far the transmission is expected to go backward/forward before the signal becomes noisy and undetectable. Equation (1) gives this summary with help of CWMax and CWmin [8].

$$\left\lfloor \frac{(MaxRange - Distance)}{MaxRange} \times (CWMax - CWMin) \right\rfloor + CWMin \quad (1)$$

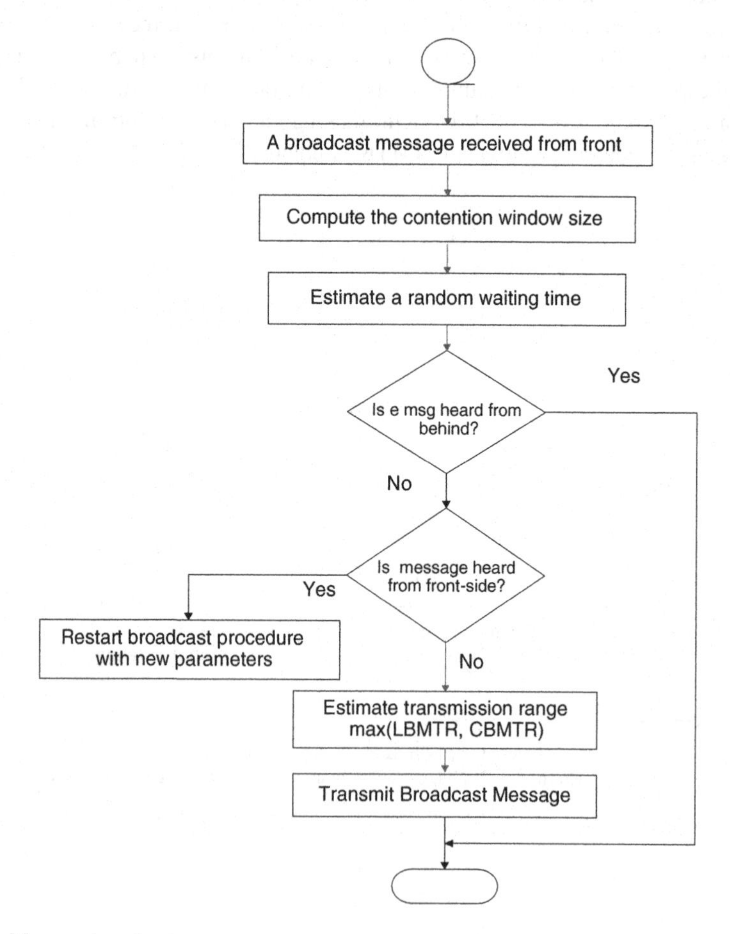

Fig. 3 Message broadcast

It is used by the cars in the targeted area to determine which one among them will become next forwarder. Ultimately because of this the number of hops are reduced. We have defined this to assure the farthest car to do this task of forwarding.

As message is received from front-side, a car uses Eq. (1) to determine CW and a random waiting time based on it.

4 Results and Discussions

We have modified the layers by changing some of the parameters according to the requirements of DSRC. The hello message generation rate is 100 ms. Figure 4 shows the throughput of the system. We have compared the results with the existing one and found they are better than the existing system. In this study, we have kept the length the network up to 6 km and vehicles are equipped with mentioned assumptions are placed in such a way that minimum distance between the two vehicles remain 20 m, thus we can say average 50–250 vehicles may be involved in the whole process.

The message delivery fraction shown in Fig. 5 and the estimated delay shown in Fig. 6, also validate the results.

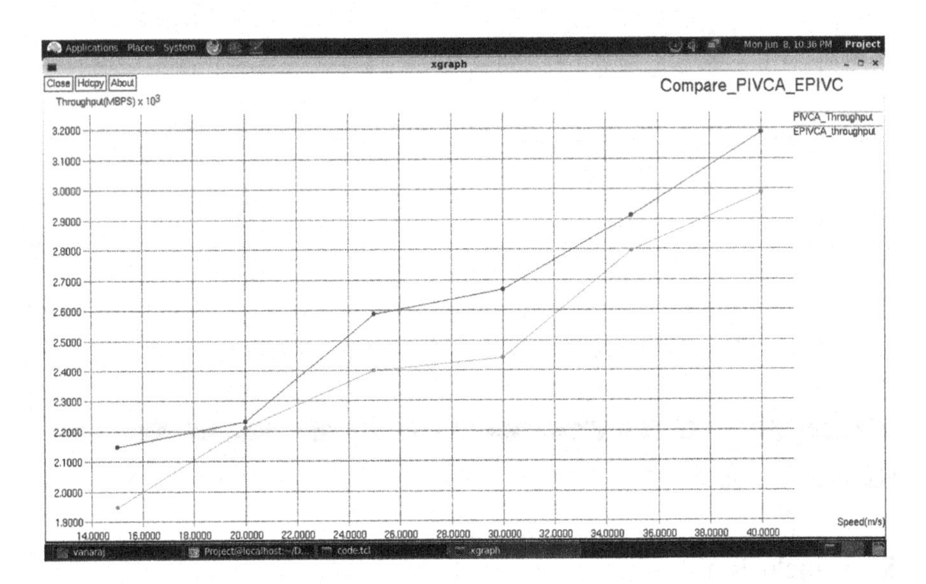

Fig. 4 Throughput

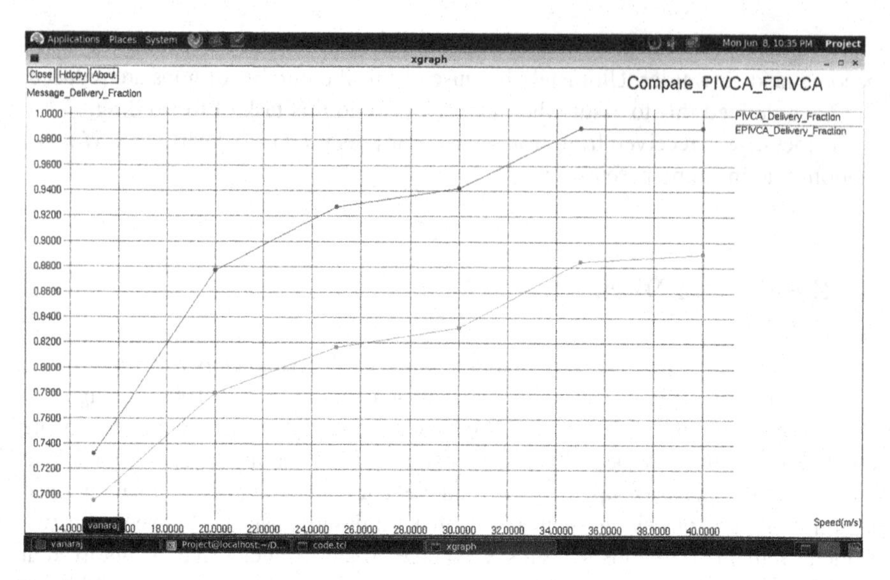

Fig. 5 Message delivery fraction

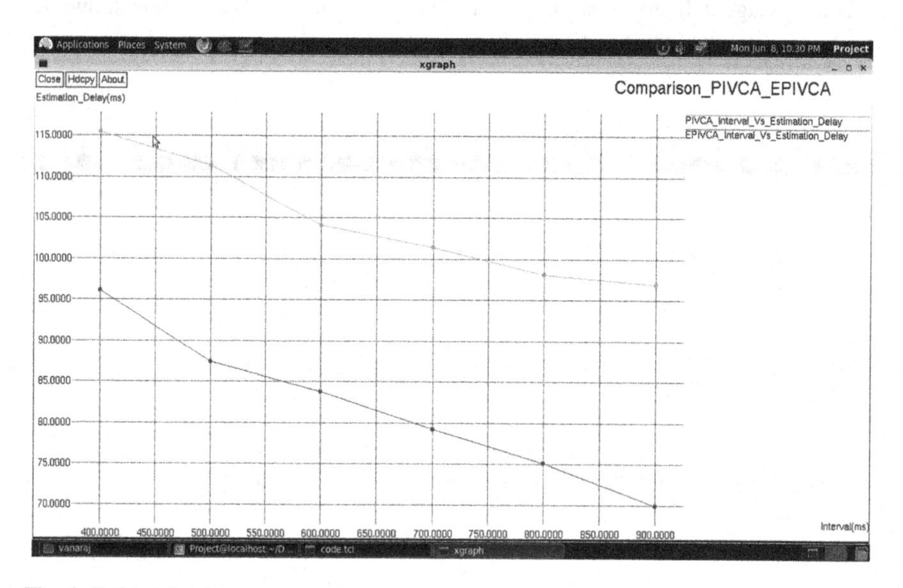

Fig. 6 Estimated delay

5 Conclusion

We proposed a broadcast protocol and provided a solution to have minimum propagation delay for real time applications and safety applications in the VANETs. Our solution can be used in 108 ambulance service in Gandhinagar city of Gujarat for vehicle to vehicle communication in urban scenario. We analyzed the network performance using Range Estimation Multi-hop Broadcast Protocol with the parameters mentioned to check the reliability of delivered messages. We have achieved different results and found that the messages reach the destination with improved message delivery fraction and throughput up to the mark.

References

1. Jafari, A.: Performance Evaluation of 802.11p for Vehicular communication networks. M. Sc. Thesis, Shaffield Hallam University (2011)
2. Zhao, J., Zhang, Y., Yeng, G.: Data pouring and buffering on the road: a new data dissemination paradigm for vehicular Ad Hoc network. IEEE Trans. Veh. Technol. **56**, 3266–3277 (2007)
3. Yang, X., Liu, J., Zhao, F., Vaidya, N.: A vehicle-to-vehicle communication protocol for cooperative collision warning. In Proceedings of Mobiquitous, Boston, MA, pp. 114–123, Aug 2004
4. Biswas, S., Tatchikou, R., Dion, F.: Vehicle-to-vehicle wireless communication proto for enhancing highway traffic safety. IEEE Commun. Mag. **44**(1), 74–82 (2006)
5. Fasolo, E., Furiato, R., Zanella, A.: Smart broadcast algorithm for inter-vehicular communication. In: Proceedings of WPMC, Aalborg, Denmark, pp. 1583–1587 Sept 2005
6. Blum, J.J., Eskandarian, A.: A reliable link-layer protocol for robust and scalable inter- vehicle communications. IEEE Trans. Intell. Transp. Syst. **8**(1), 4–13 (2007)
7. Yassein, M.B., Khaoua, O., Papanastasiou, S.: Performance evaluation of flooding in manets in the presence of multi-broadcast traffic. In: Proceedings of 11th Icpads-Workshops, Fukuoka, Japan, pp. 505–509, Jul 2005
8. Palazzi, C.E., Roccetti, M., Ferretti, S., Pau, G., Gerla, M.: Online games on wheels: fast game event delivery in vehicular ad-hoc networks. In: Proceedings of 3rd IEEE International Workshop V2V Com. IEEE IVS, Istanbul, Turkey, pp. 42–49, June 2007
9. Ratnani C., Vaghela, V.B., Shah, D.J.: A novel architecture for vehicular traffic control. In: Proceedings of IEEE International Conference on Computational Intelligence and Communication Technology, Feb 2015
10. Palazzi, C.E., Ferretti, S., Roccetti, M., Pau, G., Gerla, M.: An Inter-vehicular Communication Architecture for Safety and Entertainment. IEEE Trans. Intell. Transp. Syst., **11,** pp. 90–99 (2010)

A Scalable Model for Big Data Analytics in Healthcare Based on Temporal and Spatial Parameters

S. Hemanth Chowdary, Sujoy Bhattacharya and K.V.V. Satyanarayana

Abstract As the health care industry grows at a rapid pace, it is generating large volumes of data that needs to be stored, analyzed and acted upon by various organizations. India is a very diverse country with a large population that is having increased access to centrally managed healthcare systems. This is generating huge volumes of data, whose systematic storage and analysis for organized decision-making will be critical to the success of the industry in the coming years. This data can be classified into the realm of 'big data' for obvious reasons and appropriate technology will be required to handle it effectively. In this paper, we propose a model for analyzing historical healthcare data. Both temporal and spatial parameters have been used in this model to allow the healthcare professional different views into the information and thereby make informed judgments. Common constraints like quality, authenticity and security of the big data have also been addressed for complete effectiveness.

Keywords Big data · Healthcare data · Analytics · HBase

1 Introduction

In recent years, big data has become one of the emerging trends from research and industry perspective. At present, around a few quintillion bytes of data is generated every week, some of which is either informative or useless. This huge volume of

S. Hemanth Chowdary (✉) · K.V.V. Satyanarayana
Koneru Lakshmaiah Education Foundation (KLU),
Vaddeswaram, Guntur, India
e-mail: hemanth79s@gmail.com

K.V.V. Satyanarayana
e-mail: kopparti@kluniversity.in

S. Bhattacharya
Guide Lane, Hyderabad, India
e-mail: sujoy.b@guidelane.com

S.C. Satapathy and S. Das (eds.), *Proceedings of First International Conference on Information and Communication Technology for Intelligent Systems: Volume 1*, Smart Innovation, Systems and Technologies 50, DOI 10.1007/978-3-319-30933-0_13

data can be useful for many industries like scientific research industries, financial industries, healthcare industries and many others [1, 2]. The Healthcare industry ios one of the major contributors to this huge growth in data volumes, most of which, if properly stored and utilized, can create a revolution in the way healthcare is served. In the past, all the medical records, X-Ray reports, patient case sheets were stored in the form of hard copies only by specific institutes/providers and was not accessible outside. There was (and still is) no database maintained to store these data records in a retrievable and distributable manner. However, at present, the situation is changing and healthcare providers in India are beginning to store all the records including the patient information in the electronic format. Such storage enables easier patient health analysis with data being easily available for the healthcare professional.

IBM has characterized big data by its Volume, Velocity and Variety. Big data in health care is appealing not only because of its volume but also because of the variety and veracity of data. In healthcare industry, the big data, which comes under variety of data, includes lab reports, Electronic Patient Records, Emergency Care Data (ECD), patient Information etc., and veracity of data refers to the medical claims, cost data etc [3].

As the data is growing, it's storage in a retrievable format becomes a challenge, Typically, healthcare institutions do not invest in large data centers. So, their primary challenge lies in storage of the huge volumes of data. The best answer to this challenge is to store the data in a Non-Relational (NoSQL) database that takes care of the basic storage requirements. For example, HBase is an open source NoSQL database which provides a real-time read/write access to large datasets. HBase scales linearly to handle huge datasets with billions of rows and millions of columns and it easily combines data sources that use a wide variety of different structures and schemas.

2 Related Work

Considering the applicability of big data for handling large datasets that need analysis, it has been in wide use in multiple areas like finance, retail and image processing. But its application in the healthcare are has not been fully tapped yet. There is a necessity to produce big data solutions for the healthcare systems to get efficient analytical results based on the available data. The existing healthcare models need change because of two important factors. Firstly, for the meaningful use of the data and secondly, to enable a model where the user needs to pay for the performance. We need to apply the existing big-data analytical techniques to data related to patient health and medical records (either individual or group of population) and enable the physician to take an appropriate decision for the treatment of the patient.

The data that is captured from the sensors is continuous data, until the sensor is turned off. The big data system needs to manage this kind of data/information flow from different sources. The sources include the sensors fixed for heart, lungs etc. This continuous sensor data may contain some unwanted or irrelevant information. To prevent this, a Hidden Markov model is introduced for filtering meaningful data for the system. The data that is taken from the sensors at different conditions are analyzed for improving the quality of care [4–6]. Analytics can be performed on the other forms of data, which is taken from the social networks which contains temporal data related to the health care domain. Temporal analysis methods do not require any classifiers and they use timely techniques for different practices that are inconsistent [7]. For considering the data that is coming from different sources, the management of metadata becomes very important for correct analytics that drives decision making.

3 Architectural Framework for Healthcare System

Healthcare industry has become extremely data intensive with data coming from multiple sources. Data integration across heterogeneous data sources is the biggest challenge being identified, which would otherwise contribute to greater insights on the data available. While there are many solutions for dealing with heterogeneous data in healthcare, the model proposed in this paper gives an efficient solution. The model described is shown in the Fig. 1.

The proposed model mainly executes on two phase connected with an open source columnar database called HBase: Data Mapper Phase and Data Reducer Phase. Each phase has its own significance.

Data Mapper Phase splits the input given by the user into multiple modules like Patient Module, Disease Module and Doctor Module as shown in the figure. The input given will be split based on two parameters chosen in the spatial and temporal domains. The patient information is stored based on the spatial information, and the time stamp at which the patient is examined. By using this spatial information, it will be easy for the healthcare professionals to diagnose a regional trend of diseases.

This phase will process the input split by itself, and after processing, it will get a new key-value pair for the output. This processed output of the Mapper phase is stored in the database with different document names. This phase can handle any number of inputs and outputs, thus making the model easily scalable. Mathematically the collection of all the key-value pairs can be represented as:

$$DMP = \sum\nolimits_{k,v=0}^{n} (k, v). \tag{1}$$

In **Data Reducer Phase** retrieves the data stored in the form of key-value pairs from the healthcare database or from the hospital management system and sends the processed data to the analytical tool for formatting and display. Finally, reports are generated based on the input given by the Data Reducer to the analytical tool.

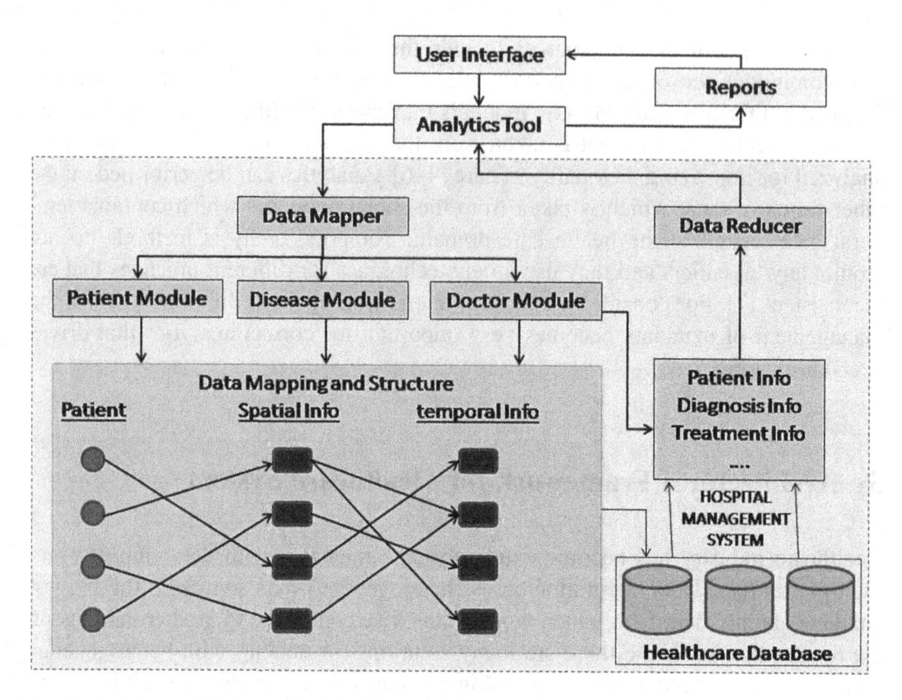

Fig. 1 Architectural framework of healthcare system

3.1 How HBase Works with Healthcare Data

HBase is an open source database, which stores the data in the form of columns. This model has the capability of simultaneously handling inputs in various formats coming from different input sources. For example, for a patient in the ICU, the database will be getting parallel inputs in the form of Blood Pressure reading from a continuous monitor, scanned images from an ultrasound scanner and heartbeat patterns from an ECG machine, The inputs are stored in the HBase after processing in the Mapper phase. HBase provides the additional facility to seamlessly add or remove a monitor in real time, thus offering supreme scalability.

Figure 2 shows how the healthcare data is proposed to be stored in the HBase database. This caselet shows how different documents coming from different sources are stored in a hypothetical medical situation. For example/source:2/allergy:345 means in this case, source:2 is the source name from where the data is coming and allergy is the category (document) name and 345 is the document number. In this format, we can store any number of documents.

The table structure of this HBase database is as shown below:

If data is received from multiple sources for the same patient, it will be stored in the same row designated for that person and identified by the Row Key for that patient. For example, if we consider patient: 1, the information coming from different sources like Source: 2, Source: 150…, is stored in Patient: 1 row only as shown in the Table 1.

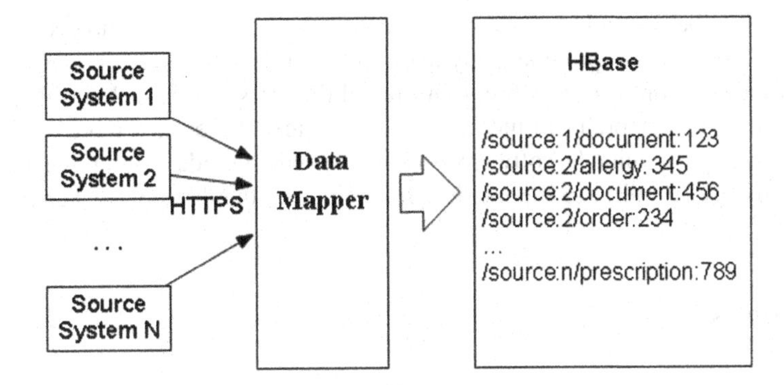

Fig. 2 Data storage in HBase

Table 1 Table structure of HBase

Row Key	Qualifiers (lease record and keys of updated items)
Patient:1	0000_LEASE, source:2/allergy:345, source:150/Body Pains:71, …
Patient:2	0000_LEASE, source:4/skin:78, source:205/hairfall:52, …
.. .	.....

4 Outcome

As healthcare systems have a wide variation from both a geographical and a diseases standpoint, the model needs to be flexible enough to cater to this variation. The inherent scalability offered by HBase effectively addresses both these requirements. The Data Mapper designed in this model effectively handles this essential requirement.

In order to ensure that the model does not just stay as a theoretical data store, the data reducer model has been added to take care of necessary reporting from the collected data. An appropriate analytical tool serves as the interface between the model and the users of the data.

5 Conclusion

Because of the explosive growth of healthcare data in India, there is an immediate need for an inclusive system that can store the data and also provide insights to the information content within the data for analysis and diagnostic purposes. The problem is further aggravated by institutions creating their own data pockets because of the non-availability of a central storage space. This issue is becoming one of national importance for efficient rollout of the various schemes that are announced by the central government. This paper proposes a model that can scale up to this challenge and act as a central data repository that can hold data from across the country and provide analytical results on the same. The model is being built to include complete flexibility and the ability to store data about all critical diseases that are important in the present day context. When this model is implemented through a suitable software package, it will provide a long-awaited and immensely useful utility to the medical fraternity and the decision makers.

References

1. Zulkernine, F., Martin, P., Ying, Z., Bauer, M., et al.: Towards cloud based analytics-as-a-service for big data analytics in the cloud. In: International Congress on Big Data, pp 62–69, (2013)
2. http://www.cloudera.com/content/cloudera/en/products-and-Services/cdh/hbase.html
3. http://www.hortonworks.com/hadoop/hbase
4. Jiang, P., Winkley, J., Zhao, C., Munnoch, R., Min, G., Tang, L.T.: An intelligent information forwarder for healthcare big data systems with distributed wearable sensors. IEEE Syst. J., pp. 1–9, (2014)
5. Raghupathi, W., Raghupathi, V.: Big data analytics in healthcare: promise and Potential. Health Inf. Sci. Syst. **2**:3, (2014). http://www.hissjournal.com/content/2/1/3
6. Vemuganti, G.: Metadata management in Big Data. Infosys lab briefing **11**, (2013)
7. Chandola, V., Sukumar, S.R., Schryver, J.: Knowledge Discovery from Massive Healthcare Claims Data, pp. 1312–1320 ACM, (2013)

Part II
Intelligent Web Mining and Knowledge Discovery Systems

Comparison of Different Techniques of Camera Autofocusing

Dippal Israni, Sandip Patel and Arpita Shah

Abstract Automatic focusing has become essential part of imaging system as the image quality matters. There have been many researches carried out for autofocusing. Autofocusing gives the benefit of high-contrast image capturing even while the scene or imaging system is moving. The mechanism adjusts focal point of imaging system so that it gives the high contrast image. Need of different autofocusing technique arises as a single autofocusing mechanism cannot serve all the application. Autofocusing is divided into (i) Active and (ii) Passive autofocusing. Active autofocusing is good choice for SLRs, but it cannot be used when using an independent light source is not possible. Passive autofocusing is the best solution in such cases which captures the scene and analyzes to determine focus and is fractioned into two sub categories (i) Contrast (ii) Phase. This paper concludes the different autofocusing techniques and its applications.

Keywords Autofocus · Phase based autofocusing · Contrast based autofocusing · Contrast measures · Focus measures · Coarse focusing · Fine focusing

1 Introduction

Autofocusing has covered a huge part in image capturing in recent days. Traditional imaging systems were using fixed focus mechanism in image capturing. While capturing an image using conventional imaging system, image may happen to be

D. Israni (✉) · S. Patel · A. Shah
Department of Computer Engineering,
CSPIT, CHARUSAT University, Anand, Gujarat, India
e-mail: dippalisrani.ce@charusat.ac.in

S. Patel
e-mail: sandip19191@gmail.com

A. Shah
e-mail: arpitashah.ce@charusat.ac.in

S.C. Satapathy and S. Das (eds.), *Proceedings of First International Conference on Information and Communication Technology for Intelligent Systems: Volume 1*, Smart Innovation, Systems and Technologies 50, DOI 10.1007/978-3-319-30933-0_14

out-focused due to many of the reasons like Moving object in still background, movement of camera, disturbance in the lens assembly of the system [1–3].

In turn to remove this problems camera went under number of various iterations to build an focusing system. Initially it absorbed a system with manual focusing which in turn moved to an autofocus mechanism.

Auto-focus has emerged through number of iterations in the past to reach to the accurateness it has now a days. Somewhere around 1960–1973, Leitz firstly patented an array of auto-focus sensor technologies. As soon as this technology immerged, various researchers focused on this era. In 1978, he presented an SLR camera with fully operational auto-focus based on his previous developments on the technology. Konica C35 AF is the first mass-produced auto-focus camera which was a simple point and shoot model released in 1977. The first 35 mm SLR developed was Pentax ME-F in 1981, which is based on focus sensors in camera body coupled with a motorized lens. F3FA in 1983 was first auto-focus camera by Nikon which was based on similar concept to the ME-F. The revolution in the auto-focus mechanism was recorded in 1985. The Minolta 7000 was the first SLR with an integrated auto-focus system, which had both the AF sensor and the driver motor were combined in camera body itself as an integrated film advance winder [2].

Modern SLR cameras use through-the-lens optical AF sensors, which provide separate sensor array which light metering. Autofocusing can be carried out in two ways: (1) Active Autofocusing and (2) Passive Autofocusing. Active Autofocus requires an independent light source which incidents light on object and sensors capture light reflected by the object and autofocusing is done. Passive autofocusing does not require independent light source. It works on the light emitted by the object itself. This is the main reason why Passive autofocusing is used in Space Imaging System [1].

Passive autofocusing can further be learnt as (1) Phase-based Autofocusing and (2) Contrast-based Autofocusing. Phase-based Autofocusing system works based on phase difference between different images.

This paper derives the survey of different focusing methods. All the autofocusing techniques are discussed in brief with their processing points. It includes basics of autofocusing and detailed description of various intermediate modules of those systems. This paper includes survey of various researches carried out by different researchers and their conclusions about those systems. It also embraces simulation results of different contrast measures [4].

2 Autofocusing Techniques

Figure below shows the basic flow of autofocusing. Image sensor captures image from light received by lens from object and some basic image processing is carried out which is optional in some cases. The Autofocusing (AF) algorithm is processed on it and decision of movement is given to driver circuit. This decision makes AF actuator to move towards the decided direction moving lens towards focus [5].

2.1 Active Autofocusing

Active AF systems identify distance to the image object independently of the optical system and then adjust the optical system to correct the focus of system. The active focusing can also be known as a technique in which a light source is made to be incident on the object and the based on the light reflected by the object, the picture is taken. The more the amount of light reflected, the better the picture quality is. Based on the reflection adjustment of lens can be done [1, 6]. Distance can be measured using various techniques including ultrasonic sound waves and infrared light.

2.2 Passive Autofocusing

In Passive AF systems, correct focus is determined by performing passive analysis of the image that is entering the imaging system. They generally do not direct any energy, such as ultrasonic sound or infrared light waves, as in active autofocusing system toward the subject. Study suggests that Passive autofocus must have light and image contrast in order to do its job. The image needs to have some detail in it that provides contrast. Passive autofocusing can be achieved by phase detection or contrast measurement.

1. Phase-based Autofocusing
   Phase-based strategy uses a phase detection sensor to decide if image captured is out-focused or in focused. A ray of light from the object falls on the lens of camera. This light passes through the lens assembly and falls on primary mirror which allows some amount of light to pass through and some light is reflected towards the Pentaprism. The light passed through the primary mirror is directed on Phase Detection Sensor with the help of secondary mirror. Phased Detection Sensor decides whether the image is in-phase or out-phased and the decision is given to the driver whether to move the lens or not to move [4].
   It may happen that the light may get split such that the light rays falling on a Phase detector may generate a wrong decision of object being out-focused and it may make lens move which actually will make lens out-focused. It also may happen that phase Detector generates correct decision to move lens and after

moving it, as there is no acknowledgment from lens to Phase Detector, it (Phase Detector) may never come to know that assembly is already focused.

2. Contrast-based Autofocusing
   In Contrast-based autofocusing system, image is captured and contrast computation is done. Based on the contrast value, decision is made if image is focused or not and accordingly autofocusing is carried out. Contrast-based Autofocusing can be done using various contrast measures. The contrast measure defines a feature of an image. These features can be divided in different categories based on various criteria like if they are in Spatial Domain or Frequency Domain, Edge Detectors or Intensity Variation based, Sharpness function or Smoothness function, Wavelet based contrast measures [7, 8].

2.3 *Comparison of Phase Based Autofocusing and Contrast Based Autofocusing*

Many of SLR today use phase detection over contrast detection in designing autofocus mechanism. The autofocusing mechanism of an SLR is very compact in size. The lens assembly used in SLRs is of a few grams and it is easy to move a lens to focus with weight in grams. But when it comes to designing a autofocus mechanism for a space imaging system, which has a lens assembly of up to some kilograms and rigidly fixed, moving such heavy lens is too tricky compared to SLRs. In such cases, moving a detector to focus is easier comparatively. Using phase detection in this case is not a good option as mentioned earlier. In situation, contrast based autofocus mechanism is designed. Table 1 shows different comparative measure of using phase based autofocusing and contrast based autofocusing [1, 4, 8].

Table 1 Phase versus contrast based autofocusing [1, 4, 8]

	Phase based autofocusing	Contrast based autofocusing
Merits	• Faster autofocusing speed • Provides much more information than contrast based autofocus	• Less complex system compared to phase based system • Immune to noise and single peak problem • Sensitive to image saturation giving better results in focusing compared to phase based
Demerits	• Complex electronics system • Prone to false focusing	• More computation • Requires patience • Better suited to fairly static subjects • Might refocus the wrong way before going the right way

3 Survey on Techniques

3.1 Active Autofocusing

An active autofocusing system suggested by B. Erbas and C.I. Underwood was designed using different focus measures. Active focusing simply refers to bringing the imaging System to a precise focus by actively controlling a component in the image structure. It involves the determination of the degree of defocusing in the system, and then moving lens in order to compensate this out-of-focus state of the system [9]. A system was designed and an experiment was carried out in order to assess the performance of the focus measures was simply to move the detector along a straight line while recording the focus measure values.

3.2 Passive Autofocusing

Passive autofocusing can be divided in two different types:

1. Phase-based Autofocusing
   Phase detection is commonly applied passive autofocus method used in single-lens reflex (SLR) cameras. Phase based auto-focusing can be done in imaging system. Figure 1 depicts the mechanism of Phase based autofocus.
   As shown in Fig. 1, it consists of a beam splitter which helps in phase detection of an image. A ray of light from the object falls on the lens of camera. This light passes through the lens assembly and falls on primary mirror which allows some amount of light to pass through and some light is reflected towards the Pentaprism. The light passed through the primary mirror is directed on Phase Detection Sensor with the help of secondary mirror. Phased Detection Sensor decides whether the image is in-phase or out-phased and the decision is given to the driver whether to move the lens or not.

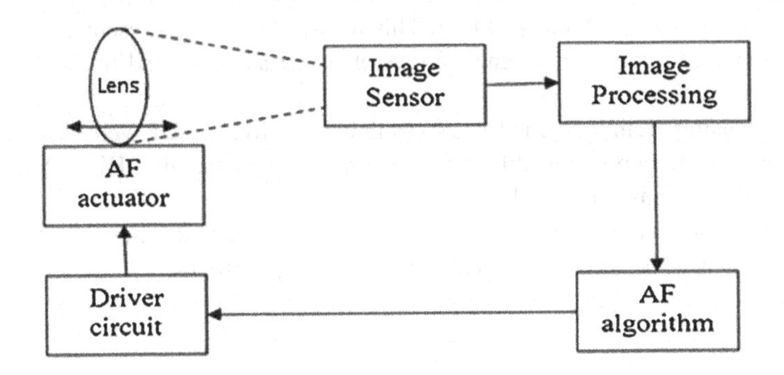

Fig. 1 Basic block diagram of autofocusing [5]

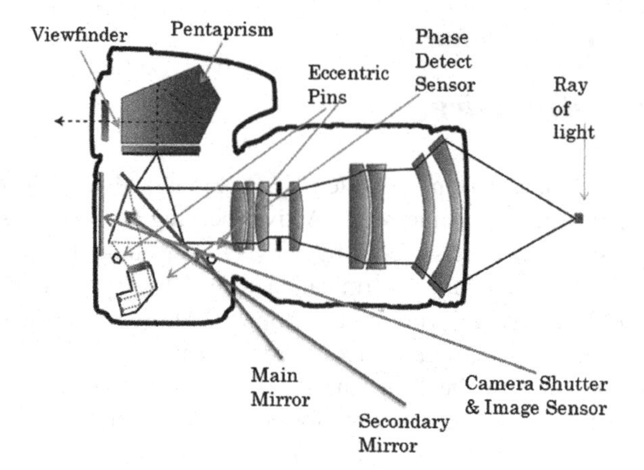

Fig. 2 Phase-based autofocusing in DSLR [2]

Figure 2 describes the mechanism of phase detection based autofocusing. Figure 2a shows the out-focused position and Fig. 2b shows the focused position of Phase based system (Fig. 3).

2. Contrast-based Autofocusing

   As explained by Said Pertuz, Domenec Puig, and Miguel Angel Garcia in "Analysis of focus measure operators for shape-from-focus", various contrast measures were simulated to check performance in accordance with the focus of an image. They concluded that every focus measure only decreases while image goes from in-focus to out-focus [7].

   Table 2 shown below describes all the focus measures suggested [3, 7, 10].

   Figure 4 shows the graph of Focus measure versus Focus Position for some of the measure above mentioned. Here focus position 1 describes full focused and 21 total out focused.

   Qinghua Zhao, Bo Liu and Zhaohui Xu used some of above mentioned focus measures to design a "Contrast Based Autofocusing System" [11]. A system was designed that captures an image and computes different focus measure and compares it to the value at previous position and generates a decision to move the lens towards the focus position. This measure can be represented in various domains like Spatial, Frequency, Wavelet based as suggested in Table 2 [12–14].

   Autofocusing technique for Inverse synthetic aperture radar were designed by various researchers using different measures of focusing like High Resolution [12, 14], Polarimetric [15, 16] etc.

   Figure 4 shows the contrast based autofocusing. It concludes that the contrast of an image is at its pick when image is completely in-focus.

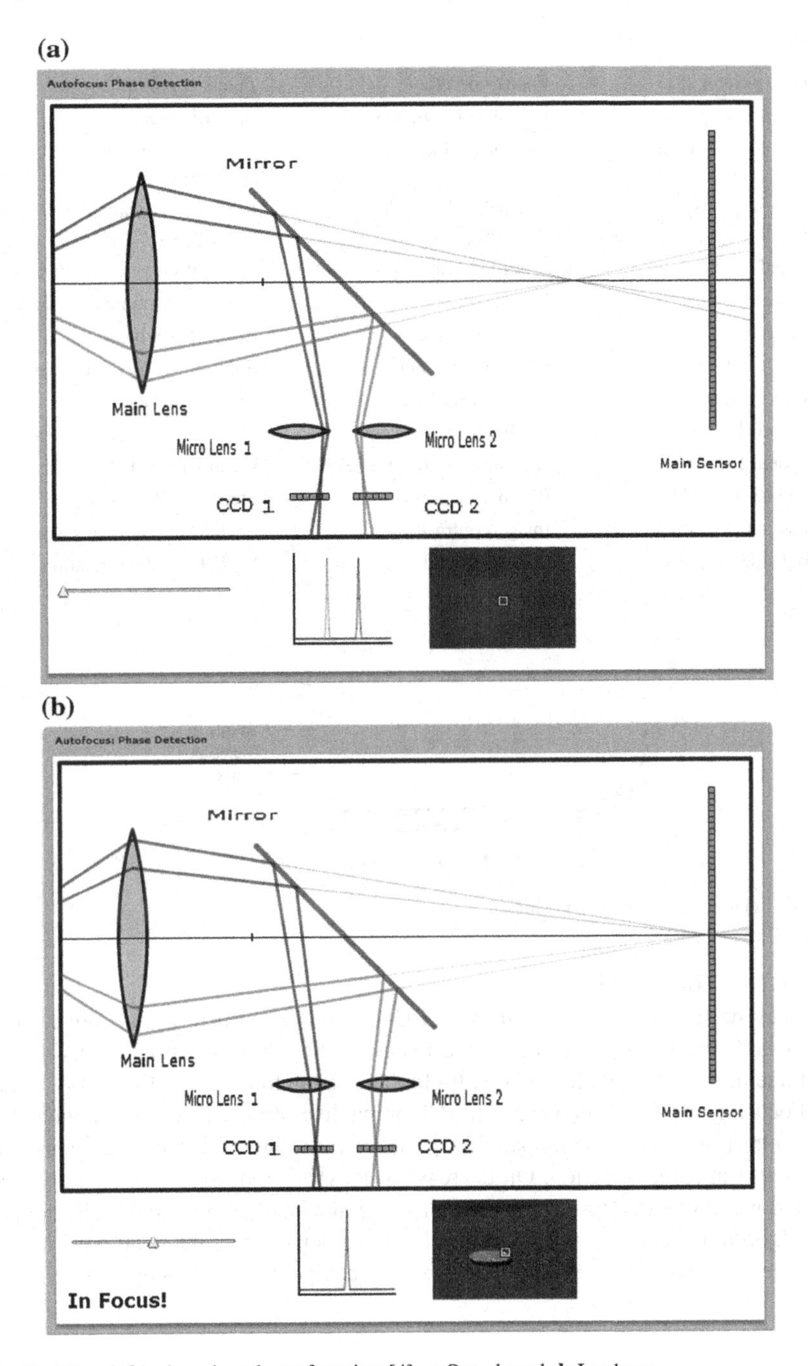

Fig. 3 (**a**) and (**b**) phase based autofocusing [4]. **a** Out-phased. **b** In-phase

Table 2 Various contrast measures

Focus operator	Focus operator	Focus operator
Gradient energy	Gray-level variance	Sum of wavelet coefficients
Gaussian derivative	Gray-level local variance	Variance of wavelet coefficients
Thresholded absolute gradient	Normalized gray-level variance	Ratio of the wavelet coefficients
Squared gradient	Modified gray-level variance	Ratio of curvelet coefficients
3D gradient	Histogram entropy	Chebyshev moments-based
Tenengrad	Histogram range	Eigenvalues-based
Tenengrad variance	DCT energy ratio	Sum of wavelet coefficients
Energy of Laplacian	DCT reduced energy ratio	Image curvature
Modified Laplacian	Modified DCT	Hemli and Scherer's mean
Diagonal Laplacian	Absolute central moment	Local Binary Patterns-based
Variance of Laplacian	Brenner's measure	Steerable filters-based
Laplacian in 3D window	Image contrast	Spatial frequency measure
Sobel Edge detector	Image Integral	Vollath's autocorrelation

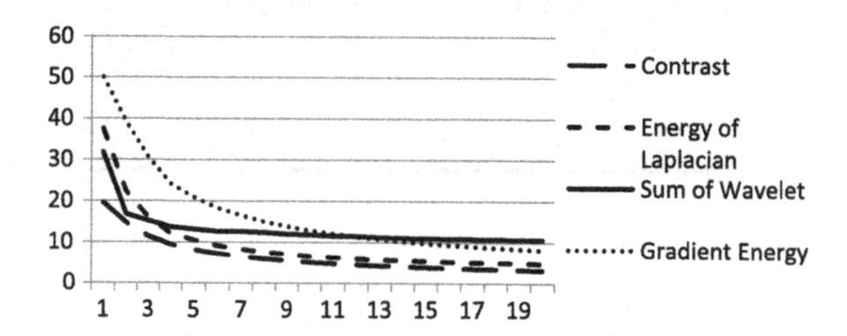

Fig. 4 Focus measure versus position

3. Moving Towards Focus
   They designed a system with a Hill-Climbing like structure that moved lens towards the focal point in iterative manner. The whole system was divided in three different modules: (1) Hill-climbing Searching Algorithm (2) Coarse Focusing (3) Fine Focusing. Hill-Climbing like structure was used such that contrast measure was compared to previous position and if increased, lens was moved in same direction. On decreasing, lens was moved to previous position and was moved with smaller steps. Figure 5 shows the Hill-Climbing Searching Algorithm. Coarse focusing was used to move lens nearby the focal point in less time. Fine focusing moved lens at exact focus point using a single step.

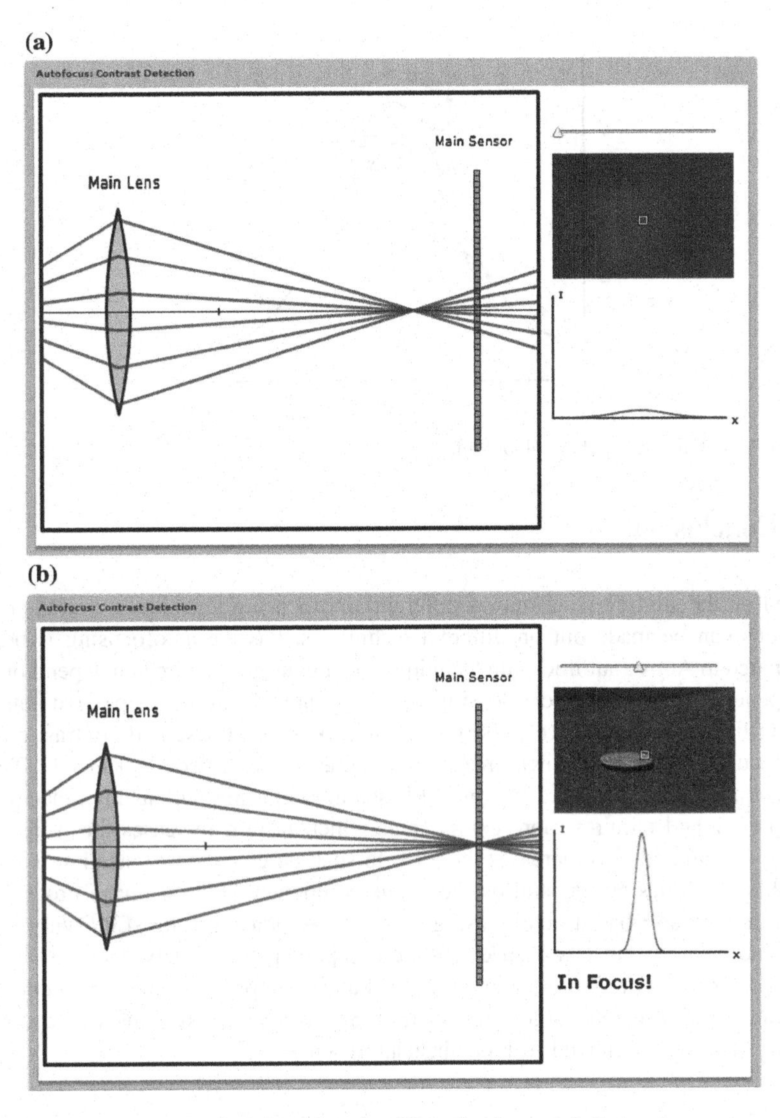

Fig. 5 (**a**) and (**b**) Contrast based autofocusing [8]. **a** Out-focused. **b** Focused

The used multiple measures to check reliability of the system. A conclusion can be drawn that a single focus measure may lead to false decision sometime as it is not mandatory that focus measure will always give the true decision as it depends on the different parameters of image like brightness, contrast, sharpness. Hence using multiple features in single system can lead to more accurateness of the system (Fig. 6).

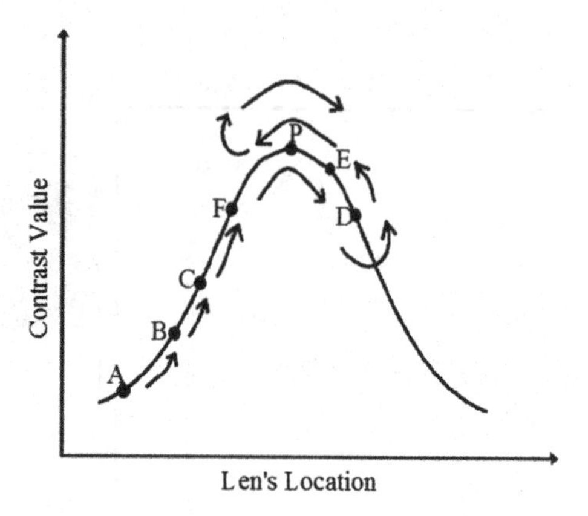

Fig. 6 Hill climbing searching algorithm [11]

4 Conclusion

Based on the study carried out on different autofocusing techniques, various conclusions can be made out on different techniques. Passive autofocusing is better compared to Active autofocusing techniques as it does not require and depend on an independent light source for focusing. But there are limitations of passive autofocusing like it does not give good results in dark areas. Phase and contrast based autofocusing have their own merits and demerits described in Table 1. Phase detection is faster compared to contrast based autofocusing but has high electronic complexity and requires more components which in turn increases the weight of whole system and also increases the cost. In such cases contrast detection is more suitable as it does not depend on electronic components but on image processing algorithms. Phase based autofocusing works more efficiently in SLRs with small lens and is faster but has high electronic component complexity. Contrast based autofocusing required more computation but is simple and more suitable for imaging system having larger lens such as Space imaging systems where image sensor (detector) is moved to focus instead of lens.

References

1. Liao, W.-S., Fuh, C.-S.: Auto-focus. In: CSIE (2003)
2. http://en.wikipedia.org/wiki/Auto-focus
3. Subbarao, M., Tyan, J.-K.: Selecting the optimal focus measure for autofocusing and depth-from-focus. In: Pattern Analysis and Machine Intelligence, IEEE (1998)
4. http://graphics.stanford.edu/courses/cs178/applets/autofocusPD.html
5. Roman G.: MEMS Focus on Cell Phone Camera Market. In: MEMS Investor Journal Inc. (2010)
6. http://electronics.howstuffworks.com/auto-focus2.htm

7. Pertuz, S., Puig, D., Garcia, M.A.: Analysis of focus measure operators for shape-from-focus. Pattern Recogn. **46**(5), 1415–1432 (2013)
8. https://graphics.stanford.edu/courses/cs178-10/applets/autofocusCD.html
9. Erbas, B., Underwood, C.I.: Active focusing system for an earth imaging reflecting telescope. In: Proceedings of 2nd International Conference on Recent Advances in Space Technologies, 2005. RAST 2005. IEEE (2005)
10. Surya, G., Subbarao, M.: Depth from defocus by changing camera aperture: a spatial domain approach. In: Computer Vision and Pattern Recognition (1993)
11. Zhao, Q., Liu, B., Xu, Z.: Research and realization of an anti-noise auto-focusing algorithm. In: 2013 5th International Conference on Intelligent Human-Machine Systems and Cybernetics (IHMSC), vol. 2. IEEE (2013)
12. Holzner, J., Gebhardt, U., Berens, P.: Auto-focus for high resolution ISAR imaging. In: EUSAR (2010)
13. Tang, J., Peli, E.: Image enhancement using a contrast measure in the compressed domain. In: Signal Processing Letters, IEEE (2003)
14. Du, X., Duan, C., Hu, W.: Sparse representation based auto-focusing technique for ISAR images. In: Geoscience and Remote Sensing, IEEE Transactions (2013)
15. Palmer, J., Martorella, M., Haywood, B.: Polarimetric ISAR auto-focusing techniques: comparison of results. In: Waveform Diversity and Design Conference (2007)
16. Martorella, M., Haywood, B., Berizzi, F., Dalle Mese, E.: Performance analysis of an ISAR contrast-based auto-focusing algorithm using real data. In: Radar Conference (2003)

An Enhanced Version of Key Agreement System with User Privacy for Telecare Medicine Information Systems

Trupil Limbasiya and Nishant Doshi

Abstract In the e-technology world, the growth of online system usage is very high. Hence it is also necessary that security and privacy of users must secure. In the telecare medical system, a remote patient and the medical server requires to authenticate each other over an insecure channel. In 2014, Li et al. proposed an enhanced scheme for telecare medicine system and claimed that it's secure against various attacks. However, in this paper, we identified that Li et al.'s scheme cannot resist against *Temporary Information, Forgery, Stolen Smart Card attack.* Therefore, in this paper proposed scheme to be secure against the mentioned attacks and also it requires less number of operations. Thus we have achieved more security with less computation overhead.

Keywords Attack · Authentication · Cryptography · Password · Smart card · Session key

1 Introduction

The development of remote user network system is helpful in medical applications such that health monitoring system, e-health care and military system, e-commerce applications etc. Storage capacity, privacy and security in communication channel and capability of processing are limited. Main concern is security and privacy of users' important information. Hence, remote user network system should be secure with respect to users' data [1].

T. Limbasiya (✉)
NIIT University, Neemrana, India
e-mail: limbasiyatrupil@gmail.com

N. Doshi
MEF Group of Institutions, Rajkot, India
e-mail: dosinikki2004@gmail.com

S.C. Satapathy and S. Das (eds.), *Proceedings of First International Conference on Information and Communication Technology for Intelligent Systems: Volume 1*, Smart Innovation, Systems and Technologies 50, DOI 10.1007/978-3-319-30933-0_15

TMIS (Telecare Medicine Information Systems) is one type of an application for remote user network system which helps to build a convenient bridge between patient/user and doctor/server. Because both are at different localities. With the rapid innovation of information technology and internet, TMIS is efficient to patients because it saves a lot of time, high expense of patient. It is also helpful for those patients which are unable to go at hospital or health care centre [2–6].

In recent times, Hao et al. introduced that Guo et al.'s scheme is not secure against user anonymity in 2013. So, Hao et al. proposed modified scheme to overcome weaknesses of Guo et al.'s authentication scheme. In 2013, Lee found that modified version by Hao et al. cannot resist session key. To challenge this issue, Lee invented newly key agreement and authentication system to provide more security. Then after, Jiang et al. pointed out that scheme of Hao et al. is not secure against stolen smart card attack. Then they proposed effective authentication and key agreement system with high security [2–6].

2 Review of Li et al.'s Scheme

In this section, we had make known to scheme of Li et al. [6] in details. The notations which are used in this paper are shown in Table 1.

Li et al. invented newly authentication and key agreement system into two mainly phase such that registration phase and authentication phase.

Table 1 Notations used in the scheme

Notations	Description
U_i	User
ID_i	Identity of U_i
PW_i	Password of U_i
SC_i	Smart card of U_i
S	Server
SCR	Smart card reader
T_2	Server side current timestamp
T_1, T_3	User side current timestamp
x	Secret key maintained by S
y	Secret number maintained by S
mk	Secret key of S
$h(\cdot), H(\cdot)$	One way hash function
$\oplus$	Bitwise XOR operator
$\|\|$	Concatenation operator

2.1 Registration Phase

There are two mainly side such that user U and server S in the registration phase. The communication channel between U and S is private channel.

1) U selects her/his identity *ID*, text password *PW* and a random number *b* & then calculates $h(ID||h(PW||b))$. *U* sends *ID* & computed parameter to *S*.
2) After receiving from *U*, *S* generates a random value *r* and calculates $IM_1 = IM_3 = h(mk) \oplus r$, $IM_2 = IM_4 = h(mk||r) \oplus ID$, $D_1 = h(ID||mk) \oplus h(ID||h(PW||b))$. Afterwards, *S* stores $\{IM1, IM2, IM3, IM4, D_1, h(\cdot), H(\cdot)\}$ into *SC* of *U*. *S* maintains status table of *U*'s registration processes. *S* sends *SC* to respective *U*.
3) After receiving *SC*, *U* calculates $D_2 = h(ID||PW) \oplus b$ and then stores D_2 into *SC*.

2.2 Authentication Phase

Authentication phase is executed between *U* and *S* to get services of designed network system.

1) *U* inserts *SC* & inputs $ID, PW.SC$ generates *u* and retrieves $b = h(ID||PW) \oplus D_2$ and then calculates $K = D_1 \oplus h(ID||h(PW||b)) = h(ID||mk), T_u(K)$ and $X_1 = h(K||IM_1||IM_2||T_u(K)||T_1)$. U sends login request $M_1 = \{IM_1, IM_2, T_u(K), X_1, T_1\}$ to *S*.
2) *S* checks timestamp ΔT holds or not. If it holds then calculates $r^{'} = IM_1 \oplus h(mk)$ and $ID^{'} = IM_2 \oplus h(mk||r')$. *S* checks $ID^{'}$ *and* r' equals or not with *ID* and *r* respectively. *S* calculates $K^{'} = h(ID^{'}||mk)$. If not equal then *S* rejects request.
3) *S* calculates $IM_1^* = h(mk) \oplus r_{new}$, $IM_2^* = ID' \oplus h(mk||r_{new})$, $sk = H(T_u(K), T_v(K^{'}), T_v(T_u(K)))$, $Y_1 = IM_1^* \oplus (SK||T_1)$, $Y_2 = IM_2^* \oplus h(SK||T_2)$ and $Y_3 = h(sk||IM_1^*||IM_2^*||T_v(K')||T_2)$.
4) *S* sends $M_2 = \{Y_1, Y_2, Y_3, T_v(K^{'}), T_2\}$ to *U*.
5) *U* verifies timestamp and it holds then calculates $sk^{'} = H(T_u(K), T_v(K^{'}), T_u(T_v(K')))$, $IM_{1new}^* = h(sk^{'}||T_1) \oplus Y_1$, $IM_{2new}^* = h(sk^{'}||T_2) \oplus Y_2$. Again *U* verifies Y_3 with computed credentials. If it holds then replaces IM_1, IM_2, IM_3, IM_4 with $IM_{1new}^*, IM_{2new}^*, IM_1, IM_2$ respectively.
6) *U* computes $X_2 = h(IM_{1new}^*||IM_{2new}^*||T_u(T_v(K'))||sk'||T_3)$ and then sends $M_3 = \{X_2, T_3\}$.
7) *S* checks timestamp If it holds then verifies earlier computed credentials with X_2. And update status table.

3 Cryptanalysis of Li et al.'s Scheme

We described in this section that how Li et al.'s scheme [6] is not secure against various attacks such that temporary and man-in-the-middle. There is also concerns regarding unnecessary storage and more number of message communication during communication. Without loss of generality, we considered a number of assumption [7, 8] centered as in earlier times remote user authentication systems and which are listed below:

1) Adversary can trace login requests of legal users during communication channel.
2) Adversary has skill to extract info which are stored in smart card of user SC_i of U_i easily.
3) Adversary can change login request message of authorized users.

We will assume that there are two legal users U_i and U_j of network system and consumer have smart card SC_i and SC_j respectively.

3.1 Temporary Information Attack

The loss of personal information for temporary time duration. Session keys (sk & sk') can be calculated based on public communication channels ($M_1 = \{IM_1, IM_2, T_u(K), X_1, T_1\}$ and $M_2 = \{Y_1, Y_2, Y_3, T_v(K'), T_2\}$) from U to S and S to U. Attacker will calculate important credentials which are as follows:

$$sk = H(T_u(K), T_v\left(K'\right), T_v(T_u(K)))$$

$$IM_1{}^* = Y_1 \oplus h(sk||T_1)$$
$$IM_2{}^* = Y_2 \oplus h(sk||T_2)$$

Y_3 will be verified by attacker for confirmation regarding calculated credentials are precise or not. Then, sk' will be computed by attacker based on public communication channels ($M_1 = \{IM_1, IM_2, T_u(K), X_1, T_1\}$ sent from $U \rightarrow S$ and sent from $S \rightarrow U$ $M_2 = \{Y_1, Y_2, Y_3, T_v(K'), T_2\}$).

$$sk' = H(T_u(K), T_v(K), T_u(T_v(K')))$$

$$IM^*_{1new} = h(sk'||T_1) \oplus Y_1$$
$$IM^*_{2new} = h(sk'||T_2) \oplus Y_2$$

Hence, attacker can modify credentials which are stored into SC such that $IM^*_{1new}, IM^*_{2new}, IM_1, IM_2$ in place of IM_1, IM_2, IM_3, IM_4. Hence, we can say that attacker has control for specific session Thus, Li et al.'s scheme is vulnerable susceptible to temporary information attack.

3.2 Forgery Attack

Identity of U can get to attacker easily. During the authentication phase of password verification and key agreement system, credentials like ID and r only will be verified. M_1 will be sent from U to S via public communication channel and M_2 is also sent from S to U in common channel. So that attacker can trace it simply. Hence, attacker can identify credentials such that $IM_1, IM_2, T_u(K), X_1, Y_1, Y_2, Y_3, T_v(K')$ etc. By using these parameters, attacker can try to forge fake login request for time T_2 into the system. S will be busy in verification process with many requests. Thus, any attacker can apply attack into system (Fig. 1).

3.3 Complex Storage & Computational

Client or user has to store three different parameters in the process of remote user authentication. More credentials (IM_3 & IM_4) into to procedure means that there will be requirement of more storage capacity and more number of operations. At the end of successful login, there will be replacement of IM_1, IM_2, IM_3 *and* IM_4 so that it will be in form of cycle again. S replaces credentials such that $IM^*_{1new}, IM^*_{2new}, IM_1, IM_2$ in place of IM_1, IM_2, IM_3, IM_4 into SC. Hence, there will be complex computations during the procedure and there is no need of using extra credentials. So cost will be high because of more storage and operations.

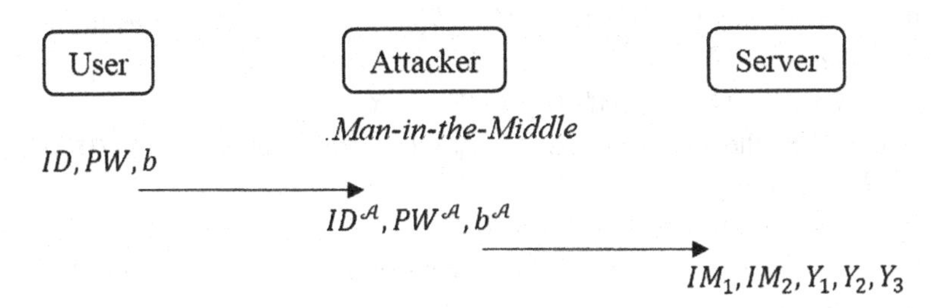

Fig. 1 Forgery attack in Li et al.'s scheme

4 Proposed Scheme

We proposed newly remote user authentication and key agreement system to overcome weaknesses of Li et al.'s scheme and to provide more security into the scheme. In our proposed scheme, there two mainly phases such that registration phase and login & authentication phase.

4.1 Registration Phase

1) ***U***: Selects ID, PW, b.
   Computes $h(ID||PW||b)$.
   Sends $\{ID, h(ID||PW||b)\}$ to S via secure channel.
2) ***S***: Generates random number value r.
   Calculates $IM_1 = h(mk) \oplus r$, $IM_2 = h(mk||r) \oplus ID$ and $D_1 = h(ID||mk) \oplus h(ID||PW||b)$.
   Stores $\{IM_1, IM_2, D_1, h(\cdot)\}$ into SC.
   Sends SC to U.
3) ***U***: Calculates $D_2 = h(ID||PW) \oplus b$.
   Updates credentials D_2 into the SC.

4.2 Authentication Phase

1) ***U***: Inserts SC and inputs ID & PW.
2) ***SCR***: Retrieves value of b from D_2.
   Generates random number r_u & timestamp T_1.
   Calculates k $= D_1 \oplus h(ID||h(PW||b))$,$T_{r_u}(k)$.
   Then computes $X_1 = h(k||IM_1||IM_2||T_{r_u}(k)||T_1)$.
   Sends $M_1 = \{IM_1, IM_2, T_{r_u}(k), X_1, T_1\}$ to S as a login request.
3) ***S***: Verifies $T_2 - T_1 < \Delta T$, ΔT is specific time duration for receiving login request.
   It holds then only retrieves $r_u^{'} = IM_1 \oplus h(mk)$ & $ID^{'} = IM_2 \oplus h(mk||r_u^{'})$.
   Computes $k^{'} = h(ID^{'}||mk)$.
   Again verifies $h(k^{'}||IM_1||IM_2||T_{r_u}(k)||T_1) = X_1$?
   If it holds then only generates r_v. Then computes $sk = h(ID^{'}||T_{rv}(k')||k'||T_{rv}(T_{r_u}(k))||T_1||T_2)$.
   Sends $M_2 = \{h(sk), T_{rv}(k^{'}), T_2\}$ to U.
4) ***SCR***: Checks $T_2 - T_1 < \Delta T$. It holds then only computes $sk' = h(ID||T_{rv}(k')||k||T_{ru}(T_{r_v}(k'))||T_1||T_2)$.
   If $sk = sk^{'}$ then only user and server are legal and continue network scheme services.

5 Security Analysis of Proposed Scheme

In this section, we will show that our proposed scheme could prevent security against Forgery attack and Temporary Information attack. There is also less requirement for communication messages transmission during login & authentication phase in our proposed scheme.

5.1 Temporary Information Attack

In our proposed scheme, we had reduced number unnecessary credentials which help to *U* for high security of confidential information & lower computational cost. There is only one way to get identity of *U* with help of both secret key *mk* and random value number *r*. At one time, both credentials may not be guessed with proper output of numbers within polynomial time. Thus, any attacker may not get *ID*.

5.2 Forgery Attack

U sends $\{IM_1, IM_2, T_{r_u}(k), X_1, T_1\}$ to S via communication channel. So, attacker may not either retrieve or get value of all three credentials such that *ID* (Identity of *U*), *PW* (Text password of *U*) and *b* (Random number) at a time. There is infrequent situation getting *ID, PW, b* because *PW, b* are not predictable at same time. Hence, there is no possibility of stolen confidential information of *U* for specific session time period.

In this section, we have concluded regarding security analysis against attacks in Table 2 and comparison regarding mathematical operations in Table 3.

We have executed one-way hash operations into JAVA. Java Development Tool Kit (Version 1.7) is also used in our execution. We have system configuration as 4 GB RAM, 500 GB HDD, Core i3 3.30 GHz processor. Our research project performance is measured on Windows 7 32-bit operating system. We have

Table 2 Security analysis against attacks

Attacks	Schemes				
	Hao et al. [3]	Lee et al. [4]	Jiang et al. [5]	Li et al. [6]	Proposed scheme
Temporary information	×	×	×	×	√
Man-in-the-middle	×	×	×	×	√
Stolen smart card	×	×	×	×	√

×: Specific scheme is not secure against certain attack
√: Specific scheme is secure against certain attack

Table 3 Comparison regarding mathematical operations

Phases	Schemes				
	Hao et al. [3]	Lee et al. [4]	Jiang et al. [5]	Li et al. [6]	Proposed scheme
Registration	$2T_h$	$4T_h$	$1T_h$	$5T_h$	$4T_h$
Authentication	$5T_h$	$15T_h$	$3T_h$	$17T_h$	$8T_h$

T_h: The execution time of one way hash function

Table 4 Average time complexity of one way hash function

Cryptographic operation	Average time complexity (ms)
$h(\cdot)$	0.0016

Table 5 Comparison of scheme execution time (ms)

Phases	Schemes				
	Hao et al. [3]	Lee et al. [4]	Jiang et al. [5]	Li et al. [6]	Proposed Scheme
Registration	0.0032	0.0064	0.0016	0.0080	0.0064
Login & authentication	0.0080	0.0240	0.0048	0.0272	0.0128

measured time complexity of one-way hash function. We have taken average time complexity of this operation after executing the project more than 100 times. Table 4 shows the average time complexity of one way hash function in milliseconds (ms). Table 5 shows execution time of different schemes based on mathematical operation h(•).

6 Conclusion

Our proposed scheme can withstand against attacks such that *Temporary Information, Forgery, Stolen Smart Card* etc. In *Li* et al.*'s* scheme, $5T_h$ are required during registration phase and $17T_h$ are required during login and authentication section. But in our proposed scheme, $4T_h$ are required during registration phase and $8T_h$ are required during login & authentication phase. Hence, we can say that our proposed scheme have low cost, less time requirement in execution and high security rather than *Li* et al.*'s* scheme. So it will be more useful in the area of medical system applications.

References

1. Lamport, L.: Password authentication with insecure communication. *Commun. ACM* **24**(11), 770–772 (1981)
2. Awasthi, A.K., Srivastava, K.: A biometric authentication scheme for telecare medicine information systems with nonce. *J. Med. Syst.* **37**(5), 1–4 (2013)
3. Hao, X., Wang, J., Yang, Q., Yan, X., Li, P.: A chaotic map-based authentication scheme for telecare medicine information systems. *J. Med. Syst.* **37**, 9919 (2013)
4. Lee, C.C., Li, C.T., Hsu, C.W.: A three-party password-based authenticated key exchange protocol with user anonymity using extended chaotic maps. *Nonlinear Dyn.* **73**(1), 125–132 (2013)
5. Jiang, Q., Ma, J., Lu, X., Tian, Y.: Robust chaotic map-based authentication and key agreement scheme with strong anonymity for telecare medicine information systems. *J. Med. Syst.* **38**(2), 12 (2014)
6. Li, C.T., Lee, C.C., Weng, C.Y.: A secure chaotic maps and smart cards based password authentication and key agreement scheme with user anonymity for telecare medicine information systems. *J. Med. Syst.* **38**(9), 77 (2014)
7. Madhusudhan, R., Mittal, R.C.: Dynamic ID-based remote user password authentication schemes using smart cards: a review. *J. Netw. Comput. Appl.* **35**(4), 1235–1248 (2012)
8. Limbasiya, T., Doshi, N.: A survey on attacks in remote user authentication scheme. In: *Computational Intelligence and Computing Research (ICCIC)*, 2014 IEEE (2014)

A Framework to Infer Webpage Relevancy for a User

Saniya Zahoor, Mangesh Bedekar and Varad Vishwarupe

Abstract The Web is a vast pool of resources which comprises of a lot of web pages covering all aspects of life. Understanding a user's interests is one of the major research areas towards understanding the web today. Identifying the relevance of the surfed web pages for the user is a tedious job. Many systems and approaches have been proposed in literature, to try and get information about the user's interests by user profiling. This paper proposes an improvement in determining the relevance of the webpage to the user, which is an extension to the relevance formula that was proposed earlier. The current work aims to create user profiles automatically and implicitly depending on the various web pages a user browses over a period of time and the user's interaction with them. This automatically generated user profile assigns weights to web pages proportional to the user interactions on the webpage and thus indicates relevancy of web pages to the user based on these weights.

Keywords User profiling · Web personalization · Implicit user behavior modeling · Client side analysis

1 Introduction

In recent years, there has been a tremendous growth in the web and its usage, so much so that today many users find it difficult to get information that is relevant to them. Moreover, the behavior of the user is dynamic which makes it difficult to

S. Zahoor · M. Bedekar (✉)
Computer Engineering Department, MAEER's MIT Kothrud, Pune, India
e-mail: mangesh.bedekar@gmail.com

S. Zahoor
e-mail: saniya.zahoor@yahoo.com

V. Vishwarupe
IT Engineering Department, MIT College of Engineering, Kothrud, Pune, India
e-mail: varad44@gmail.com

S.C. Satapathy and S. Das (eds.), *Proceedings of First International Conference on Information and Communication Technology for Intelligent Systems: Volume 1*, Smart Innovation, Systems and Technologies 50, DOI 10.1007/978-3-319-30933-0_16

track his current interests and changes in his interests. If the user's interests are asked explicitly, most users tend to either ignore giving information or fill in wrong/incomplete information. For example, when asked for rating webpage's, users at times tend to either give wrong information, or incomplete information which is also learned as it is stated explicitly. So, there is a need to learn the user implicitly, whereby the system learns about the user and his interests in a transparent manner. The user does his usual web activities and no questions are asked. Various activities that the user does on his browser are tracked to infer interest. All this is done to obtain knowledge about user's behavior of web and thus learn about his interests. The User Profile, thus built contains entities found interesting by the

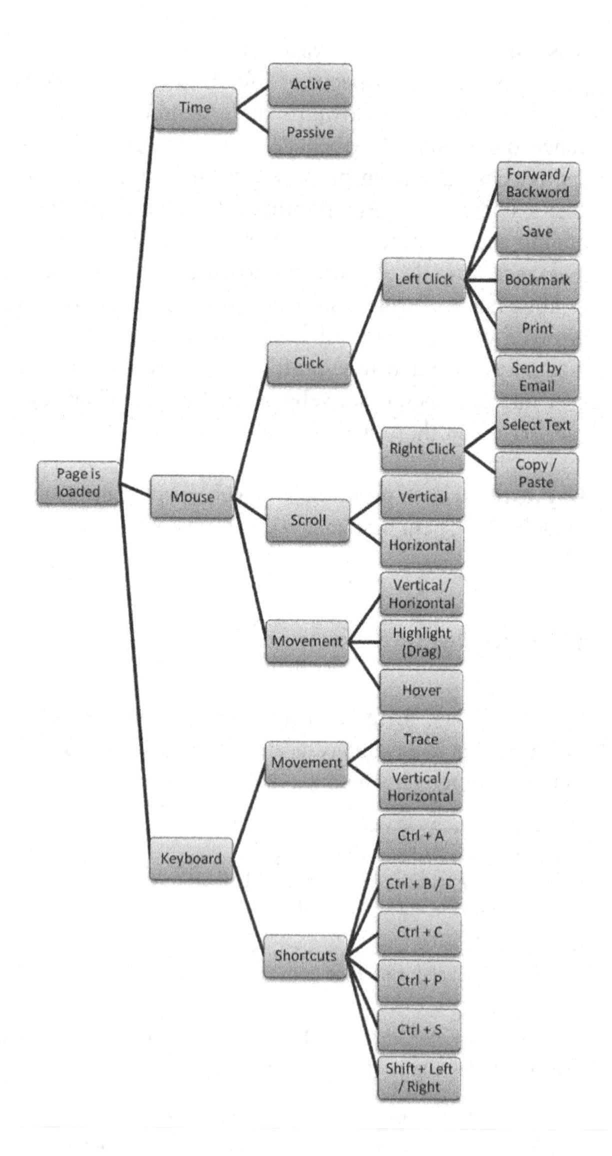

Fig. 1 Various implicit interest indicators

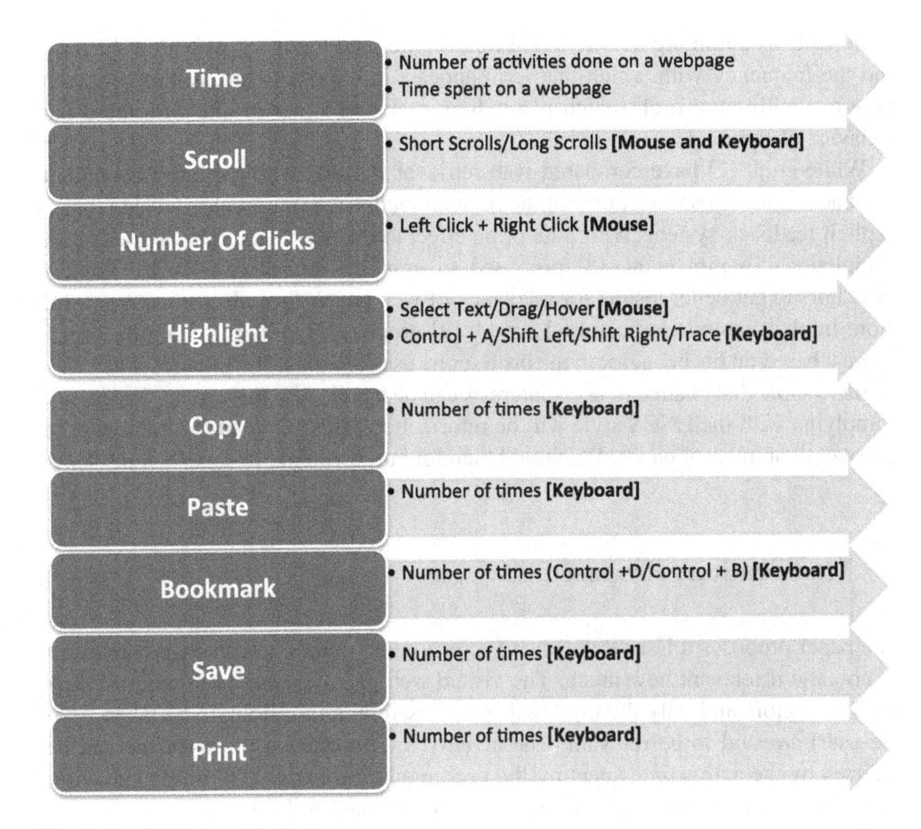

Fig. 2 Implicit interest indicators

user [1–4]. Zahoor et al. [5] presented an comparative survey of various implicit interest indicators used for User Profiling as shown in Fig. 1.

Similarly, select text, control all, drag and so on do the highlighting of text and should have the same weight. The regrouped interest indicators are as indicated in Fig. 2.

In this paper, the focus is on the above set of implicit interest indicators (9) to propose a measure that will generate user profiles automatically depending on the various web pages a user browses over a period of time and thus indicates relevancy of web pages for a particular user.

2 Related Work

Researchers have done work of allotting weights to the pages visited. This was primarily done on relating the number of pages a user visits and number of pages he finds relevant for a given search. The method proposed by Teevan et al. [6] ranks documents by summing over terms of interest the product of the weight of the term and the frequency with which that term appears in the document. Rating any web page explicitly after each search is a tedious task and is not in line with the usual behavior of the user.

White et al. [7] have compared web retrieval systems with explicit v/s implicit feedback. They go on to show that the implicit feedback systems can indeed replace explicit feedback systems with little or no effect to the users search behavior or task completion. Shapira et al. [8] proposed to combine some well known interest indicators to get better results for relevancy of web pages, they also propose several more implicit interest indicators. Li et al. [9] proposed a system to identify users' interest based on his behavior in the browser he uses to access the Internet. They also go on to show that change in users interests can be handled by the system. Papers not complying with the LNCS style will be reformatted. This can lead to an increase in the overall number of pages. We would therefore urge you not to squash your paper.

3 Modifications Proposed

The paper proposes a User Relevance Factor to determine if a web page is relevant or not, and if relevant how much. The visited web pages are sorted in order of their relevant factors and only the top few pages (based on a threshold, to be taken from the user) are said to be relevant. The activity by the user on the webpage can be inferred by the active time spent by the user, mouse movement, scrolling behavior, save, bookmark and print on the webpage as indicated in [10].

The relevance factor, Rf is, calculated by the following formula,

$$Rf = \log\left(\frac{(time) \times (movement)}{scroll}\right) * Save * Print * Bookmark$$

The formula included 6 factors in it and it was based on the concept that increase in the implicit interest indicators increases the relevancy of the search results for a particular user but there were following issues with the formula,

SCROLL: If a user is scrolling a lot, then it nullifies the effect of spending time on the webpage and performing actions on it. Placing Scroll at the denominator changes the effect on the other activities on the webpage. A concept of ratio of short scrolls and long scrolls (S/L) is introduced to handle this problem. Short scrolls means that user is reading something on the webpage, and the small scrolls done more number of times indicates the user is positively interested in the webpage while as long scrolls means that user is seeing the page quickly and the long scrolls done less number of times indicates less interest of the user on the webpage. The relevance factor R will be directly proportional to short scrolls (S) and inversely proportional to long scrolls (L).

TIME ISSUE: Time is taken directly proportional in the formula. There are some issues in the following cases:

- user spends more time on the webpage, does less activity;
- user spends small time and does a lot of activity.

The formula gives higher priority to the former which should not occur. The time factor in the equation was modified to ratio f, where f equals the number (N) of activity done on the webpage divided by the time (T) spent on the webpage which shows the user's engagement on the webpage. There can be four cases for f which are shown in Table 1.
SAVE, PRINT, BOOKMARK: what if a user performs these actions n number of times that too should have an effect on the relevancy for a user. If a user prints a page n number of times, that means the page has higher relevancy for him as compared to the case if he prints the page once. Thus, capturing actions and capturing the number of times the action can be included in the formula as well.

Factors included are Highlight, Copy, Paste, Bookmark, Save, Print, Number of Clicks on a Webpage (Left Click & Right Click), Number of Activity done on the Webpage divided by time spent on the Webpage (N/T) and Number of Short Scrolls divided by Number of Long Scrolls (S/L) as the implicit interest indicators in relevance factor as mentioned below,

$$R = \frac{N}{T} + (Sa * n^{s} + B * n^{b} + Pr * n^{Pr} + C * n^{c} + Pa * n^{Pa} + H * n^{h} + Cl * n^{Cl}) + \frac{S}{L}$$

where
N Number of activities on a particular webpage during a particular session
T Time spent on the webpage during a particular session
n number of times the activity is performed during a particular session
Cl clicks = leftclicks + rightclicks
Sa Save,
B bookmark,
Pr Print,
Pa Paste
H Highlight
C Copy,
L Long scrolls,
S Short scrolls.

Table 1 Activity V/S time

Activity	Time	
Cases	More time	Less time
More activity	Medium value	High value
Less activity	Less value	Medium value

Table 2 Initial weights of interest indicators

Click	Highlight	Copy	Paste	Bookmark	Save	Print
0.33	0.20	0.15	0.12	0.10	0.07	0.03

After various iterations, the values mentioned in table gave best results

Among the Nine implicit interest indicators that are in the formula, Time (seconds) and Scroll (distance covered) are taken directly as value and the rest of the seven interest indicators have the ordering as,

$$\text{Click} > \text{Highlight} > \text{Copy} > \text{Paste} > \text{Bookmark} > \text{Save} > \text{Print}$$

Click is placed at the initial position, as it is the most frequent action that a user does on the webpage. It also indicates about which part of the webpage (DOM) a user is interested. To highlight text on a webpage, the user needs to first click then highlight. In order to paste 'copied' text, the user first needs to copy the text then highlight and then paste. Click, highlight, copy and paste tells us which part of the webpage a user is interested and that will have higher weight as compared to bookmark which tells us about the URL the user is interested in. The order followed for Bookmark, save and print is the same as in [11]. The weights for these implicit activities are as indicated in Table 2.

4 Implementation Details

The implementation starts with installation of XAMPP server on the client's machine and creation of database with tables containing specific columns that represent a lot of useful information. One of the tables contains the two columns, one for the URL and the other for the time spent. Likewise there are other tables which contain information about user's interest on the webpage. Once the server has been installed with the required database and tables, every time the user uses a browser he needs to switch on the server after which the data starts getting stored in the database.

JavaScript (JS) is an interpreted computer programming language. Mozilla Firefox Browser is free and open source. One of the main reasons why Mozilla Firefox was used was because of its unique add-on Greasemonkey. Once all these user initiated events are captured, this data along with the URL of the web page is stored into the database. The relevance factor script is then invoked using Greasemonkey which monitors the users' behavior on the web page. In this way all the necessary data values required for the method are captured and stored. Since the databases are stored on XAMPP server.

Table 3 URL's and corresponding values

Case	URL	Cl	H	C	P	B	Sa	P	S	L	T
1	URL1	3	2	1	1	1	2	1	44	19	189
2	URL2	4	4	1	1	1	1	1	103	51	96

5 Mathematical Proof of the Formula

Consider the case of a user who visits two URLs with the following values (Table 3).
Case 1:

$$f = (3+2+1+1+1+2+1)/189 = 11/189 = 0.0582.$$
$$\mathrm{Ne} = 63.\ \text{Average Scroll(i)} = 4805.134/63 = 76.27.$$
$$\mathrm{S} = 44, \mathrm{L} = 19, \mathrm{S/L} = 44/19 = 2.3158.$$
$$\mathrm{R}(1) = (0.0582) + (0.33 * 3 + 0.20 * 2 + 0.15 * 1 + 0.12 * 1 + 0.10 * 1 + 0.07 * 2 + 0.03 * 1) + (2.3158) = 4.304.$$

Case 2:

$$f = (4+4+1+1+1+1+1)/96 = 13/96 = 0.1354.$$
$$\mathrm{Ne} = 154.\ \text{Average Scroll (i)} = 37785.53/154 = 245.360.$$
$$\mathrm{S} = 103, \mathrm{L} = 51, \mathrm{S/L} = 103/51 = 2.019608.$$
$$\mathrm{R}(2) = (0.1358) + (0.33 * 4 + 0.20 * 4 + 0.15 * 1 + 0.12 * 1 + 0.10 * 1 + 0.07 * 1 + 0.03 * 1) + (2.01960) = 4.745.$$

As can be observed, R(2) has higher value than R(1) i.e.; Case 2 URL is more relevant to the user as compared to the Case 1 URL. Subjectively, the user was asked which URL is more relevant to him—Case 1 or Case 2, and the answer was similar.

6 Conclusion

User's behavior on a webpage can reveal a lot about his interests on the web. The actions that the user performs and his usage on the web can be captured and can help in understanding the user very well. Almost all actions of the user, which are done in the browser, can be captured via the Mouse and the Keyboard. Only Keyboard and Mouse actions are not sufficient to identify the user's interests. The proposed framework handles the problem well by considering the ratio of short to long scrolls to get a proper interest measure. Time spent, considered as a single entity is a misnomer to indicate relevancy which has to be handled too.

The framework ranks all the visited web pages according to its relevancy to the user hence this will be used in giving relevant search results to the user. For each search term the pages browsed by the user are recorded and ranked according to the users profile. Once a concrete database gets created over a period of time, as soon as the user searches any term he will get a list of web pages visited by him for similar search done earlier which would be ranked according to the user's personal relevance.

References

1. Faucher, J., McLoughlin, B., Wunschel, J.: Implicit web user interest, Technical Report MQP-CEW-1101, Worcester Polytechnic Institute, Spring (2011)
2. Hauger, D., Paramythis, A., Weibelzahl, S.: Using browser interaction data to determine page reading behavior. In: UMAP'11, Proceedings of the 19th International Conference on User Modeling, Adaption, and Personalization, pp. 147–158. Girona, Spain, 11–15 July 2011
3. Kríž, J.: Keyword extraction based on implicit feedback. Inf. Sci. Technol. Bull. ACM Slovakia, **4**(2), 43–47
4. Leiva Torres, L.A., Hernando, R.V.: A gesture inference methodology for user evaluation based on mouse activity tracking. In: IHCI 2008, Proceedings of the IADIS International Conference on Interfaces and Human Computer Interaction, Amsterdam, The Netherlands, 25–27 July 2008
5. Zahoor, S., Bedekar, M., Kosamkar, P.: User implicit interest indicators learned from the browser on the client side. In: International Conference on Information and Communication Technology for Competitive Strategies, Udaipur, Rajasthan, India, 14–16 Nov 2014
6. Teevan, J., Dumais, S., Horvitz, E.: Personalizing search via automated analysis of interests and activities. In: Proceedings of the 28th Annual International ACM SIGIR Conference on Research and Development in Information Retrieval (SIGIR '05), pp. 449–456. ACM, New York
7. White, R., Ruthven, I., Jose, J.M.: The use of implicit evidence for relevance feedback in web retrieval. In: Proceedings of the Twenty-Fourth European Colloquium on Information Retrieval Research (ECIR '02). Lecture Notes in Computer Science, pp. 93–109. Glasgow (2002)
8. Shapira, B., Taieb-Maimon, M., Moskowitz, A.: Study of the usefulness of known and new implicit indicators and their optimal combination for accurate inference of users interests. Proceedings of SAC '06, pp. 1118–1119
9. Li, F., Li, Y., Wu, Y., Zhou, K., Li, F., Wang, X.: Discovery of a user interests on the internet. In: Proceedings of the IEEE/WIC/ACM, International Conference on Web Intelligence and Intelligent Agent Technology, pp. 359–362 (2008)
10. Zahoor, S., Dr. Bedekar, M.: Implicit client side user profiling for improving relevancy if search results, CCSEIT-2014. In: Proceedings of Fourth International Conference on Computational Science, Engineering and Information, Technology, Army Institute of Technology, Pune, India, 8–9 Aug 2014
11. Zahoor, S., Rajput, D., Bedekar, M., Kosamkar, P.: Capturing, understanding and interpreting user interactions with the browser as implicit interest indicators. In: ICPC 2015, International Conference on Pervasive Computing, Sinhgad College of Engineering, Pune, 8–10 Jan 2015

Comparative Study of Various Features-Mining-Based Classifiers in Different Keystroke Dynamics Datasets

Soumen Roy, Utpal Roy and D.D. Sinha

Abstract Habitual typing rhythm or keystroke dynamics is a behavioural biometric characteristic in Biometric Science relates the issue of human identification/authentication. In 30 years of on-going research, many keystroke dynamics databases have been created on various pattern of strings ("greyc laboratory", ".tie5Roanl", "the brown fox", …) taking various combination of keystroke features (flight time, dwell time) and many features-mining classification algorithms have been proposed. Many have obtained impressive results. But in evaluation process, a classifier's average Equal Error Rates (EERs) are widely varied from 0 to 37 % on different datasets ignoring typographical errors. The question may arise, which classifier is best on which pattern of keystroke databases? To get the answer, we have started our experiment and created our own five rhythmic keystroke databases on different daily used common pattern of strings ("kolkata123", "facebook", "gmail.com", "yahoo.com", "123456") and executed various classification algorithms in R statistical programming language, so, we can compare the performance of all the classification algorithms soundly on different datasets. We have executed 22 different classification algorithms on collected data considering various keystroke features separately. In the observation, obtained best average EER of the classifier Lorentzian is 1.86 %, where 2.33 % for Outlier Count, 3.69 % for Canberra, 5.3 % for Naïve Baysian and 8.87 % for Scaled Manhattan by taking all five patterns of strings and all combination of features in the consideration. So the adaptation of keystroke dynamics technique in any existing system increases the security level up to 98.14 %.

S. Roy (✉) · D.D. Sinha
Department of Computer Science and Engineering, University of Calcutta,
92 APC Road, Calcutta 700009, India
e-mail: soumen.roy_2007@yahoo.co.in

D.D. Sinha
e-mail: devadatta.sinha@gmail.com

U. Roy
Department of Computer and System Sciences, Visva-Bharati,
Santiniketan 731235, India
e-mail: roy.utpal@gmail.com

S.C. Satapathy and S. Das (eds.), *Proceedings of First International Conference on Information and Communication Technology for Intelligent Systems: Volume 1*, Smart Innovation, Systems and Technologies 50, DOI 10.1007/978-3-319-30933-0_17

Keywords Keystroke dynamics · EER · Canberra · Chebyshev · Czekanowski · Gower · Intersection · Kulczynski · Lorentzian · Minkowski · Motyka · Ruzicka · Soergel · Sorensen · Wavehedges · Manhattan distance · Euclidean distance · Mahanobolis distance · Z Score · KMean · SVM · Naïve Baysian

1 Introduction

Knowledge-based user authentication technique is very popular for its simplicity characteristics and users are very comfortable on it. But today, passwords or PIN is not limited due to brute-force, shoulder surfing or key logger attack. It demands higher level of security keeping simplicity with giving better performance. Some of the words we type daily like our name, address, email ID, password, … and we are habituated to type it in same rhythm which is unique just like our signature as a behavioral biometric characteristics and can be used in human identification or verification technique. It cannot be lost or stolen or mimicry in addition with no extra security apparatus is needed. The accuracy level of keystroke dynamics characteristics is not much promising in practice. It demands higher level of synthesis, so this technique can be effectively implemented in real life.

In our experiment, we have implemented Java Applet program to get the raw data of key press and release times of various pattern of strings ("kolkata123", "facebook", "gmail.com", "yahoo.com", "123456"), we have not taken password type strings or hard strings because users are not habituated to type hard strings in same rhythm. So, in the training session, we have considered some daily used words where all the users are habituated to press the words in same rhythm. Here getTime() function return the time of key press and release events in ms unit. Then we have calculated the following features of keystroke dynamics also in ms unit: time duration between key press and release of single key or key hold time or dwell time or key duration time (KD), time duration between two subsequent release or up-up key latency (UU), time duration between release of a key and press of next key or up-down key latency (UD), time duration between press of a key and release of next key or di-graph time or down-up key latency (DU), time duration between two subsequent press or down-down key latency (DD), total time (ttime), tri-gap time (trigap) and four-gap time (4gap).

Rhythmic keystroke is a behavioral biometric characteristics measured in Keystroke Dynamics methods is not a new. In the year 1897, Bryan and Harter investigated keystroke dynamics. In 1975, Spillane described the concept of keystroke dynamics. After that many researchers collected keystroke dynamics databases and evaluated by different classification algorithms, where statistical methods are common, many distance-based algorithms and many machine-learning methods have been also applied, many have obtained impressive results. But in evaluation process the classification's average Equal Error Rates are varied widely.

We have collected press and release time of 12096 keystrokes of 1440 samples of patterns from 12 different individuals in 4 different sessions with minimum of 1 month interval for five different common words ("kolkata123", "facebook", "gmail.com", "yahoo.com", "123456") in our experiment. Then we have considered 8 different features and combination of features then we have executed 22 different classifiers on that collected data. In our observation, obtained best average EER of the classifier Lorentzian is 1.86 %, where 2.33 % for Outlier Count, 3.69 % for Canberra, 5.3 % for Naïve Baysian, 8.87 % for Scaled Manhattan by taking all five patterns of strings and all combination of features in the consideration. So the adaptation of keystroke dynamics technique in any existing system increases the security level up to 98.14 %.

2 Keystroke Dynamics

2.1 Basic Idea

Keystroke dynamics is a technique work on set of some timing or pressure data of keystroke which is generated at typing on keyboard which is unique and can be used to classify the users.

2.2 Science and Features Selection

Placement of fingers on keyboard, hand weight, length of finger, neuro-physiological factors are made typing style unique. We have calculated key press and release time P_i and R_i for key K_i which represent entered character set, where $6 \le i \le$ length of the entered pattern. The features of the keystroke dynamics as follows [1]:

$$\text{Key Duration}(T_1) = R_i - P_i \tag{1}$$

$$\text{Up Up Key Latency }(T_2) = R_{i+1} - R_i \tag{2}$$

$$\text{Down Down Key Latency}(T_3) = P_{i+1} - P_i \tag{3}$$

$$\text{Up Down Key Latency}(T_4) = P_{i+1} - R_i \tag{4}$$

$$\text{Down Up Key Latency}(T_5) = R_{i+1} - P_i \tag{5}$$

$$\text{Total Time Key Latency}(T_6) = R_n - P_1 \tag{6}$$

$$\text{Tri-graph Latency } (T_7) = R_{i+2} - P_i \tag{7}$$

$$\text{Four-graph Latency } (T_8) = R_{i+3} - P_i. \tag{8}$$

2.3 *Keystroke Dynamics as User Authentication*

There are different ways in which a user can be authenticated. However all of these ways can be categorized into one of three classes: “Something we know” e.g. password, “Something we have” e.g. token, “Something we are” e.g. biometric property. Here, keystroke dynamics is the combination of three, something we know that is the pattern of strings, and something we have that is our finger tips size or our hand weight and something we are that is our typing style what we have learned in our life.

2.4 *Factors Affecting Performance*

Some of the factors which affect the way of keystroke Dynamics as follows: Text length, sequences of character types, word choice, and number of training sample, statistical method (mean or median) to create template, mental state of the user, tiredness or level of comfort, keyboard type, keyboard position and height of the keyboard, hand injury, weakness of hand mussel, shoulder pain, education level, computer knowledge, and category of users.

2.5 *Algorithms*

Following features mining algorithms [2, 3] can be applied on keystroke dynamics database. Here, P refers to the training set and Q refers to the test set. Mean and standard deviation is represented by μ and α respectively.

Canberra: $D_{car} = \sum_i^n \frac{\lvert P_i - Q_i \rvert}{P_i + Q_i}$	Minkowski: $D_{mink} = \sqrt[p]{\sum_i^n \lvert P_i - Q_i \rvert^p}$	Euclidean Distance: $D_{eu} = \sqrt[2]{\sum_i^n (\lvert P_i - Q_i \rvert)^2}$
Chebyshev: $D_{cheb} = \sum_i^n \max\lvert P_i - Q_i \rvert$	Motyka: $D_{mot} = \frac{\sum_i^n \max(P_{i,}) \rvert}{\sum_i^n (P_i + Q_i)}$	Mahanobolis Distance: $D_{maha} = \sqrt[2]{\sum_i^n ((\lvert P_i - Q_i \rvert)/\alpha_i)^2}$
Czekanowski: $D_{cze} = \frac{\sum_i^n \lvert P_i - Q_i \rvert}{\sum_i^n (P_i + Q_i)}$	Ruzicka: $D_{ruz} = 1 - \frac{\sum_i^n \min(P_i, Q_i)}{\sum_i^n \max(P_i, Q_i)}$	Z Score: $D_z = \sum_{i=1}^n (\lvert P_i \rvert - \mu(\lvert Q_i \rvert))/\alpha_i$

(continued)

(continued)

Gower: $D_{gow} = \frac{1}{n}\sum_i^n \lvert P_i - Q_i \rvert$	Soergel: $D_{soe} = \frac{\sum_i^n \lvert P_i - Q_i \rvert}{\sum_i^n \max(P_i, Q_i)}$	Lorentzian: $D_{lor} = \sum_i^n \ln(1 + \lvert P_i - Q_i \rvert)$
Intersection: $D_{ins} = \frac{1}{2}\sum_i^n \lvert P_i - Q_i \rvert$	Sorensen: $D_{sor} = \frac{\sum_i^n \lvert P_i - Q_i \rvert}{\sum_i^n (P_i + Q_i)}$	Manhattan Distance: $D_{man} = \sum_{i=1}^n (\lvert P_i - Q_i \rvert)$
Kulczynski: $D_{kuld} = \frac{\sum_i^n \lvert P_i - Q_i \rvert}{\sum_i^n \min(P_i, Q_i)}$	Wavehedges: $D_{wv} = \frac{\sum_i^n \lvert P_i - Q_i \rvert}{\sum_i^n \max(P_i, Q_i)}$	

Table 1 Background of keystroke dynamics

Authors	Classifiers	Length of the pattern	Features	EER (%)
Joyce and Gupta [4]	Manhattan	33	UD	0.25–16.36
Bleha et al. [5]	Euclidian	11–17	UD	2.8–8.1
Haider et al. [6]	Nural Network	7	UD	16.1
Yu and Cho [7]	SVM	6–7	UD	10.2
Killourly S. [8]	Manhattan (Scaled)	10+	UD	9.6
Kang et al. [9]	K mean	7–10	KD, UD	3.8
Giot et al. [10]	SVM	100	KD, UD	15.28

3 Background Details

In 30+ years of experience, many researchers have proposed their algorithms, taking various features and various length of pattern strings (Table 1).

4 Experimental Setup and Database

We have collected key press and release time as raw data in ms unit of all the entered keys for all five patterns using Java Applet program from 12 individuals during 12 months. Then we have calculated 8 different keystroke features using Eqs. 1–8. Then we have implemented 22 different classification algorithms (Canberra, Chebyshev, Czekanowski, Gower, Intersection, Kulczynski, Lorentzian, Minkowski, Motyka, Ruzicka, Soergel, Sorensen, Wavehedges, Manhattan Distance, Euclidean Distance, Mahanobolis Distance, Z Score, K-Mean, SVM, Naïve Baysian) in R Statistical language and evaluated the performance on the databases of the different pattern of strings “kolkata123”, “facebook”, “gmail.com”, “yahoo.com”, “123456” and combination of all strings.

5 Experimental Results

The following line chart represents the time stamps of 6 sample of same subjects and we can see the ratio of time tamps are almost same (Fig. 1).

The following line chart represents the 6 sample of time stamps of 6 different subjects and we can see the different time tamps ratios for 6 different subjects (Fig. 2).

In our experiment, we have seen that most of the time some distance-base algorithms achieved impressive results given bellow. Here, EERs are calculated in % (Table 2).

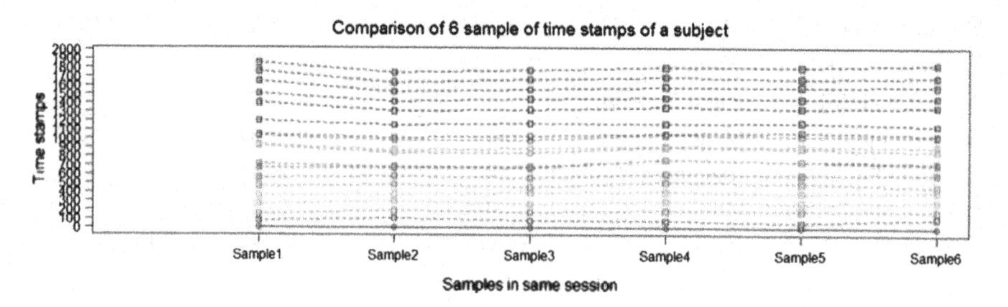

Fig. 1 Comparisons of 6 samples of time stamps of a subject

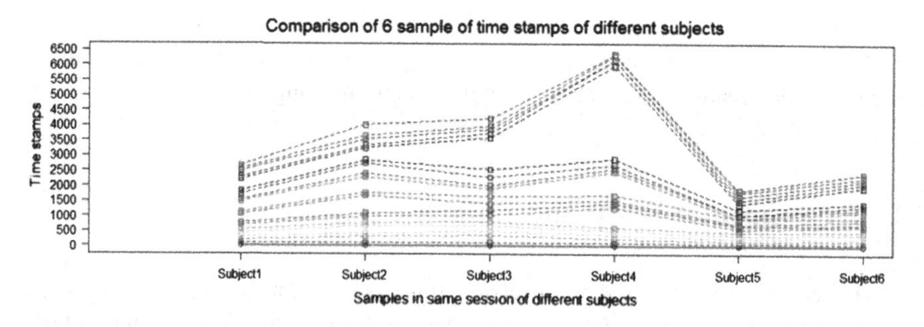

Fig. 2 Comparisons of 6 ample of time stamps of different subjects

Table 2 Obtained average EERs of different patterns for different classification algorithms

Classifiers	All strings	kolkata123	Facebook	gmail.com	yahoo.com	123456	.tie5Roanl [8]
Canberra	**3.69**	**8.93**	**10.83**	**12.47**	**12.92**	**15.18**	**29.74**
Cheby	12.53	13.48	15.85	16.16	20.04	19.24	25.68
Czekanowski	11.40	14.36	14.61	15.88	19.26	18.53	30.77
Gower	51.36	52.56	53.25	53.82	53.25	49.94	62.32
Intersection	60.21	63.67	54.04	54.67	55.49	59.12	41.44
Kulczynski	11.40	14.36	14.61	15.88	19.26	18.53	30.77
Kulczynskis	11.40	14.36	14.61	15.88	19.26	18.53	30.77
Lorentzian	**1.86**	**9.09**	**9.94**	**10.64**	**13.10**	**16.38**	**25.27**
Minkowski	23.99	18.78	19.07	22.89	21.40	20.49	31.99
Motyka	11.40	14.36	14.61	15.88	19.26	18.53	30.77
Ruzicka	11.40	14.36	14.61	15.88	19.26	18.53	30.77
Soergel	11.40	14.36	14.61	15.88	19.26	18.53	30.77
Sorensen	11.33	14.36	14.61	15.88	19.26	18.53	30.72
Wavehedges	11.40	14.36	14.61	15.88	19.26	18.53	30.77
Euclidean	22.82	16.00	18.88	19.07	22.38	20.20	29.11
Manhattan	12.53	13.48	15.85	16.16	20.04	19.24	25.68
ScaledManh	**8.87**	**11.77**	**9.75**	**11.90**	**14.74**	**17.61**	**15.45**
OulierCount	**2.33**	**9.85**	**9.97**	**13.42**	**12.15**	**16.30**	**17.12**
Mahalanobis	26.17	15.40	16.29	16.32	26.77	31.72	26.77
KMeans	18.40	15.21	13.19	13.89	16.98	17.96	16.62
SVM	18.18	15.06	11.30	14.49	16.24	16.82	17.16

6 Evaluation, Analysis and Comparison

In this subsection, comparisons on the basis of average EERs have been made between top most distance-based classification algorithms on different patterns of strings. Lorentzian classification algorithm achieved impressive results on 5 patterns of strings. It achieved only 1.86 % of average EER. Lorentzian algorithm represented in red colour in the following bar chart (Figs. 3 and 4).

In the following line chart, average EERs for different features-mining techniques for different patterns of strings have been represented to get the answer, which pattern of strings are suitable and which classification algorithms can be applied on keystroke database. Line in blue colour in Fig. 5 represents all strings and most of the time, it achieves optimum results. Twenty two different classification algorithms have been evaluated on different datasets but some of that algorithms achieved impressive results. In the following line chart, we can see that Lorentzian achieved 1.86 % of average EER, Outlier Count achieved 2.33 % and Canberra achieved 3.69 %.

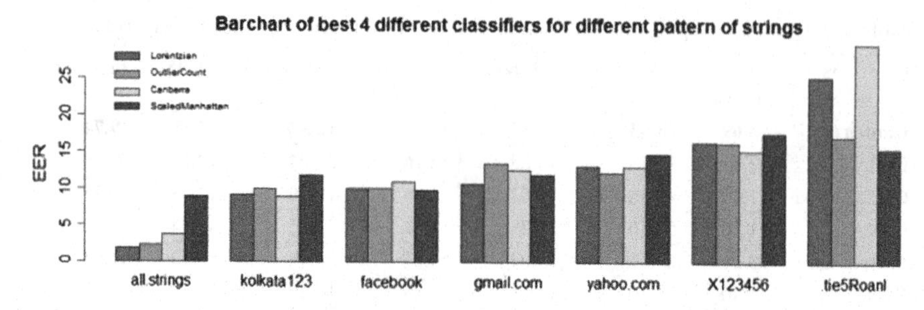

Fig. 3 Comparisons between top 4 classification algorithms on different patterns

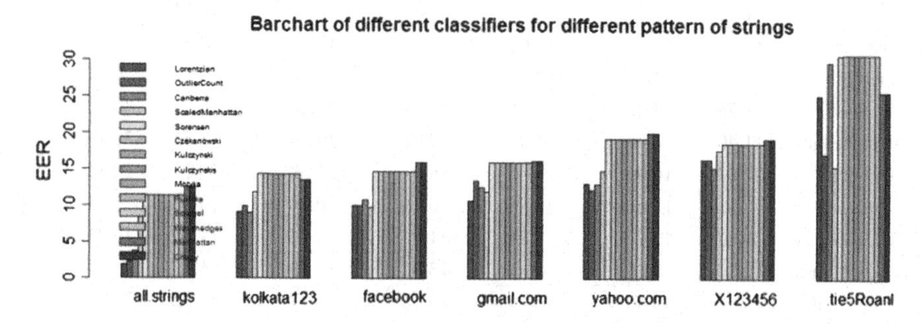

Fig. 4 Comparisons between top classification algorithms basis on average EER

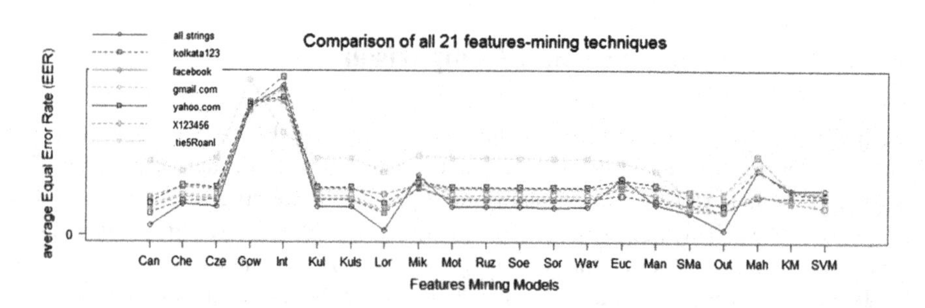

Fig. 5 Line chart of all 22 classifiers on different patterns

In the bellow figure, we have represented the EERs for different patterns of strings for all classification algorithms in histogram and we have seen except 2–4 classification algorithms most of the time we got impressive results (Fig. 6).

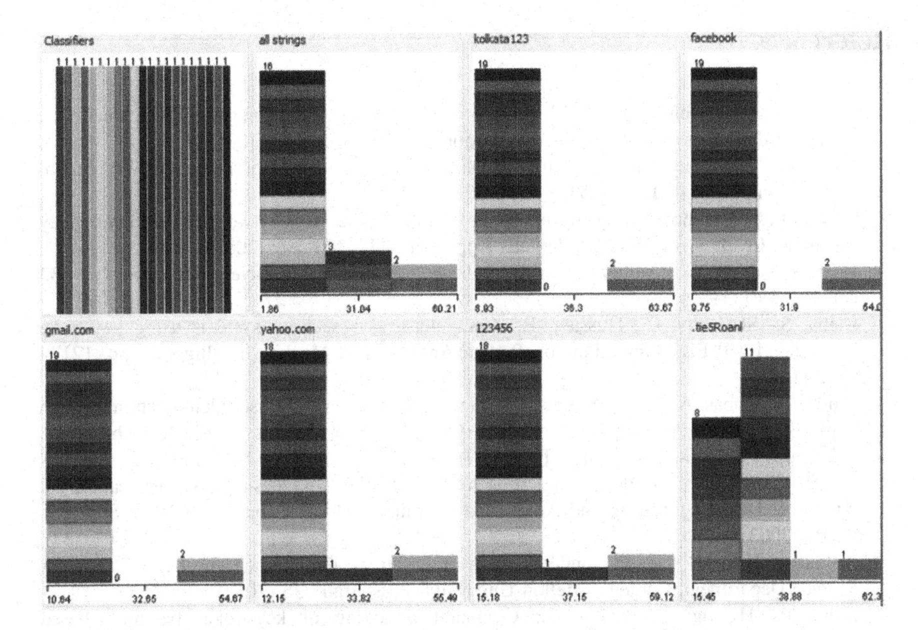

Fig. 6 Histogram of EERs of different patterns of strings

7 Conclusion

We have evaluated 22 different classification algorithms and we have seen most of the time obtained results are impressive. But Lorentzian classification algorithms achieved best average EER only 1.86 %, where Ourlier Count achieved only 2.33 %. In our evaluation process, we have worked on different databases on keystroke dynamics for verification but we have seen hard type of password type patterns are not suitable to train the system, we have to use some daily-used common words to train the system. We have also seen the pattern of string like "123456" are also not suitable in keystroke dynamics. Keystroke dynamics characteristics may change over time depending on mental state or muscle pain or tiredness of the person. Length of the string or PIN, which is used in authentication and type of that string, affect the way of regular typing rhythm. Position of the keyboard and type of key board also affect the way of typing style. Need much more experiment on it, like key pressure, finger placement etc. can also be calculated and considered to optimize the performance in accuracy of keystroke dynamics.

References

1. Roy, S., Roy, U., Sinha, D.D.: Enhanced knowledge-based user authentication technique via keystroke dynamics. Int. J. Eng. Sci. Invention (IJESI) **3**(9), 41–48 (2013)
2. Monrose, F., Rubin, A.D.: Keystroke dynamics as a biometric for authentication. Future Gener. Comput. Syst. **16**(4), 351–359 (2000)
3. Cha, S.H.: Comprehensive survey on distance/similarity measures between probability density functions. Int. J. Math. Models Methods Appl. Sci. **4**(1), 300–307 (2007)
4. Joyce, R., Gupta, G.: Identity authorization based on keystroke latencies. Commun. ACM **33** (2), 168–176 (1990)
5. Bleha, S., Slivinsky, C., Hussien, B.: Computer-access security systems using keystroke dynamics. In: IEEE Transactions on Pattern Analysis and Machine Intelligence, pp. 1217–1222 (1990)
6. Haider, S., Abbas, A., Zaidi, A.K.: A multi-technique approach for user identification through keystroke dynamics. In: IEEE International Conference on Systems, Man, and Cybernetics. vol. 2, pp. 1336–1341. Nashville, TN (2000)
7. Yu, E., Cho, S.: Novelty detection approach for keystroke dynamics identity verification. In: Intelligent Data Engineering and Automated Learning, vol. 2690, pp. 1016–1023. Springer, Berlin (2003)
8. Killourhy, K.S.: A scientific understanding of keystroke dynamics, PhD thesis, Computer Science Department, Carnegie Mellon University, Pittsburgh, US (2012)
9. Kang, P., Hwang, S.S., Cho, S.: Continual retraining of keystroke dynamics based authenticator. In: Advances in Biometrics, Proceedings, vol. 4642, pp. 1203–1211. Springer, Berlin (2007)
10. Giot, R., El-Abed, M., Rosenberger, C.: GREYC keystroke: a benchmark for keystroke dynamics biometric systems. In: Proceedings of the IEEE 3rd International Conference on Biometrics: Theory, Applications and Systems (BTAS '09), pp. 1–6 (2009)

A Novel Leakage Reduction Technique for Ultra-Low Power VLSI Chips

Vijay Kumar Magraiya, Tarun Kumar Gupta and Krishna Kant

Abstract The modern portable devices demand ultra-low power consumption due to the limited battery size. With each new generation, the need of more transistors on the same chip is increasing due to the increased functionality. The leakage causes static power consumption is exceeding the dynamic power in the sub-nanometer designs. Therefore, effective leakage reduction technique is required to minimize the power consumption. In this paper, we have explored the existing leakage reduction techniques and propose a new leakage reduction technique that provides significant reduction in the leakage without significant area/power overhead. The simulation results on Synopsys HSPIC shows that that proposed leakage reduction technique provides 10 % reduction in leakage over the existing leakage reduction technique in the literature.

Keywords Low power design · Leakage reduction · Integrated circuits · VLSI

1 Introduction

The power consumption in the portable devices has become a severe challenge in the modern portable device and demands significant effort to address it. With the technology scaling [1–3], the supply voltage is also scaling to limit and maintain the field. The reduction in the supply voltage requires the reduction in threshold voltage

---

V.K. Magraiya (✉) · T.K. Gupta
ECE Department, Maulana Azad National Institute of Technology, Bhopal, India
e-mail: vijay.magraiya@gmail.com

T.K. Gupta
e-mail: taruniet@rediffmail.com

V.K. Magraiya · K. Kant
ECE Department, ShriRam College of Engineering and Management, Banmore, Morena, India
e-mail: krishna.kant153@gmail.com

S.C. Satapathy and S. Das (eds.), *Proceedings of First International Conference on Information and Communication Technology for Intelligent Systems: Volume 1*, Smart Innovation, Systems and Technologies 50, DOI 10.1007/978-3-319-30933-0_18

to maintain the logic levels. With each new technology up to 5X increase in the leakage, that leads to the significant static power consumption. This necessitates the leakage reduction technique that is able to reduce the leakage to acceptable level. Among the several leakage reduction technique, some of the technique demands device modification whereas other technique demands modification in the existing circuit or architecture to reduce leakage current [4]. The device level techniques are costly and cannot be adopted for most of the design, whereas the circuit and architectural level techniques are very flexible and can be easily adopted.

Significant efforts have been devoted to achieve efficient leakage reduction at the circuit level. The transistor stacking techniques [5] reduces the leakage current of the series connected transistors. The stacking causes increase in the source voltage that reduces voltage on gate-to-source, drain-to-source and create body-bias. The dual-threshold (DT) and multi-threshold (MT) approach of leakage reduction [6] reduces leakage by adopted high threshold transistors for non-critical path whereas low threshold transistor for the critical path to avoid increase in delay. Some new techniques such as LECTOR [7], ONOFIC, super cutoff, etc have also shown as good approach to reduce leakage current. The LECTOR approach utilizes two extra transistors in series between pull-up and pull-down network whereas ONOFIC approach utilizes parallel connected nMOS and pMOS transistor between pull-up and pull-down network. In the super cutoff approach as reported in [8], utilizes on pMOS transistor to gate leakage current and biased in supper cutoff condition. Although, the leakage reduced by supper cutoff is very high but the variable bias requirement increases its complexity.

This paper proposes a novel leakage reduction technique that reduces leakage significantly without power/area and performance overhead. The simulation results shows that proposed approach significantly reduces leakage current over the existing. Thus, proposed leakage reduction can be effectively utilized in portable devices.

2 Sources of Leakage and Different Leakage Reduction Techniques

Significant efforts have been devoted to reduce leakage. Before getting detailed discussion on the different techniques, the first subsection shows sources of leakage.

2.1 Sources of Leakage

The leakage current [9] within a MOS can be expressed by the Eq. (1).

$$I_{leak} = I_{sub} + I_{gate} + I_{BTBT} + I_{GIDL} \tag{1}$$

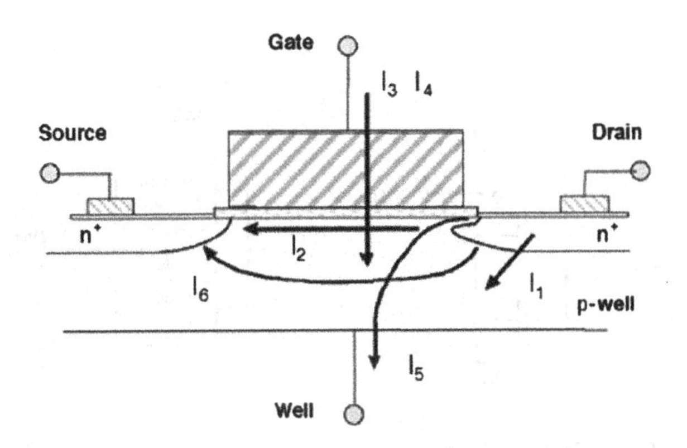

Fig. 1 Sources of leakage in a transistor

There are six major sources of leakage in the scaled device as shown in Fig. 1. I_1 is the reverse bias pn-junction leakage whereas I_2 is the sub-threshold leakage. I_2 is the I_{DS} when the gate is biased below V_{th}. I_3 is the oxide tunneling current which arises due to large electric field and can be reduced by using high-K material. I_4 is the leakage due to hot carrier injection, whereas I_5 is gate induced drain leakage. Finally the I_6 represent punch through current which is excessive and may burn out the MOS.

2.2 *Leakage Reduction Techniques*

2.2.1 Transistor Stacking

The sub-threshold current in the transistor reduces significantly when two or more transistor in series is in OFF condition. In this condition, source voltage of the stacked transistor [5] increases that reduces V_{GS}, V_{DS} and increases the V_{BS} as shown in Fig. 2a. All these effects result in significant reduction in the leakage current. To further reduce the leakage, more series connected transistors are turned off by applying input. This approach is sometimes called as forced stack technique.

2.2.2 Super Cut-Off Technique

In this approach a sleep pMOS transistor is biased in super cut-off region [8] as shown in Fig. 2b. It overcomes the high threshold requirement of the dual threshold logic. When the pMOS transistor is kept at high logic, it operates in OFF mode and draws leakage current. This leakage current can be further reduced if the V_{gs} is kept above V_{dd}. This method significantly reduces leakages as the Subthreshold current reduces exponentially with increasing V_{GS} of sleep transistor.

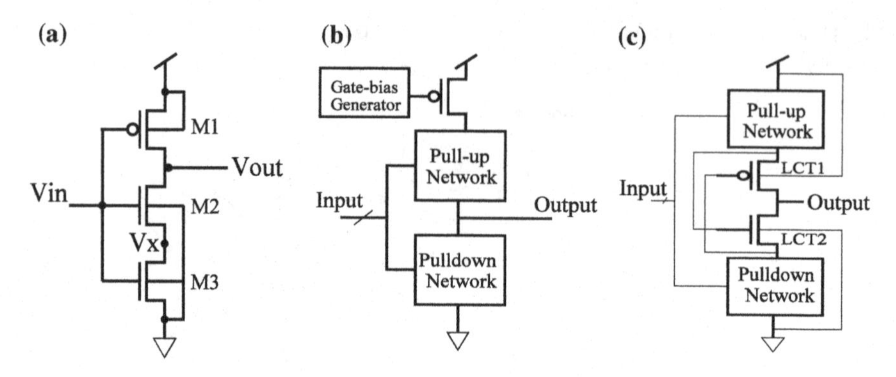

Fig. 2 Different leakage reduction techniques: **a** Transistor stacking. **b** Super cut-off. **c** LECTOR

2.2.3 Leakage Control Transistor (LECTOR) Technique

The LECTOR [7] technique utilizes two transistors a PMOS and an NMOS which are connected as shown in Fig. 2c. In this approach, each controlling transistor controls the other LCT. These controlling transistors are in near cut-off thus reduces leakages current or increases resistance in the path of supply to ground. Further, this approach uses single threshold and is input pattern independent approach. It can be seen that the LECTOR approach is also using the concept of transistor stacking where inserted controlling transistor reduces leakage in supply ground path. The large resistance between is seen corresponding large leakage reduction when one or more transistors are in cut-off mode. This approach reduces leakage current significantly in both active and standby modes.

2.2.4 Body Biasing Approach

This approach utilizes the substrate biasing to reduce the leakage current as shown in Fig. 3a. As the reverse biasing of body bias increases it increases the threshold voltage which in turn reduces the leakage current. Thus, the application of the high reverse bias in the non-critical path to reduces leakage while keeping the lower threshold in the critical path to maintain the performance of the whole design. The additional substrate bias circuit generates appropriate body bias voltage [10, 11] for both NMOS and PMOS when the logic circuit works in the active or standby mode. The reverse body biasing in the standby mode increases resistance of the leakage path thus reduces leakage current significantly. The threshold voltage of the device depends on body to source voltage. Threshold voltage varies according to the relation given in Eq. (2).

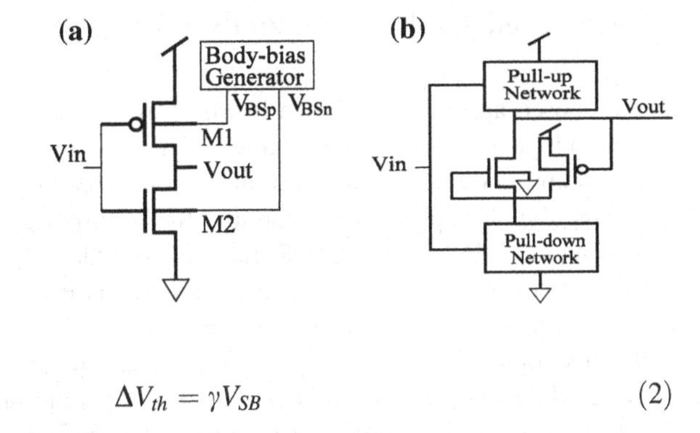

Fig. 3 Different leakage reduction techniques: **a** Body bias and, **b** ONOFIC

$$\Delta V_{th} = \gamma V_{SB} \quad (2)$$

where γ is the body bias coefficient.

2.2.5 ONOFIC Approach

The ONOFIC approach [4] as shown in Fig. 3b is a circuit level approach and utilizes single threshold to reduce leakage current. The ONOFIC approach turns ON/OFF the addition transistor such that it reduces leakage current significantly without affecting the performance of the circuit. This reduction in leakage current significantly reduces the power consumption. Moreover in this approach as body terminal of the both PMOS and NMOS are in connected to Vdd and Gnd respectively it increases the stacking effects and reduces leakage current. The NMOS/PMOS transistors must be in cut-off or in linear mode depending on the output logic. The existing approaches provide reduction in leakage at the cost of large area overhead. In order to reduces leakage within limited area/power overhead we propose a novel leakage reduction technique which is discussed in the next section.

2.2.6 Dual Threshold (DT) Technique

In this technique, transistors of non-critical path are kept at higher threshold to reduce the leakage whereas the transistors in critical path designed with low threshold as shown in Fig. 4. The lower threshold in the critical path reduces the delay of the design whereas higher threshold on non-critical path allows lower leakage of the circuit. Thus, DT technique [12] provides lower leakage without disturbing the performance. The major limitation of the DT technique is the higher fabrication cost of DT transistors which reduces its popularity.

3 Proposed Leakage Reduction Technique

The proposed approach as shown in Fig. 5 is a circuit level approach uses single threshold logic to reduce leakage current. This approach significantly reduces active and standby mode of leakage current without affecting performance thus reduces static and dynamic power consumption. In the proposed logic an additional logic consisting of NMOS and a PMOS transistor is added to reduce the leakage current where the connection of these transistors is in series. The transistors in proposed logic are connected as shown in Fig. 8.

In this approach different input pattern at the inserted transistors M1 and M2 provides different control to the leakage current. In standby condition both the transistors are kept in cut-off mode to reduce the leakage current whereas in normal active mode both are kept in ON conditions.

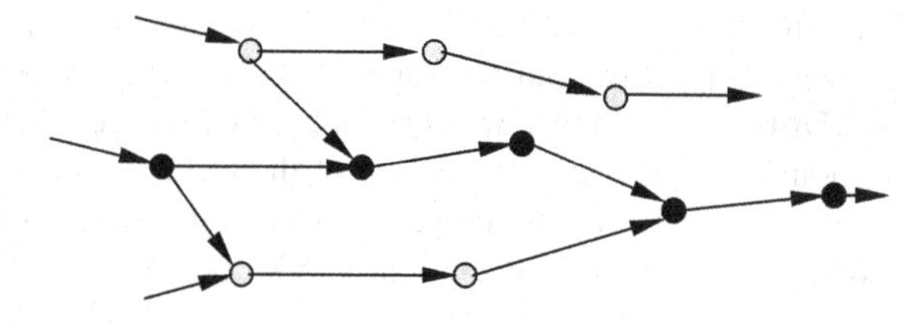

Fig. 4 Critical path with *dark dots*

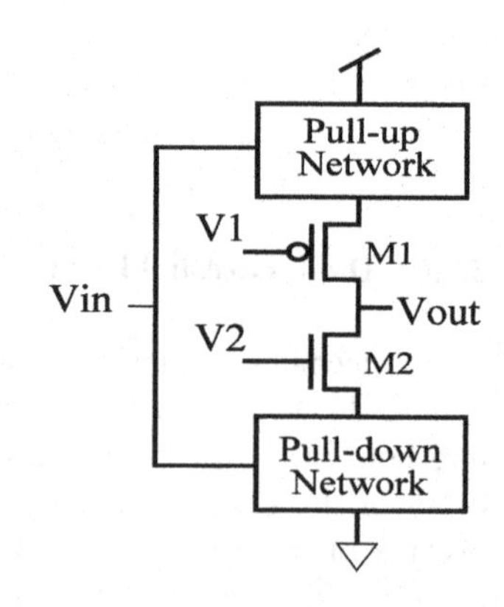

Fig. 5 Proposed approach of leakage reduction

4 Simulation Results and Analysis

This section first provides simulation environment and then show design metrics to evaluate the proposed design.

4.1 Simulation Environment

All the approaches of leakage reduction are first implemented in Tanner 14.1 with similar sizing of the transistors. The 3-input NAND gate is used as design for applying leakage reduction techniques. From the schematic diagram, the netlist is created and simulated with 32 nm technology PTM files using Hspice simulator from Synopsys. The all possible combination of the input patterns are applied and corresponding leakage is evaluated and compared.

4.2 Comparative Analysis

The simulation results of various leakage reduction techniques on 32 nm technology node are illustrated in Table 1. From the simulation results it can be seen that for different input pattern the leakage current is different.

It can be seen that the leakage is small for '000' all zero input pattern whereas higher for '111' input pattern.

From Fig. 6 can be seen that proposed approach reduces more leakage current over ONOFIC approach whereas it requires same area as shown in Fig. 7. Finally the comparison for leakage power for different approach is compare in Fig. 8.

Table 1 Leakage current (nA) using various leakage reduction techniques on 3-input NAND gate at 32 nm node

Technique	Input pattern							
	000	001	010	011	100	101	110	111
NAND3 GATE	0.181	0.183	0.187	1.52	0.144	0.15	0.177	4.56
DT_NAND 3	0.195	0.2	0.199	0.2	0.141	0.146	0.158	0.175
DT1_NAND 3	0.227	0.228	0.101	0.232	0.168	0.172	0.192	3.92
DT2_NAND 3	0.151	0.158	0.158	0.16	0.119	0.125	0.121	0.143
BB_NAND3	0.176	0.176	0.186	1.521	0.013	0.013	0.009	4.45
LECTOR_NAND3	0.053	0.056	0.056	1.1	0.065	0.071	0.079	1.396
SC_NAND 3	0.005	0.007	0.013	0.01	0.008	0.011	0.014	3.569
ONOFIC_NAND3	0.04	0.042	0.045	1.086	0.047	0.053	0.056	7.65
Proposed	0.053	0.054	0.056	1.1	0.058	0.064	0.066	6.331

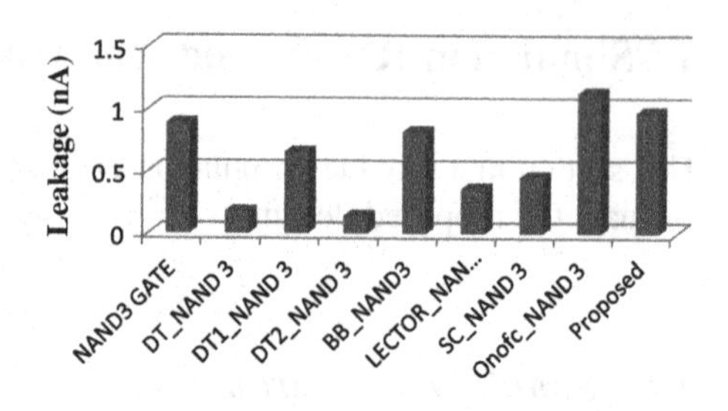

Fig. 6 Comparison of leakage current with various Leakage reduction techniques

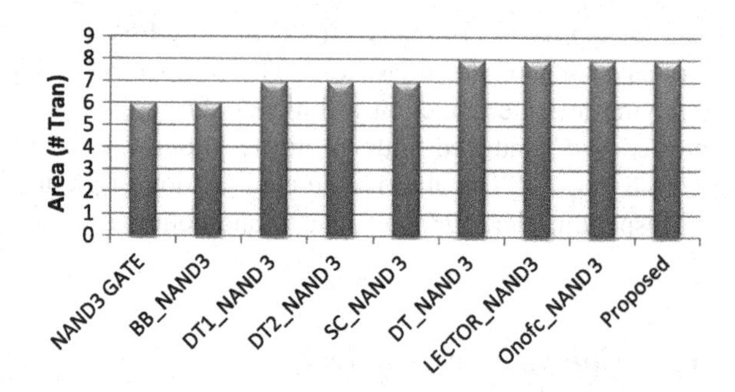

Fig. 7 Area overhead of various leakage reduction techniques

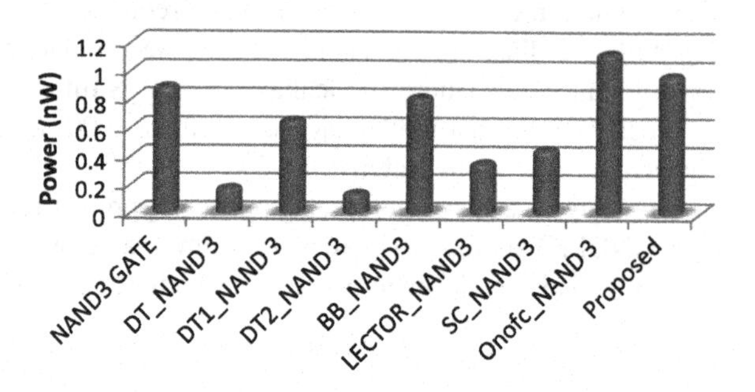

Fig. 8 Leakage power consumption with reduction techniques

5 Conclusion

This paper addresses the prime concern of high leakage power consumption of modern portable devices by proposing novel leakage reduction technique. In this paper we have first analyzed different leakage reduction techniques existing in the literature and the compare the by implementing and simulating using Tanner and Hspice tool with 32 nm technology node. The simulation results using 3-input NAND gate show that proposed approach reduces leakage current significantly over the existing approaches with large area, power and delay overhead.

References

1. Dhar, S., Pattanaik, M., Rajaram, P.: Advancement in nanoscale CMOS device design en route to ultra-low-power applications. In: VLSI Design (2011)
2. Moore, G.E.: Progress in digital integrated circuits. In: Proceedings of the IEEE International Electron Devices Meeting pp. 11–13 (1975)
3. Mohamed, K.S.: Work around Moore's Law: current and next generation technologies. Appl. Mech. Mater. **110**, 3278–3283 (2012)
4. Tajalli, Leblebici, Y.: Design trade-offs in ultra-low-power digital nanoscale CMOS. In: IEEE Transactions on Circuits and Systems I: Regular Papers vol. 58. pp. 2189–2200 (2011)
5. Mukhopadhyay, S., Neau, C., Cakici, R.T., Agarwal, A., Kim, C.H., Roy, K.: Gate leakage reduction for scaled devices using transistor stacking. IEEE Trans. VLSI Syst. **11**, 716–730 (2003)
6. Pakbaznia, E., Pedram, M.: Design of a tri-modal multi-threshold CMOS switch with application to data retentive power gating. IEEE Trans. VLSI Syst. **20**, 380–385 (2012)
7. Hanchate, N., Ranganathan, N.: LECTOR: a technique for leakage reduction in CMOS circuits. IEEE Trans. VLSI Syst. **12**, 196–205 (2004)
8. Kawaguchi, H., Nose, K., Sakurai, T.: A super cut-off CMOS (SCCMOS) scheme for 0.5-V supply voltage with pico-ampere stand-by current. IEEE J. Solid-State Circuits **35**, 1498–1501 (2000)
9. Roy, K., Mukhopadhyay, S., Meimand, H.M.: Leakage current mechanisms and leakage reduction techniques in deep-submicrometer CMOS circuits. Proc. IEEE **91**, 305–327 (2003)
10. Agostinelli, M., Alioto, M., Esseni, D., Selmi, L.: Leakage-delay tradeoff in Finfet logic circuits: a comparative analysis with bulk technology. IEEE Trans. VLSI Syst. **18**, 232–245 (2010)
11. Kang, S.M., Leblebici, Y.: CMOS Digital Integrated Circuits Analysis and Design. McGraw-Hill, New Delhi (2003)
12. Kao, J.T., Chandrakasan, A.P.: Dual-threshold voltage techniques for low-power digital circuits. IEEE J. Solid-State Circuits **35**, 1009–1018 (2000)

Techniques for Designing Analog Baseband Filter: A Review

Sandeep Garg and Tarun Kumar Gupta

Abstract Different Design Techniques for implementing the Analog Baseband filter are presented in this paper. The baseband filter is classified on the basis of approximation techniques, type of component, technology and type of signal used. These techniques are compared on the basis of CMOS technology, minimum cutoff frequency, filter order, power consumption, supply voltage and integrating capacitance. The filters compared in this paper are active continuous time analog filters.

Keywords Analog Baseband filters · Channel select filters · CMOS · Biquadratic

1 Introduction

An analog Baseband filter is a three terminal device which is used in telecommunication systems as a low pass filter [1]. It is used in Zero IF Receiver also called Direct Conversion Receiver for selecting the desired message signal. It is connected at the output of the Radio Frequency Mixer. It detects the baseband signal without any distortion and also eliminates the out-of-band signal. So it is also called channel select filter. It is used for filtering the lower frequency contents in the baseband signal. In the case of super heterodyne receivers, an IF amplifier is used. This filter eliminates the use of this IF amplifier in direct conversion receiver reducing overall cost and complexity.

In the receiver shown in Fig. 1, the modulated signal is received at antenna. It is then passed through bandpass filter for selecting the desired Amplitude Modulated signal. The Radio Frequency low noise amplifier reduces the noise content and increases the power level of the signal. The mixer stage is used to convert all

S. Garg (✉) · T.K. Gupta
Electronics and Communication Engineering Department,
Maulana Azad National Institute of Technology, Bhopal, India
e-mail: sandeepgargvlsi@gmail.com

T.K. Gupta
e-mail: taruniet@rediffmail.com

S.C. Satapathy and S. Das (eds.), *Proceedings of First International Conference on Information and Communication Technology for Intelligent Systems: Volume 1*,
Smart Innovation, Systems and Technologies 50, DOI 10.1007/978-3-319-30933-0_19

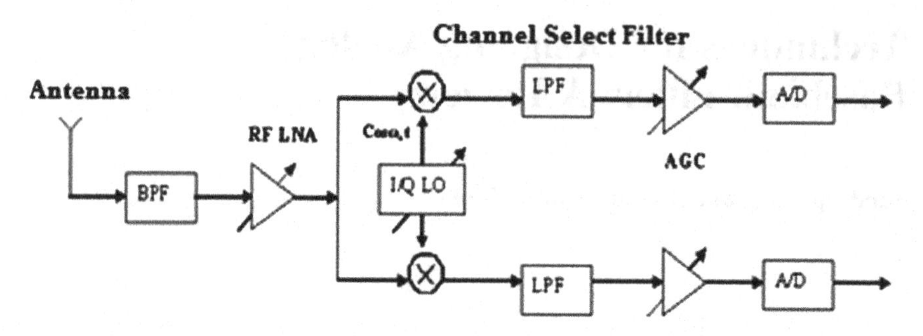

Fig. 1 Architecture of direct conversion receiver

income signals to same intermediate frequency and to remove the image signal. Analog baseband filter is connected after the mixer stage to recover the baseband or message signal. For faithful reproduction of signal, the channel selectivity and dynamic range of the filter must be high.

2 Review of Baseband Filter

The design of baseband filter depends on process technology and approximation technique[2]. Depending on the frequency response of filter, there are three types of approximation techniques—Butterworth, Chebychev and Bessel filters. Butterworth and Chebychev are amplitude filters while Bessel filter is a linear phase filter. In Butterworth approximation, a maximally flat amplitude response is obtained in passband but the roll-off from passband to stopband is slow. Chebychev approximation provides faster roll off than Butterworth filter but produces ripples in the passband. Bessel filter is a linear phase filter which shifts the phase of all frequency components present in the signal by equal amount, preserving the shape of the signal at the output. The three approximation techniques are shown in Fig. 2.

The roll off from passband to stopband also depends on order N of the filter where N is the number of poles. As order of filter increases, the slope becomes steeper as shown in Fig. 3.

On the basis of components used, the analog filters can be classified as Active and Passive filters[3]. In Passive filters, passive components like resistance, inductance and capacitance are used. Since Passive filters have no amplifying component, so they do not amplify the signal. They have a gain less than one. Active filters use op-amp as active device which has high amplifying gain. So they increase signal strength.

The analog filters can also be classified as continuous and discrete time filters. The continuous time filters are used for signal conditioning of continuous signals. Baseband filters are used in wired and wireless communication system and in high

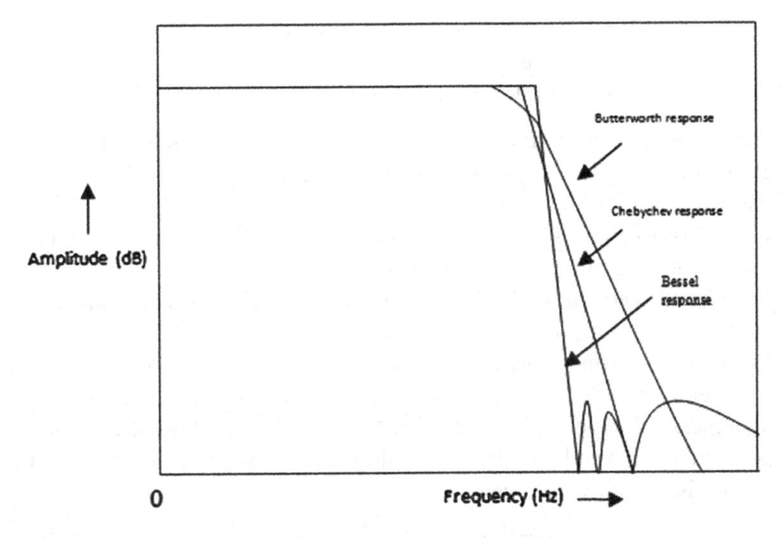

Fig. 2 Frequency response comparison for various approximation techniques

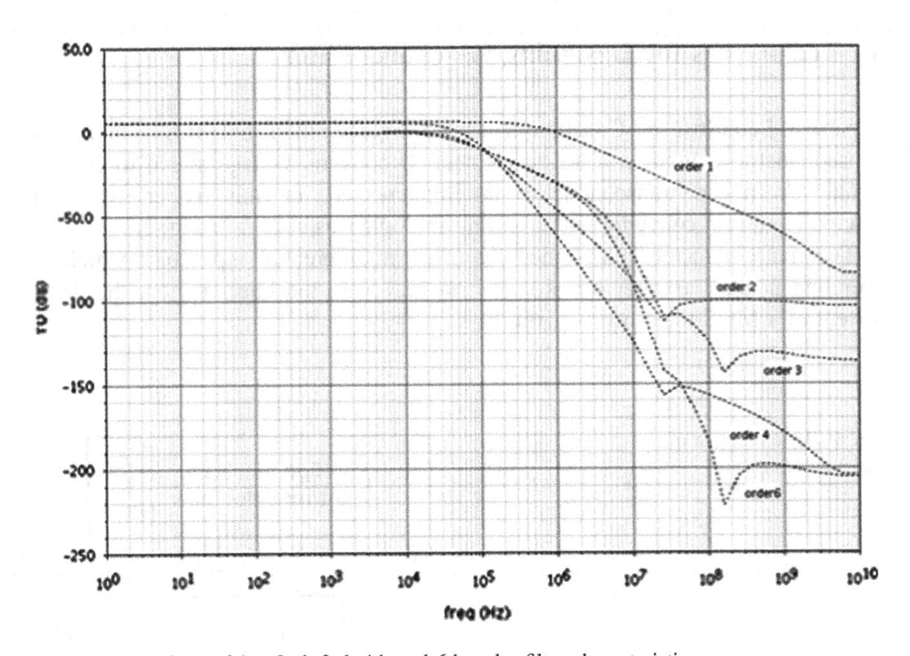

Fig. 3 Comparison of 1st, 2nd, 3rd, 4th and 6th order filter characteristics

data-rate hard disks. Discrete time filters are used for conditioning of discrete signals using DSP algorithms.

The parameters used in filter design are[4]:

(1) **Pass band attenuation**: It is the minimum signal attenuation in the stop-band.
(2) **Pass band ripple**: It is the maximum deviation from desired gain in pass-band.
(3) **Cut-off frequency (f_c)**: It is the frequency at which gain of filter reduces by 3 dB of its maximum gain. It is also called −3 dB point.
(4) **Stop band frequency (f_s)**: It is the frequency at which minimum stop band attenuation is achieved.
(5) **Steepness**: It defines how fast the filter switches from pass band to stop band. It depends on no. of poles in the filter transfer function. As no. of poles increases, the slope becomes steeper. A slope of −20 dB per decade or −6 dB per octave is given by each pole. A slope of +6 dB per octave or +20 dB per decade is given by each zero.

In this paper, various techniques for designing continuous active time filters are discussed.

3 Techniques for Designing Baseband Filter

In paper [5] and [6], Gm-C technique is used to realize filter. In Gm-C technique, a transconductor is used as a active element. It converts voltage signal to current signal. A linear transconductor is implemented in this paper using a voltage-to-current converter and a current multiplier as shown in Fig. 4.

The transconductor is designed by using a differential input pair which is operating in the linear region as shown in Fig. 5. In the linear region, the drain current of a MOS transistor is given by

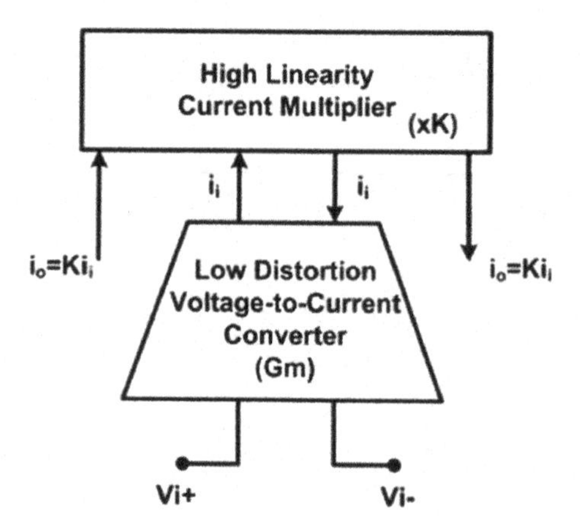

Fig. 4 Basic diagram of transconductor

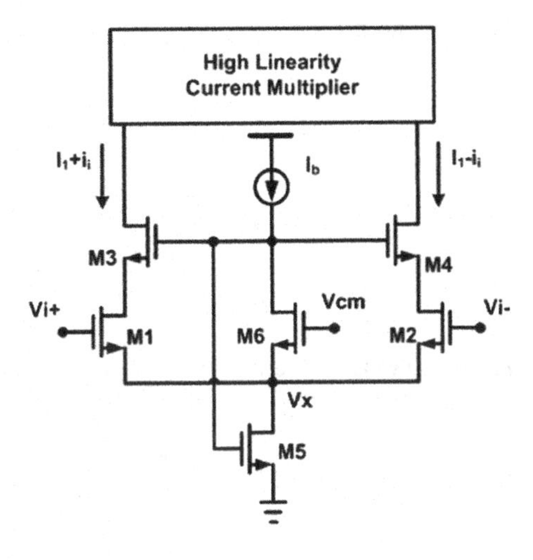

Fig. 5 Transconductor with current multiplier

$$I_D = \mu_n C_{ox} \frac{W}{L} \left[(V_{GS} - V_t) V_{DS} - \frac{1}{2} V_{DS}^2 \right]$$

where V_{th} = threshold voltage of MOS device, V_{DS} = drain to source voltage, V_{GS} = gate to source voltage, μ_n = mobility of electrons, C_{ox} = oxide capacitance, W = width of channel, L = length of channel.

The transconductance of the device can be obtained as:

$$g_m = \frac{\partial I_D}{\partial V_{GS}} = K V_{DS}$$

Using this transconductor, a third order low-pass Butterworth filter is designed. The order of filter is taken so as to meet the design specifications. The use of the transconductor allows the cutoff frequency of filter to be tuned from 135 kHz to 2.2 MHz. This range of frequency is used in several mobile applications.

In paper [7], Active RC technique is implemented as shown in Fig. 6. In this technique, resistance and capacitance are used as passive elements and operational amplifier is used as an active element for filter realization. The main disadvantage of this technique is that as the order of filter increases, the no. of opamps required will also increase. So the power consumption will increase heavily as no. of opamp increases.

Linearity of Active RC filters is high. In this technique, the bandwidth requirement of opamp is much higher than the filter cut-off frequency. This causes higher power consumption.

In paper [8], Active-G_m-*RC* approach is used to design baseband filter. In this approach, the filter is designed with opamp as active element and R and C as passive elements as shown in Fig. 7.

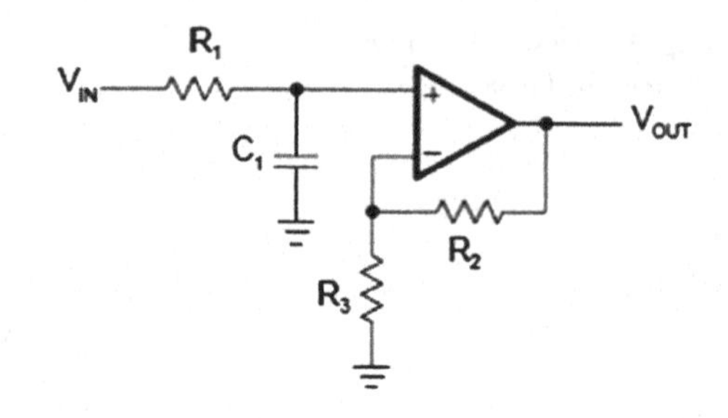

Fig. 6 Active RC cell

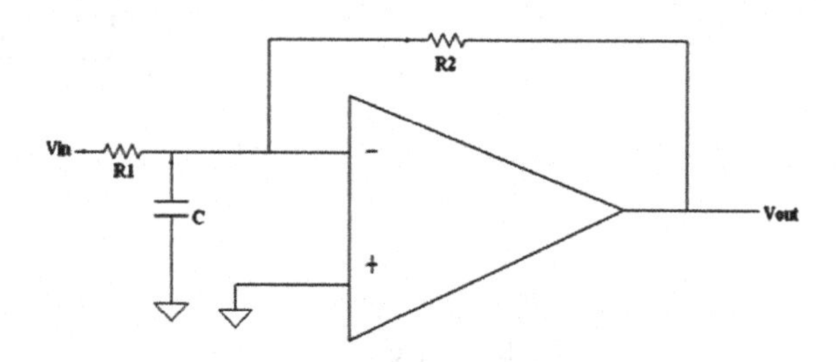

Fig. 7 Active Gm-RC cell

In Active-RC Biquadratic cells, two opamps are used for implementing a second order transfer function whereas this technique uses a single op-amp due to which the power consumption becomes half. The filter frequency response is synthesized using the frequency response of opamp. Due to this reason, the op-amp designed does not require to have the unity gain frequency ω_u much larger than the frequency response of filter pole to be synthesized.

The Transfer function of the Biquad cell is given by

$$T(s) = \frac{G}{\frac{s^2}{\omega_0^2} + \frac{s}{\omega_0 Q} + 1}$$

where G = DC gain of cell, Q = Quality factor of cell, ω_0 = cut-off frequency of cell, ω_u = unity gain frequency of the opamp.

In paper [9], a universal current-mode biquadratic filter is implemented as shown in Fig. 8. Using this topology, biquadratic functions of all the types can be realized. Using controllable CMOS switch pattern, the programming of the different filters can be done digitally. CMOS technology operating on current mode is used for implementing the filter. The filter proposed here can be used in a mixed (digital–analog) signal Low Scale Integration technique and can be used for high frequency low power applications.

For full IC design, the circuit uses two grounded capacitors. By varying the bias current, the frequency of the filter can be tuned .

In paper [10], a low-pass filter of first order is implemented having power consumption of 5 nW and minimum cut-off frequencies of 2 MHz. The circuit is used for signal conditioning applications. The filter is designed to be used in

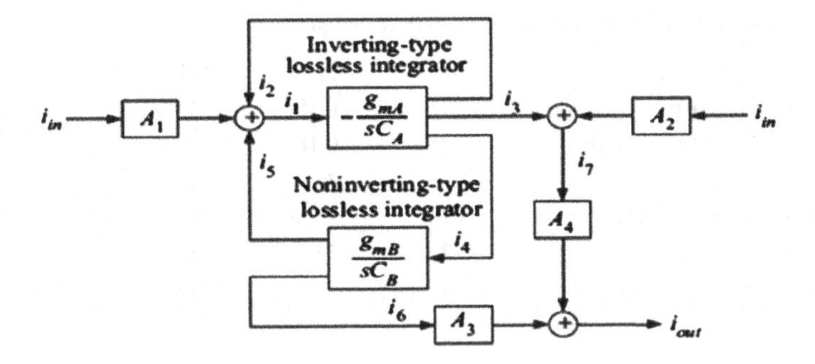

Fig. 8 Universal biquadratic filter using current mode

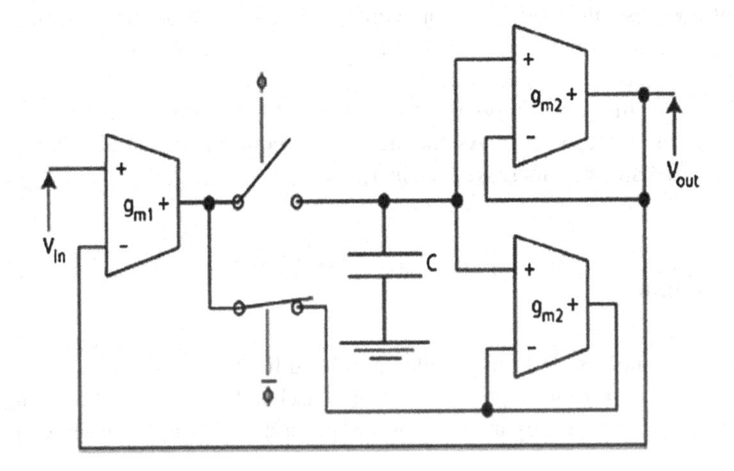

Fig. 9 Low-pass filter for ultra low power application with two buffers

medical equipment based on very low frequency applications. For filtering very low frequency signals, a transconductance having a clock signal is used in filter. It gives high bias current level with low transconductance. The technology used is 0.35 μm and supply voltage is 1 V.

The noise measured is 32-μVRMS. The power consumption in this filter is lower as compared with previously reported filters of same or higher order. 3-dB frequency of the proposed filter is better than the previous filters described in the paper.

In dissertation [11], for implementing the filter, OTA (Operational Transconductance amplifier) is used as basic building block as shown in Fig. 9. In a G_m-C filter, to increase the output resistance and to reduce transconductance, a switch is used at the output of transconductor. Very low value of transconductance can be obtained by this technique. Digital tuning can be easily done using clock signal.

Table 1 Comparisons of different low pass filter techniques

References	[2]	[3]	[4]	[5]
CMOS process (μm)	0.35	0.50	1.20	0.80
Minimum cut-off frequency (Hz)	0.25	0.18	0.17	0.1
Filter order	1	3	2	1
Power consumption (Watt)	8.2n	230n	5n	4.712n
Supply voltage (V)	2.7	3.0	5.0	1.0
Integrating capacitance (F)	29p	15p	10n	17p

4 Comparison of Different Low Pass Filters

Different low pass filter topologies are compared in Table 1 on the basis of process technology, cutoff frequency, filter order, power consumption of filter, supply voltage and integrating capacitance.

In the table, first row shows the process technology (channel length) of MOS device used in various papers. As channel length reduces, area on die also reduces. So packing density will increase (Table 1).

5 Conclusion

Various techniques for designing analog baseband filter have been discussed in this paper. Depending on the requirement of the baseband filter (power consumption, supply voltage, linearity, dynamic range), appropriate technique can be chosen.

References

1. D'Amico, S., Baschirotto, A.: Active Gm-RC continuous-time biquadratic cells. In: Analog Integrated Circuits and Signal Processing, vol. 45, pp. 281–294. Springer (2005)
2. Timar, A., Rencz, M.: Design issues of a low frequency low-pass filter for medical applications using CMOS technology. In: IEEE, Design and Diagnostics of Electronic Circuits and Systems, pp. 1–4 (2007)
3. Solis-Bustos, S., Martinez, J.S., Maloberti, F., Sanchez-Sinencio, E.: A 60-dB dynamic-range CMOS sixth-order 2.4-Hz low-pass filter for medical applications. In: IEEE Transactions on Circuits and Systems, vol. 47(12), pp. 139–1398 (2000)
4. Jung, W., Zumbahen, H.: Op-amp Application Handbook, Newens Publications (2005)
5. Lo, T.-Y., Hung, C.-C.: Multimode $G_m - C$ channel selection filter for mobile applications in 1-V supply voltage. In: IEEE Transactions on Circuits and Systems-II, vol. 55(4), pp. 314–318 (2008)
6. Lo, T.-Y., Chih Hung, C.: A 1-V Gm-C low-pass filter for UWB wireless application. In: IEEE Solid-State Circuits, pp. 277–280 (2008)

7. D'Amico, S., Giannini, V., Baschirotto, A.: A 4th-order active-Gm-RC reconfigurable (UMTS/WLAN) Filter. IEEE J. Solid-State Circuits **41**(7), 1630–1637 (2006)
8. Krummenacher, F.: High-voltage gain CMOS OTA for micropower SC filters. In: IEEE Electronics Letters, vol. 17, pp. 160–162 (1981)
9. Tangsrirat, W.: Low-voltage digitally programmable current-mode universal biquadratic filter. Int. J. Electron. Commun. **62**, 97–103 (2008)
10. Rodriguez Villegas, E., Casson, A.J., Corbishley, P.: A subhertz nanopower low-pass filter. In: IEEE Circuits and Systems, vol. 58(6), pp. 120–25 (2011)
11. Bajpayee, N.: Ultra low power low pass active CMOS filter design for biomedical application. In: Dissertation, Department of Electronics & Communication Engineering, Thapar University, Patiala (2012)

DAREnsemble: Decision Tree and Rule Learner Based Ensemble for Network Intrusion Detection System

Dwarkoba Gaikwad and Ravindra Thool

Abstract The Intrusion detection system is a network security application which detects anomalies and attackers. Therefore, there is a need of devising and developing a robust and reliable intrusion detection system. Different techniques of machine learning have been used to implement intrusion detection systems. Recently, ensemble of different classifiers is widely used to implement it. In ensemble method, the appropriate selection of base classifiers is a very important process. In this paper, the issues of base classifiers selection are discussed. The main goal of this experimental work is to find out the appropriate base classifiers for ensemble classifier. The best set of base classifier and the best combination rules are identified to build ensemble classifier. A new architecture, DAREnsemble, have proposed for intrusion detection system that consists of unstable base classifiers. DAREnsemble is formulated by combining the advantages of rule learners and decision trees. The performance of the proposed ensemble based classifier for intrusion detection system has evaluated in terms of false positives, root mean squared error and classification accuracy. The experimental results show that the proposed ensemble classifier for intrusion detection system exhibits lowest false positive rate with higher classification accuracy at the expense of model building time and increased complexity.

Keywords Intrusion detection system · Random forest · Combination rule · Ensemble · False positive

D. Gaikwad (✉) · R. Thool
Computer Department, SGGSIO Engineering and Technology,
Vishnupuri, Nanded, India
e-mail: dpgaikwad@aissmscoe.com

R. Thool
e-mail: rcthool@yahoo.com

S.C. Satapathy and S. Das (eds.), *Proceedings of First International Conference on Information and Communication Technology for Intelligent Systems: Volume 1*,
Smart Innovation, Systems and Technologies 50, DOI 10.1007/978-3-319-30933-0_20

1 Introduction

The applications of the Internet help society in many areas such as electronic communication, teaching, commerce and entertainment. In business, the Internet is used as an important component for communication and money transfer. Therefore, it has become a part of daily life of the people. Due to huge usage of the Internet, any user in the network can remotely access the information, manipulate information, and render a computer system unreliable. Such type of unauthorized access to software systems or computer in a network called an intrusion. Intrusion detection system can resist external attack on system in the network. It also can be used to detect intruders, malicious activities, malicious code and unwanted communications over the Internet. The usual firewall and routers helps to protect the system in network in different ways, but they cannot perform the task of intrusion detection system [1].

In general, intrusion detection systems are divided into two types based on detection technique; anomaly and misuse detection. Misuse detection technique is a signature-based detection which is used to identify well-known attacks. Whereas, anomaly-based intrusion detection system is based on pattern matching that inspects on-going traffic, transaction, activity, and behaviour in order to identify intrusions. It has abilities to detect new attacks and outlier [2]. But, it exhibits a high false positive rate. A host-based intrusion detection system requires small programs to be installed on individual systems to be monitored [3, 4]. In recent years, the soft computing techniques such as Genetic algorithm, Fuzzy logic and Neural Networks and are mostly used to implement anomaly-based intrusion detection system. The lack of interpretability is the main drawback of soft computing techniques [5]. To overcome this problem, the ensemble of different classifiers has been used. The ensemble of similar or heterogeneous classifiers can reduce the bias and variance on the different training data set. There are many methods of ensemble such as Bagging, Boosting, Random forest, and Stacking of the classifiers. In this paper, the ensemble of decision trees and rule learner have implemented with voting rule combination method.

The reminder of the paper is ordered as follows. In Sect. 2 is dedicated to describe the elated works. In Sect. 3, the proposed architecture is described. In Sect. 4, the experimental results are discussed. Finally, Sect. 5 is dedicated to conclude the paper.

2 Related Work

In this section, the related works are discussed in short. Arun Raj Kumar and Selvakumar [5] have proposed NFBoost ensemble algorithm for DDoS attack classification. The subsystem of NFBoost is implemented using hybrid Neuro-fuzzy adaptive systems. Krawczyk et al. [6] have proposed an OCClustE architecture that is the ensembles of one class classifier. In this implementation, the clustering method is used to partition training data set into small data sets. The clusters are used to train a one-class classifier. This method is used to form a pool of M classifiers for each of the target classes. Chebrolu et al. [7] have applied CART and Bayesian networks to select the significant features. The CART and Bayesian networks have used to classify the training dataset. Due to reduction of features, the performance of Intrusion detection has been improved. Mukkamalaa et al. [8] have addressed different ensemble methods using soft computing and hard computing techniques. In this contribution, authors have studied the performances of Support Vector Machines, Multivariate Adaptive Regression Splines and Artificial Neural Networks for intrusion detection system. Authors have observed that the performance of ensemble of these three is superior in the individual's performance for intrusion detection applications. Menahem et al. [9] have suggested an ensemble method for detecting the malware in the network. In the ensemble method, authors have used five base classifiers that belong from different families of classifiers. It combines the results of individual classifiers to build one final classifier. Liu et al. [10] have proposed the ensemble classifier of SVM base classifier. It is a new approach for learning from imbalanced datasets. It is used with a combination of under-sampling and over-sampling techniques. It integrates the classification results of weak classifiers which are constructed individually on the processed data.

Lin et al. [11] have proposed a CWV scheme to leverage the different domain knowledge among multiple intrusion detection systems. It reduces false positives and false negatives. It helps to increases the efficiency of alert post-processing. Obimbo et al. [12] have introduced and implemented ranking system based on SOFM. It creates many SOFMs trained separately instead of one to improve the precision on classification rate. It also used to reduce the false positive rate on each type of attacks. The system is generated which ranks of each SOFM. Elbasiony et al. [13] have proposed an intrusion detection framework in which the hybridization of classification and clustering technique is used. The Random Forest classification algorithm has used in misuse detection method and weighted-mean clustering algorithm used in Anomaly detection method. Panda et al. [14] have proposed a hybrid smart approach for Intrusion detection system. Un-supervised and supervised filter are used to filter the original training data set.

3 Proposed Ensemble Classifier for Intrusion Detection System

In this section, the overall working principle of the proposed architecture of intrusion detection system is described in detail. In particular, the combination of Decision Trees and Rule leaner as base classifiers of the ensemble is investigated and applied. In ensemble method, a set of base classifiers are constructs from training dataset and classification is performed by taking a vote on the predictions made by each base classifier [15, 16]. The main condition for an ensemble classifier is that the base classifiers should be unstable and independent of each other to perform better than a single base classifier. To get more accurate classification, selection of base classifier is very important step in ensemble technique. In this paper, the combination of unstable classifiers is used to get more classification accuracy. We are aiming at a more accurate classification decision at the expense of increased complexity. We look for the best set of base classifiers and best combination method. Rule Learners, and Decision tree they provide low generalization error and exhibit higher classification accuracy [17, 18]. Because of these facts, we have chosen these classifiers as base classifiers.

Figure 1 depicts the logical view of the proposed DAREnsemble which is divided into different levels. In the first step, training datasets D_1, D_2, D_3 is created from the original NSL_KDD dataset D. The size of each dataset is same but distributions of samples are not identical. A base classifier C_1, C_2 and C_3 are then constructed on each training set. The ensemble classifier or combiner can be obtained by using different combining rules such as minimum probability, maximum probability, product of probabilities and majority voting. The idea behind using different rule methods for experiments is to determine the best combination (rule). For combination of the predictions of base classifiers Eq. (1) is used. The

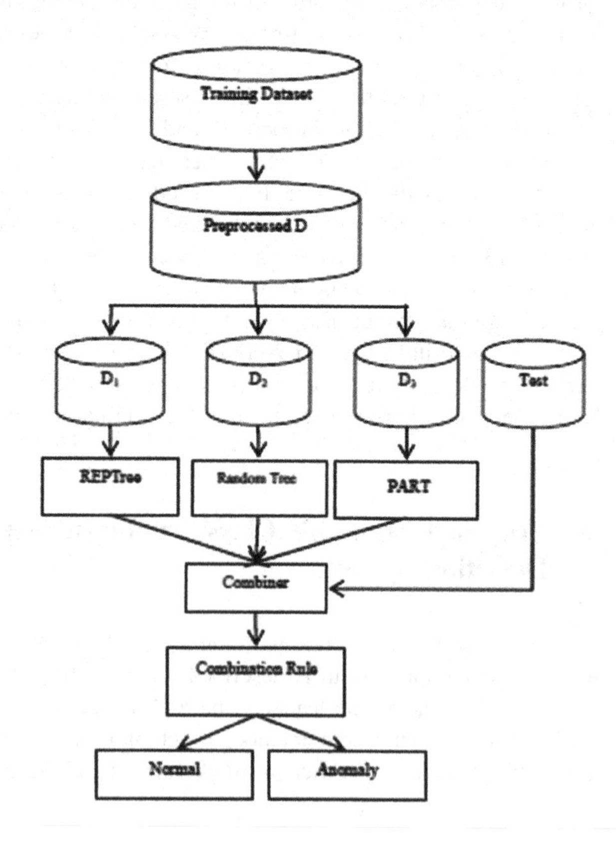

Fig. 1 The logical view of the proposed DAREnsemble

equation is used to take a majority vote on the individual. All trained classifiers are combined together to predict the class of test example by using different combination rules. The majority vote will result in an ensemble decision for class ω_k, where k is number of classifiers and in our case the value of k is 3.

$$C^*(x) = \text{combined vote}(C_1(x), C_2(x), C_3(x)). \quad (1)$$

The algorithm (1) describes the procedure for an ensemble method using decision tree and rule based classifier. This idea is inspired by the general procedure of ensemble of multi classifier. The algorithm starts with the division process of the pre-processed NSl_KDD dataset. Then, each base classifier is trained on each dataset D3 in parallel. On testing mode, each classifier is applied to test the x, tuple and predict its class. Finally, the output of all classifiers is combined using the combination rule to predict the class of x set tuple.

Algorithm 1 The proposed ensemble method.
Input: Original Training dataset D (NSL_KDD), and T test data set Start:
1. For i = 1 to 3
2. Create $D_1 \ldots D_3$ data sets to train the classifier.
3. Build REPTree, RandomTree and PART (as base classifier) from D.
4. End For
5. For each test example $x \in T$ Do
6. $C^*(x)$ = combined vote $(C_1(x), C_2(x), C_3(x))$.
7. End For
END. (Where Vote= combination rule (E.g. Majority vote, Product of Probability, Maximum Probability etc.)
Output: Ensemble Classifier C*.

4 Experimental Result

The experimental results of proposed ensemble methods for intrusion detection system have analyzed in this section. The performance are evaluates in terms of classification accuracy, false positive rate and RMSE on two sets of base classifiers. At the same time, we compare it with the other ensemble methods such as Bagging, Boosting, and Random Forest. For experimental, the Intel (R) CORE™ i5-3210 M CPU @ 2.50 GHz, Installed 8 GB RAM, and Lenovo Laptop with 32 bit operating is used. Firstly, the proposed DAREnsemble classifier with three base classifiers has been implemented for intrusion detection system. This system is evaluated using different combination of rules. Table 1 shows the performance results of classifier on five combination rules using cross validation. Figure 2, depict the performances of the proposed DAREnsemble with Ridor, REPTree, RandomTree base classifiers. It exhibits highest accuracy on average of probability and product of probability

Table 1 Classification accuracy of the Ensemble of Ridor, REPTree, RandomTree using cross validation

Sr. No	Combination rule	False positive	RMSE	Classification accuracy
1	Average of probability	0.001	0.0351	99.88
2	Product of probability	0.002	0.0382	99.88
3	Majority voting	0.003	0.0526	99.72
4	Minimum probability	0.002	0.0382	99.20
5	Maximum probability	0.005	0.0541	99.50

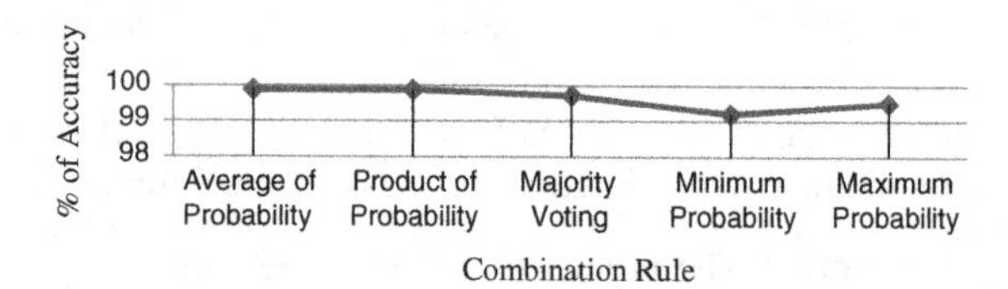

Fig. 2 Classification accuracy of the Ensemble of Ridor, REPTree, RandomTree using cross validation

combination rule and lowest accuracy with the minimum probability combination rule using cross validation. It shows that the false positive is lower when average of probability is used with average RMSE.

Table 2 depicts the performances of DAREnsemble on test data set. According to Fig. 3, the proposed DAREnsemble with Ridor, REPTree, RandomTree as a base classifiers exhibits highest accuracy on Average of Probability combination rule and lowest accuracy with product of probability combination rule on the test data set. The minimum probability of combination rule shows lower false positive and RMSE.

Table 2 Classification accuracy of Ensemble of Ridor, REPTree, and RandomTree on test data set

Sr. No	Combination rule	False positive	RMSE	Classification accuracy
1	Average of probability	0.179	0.404	78.88
2	Product of probability	0.127	0.380	75.30
3	Majority voting	0.179	0.460	78.82
4	Minimum probability	0.126	0.379	75.30
5	Maximum probab ility	0.173	0.388	78.03

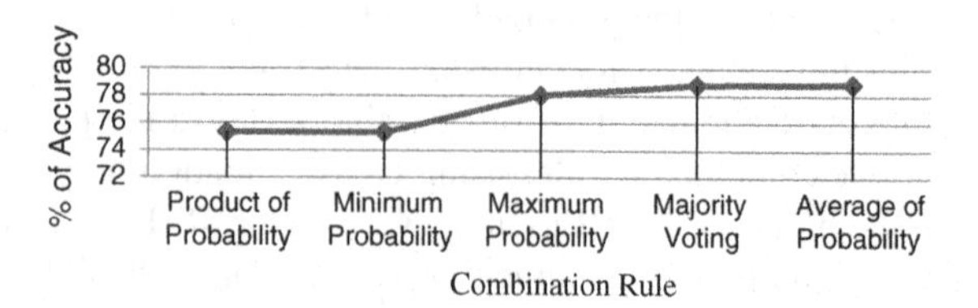

Fig. 3 Classification accuracy of the Ensemble of Ridor, REPTree, and RandomTree on test data set

Table 3 Classification accuracy of ensemble on cross validation and test dataset

Sr. No.	Combination rule	Accuracy on cross validation	Classifier accuracy of test dataset
1	Naïve Bayes	90.22	76.68
2	AdaBoost (Base Classifier RandomTree)	99.75	79.01
3	Bagging (Base Classifier Random Tree)	99.75	79.87
4	Random Forest	99.75	79.87
5	DAREnsemble (Ridor, REPTree, RandomTree)	99.88	78.88

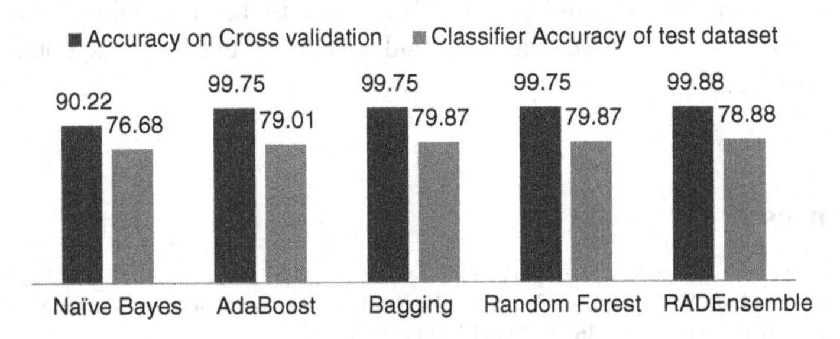

Fig. 4 Classification accuracy of ensemble on cross validation and test dataset

The results of DAREnsemble classifier and existing ensembles such as Bagging, Boosting and Random Forest are shown in Table 3. The DAREnsemble is an ensemble method in which the base classifiers are from different families of the classifier. DAREnsemble make use of Rule Learner and Decision Trees which from two families of the classifier. The DAREnsemble with three base classifiers are tested using different combination rules to find out the best combination rule. The accuracy is evaluated on test data set and using cross validation method. According to Table 3 and Fig. 4, the proposed DAREnsemble with RIDOR, RandomTree and REPTree base classifiers exhibit highest classification accuracy on Cross validation and average accuracy on the test dataset.

5 Conclusion

In this paper, the related works are presented and it is found that the ensemble techniques provide higher classification accuracy and lower false positive (FP) rate. In ensemble method, unstable classifiers are used for increasing classification accuracy. In this paper, the DAREnsemble classifier with rule learner and decision trees base classifiers for intrusion detection system have presented. Rule Learner and Decision Trees have used because they provide low training and generalization error. The Riple down Rule learner and two decision trees (REPTree, RandomTree) have used as base classifiers of ensemble classifier. The proposed system has evaluated in terms of false positive (FP) and accuracy on test dataset and cross validation. The experimental result shows that the system exhibits highest classification accuracy when the Average of Probabilities voting method of combination rule is used. The system provides 99.88 and 78.88 % classification accuracy, by cross validation and on the test data set respectively. The proposed DAREnsemble classifier have compared with other ensemble methods such as Bagging, Booting, and Random Forest. It is found that on cross validation DAREnsemble provides higher accuracy as compared to other ensembles when the Average Probability combination rule is used. It is also shown that the DAREnsemble provide higher accuracy on test dataset when the maximum probability combination rule is used. The drawback of the proposed ensemble classifier is that it requires more model building time for each base classifier and ensemble classifier as compared to existing methods.

References

1. Tsai, C.-F., Hsu, Y.-F., Lin, C.-Y., Lin, W.-Y.: Intrusion detection by machine learning: a review. Expert Syst. Appl. **36**, 11994–12000 (2009)
2. Bhuyan, M.H., Bhattacharyya, D.K., Kalita, J.K.: Network anomaly detection: methods, systems and tools. In: IEEE Communications Survey and Tutorials, vol. 16(1), First Quarter (2014)
3. Basics of Intrusion detection system, www.sans.org/readingroom/whitepapers/detection
4. Major Types of IDS, http://advancednetworksecurity
5. Arun Raj Kumar, P., Selvakumar, S.: Detection of distributed denial of service attacks using an ensemble of adaptive and hybrid neuro-fuzzy systems. Comput. Commun. **36**, 303–319 (2013)
6. Krawczyk, B., Wozniak, M., Cyganek, B.: Clustering-based ensembles for one-class classification. Inf. Sci. **264**, 182–195 (2014)
7. Chebrolu, S., Abraham, A., Thomas, J.P.: A feature deduction and ensemble design of intrusion detection systems. Comput. Secur. **24**, 295–307 (2005)
8. Mukkamalaa, S., Sunga, A.H., Abrahamb, A.: Intrusion detection uses an ensemble of intelligent paradigms. J. Network Comput. Appl. **28**, 167–182 (2005)
9. Menahem, E., Shabtai, A., Rokach, L., Elovici, Y.: Improving malware detection by applying multi-inducer ensemble. Comput. Stat. Data Anal. **53**, 1483–1494 (2009)
10. Liu, Y., Yu, X., Huang, J.X., An, A.: Combining integrated sampling with SVM ensembles for learning from imbalanced datasets. Inf. Process. Manage. **47**, 617–631 (2011)
11. Lin, Y.-D., Lai, Y.-C., Ho, C.-Y., Tai, W.-H.: Creditability based weighted voting for reducing false positives and negatives in intrusion detection. Comput. Secur. **39**, 460–474 (2013)
12. Obimbo, C., Zhou, H., Wilson, R.: Multiple SOFMs working cooperatively in a vote-based ranking system for network intrusion detection. In: Procedia Computer Science, vol. 6, pp. 219–224, Complex Adaptive Systems, vol. 1 (2013)
13. Elbasiony, R.M., Sallam, E.A., Eltobely, T.E., Fahmy, M.M.: A hybrid network intrusion detection framework based on random forests and weighted k-means. Shams Eng. J. Shams Univ. **4**, 753–762 (2013)

14. Pandaa, M., Abraham, A., Patra, M.R.: A hybrid intelligent approach for network intrusion detection. In: International Conference on Communication Technology and System Design, Procedia Engineering, vol. 30(2012), pp. 1–9 (2011)
15. Tan, P.-N., Steinbach, M., Kumar, V.: Introduction to data Minin. Published by person, Indian subcontinent version, ISBN-978-93-325-1865-0 (2006)
16. Sharma, P., Ripple-down rules for knowledge acquisition in intelligent system. J. Technol. Eng. Sci. **1**(1) January–June (2009)
17. Gaikwad, D.P., Thool, R.C.: Intrusion detection system using ripple down rule learner and genetic algorithm. Int. J. Comput. Sci. Inf. Technol. (IJCSIT) **5**(6), 6976–6980 (2014)
18. Gaikwad, D.P., Thool, R.C., Intrusion detection system using bagging ensemble method of machine learning. In: International Conference on Computing Communication Control and Automation (2015)

Scalable Design of Open Source Based Dynamic Routed Network for Interconnection of Firewalls at Multiple Geographic Locations

Chirag Sheth and Rajesh A. Thakker

Abstract A single firewall becomes traffic bottleneck depending on network expansion, number of connections and throughput required. The present paper describes a method of interconnecting the firewalls at multiple geographic locations through Open source network. The placement of firewall and routing device along with protocols will form the proposed optimized system to improve overall network security and scalability. To evaluate the performance of the approach, authors carried out performance testing under laboratory setup. OpenBSD PF firewalls processed 360 kpps of traffic with 80 % CPU load. Further, design for site requiring higher capacity is proposed and forwarding performance of firewalls is tested with different packet sizes under laboratory traffic by changing maximum transmission unit (MTU). Dynamic extension of design is proposed to connect to other networks using dynamically routed interconnections.

Keywords Network security · Network firewall · Network management · Interconnection of firewalls at multiple locations · OpenBSD packet filter

1 Introduction

A firewall protects one or more inside networks, connects inside protected network to the internet and also connects to other internal geographically separated firewalls for access to the rest of the network. Today, with growth and expansion, most companies are geographically spread all over the world. There is a need for increased network security when connecting one inside protected network with another inside

C. Sheth (✉)
Tata Consultancy Services Limited, Garima Park, Gandhinagar, India
e-mail: chirag.sheth@tcs.com

R.A. Thakker
Electronics and Communication Department, Vishwakarma Government Engineering College, Chandkheda, India
e-mail: rathakker2008@gmail.com

S.C. Satapathy and S. Das (eds.), *Proceedings of First International Conference on Information and Communication Technology for Intelligent Systems: Volume 1*, Smart Innovation, Systems and Technologies 50, DOI 10.1007/978-3-319-30933-0_21

protected network in different geographical location, specifically over internet [1]. One of the known approach to avoid sending data in plain text over internet is to establish site to site virtual private network (VPN) using firewall at both locations. VPN will ensure data security by encrypting the traffic [2]. However, major challenge with site to site VPN is that it provides static network routing. In case of multiple sites, each site needs dedicated site to site VPN connectivity with all other sites. This increases network complexity and bottleneck in case of any site or link failure [3]. Considerable amount of work has been done to study encryption algorithm on site-to-site IP Security (IPsec) VPN [4], Border Gateway Protocol (BGP) and Multi-Protocol Label Switching (MPLS) performance comparisons [5, 6] and setup of virtual cluster cloud [7] for inter-site connectivity. Some of the proprietary firewall products like Checkpoint does provide route based VPN solution using Virtual Tunnel Interface [8], however additional SPLAT Pro Advanced Routing Suite license is required to use the feature. The proposed design presents Open source system of interconnection of firewalls to achieve dynamic routing. Apart from increased security, it also helps in improving scalability.

2 Description of Proposed Design

The proposed network design is for global or wide area network that interconnects all sites at different geographical location and also interconnects all the networks within the same site. With the proposed network design, the way connectivity is achieved between the sites is irrelevant to the site firewalls. Firewalls only need dedicated network interfaces to connect to different networks as shown in Fig. 1.

Direct communication between devices at separate sites requires a network that connects the sites without translating addresses. The proposed network provides connectivity between internal networks. It does not provide internet connectivity. For internet connectivity, firewall will use dedicated internet interface to route all internet traffic. Depending on performance needs, same firewall could also act as

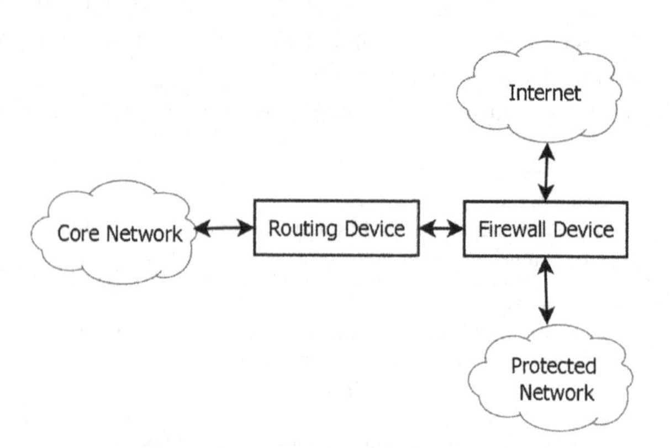

Fig. 1 Single site layout

routing device, if there are no high performance requirements. In the proposed setup mentioned in Fig. 1, a standard firewall has the following setup:

(1) Internet interface for connecting to internet.
(2) Interface for connecting to routing device of proposed network.
(3) Inside interface(s) for connecting to protected network.

All communication to internal networks that is not behind the local firewall is routed via the proposed network interface. The remote internal network could be in the same datacenter in nearby rack, on the other side of the internet. Remote network connectivity is irrelevant to the firewalls as that part is handled by the proposed network. Scaling the network becomes a simple process of connecting new devices to the proposed network. The proposed network does not perform any detailed filtering of traffic. It encrypts all traffic between sites and anti-spoof all traffic entering the proposed network.

As shown in Fig. 2, the proposed standalone network consists of the device used for routing and the links connecting them. For inter-site connectivity the links between routing devices is proposed to use internet protocol security (IPsec) encapsulated generic routing encapsulation (GRE) tunnels. Within a site, it is

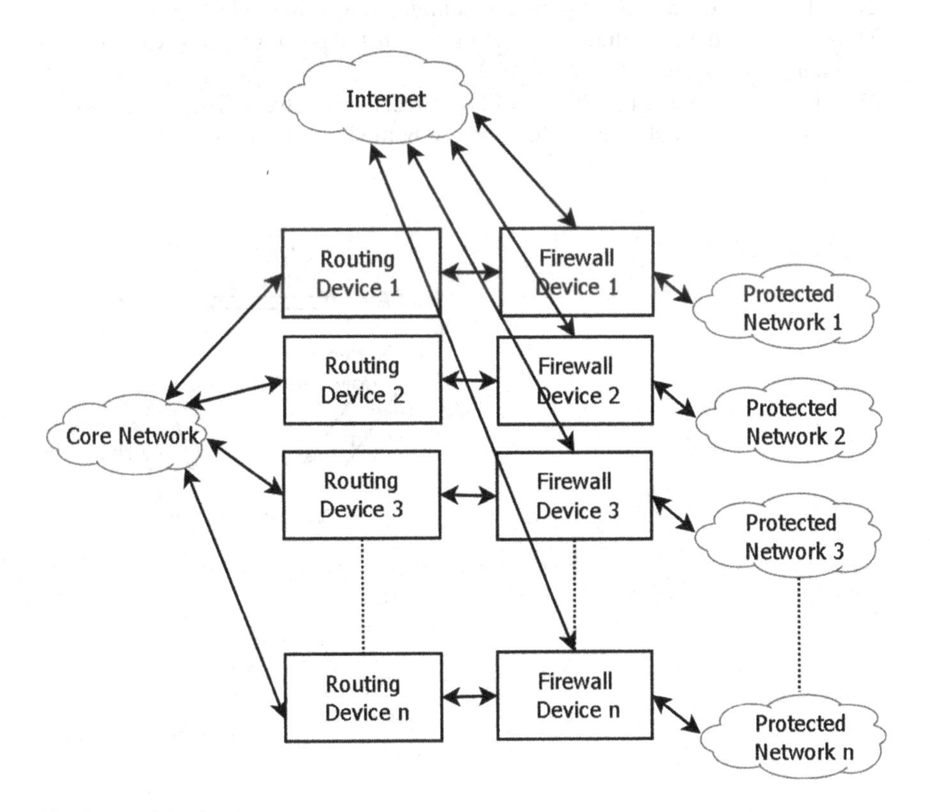

Fig. 2 Multiple sites layout

proposed to be direct links or VLAN sub-interfaces. Inter-site traffic is always encrypted. Inside the proposed network there is no filtering of traffic. Dynamically routed network uses Border Gateway Protocol (BGP) for routing. Also, the network is split up into multiple autonomous systems (AS) to stabilize the routing. Stabilization of traffic will help preventing BGP breakdown. The design uses one backbone AS and multiple leaf AS. The leaf AS connects the backbone AS for connectivity to the rest of the network.

The backbone AS advertises all prefixes to leaf AS as shown in Fig. 3. A leaf AS only advertises its own prefixes on peering with other AS. A leaf AS may have peering with other leaf AS, but it does not provide transit between AS. All traffic between AS in the network will go either via the backbone AS or via a direct peering between the AS. Inside each AS a BGP-only setup is used, with the Multi Exit Discriminator (MED) attribute used for path selection. The MED provides a dynamic way to influence another AS in the way to reach a certain route when there are multiple entry points for that AS.

Below are some of the characteristics of proposed network for path selection:

(1) All routing device should have internal BGP (iBGP) peering with only directly connected peers.
(2) All routing device iBGP peers are configured as route-reflectors.
(3) All routing device will re-write next-hop in iBGP peering so it is reachable via connected routes.
(4) All routing device increment MED on received prefixes. The value used to increment MED should be the same on both sides of a peering.

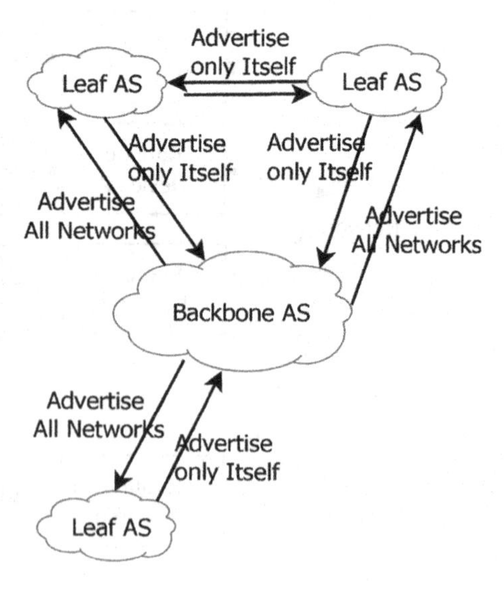

Fig. 3 BGP setup

Routing device has three types of interfaces:

a) Core interface—The core interface connects to other routing devices. Access list on core interfaces is set to no filtering.
b) Edge interface—The edge interface connects to firewall edge device. Traffic filtering on edge interfaces is set to stateless filtering with unicast reverse path forwarding (uRPF).
c) Uplink interface—The uplink interface connects to a packet switched network, like the internet, and is used to provision tunnels on it. An internet uplink interface is also used for management of the routing device over the internet. Traffic filtering on uplink interfaces is set to stateful filtering and only permit communication with peers.

The routing device used in setup is standard server hardware using OpenBSD operating system, without any additional software being used. It permits easy management as only base operating system needs to be patched or upgraded [9, 10]. To enhance the security, the uplink interface uses stateful filtering and permit traffic to/from VPN peers only. One of the uplink interfaces is also used as internet management interface. Only internet control message protocol (ICMP), internet security association and key management protocol (ISAKMP) and internet protocol security (IPsec) traffic is permitted to/from peers. It means that no communication is possible with a VPN peer if IPsec fails. The Do not Fragment (DF) bit is cleared on the IPsec packets to handle path maximum transmission unit (MTU) issues on the network. OpenBSD copies the DF bit from the original packet to GRE and then to encapsulating security payload (ESP). Further, interface-bound state policy is used to tighten security. The routing devices are configured to use single routing table. No default route is used. All routing devices peer with the directly connected routing devices, MED is incremented on all received prefixes, and next-hop is re-written on all advertised prefixes. The iBGP and eBGP configuration differs by one line depending on if it's for a routing device in backbone AS or leaf AS; backbone routing device have "announce all" and leaf routing device have "announce self."

2.1 *Performance Measurement of OpenBSD Firewall*

Tests are carried out to measure performance of OpenBSD Firewalls in the proposed design with two sites [11]. AB (ApacheBench) and NGNIX are used for generating and receiving the traffic. AB is a single-threaded command line computer program which loads the test servers by sending an arbitrary number of concurrent requests. Nginx is an open-source, high-performance HTTP server. The testing setup uses:

(a) 2 × OpenBSD PF Firewall, running two VLANS with rulebase having two sections of rules for each vlan-network.

(b) 2 × OpenBSD Routing Devices used to forward packets.
(c) 2 × Target Servers running Nginx
(d) 2 × Generator Servers running AB
(e) MTU on all servers 256 bytes

The test is successful and generated http traffic to and from the VLANs. In the laboratory setup described, output obtained is 360 kpps with about 80 % CPU load on the OpenBSD Firewall. To the best of author's knowledge, no standard literature available mentions about discussed performance parameters. As a result, there is no benchmark to compare the results obtained.

3 High Capacity Site Design

At backbone sites the routing devices also need to process traffic between remote sites. The IPsec processing is a lot more CPU intensive than just forwarding. When the device needs to handle IPsec traffic, local forwarding performance is greatly reduced. To increase the capacity of the site, a new type of device is proposed, which is dedicated to routing traffic inside a site. It takes over the task of interconnecting all the network core and edge devices at a site from the routing device. With this new device installed, the routing devices are no longer need to process traffic between local devices, they only need to handle VPN traffic. As a result, VPN capacity will be increased by adding additional routing devices to the site. The new devices will take care of the routing between all devices. As a bi-product, the capability to add additional routing devices at a site also provides a way to overcome the issues with link handling. A typical setup has site routing devices usually in pairs for high availability is shown in Fig. 4.

Site may have as many site routes as needed to build the network. Some of the other characterizes of typical setup are:

(1) Routing device should connect to at least two site routing device, each connection on dedicated link or VLAN.
(2) Interconnect routing device should connect to at least two site routing devices, each connection on dedicated link or VLAN.
(3) Edge devices should connect to at least two site routing devices, using one shared VLAN between site routing devices and Edge device(s).

As we add dynamically routed firewall edge devices and interconnect routing device, we will use the site routing with less traffic as primary path. It will achieve some basic load-sharing across site routing device and routing device. The difference from the routing device is that the site routing device has no uplink Interfaces. This also means that management of the devices cannot be done over the Internet from the Internet management networks. The preferred layout is to use two physical

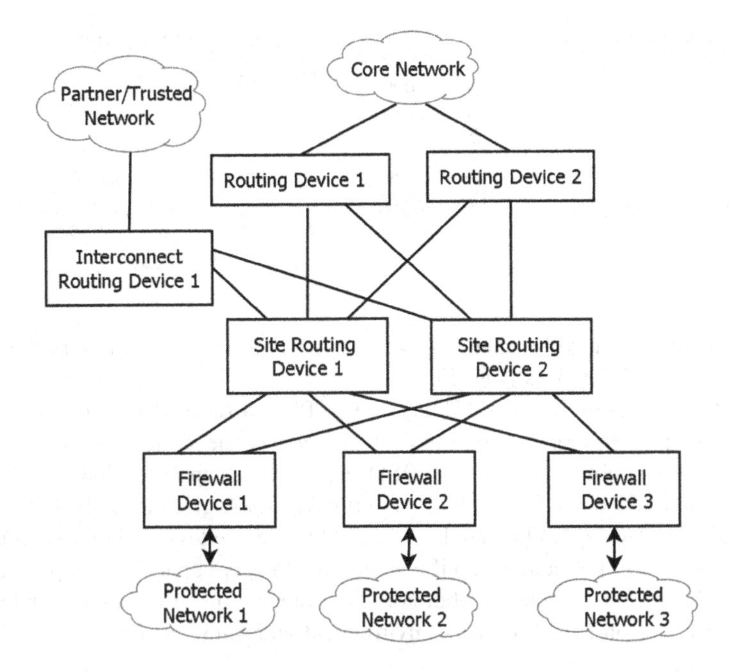

Fig. 4 Design layout with site and interconnect routing devices

interfaces configured as a fail-over trunk to connect to each switched network. The interfaces in one fail-over trunk should connect to separate switches.

3.1 Performance Tests for High Capacity Site Design

Tests are carried out to measure forwarding performance with different packet sizes. A single Site Router with two Edge interfaces and one core interface is used for testing. Two Gigabit Ethernet interfaces are configured as a failover trunk, with VLAN-interfaces configured on the trunk. MTU are configured on generator and receiver interface to test with different packet sizes. Systems configuration are similar to that used during test above as per Table 1. In the laboratory setup, below

Table 1 System configurations

Configurations	OpenBSD PF & routing device	Generator server	Target server
Hardware platform	Supermicro X9SRD-F		
Operating system	Open BSD 5.5	Kali Linux	Kali Linux
Product architecture	Multi-processor, Multi-core		
Processing cores	1		

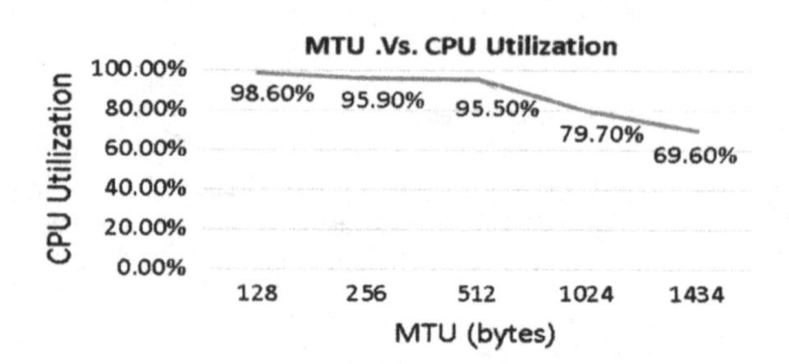

Fig. 5 MTU Vs. CPU utilization

are the results obtained by changing MTU in order to test forwarding performance with different packet sizes (Fig. 5).

The tests indicates that as MTU increases, CPU utilization decreases as shown in Fig. 5. No packet drop is observed which reflects the correctness of proposed design. It is also observed that OpenBSD site router is capable to handle 380 kpps with MTU of 256 bytes before it is fully utilized, which gives us around 700 Mbps of bandwidth. Hence, under standard size MTU (256 bytes), authors recommends to use OpenBSD Site router setup if it is required to move up to 700 Mbps of traffic. Results obtained are limited to laboratory environment under discussion and it may vary depending on actual network environment and traffic conditions.

4 Dynamic Network Extension

This design is an extension to the proposed design and specifies how to connect the presented network to other networks using dynamically routed interconnections. This allows interconnections in multiple locations to achieve redundancy and also improve routing between the networks. It supports both local and VPN connectivity between networks. This design will also enable routing device to have dynamically routed interconnections with other networks. The Interconnections should be possible over direct physical interconnect and also encrypted VPN over Internet or other third party network. The Interconnections support source address verification. The Interconnections also support filtering of prefixes in the routing protocol.

To handle the interconnections, interconnect routing device is used as shown in Fig. 4. It is dedicated to the task of interconnecting with other networks. This setup is used when we need to connect the network to other networks at more than one location using dynamic routing. It also means that any form of address translation is not supported. Internal networks use the same address space, as a result, interconnecting them always requires some form of address translation. Most networks have a VPN interconnect design to address these issues. BGP is the routing protocol supported for network interconnections. Under normal conditions, the proposed network will advertise the same networks at all interconnect locations and the other network is expected to do the same towards proposed network. Interconnect routing device connects to the routing device for connectivity to the proposed network.

The interfaces used are core interfaces and filtering of traffic is not done on either device. Since the interconnect routing device is a modified version of proposed routing device, the routing configurations have the same setup, with some modifications.

5 Conclusion

Authors proposed system and method comprising Open source network for interconnection of firewalls at multiple geographical locations connected through wide area network and also interconnection of all the networks within the same site. It applies to all communication to internal networks that is not behind the local firewall. All such communication will be routed through proposed Open source network which is connected to firewall via dedicated interface. Performance tests carried out on PF firewalls confirms that firewalls are processing 360 kpps of HTTP traffic with about 80 % CPU load on the OpenBSD Firewall. There could be variation in discussed performance parameters and results on different make and model of hardware. Network design for sites which needs higher capacity as compared with other network sites is also proposed. Performance tests of the proposed setup under laboratory traffic infers to use Site router setup if it is required to support up to 700 Mbps of traffic. Authors also proposed extension of the design in order to connect the proposed network to other external networks using dynamically routed interconnections.

References

1. Sheth, C., Thakker, R.: Performance evaluation and comparative analysis of network firewalls. In: IEEE International Conference on Devices and Communications (ICDeCom), pp. 1–5. IEEE Press (2011)
2. Bush, R., Griffin, T.G.: Integrity for virtual private routed networks. In: 22nd Conference of the IEEE Computer and Communications (INFOCOM), pp. 1467–1476. IEEE Press (2003)
3. Hamed, H., Al-Shaer, E., Marrero, W.: Modeling and verification of IPSec and VPN security policies. In: 13th IEEE International Conference on Network Protocols, pp. 259–278. IEEE Press (2005)
4. Zaharuddin, M.H.M., Rahman, R.A., Kassim, M.: Technical comparison analysis of encryption algorithm on site-to-site IPSec VPN. In: IEEE International Conference on Computer Applications and Industrial Electronics, pp. 641–645. IEEE Press (2010)
5. Mai, J., Du, J.: BGP performance analysis for large scale VPN. In: IEEE International Conference on Information Science and Technology, pp. 722–725. IEEE Press (2013)
6. Almandhari, T.M., Shiginah, F.A.: A performance study framework for Multi-Protocol Label Switching (MPLS) networks. In: IEEE 8th GCC Conference and Exhibition, pp. 1–6. IEEE Press (2015)
7. U-chupala, P., Uthayopas, P., Ichikawa, K., Date, S., Abe H.: An implementation of a multi-site virtual cluster cloud. In: 10th IEEE International Joint Conference on Computer Science and Software Engineering, pp. 155–159. IEEE Press (2013)

8. Check Point Software Technologies Ltd., https://sc1.checkpoint.com/documents/R76/
9. Lencse, G., Repas, S.: Performance analysis and comparison of different DNS64 implementations for Linux, OpenBSD and FreeBSD. In: IEEE 27th International Conference on Advanced Information Networking and Applications, pp. 877–884. IEEE Press (2013)
10. Attebury, G., Ramamurthy, B.: Router and firewall redundancy with OpenBSD and CARP. In: IEEE International Conference on Communications, pp. 146–151. IEEE Press (2006)
11. PF: The OpenBSD Packet Filter, http://www.openbsd.org/faq/pf/

Fault Location Estimation in HVDC Transmission Line Using ANN

Jenifer Mariam Johnson and Anamika Yadav

Abstract This paper presents a simple, yet accurate method for fault location in HVDC transmission lines using Artificial Neural Networks. The ±500 kV HVDC system has been modelled using PSCAD/EMTDC and further analysis has been done using MATLAB. Single-end AC RMS voltage and DC voltage and current have been used to identify the location of the fault. This method is relatively simple because the standard deviation of the fault data alone gives acceptably accurate results while using ANN.

Keywords ANN · Fault location · HVDC transmission line · PSCAD/EMTDC

1 Introduction

High Voltage DC transmission systems are being extensively used now owing to the ever increasing need for electric power and hence their bulk transmission over often very long distances [1]. Such transmissions may even involve transfer of power between asynchronous systems. The HVDC technology comes as an economic solution, in comparison to HVAC, to both these issues for transmissions beyond a certain break-even distance [2–4].

Faults in the transmission line are frequent and at the same time need to be cleared as soon as possible to ensure the continuity of power transfer and to avoid economic losses. Since HVDC technology is preferably used for transmission line lengths above the break-even distance, which is as high as 800 km, the accurate and fast location of faults, once they have occurred, is very important for the reliability and efficiency of the system.

J.M. Johnson · A. Yadav (✉)
Department of Electrical Engineering, National Institute of Technology,
Raipur, Chhattisgarh 492010, India
e-mail: ayadav.ele@nitrr.ac.in

J.M. Johnson
e-mail: jeniferjohns91@yahoo.com

S.C. Satapathy and S. Das (eds.), *Proceedings of First International Conference on Information and Communication Technology for Intelligent Systems: Volume 1*,
Smart Innovation, Systems and Technologies 50, DOI 10.1007/978-3-319-30933-0_22

The present-day primary protection schemes for HVDC transmission lines employ travelling wave based methods, whereas the back-up protection is based on dc minimum voltage and dc line differential protection methods. These schemes, however, are not sufficiently accurate or sensitive and hence are less reliable. The present scenario hence demands better line protection techniques that would ensure better reliability that is a key issue due to the extensive usage of the HVDC technology these days.

The travelling wave algorithms estimate the fault location based on the time taken by the fault generated travelling wave to propagate along the transmission line [5–8]. In methods requiring two-terminal data, the global positioning system (GPS) is usually employed to keep the measurements synchronized. The travelling wave theory follows that the transients that are generated as a result of faults or switching procedures are composed of travelling waves that continue to bounce back and forth between the fault point and the terminals until a post-fault steady state is reached. The signals intended to be used need to be synchronized before the travelling wave algorithm is applied. Once the signals have been synchronized, the travelling wave algorithm is used to estimate the travel time of forward and backward fault transients along the concerned transmission line between the fault point and the location of the relay. The fault location is then estimated using these travel times. Although these methods have fast response and high accuracy, the accuracy is affected by the accurate detection of the surge arrival time. However, in methods requiring only single terminal data, there is no need of GPS and these methods are hence more economical [9–12]. But they require the detection of secondary reflection wave as well.

A single ended fault detection and location method for HVDC transmission systems using DWT was first proposed in [9]. A two-ended travelling wave-based fault location method for an HVDC transmission system was put forth and implemented in [10]. A two-ended fault location method for overhead HVDC transmission line using steady-state voltages and currents was proposed in [12].

In this paper, a relatively simple fault location method has been proposed that uses the AC RMS voltage and DC current and voltage at a single end of the HVDC transmission line. A ±500 kV HVDC system was modelled using PSCAD/EMTDC and a single line to ground fault was simulated on the positive line at different locations from the rectifier end to the inverter end and the data was collected and recorded. This data was analyzed in MATLAB and it was found out that the rectifier end AC RMS voltage, DC current and DC voltage of the faulted line alone were sufficient to determine the location at which the fault occurred. The standard deviations of these measurements for each location were then used to train the Artificial Neural Network. The accuracy of this technique was then verified by testing the network with data obtained with different fault locations than those given for training. The accuracy obtained was ±2 km (0.21 %).

2 HVDC Transmission

A 936 km long ±500 kV transmission system has been modelled in PSCAD/EMTDC for further analysis. The CIGRE Benchmark HVDC model which is monopolar was modified to make it bipolar as in Fig. 1. The tower structure of the ±500 kV HVDC transmission line is also depicted in Fig. 2, which has been adopted from [13].

2.1 DC Transmission Line Fault

The common types of faults that are likely to occur in HVDC transmission lines are the single line to ground and line to line fault. Moreover, the single line to ground fault can occur either on the positive or on the negative polarity transmission line. For our analysis, a single line to ground fault has been simulated on the positive polarity transmission line, assuming the remaining two fault characteristics to be of similar nature.

A permanent fault is simulated on the HVDC transmission like at every kilometer from 1 to 935 km. A half of this data is utilized for training the neural network and the remaining to test the same network.

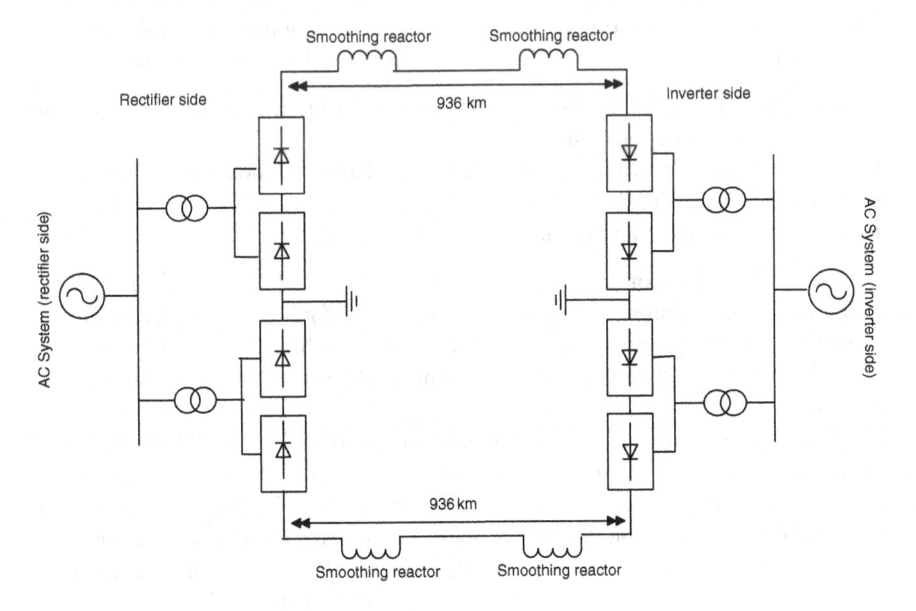

Fig. 1 Bipolar HVDC system model

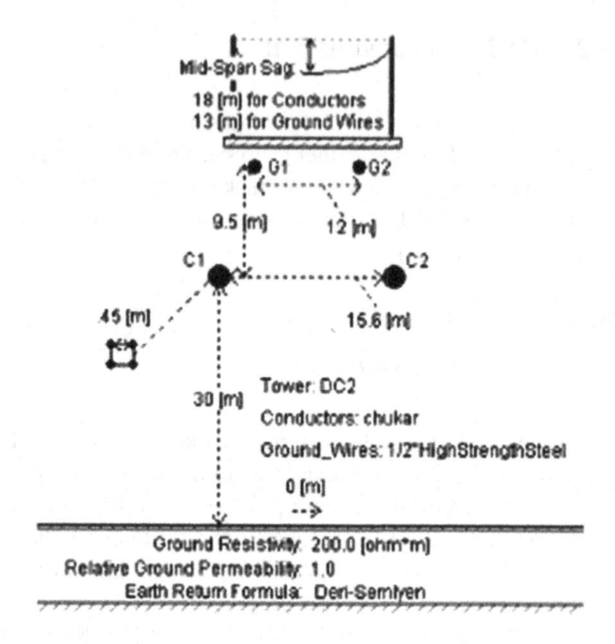

Fig. 2 Tower structure for ±500 kV HVDC lines

3 Artificial Neural Network

An ANN is an artificial (man-made) system motivated by research into the human brain. It mimics the operation of the brain in performing a particular task [14]. The brain has many features that are desirable in Artificial Intelligence systems:

- it is robust and fault tolerant; nerve cells in the brain die every day without affecting its performance significantly
- it can deal with information that is fuzzy, probabilistic, noisy or inconsistent
- it is highly parallel
- it is small, compact and dissipates very little power

The building block of any ANN is a very simple model of the fundamental cell of the brain—the neuron. Neurons are connected to form networks with different structures to perform certain functions. The manner in which the neurons of a neural network are structured depends on the learning abilities and functions that we want to achieve.

The data extracted from the PSCAD/EMTDC model of the HVDC transmission system is used as the input and target data for training the neural network. In particular, the rectifier end AC RMS voltage and the DC voltage and current values, also on the rectifier end, on the faulted line are used here. The standard deviation of these fault data over a sufficient pre-fault, during-fault and post-fault duration are fed as the input data to the neural network in our method. The corresponding distances act as the target for the network. The neural network is trained using the Levenberg-Marquardt backpropagation algorithm.

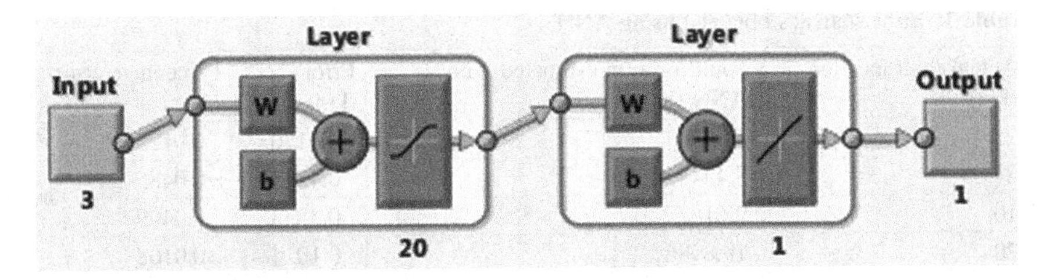

Fig. 3 Neural network architecture used

The main goal of designing the neural network is to accurately estimate the fault location along the transmission line under consideration. For this, the network is initially trained with a set of training data set consisting of extracted samples of pre-fault and post-fault AC RMS voltage, DC voltage and current measurements taken at the rectifier end of the concerned transmission line from different fault case simulation, along with the corresponding targets, being the targeted fault location. The neural network architecture used in this paper consists of two layers (one hidden and one output) of 20 and 1 neuron respectively, as shown in Fig. 3. The activation functions used for the layers are tangent sigmoid and linear function respectively. The three inputs given to the network are the AC RMS voltage, DC voltage and current at the rectifier end. The output is the fault location corresponding to the input data given. Once trained, the network is now ready to predict the fault location for any data including those for which it has not been trained with.

4 Results

The fault data that had earlier not been used for training the neural network is used to test the same network to determine how well the neural network used can identify the fault location. The test data consist of total 468 fault cases simulated at every 2 km from 2 to 934 km length. It is found out that the network that predicts the fault location with an accuracy of ±2 km (0.21 %), which is quite acceptable with respect to the length of the transmission line under consideration.

Table 1 illustrates few test results along with the error statistics of proposed fault location algorithm implemented using ANN. The error plot has been depicted in Fig. 4 which shows the difference between the estimated and target fault location for 468 test fault cases. It can be observed that the error is more at the sending end from where voltage and current measurements have been made. The minimum error obtained is 4.633e-5 (4.95e-6 %) for fault at 164 km whereas the maximum error obtained is −1.9908 (0.213 %) for fault at 54 km.

$$Percentage\ error = \frac{Actual\,fault\ location - Estimated\ location}{Length\ of\ the\ transmission\ line} \times 100 \quad (1)$$

Table 1 Error statistics obtained using ANN

Actual fault location (km)	Fault location estimated with ANN (km)	Error (km)	Percentage error (%)
2	1.883	0.1170	0.0125
8	8.4543	−0.4543	−0.0485
10	9.6167	0.3833	0.0409
20	19.8986	0.1014	0.0108
42	43.1579	−1.1579	0.1237
50	50.2751	−0.2751	−0.2939
54	55.9907	−1.9907	0.2126
62	61.1674	0.8326	0.0889
100	100.0509	−0.0509	−0.0054
150	149.9481	0.0519	0.0055
200	200.0856	−0.0856	−0.0091
300	299.927	0.0730	0.0078
400	400.0856	−0.0856	−0.0091
500	500.0427	−0.0427	−0.0045
600	599.927	0.0730	0.0077
700	699.8462	0.1538	0.0164
800	800.0639	−0.0639	−0.0068
900	900.0382	−0.0382	−0.0041
910	909.8746	0.1254	0.0134
920	920.1917	−0.1917	−0.0204
930	929.8417	0.1583	0.0169
934	933.9065	0.0935	0.0099

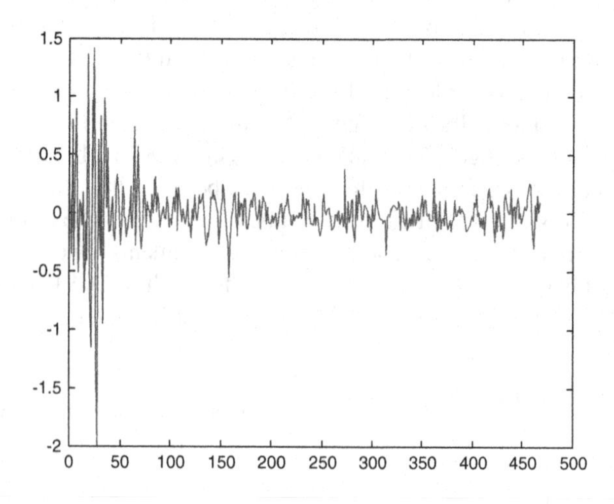

Fig. 4 Error plot obtained using ANN

5 Conclusion

The proposed method is thus a relatively simple method for the prediction of fault location in a bipolar HVDC transmission system. The simplicity lies in the fact that the voltages and current data obtained can be used in the neural network by just taking the standard deviation of the fault data for fault at various locations. The accuracy is ±1 km which can be further improved if we take the inverter end data along with the rectifier end data. Although this option provides better accuracy, it has been avoided owing to the additional communication channel requirement that would be involved if the receiving end and sending end data were both required for the fault location algorithm.

References

1. Arrillaga, Y.H.L.J., Watson, N.R.: Flexible Power Transmission: The HVDC Option. Wiley, New York (2007)
2. Willis, L., Stig, N.: HVDC transmission: yesterday and today. IEEE Power Energy Mag. **5**, 22–31 (2007)
3. John, H.G.: Economic aspects of D-C power transmission, Power App. Syst., Part III. Trans. Amer. Inst. Elect. Eng. **76**, 849–855 (1957
4. Perrin, J.F.: High-voltage DC power transmission. J. Inst. Elect. Eng. **7**, 559-31 (1961)
5. Dewe, M.B., Sankar, S., Arrillaga, J.: The application of satellite time references to HVDC fault location. IEEE Trans. Power Del. **8**, 1295–1302 (1993)
6. Nnanyakkara, K., Rajapakse, A.D., Wachal, R.: Fault location in extra HVDC transmission line using continuous wavelet transform. In: IPST Conference, Delft, the Netherlands (2011)
7. Magnago, F.H., Abur, A.: Fault location using wavelets. IEEE Trans. Power Del. **13**, 1475–1480 (1998)
8. Evrenosoglu, C.Y., Abur, A.: Travelling wave based fault location for teed circuits. IEEE Trans. Power Del. **20**, 1115–1121 (2005)
9. Shang, L., Herold, G., Jaeger, J., Krebs, R., Kumar, A.: High-speed fault identification and protection for HVDC line using wavelet technique. In Proceedings of IEEE Porto Power Tech Conference Porto, Portugal, pp. 1–4 (2001)
10. Chen, P., Xu, B., Li, J.: A travelling wave based fault locating system for HVDC transmission lines. In Proceedings of International Conference on PowerCon, Chongqing, China, pp. 1–4 (2006)
11. Ancell, G.B., Pahalawaththa, N.C.: Maximum likelihood estimation of fault location on transmission lines using travelling waves. IEEE Trans. Power Del. **9**(2), 680–689 (2004)
12. Suonan, J., Gao, S., Song, G., Jiao, Z., Kang, X.: A novel fault-location method for HVDC transmission lines. IEEE Trans. Power Del. **25**, 1203–1209 (2010)
13. Yuangsheng, L., Gang, W., Haifeng, L.: Time-domain fault-location method on HVDC transmission lines under unsynchronized two-end measurement and uncertain line parameters. IEEE Trans. Power Del. **30**(3), 1031–1038 (2015)
14. Aggarwal, R., Song, Y.: Artificial neural networks in power systems. Part I. General introduction to neural computing. IEEE Trans. Power Eng. **11**, 129–134 (1997)

Cellular Radiations Effect on Human Health

Saloni Bhatia, Sharda Vashisth and Ashok Salhan

Abstract This paper includes the various past and present researches which involve the study of the cellular radiations on the human cells. The effect of the weak electromagnetic fields from various sources like cell phones, cordless phones and Wi-Fi can be hazardous over long term exposures and can cause various health problems. The technique of superimposition of incoherent noise field can prove advantageous in suppressing the biological effects of radiofrequency (RF) electromagnetic fields.

Keywords Reactive oxygen species · Specific absorption rate · Deoxy-Ribonucleic acid

1 Introduction

The man had created various technologies which have become an integral part of human life and among them the most important is the creation of cell phones which enable us to interact with others around the globe. But every technology, along with advantages is also linked with disadvantages. The cellular companies guarantee no harmful effects of the cellular radiations on human health but this safety level is accessed considering only the heating effect which is almost insignificant due to the weak radiations. However, there are non-thermal effects also which are linked with direct electrical effect on our cells, organs and tissues which do far more damage to our health at energy levels [1].

The effects of EM radiations includes damages to the glands which results in obesity and various other disorders, loss of fertility, cancer, autism, increase in allergies, memory loss and in severe cases DNA damage also. However, the effects are observed over the long term exposure but are alarming to the living species. The report released by the government agencies around the world on wireless

S. Bhatia (✉) · S. Vashisth · A. Salhan
Department of EECE, Institute of Technology and Management, Gurgaon, Haryana 122001, India
e-mail: salonibhatia13@gmail.com

S.C. Satapathy and S. Das (eds.), *Proceedings of First International Conference on Information and Communication Technology for Intelligent Systems: Volume 1*, Smart Innovation, Systems and Technologies 50, DOI 10.1007/978-3-319-30933-0_23

telecommunications clearly indicate and elaborates the harmful effects of these radiations [2].

The various works done in past had adopted different procedures to check on the radiations effect on both in vivo and in vitro species but the simplest work was explained by Litovitz whose theory was adopted for the experimental purposes. The main theory, involves addition of random electromagnetic fields (ELF) which is generally white noise to a regularly repeating field of cell phones [1]. The technique like a cell culture is then adopted to check out the effect of the cellular radiations on cell linings of various cells. The cells are kept for certain hours under the electromagnetic field exposure provided by cellular radiations along with the ELF magnetic field and the effect of these radiations on cell lining is analysed. The observations and conclusions of various studies are discussed in the following sections.

2 Biological Effects

The electromagnetic radiations are classified in two categories, Ionizing Radiations and Non-Ionizing Radiations. In Ionizing radiations, the charged ions are produced by the removal of valence electrons. These radiations lie in the high end of the electromagnetic spectrum like gamma rays, X-rays and higher region of ultraviolet radiations. They can cause nuclear effects, chemical effects, electrical effects and heating effects. The exposure to these radiations is chronic to our health [3] whereas, the Non-Ionizing radiations do not have enough energy to create ions and lies in the lower region of electromagnetic spectrum. These include microwave radiations, radiofrequency radiations, infrared, visible light and lower region of ultraviolet radiations. The effects of these radiations are not disastrous like ionic radiations but there long term exposure has adverse effect [4].

The radiations of cellular technology are non-ionizing in natures which do not cause significant heating but the electrical effects of these radiations are to be analysed. These non-ionizing electromagnetic radiations might result in the leakage of cell membrane in the way that they generate alternating fields which results in the production of alternating currents that flow through cells and tissues and remove some of the structurally important calcium ions which results in leakage of these ions [1].

There are various tight junction barriers in our body which protect us from various toxins and allergies to enter our sensitive parts like brain, heart etc. However, due to the inward leakage of these calcium ions, the junction barriers gets open up which can be fatal in some cases for e.g. the opening of brain barrier can cause destructions of some of the neurons and may result in Alzheimer's disease. Similarly, opening of epithelial respiratory barrier may cause asthma; opening of gut barrier may results in the destruction of autoimmune system etc. The electrical energy generated by the cellular radiations interacts with the electrical signals of neurons and this interaction may damage the functioning of brain and can lead to loss of memory [1].

The microwave radiations are observed to be more harmful than other non-ionizing radiations since the cell membranes provides high resistance to the passage of direct currents but due to their thinness for alternating currents they behave like a capacitor. The effective resistance of the capacitor is inversely proportional to the frequency, microwave being higher in frequencies experience less resistance and thus, passes more easily through cell membranes and tissues [1].

The SAR (specific absorption rate) specifies the amount of radiations absorbed by human head which lies in between 1.6 and 2 W/kg for 1 g tissues and mathematical given as [5]:

$$SAR = \frac{\sigma E^2}{\rho} \quad (1)$$

Where, σ is conductivity, E is electric field density of the tissue and ρ is density of the tissue.

Both the thermal and non- thermal effects are caused if SAR exceeds the specified limit. In this case, more effect occurs on head, ear pinna, internal ear and brain. Figure 1 illustrates the amount of absorption rate for people lying in different age group and the penetration rate of electromagnetic radiations inside the head and brain. It can be analysed that absorption rate and amount of penetration of radiations are more for children than for adults since the children have less skull thickness. Thus, people belonging to lower age group are prone more to these problems [6].

The fields which cause the damage lies in the low frequency range and hence, for the removal or reduction of these effects, the low magnetic field is provided.

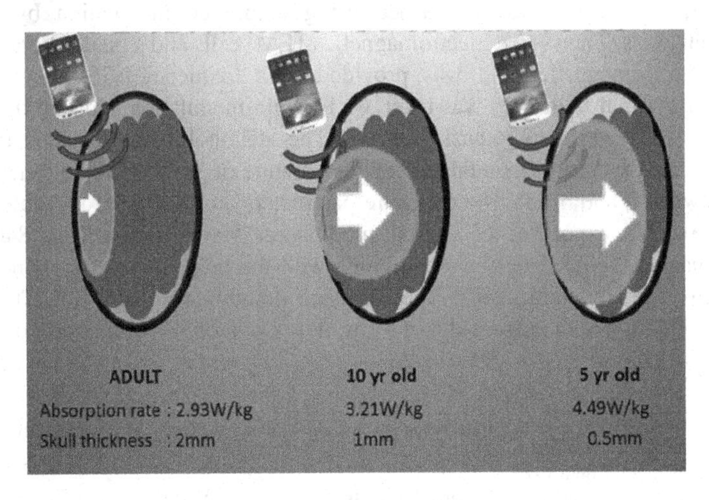

Fig. 1 The figure illustrating the depth of penetration of electromagnetic waves emitted by cell phones on people belonging to different age groups [6]

3 Methods and Techniques

Different methods were explained by different authors, experiments were performed and various results were obtained, which are discussed in this section. The concept adopted in the detection and reduction of electromagnetic radiations effect, is superposition of incoherent electromagnetic noise fields over the coherent fields generated by cellular radiations [7]. The study was undertaken to analyse the superposition of an incoherent magnetic field over RF induced effect. The SP2/0 cells were cultured and divided into three groups. First group cells were kept under the RF radiations exposure produced by mobile phone simulator. The second group was exposed only to the incoherent magnetic noise field whereas; the third group was kept under the exposure of RF radiation along with incoherent noise field. The duration of exposure was kept 2 h. It was recorded that the Reactive Oxygen Species (ROS) production during the exposure of 900 MHz RF radiation increases whereas; there was no significant increase in the ROS production in the cells belonging to the third group.

The above concept was proposed earlier by Litovitz by analysing the ornithine decarboxylase (ODC) activity in L929 cells [1]. The experiments were performed on microwaves which are modulated in amplitude, frequency, square wave and digital and analog modulation schemes adopted in cellular communication. The 60 Hz amplitude modulated wave was considered with modulation index of 0.23. The incident power was kept at 1 W to get nominal specific absorption rate (SAR) of 2 W/kg. The digital modulation model involves transmitting of pseudo-random test sequence in TDMA (Time Division Multiple Access) mode with each burst lasting for nearly 7 ms with repetition rate of 50 Hz and duty cycle of 33 %.

The ELF band–limited noise was generated with nominal bandwidth between 30 and 100 Hz [8, 9].

The first exposure was performed using amplitude modulation by use of Helmholtz coils, Transverse Electromagnetic (TEM) cells and signal generator. The guided electromagnetic field was provided with frequency 835 MHz and the modulating signal frequency was taken 60 Hz with modulation depth of 0.23. The exposure was given continuously for 8 h. The Helmholtz was installed inside a water jacketed incubator maintained at a temperature of 37 °C. The noise level was kept between 0 and 4 μT. The exposure parameters are illustrated in Table 1.

The second exposure for digital modulation was performed at 840 MHz Electromagnetic guided field was provided with the help of cellular phone, TEM cells and with Helmholtz coils. The exposure duration was kept 8 h. The Pulse width modulation with pulse width of 7 ms, duty cycle of 33 % and pulse repetition

Table 1 Exposure parameters without the effect of external magnetic field [9]

Parameter	Value	Method
Power	1 W	Measured
Electric field strength	1 V/cm	Calculated
SAR	2.5 W/kg	Measured

Table 2 Exposure parameter for superimposed magnetic field [9]

Parameter	Value
Magnetic flux density	0.25, 0.5, 1, 2 and 4 μT superimposed on EM field in exposure 1
Magnetic flux density	1, 2, 5 and 10 μT superimposed on EM field in exposure 2

rate of 50 Hz was provided by D-AMPS random test sequence in TDMA mode. The exposure parameters were same as that of Table 1 with peak power of 3.8 W.

The third exposure was performed using magnetic field (AC field) with random amplitude variations in time created using noise generator which was superimposed on electromagnetic field given in first and second exposure. The duration of exposure was taken 8 h. The exposure parameters are illustrated in Table 2.

It was observed that, when amplitude modulated (AM) microwave radiations were superimposed with ELF noise above 0.5 μT, the induced ornithine decarboxylase (ODC) enhancement due to microwave radiations decreases with rms amplitude of ELF noise and complete inhibition in ODC activity was achieved with noise fields above 2 μT. The noise level for digital modulation was kept between 0 and 10 μT. It was analysed that, inhibition in ODC activity starts when noise level was above 5 μT and the complete inhibition was achieved above 10 μT [8, 10].

The EM radiations effect on DNA due to cellular phone was also illustrated in one of the study along with the reduction in the damage caused by the use of Electromagnetic noise. The influence of Global System for Mobile Communication (GSM) with radio frequency field of 1.8 GHz causes the DNA damage and results in human lens epithelial cell the formation of intracellular reactive oxygen species. The alkaline comet assay was performed and cells for 24 h are exposed at the specific absorption rate (SAR) of 1, 2, 3 and 4 W/kg. The cells were broadly categorized in 4 groups, with first group kept under sham exposure, second group placed in contact of microwave radiations with SAR value of 1, 2, 3 and 4 W/kg. The third group was exposed to 2 μT magnetic fields, and last group was exposed simultaneously to 2 μT magnetic field and microwave radiations [9, 11].

It was observed that there is no significant difference in DNA damage of microwaves exposure with SAR value of 1 and 2 W/kg at SAR value of 3 and 4 W/kg there is significant increase in DNA damage. The 4 W/kg has more effect than 3 W/kg.

The similar effect was studied over the chick embryos. The experiment was conducted in two halves. The first half was exposed to 100 Hz magnetic field in the form of pulses with pulse width of 500 μs and rise and fall time of 2 μs. The magnetic field strength was kept at 1 μT. The second half was exposed to 100 Hz magnetic field superimposed with noise. The parameters were kept same as earlier

Table 3 The Sar value For 900 and 1800 MHz frequency band for monopole antenna positioned externally [5]

Frequency (MHz)	Absorbed power (W)	SAR (W/kg)
900	0.0381	0.000116735
1800	0.0153	4.697e-005

Table 4 The Sar value For 900 and 1800 MHz frequency band for monopole antenna positioned internally [5]

Frequency (MHz)	Absorbed power (W)	SAR (W/kg)
900	0.0519	0.000158886
1800	0.0360	0.000110281

but exposure duration was increased to 48 h. It was observed that the damage to chick embryos during first half of the exposure was more as compared to the second half of the exposure [9, 12].

The amount of penetration of Electromagnetic (EM) radiations depends upon the frequency of the field and thus, is greater for lower frequencies. The experiment was conducted to demonstrate the EM radiations effect on the human head with the help of the radiation pattern of monopole antenna which is kept at internal as well as in external position to human head.

The mobile phone was constructed which was having monopole antenna internally as well as externally. The experiment was carried out at both 900 and 1800 MHz GSM phone radiating 0.6 W powers. The results are illustrated in Tables 3 and 4.

From Tables 3 and 4 it is observed that the SAR rating for 900 MHz band is more as compared to 1800 MHz band which is due to the skin effect which is inversely proportional to the frequency. The skin effect described the depth of penetration of EM waves for a given material. The SAR values for the antenna positioned internally have greater values than the antenna positioned externally which is due to the fact that the antenna in internal position is more close to ear and head [5].

4 Conclusion

Various studies conducted over different cell lining shows different amount of damage. The Electromagnetic effect was analysed by the inhibition of certain enzyme activities and the stimulation of species of oxygen ions which have adverse effect over the human health. The non- thermal radiations do not show any kind of adverse effect or any kind of DNA damage in the short run exposure, but with increase in the exposure time the damage can be observed. It is concluded that in the long run the electromagnetic radiations emitted through cell phones, Wi-Fi etc. may cause these serious problems. The superimposition of magnetic field strength of lower values like 2 μT and above 5 μT over the cellular electromagnetic radiations shows less damage. A device may be build which can protect the human body against these harmful radiations. The device may be equipped with the white noise circuit which will be operated at lower frequencies so that, the interference of this noise occurs with these lower frequencies which causes non-thermal effects whereas, the higher frequencies which carries the information will not get interfered and the message can be transmitted without any information loss.

References

1. Goldsworthy, A.: The biological effects of weak electromagnetic fields—problems and solutions. March 2012 (Online). Available: http://www.cellphonetaskforce.org/
2. Trower, B.: Address to Welsh assembly on wireless telecommunication (Online). Available: http://www.mastsanity.org/
3. Non-Ionizing Radiations.: (Online). Available: http://en.wikipedia.org/
4. Ionizing Radiations.: (Online). Available: http://en.wikipedia.org/
5. Lias, K., Mat, D.A.A., Kipli, K.,Marzuki, A.S.W.: Human health implications of 900 MHz and 1800 MHz mobile phones. In: IEEE 9th Malaysia International Conference on Communications, Dec 2009
6. How mobile phone radiations penetrate the brain.: (Online). Available: http://www.electroschematics.com/
7. Kazemi, E., Mortazavi, S,M.J., Ghanbari, A.A., Mozdarani, H., Zahed, S.S., Mostafavi, Z.: The effect of superposition of 900 MHz and incoherent noise electromagnetic fields on the induction of reactive oxygen species in SP2/0 Cel line. In: IJRR (2014)
8. Litovitz, T.A., Penafiel, L.M., Farrel, J.M., Krause, D., Meister, R., Mullins, J.M.: Bioeffects induced by exposure to microwaves are mitigated by superposition of ELF noise. Bioelectromagnetics **18**, 422–430 (1997)
9. http://www.emf-portal.de/
10. Litovitz, T.A., Krause, D., Montrose, C.J., Mullins, J.M.: Temporally incoherent magnetic fields mitigate the response of biological systems to temporally coherent magnetic fields. Bioelectromagnetics **15**(5), 399–409 (1994)
11. Yao, K., Wu, W., Wang, K., Ni, S., Ye, P., Yu, Y., Ye, J., Sun, L.: Electromagnetic noise inhibits radiofrequency radiation-induced DNA damage and reactive oxygen species increase in human lens epithelial cells. Mol. Vis. **14**, 964–969 (2008)
12. Litovitz, T.A., Montrose, C.J., Doinov, P., Brown, K.M., Barber, M.: Superimposing spatially coherent electromagnetic noise inhibits field-induced abnormalities in developing chick embryos. Bioelectromagnetics **15**(2), 105–113 (1994)
13. Lai, H., Singh, N.P.: Single- and double-strand DNA breaks in rat brain Cells after acute exposure to radiofrequency electromagnetic radiation. Int. J. Radiat. Biol. **69**(4), 513–521 (1996)

Micro-interaction Metrics Based Software Defect Prediction with Machine Learning, Immune Inspired and Evolutionary Classifiers: An Empirical Study

Arvinder Kaur and Kamadeep Kaur

Abstract Software developer's pattern of activities, level of understanding of the source code and work practices are important factors that impact the defects introduced in software during development and its post-release quality. In very recent previous research (Lee et al. in Micro interaction metrics for defect prediction, pp 311–321, 2011), process metrics and micro-interaction metrics (Lee et al. in Micro interaction metrics for defect prediction, pp 311–321, 2011) that capture developer's interaction with the source code have been shown to be influential on software defects introduced during development. Evaluation and selection of suitable classifiers in an unbiased manner is another conspicuous research issue in metrics based software defect prediction This study investigates software defect prediction models where micro-interactions metrics (Lee et al. in Micro interaction metrics for defect prediction, pp 311–321, 2011) are used as predictors for ten Machine Leaning (ML), fifteen Evolutionary Computation (EC) and eight Artificial Immune recognition system (AIRS) classifiers to predict defective files of three sub-projects of Java project Eclipse. They are -etc, mylyn and team. While no single best classifier could be obtained with respect to various accuracy measures on all datasets, we recommend a list of learning classifiers with respect to different goals of software defect prediction (SDP). For overall better quality of classification of defective and non-defective files, measured by F-measure, ensemble methods-Random Forests, Rotation Forests, a decision tree classifier J48 and UCS an evolutionary learning classifier system are recommended. For risk-averse and mission critical software projects defect prediction, we recommend logistic, J48, UCS and Immunos-1, an artificial immune recognition system classifier. For minimizing testing of non-defective files, we recommend Random Forests, Rotation Forests, MPLCS (Memetic Pittsburgh Learning Classifier) and Generational Genetic Algorithm (GGA) classifier.

A. Kaur · K. Kaur (✉)
USICT, GGS Indraprastha University, Sector 16-C, Dwarka, Delhi 110078, India
e-mail: kdkaur99@gmail.com

A. Kaur
e-mail: arvinderkaurtakkar@yahoo.com

S.C. Satapathy and S. Das (eds.), *Proceedings of First International Conference on Information and Communication Technology for Intelligent Systems: Volume 1*, Smart Innovation, Systems and Technologies 50, DOI 10.1007/978-3-319-30933-0_24

Keywords Machine learning (ML) · Evolutionary classifier (EC) · Artificial immune recognition system (AIRS) · Software defect prediction (SDP)

1 Introduction

Software defect prediction (SDP) has been recognized as a significant research area in software quality engineering which can contribute to development of better quality software, within budget and time, as it leads to judicious utilization of scarce testing resources and time [1–5]. There has been significant and seminal empirical research in development of software metrics based on static code attributes for early prediction of defect-prone modules [1–3]. The software metrics based on static code attributes such as complexity, cohesion, size and coupling are termed as static code metrics (SCM). Recently, some researchers have introduced a new concept that software developers' coding habits and the way they handle coding tasks, greatly affect the number of defects introduced into a software module [6]. This is termed as developer's interactions with source code or micro-interactions with the source code [6]. Software metrics which measure developer's interactions with source code are termed as Micro-Interaction metrics (MIM) [6]. Another conspicuous research direction in software defect prediction (SDP) demonstrates that the selection of a learning method is highly important in predicting software defects [2]. Menzies et al. [2] established a baseline experiment by utilizing rich developments in machine learning and data mining to demonstrate that the selection of a learning method greatly affect the accuracy of a software defect prediction (SDP) model. Many researchers have explored Machine Learning (ML) methods in SCM based SDP [7]. Further, some researchers have also explored evolutionary computation (EC) [8] and artificial immune recognition systems (AIRS) [9, 10] in static code metrics(SCM) based SDP [5, 9–12]. Recently, Shepperd et al. [7, 13], have shown that the use of machine learning techniques in SDP is affected by researcher bias. They recommend inter-group studies to eliminate researcher bias. In our opinion conceptual replication of machine learning models across different software metrics, data sets and benchmarking of models [4] may also be a solution to the problem of eliminating researcher bias. Most of the empirical studies investigate machine learning (ML), evolutionary computation (EC) and artificial immune recognition systems (AIRS) methods in static code metrics (SCM) based software defect prediction [1–5]. Lee et al. [6] have already shown that MIM based SDP is more accurate than SCM based defect SDP. However, they do not investigate ML, EC and AIS conjunctively in MIM based SDP [6]. To the best of our knowledge, at present no empirical study has systematically compared ML, EC and AIS in MIM based SDP. In this paper, we evaluate ten ML, ten EC and eight AIS methods in MIM based software defect prediction. We utilize three publicly available MIM based defect data sets for this purpose [6]. Our results show that, although there are no silver-bullet solutions to prediction and learning problems in software engineering and it is futile to search for a single learning system that would handle all the problems, but it is possible to select learning systems that frequently performs well across a variety of data sets and prediction problems [14]. Further, in software defect prediction, optimization of different accuracy measures is related to achievement of different goals in model-based reasoning [14]. If the goal is to maximize recall or probability of detection of defective files in mission critical software projects, then a classifier with higher recall should be selected for defect prediction, even if it has lower precision. If the goal is minimize testing of non-defective files, then a classifier with higher precision should be used. If maximization of overall quality of classification is the goal the a classifier with higher F-measure should be selected [14]. The importance of our work in this paper is that we present experimental results with respect to different goals of prediction. The rest of the paper is organized as follows: Section 2 presents related work on ML, EC and AIS methods used in this study and our contribution. Section 3 presents the empirical data and modeling procedure used in this study. Sect. 4, presents the experimental results and discussion. Section 5 presents conclusions.

2 Related Work on ML, EC and AIRS Classifiers in SDP

Machine Learning (ML) means—learning from data through algorithms that build models from training data examples. In software defect prediction, supervised machine learning is commonly employed, whereby some algorithm learns from data and maps example inputs to outputs. Most common supervised machine learning approaches in SCM based defect prediction are Random Forests, Bagging, Boosting, Decision trees, Naïve Bayes, Bayes Networks, logistic regression and rule based [2, 3, 5]. Because of availability of large number of supervised machine learning approaches and their applicability in software defect prediction [2], researchers have tried to benchmark and rank these approaches in software defect prediction [4]. Lessman et al. [4] found that Random Forests were top performing machine learning classifiers across 10 NASA software projects. NASA metrics data sets pertain to static code metrics(SCM) of software developed in C and C++. Recently, we have shown that Random Forests, logistic regression, Kstar, Naïve Bayes and Bayesnet are most accurate single learner based approaches across large number of Java based software SCM data sets [15] and Rotation Forests, Bagging and Boosting are useful ensemble approaches for the same [16].

Evolutionary computation (EC)—refers to the computing paradigm that draw insights from various natural and biological processes like Darwinian evolution, survival of fittest and swarm intelligence. Two popular basic Learning based Classifier systems (LCS) in this context are Michigan Learning classifier system (MLCS) and Pittsburgh Learning classifier system(PLCS) [17]. In a Michigan-style LCS there is only a single set of rules in a population and the best classifiers are selected within that set using evolutionary algorithms [17]. In a Pittsburgh-type LCS (PLCS) the population has many rule set and evolutionary algorithm recombines and reproduces the best of these rule sets [17]. Recently, Malhotra et al. [8]

Table 1 List of ML, EC and AIRS techniques

Name of classifier	Category	Name of classifier	Category	Name of classifier	Category
Random Forest [18]	ML	GFS-MAX-LOGITBOOST [17]	EC, genetic fuzzy rules	GANN [17]	EC, neural network
Rotation Forest [18]	ML	GFS-Adaboost [17]	EC, genetic fuzzy rules	HIDER [17]	Hierarchical decision rules
Bagging [18]	ML	GFS-SP [17]	EC, genetic fuzzy rules	DT-GA-C [17]	Decision tree-genetic Algorithm classifier-C
AdaBoost [18]	ML	XCS [17]	EC, MLCS	AIRS1 [19]	AIRS
BayesNet [18]	ML	UCS [17]	EC, MLCS	AIRS2 [19]	AIRS
Logistic [18]	ML	CPSO [17]	EC, MLCS	AIRSParallel [19]	AIRS
Kstar [18]	ML	REPSO [17]	EC, MLCS	Immunos1 [19, 20]	AIRS
REPTree [18]	ML	GAS-ADI [17]	EC, PLCS	Immunos2 [19]	AIRS
J48 [18]	ML	GAS-I [17]	EC, PLCS	Immunos99 [19]	AIRS
Naïve Bayes [18]	ML	MPLCS [17]	EC, PLCS	CLONALG [19]	AIRS
GFS-LOGITBOOST [17]	EC, genetic fuzzy rules	NNEP [17]	EC, neural network	CLONALG.CSCA [19]	AIRS

studied 15 evolutionary algorithms on five static code metrics (SCM) data sets and found that MPLCS was most accurate on two SCM data sets, NNEP was most accurate on another two datasets [8]. However, they do not compare the results with non-evolutionary machine learning (ML) and they do not perform goal based evaluation of various algorithms. Our contribution is that we perform goal based evaluation of algorithms with micro-interaction metrics as predictors.

Artificial immune systems (AIS)—are computing paradigms that draw insights from immune systems of vertebrates. After conducting extensive literature review we found that software defect prediction literature has some studies on Artificial immune recognition systems (AIRS) in SCM based defect and change prediction [5, 9–12]. Abaei and Selamat [5] found that immunos99 performs well among AIS classifiers when feature selection technique is applied, and AIRSParallel performs better without any feature selection techniques but Random Forests perform well on small and large data sets. There are also some studies involving AIRS in SCM based change prediction [11, 12]. To the best of our knowledge, after extensive literature search we found out that there are no studies that compare and analyze ML, EC and AIRS in micro-interaction metrics (MIM) based software defect prediction (SDP). Table 1 presents the list of ML, EC and AIS techniques used in this paper for MIM based defect prediction.

3 Empirical Data Modeling

In this study, to rank the relative importance of Machine learning, Evolutionary Computation and Artificial Immune Systems in Micro-interaction (MIM) metrics [6] based defect prediction and for comparison of ML, EC and AIS in MIM-SDP, a publicly available MIM metrics data set collected by Lee et al. [6] is used. This data set pertains to three sub-projects of an open source Java Project Eclipse. The subprojects are—Mylyn, etc. and Team. Eclipse is a Java based Integrated Development Environment. The data set consists of 1042 (985 non-defective and 57 defective) instances or files in etc., 1061 (909 non-defective and 152 defective) instances or files in MyLyn and 239 (154 non-defective and 85 defective) instances in Team sub-projects. Each instance consists of 56 attributes or values of Micro-interaction metrics at file and task level. MIM measure developers interactions with source code files such as effort spent on a given file, degree of developer's interest in a given file, time intervals between events on a given file, sequential event patterns in tasks, time spent on a task by a developer. The details of MIMs can be referred in detail in [6], however the partial list is provided in Table 2 for convenience of readers [6].

Experiments for data modeling were conducted using WEKA [18] for ML, AIRS WEKA plugin [21] and KEEL [17] for evolutionary classifiers. In consistence with empirical studies practices in software engineering, each of the data set was divided into ten bins and 10-cross validation procedure was repeated hundred times [6]. The MIM listed by Lee et al. in [6] were used as predictor variables. If a

Table 2 Partial list of micro-interaction metrics [6] or predictor variables

Name of MIM	Description
Recurrence-of-File-Selection [6]	Number of times a file is selected in a given task
Recurrence-of-File-Editing [6]	Number of times a file is edited in a given task
Recurrence-of-manipulation-Events [6]	Number of times a file is manipulated in a given task
Recurrence-of-propagation-events [6]	Number of events a file is propagated in a given task
Mean-Degree of-interest-in-a-File [6]	Average of Developer's interest in a file
Variance of-Degree-of-interest-in-a-File [6]	Variance of Developer's interest in a file
Average-Time-between-Edits [6]	Time passed between two edit events on a file
Average-Time-between-Selections [6]	Time passed between two Selection events on a file
Average-Time-between-selection and Edit [6]	Time passed between a selection and edit events on a file
Number-of-Distinct-selected-Files [6]	Number of distinct or non-matching files selected in a task
Number-of-Distinct-edited-Files [6]	Number of distinct or non-matching files edited in a task
Time-consumption [6]	Total time consumed on a task
Percent-non-Java-edit [6]	(Number of non java edit events/total edit events) * 100

Java file had one or more defects, it was labeled as defective, other-wise, it was labeled as non-defective. The defectiveness of a file was considered as the response variable. The accuracy evaluation measures used for evaluating ML, AIS and EC classifiers in MIM—SDP in this study are:-

Precision—It is the proportion of defective files correctly predicted out of the all files predicted as defective by a software defect prediction model. This measure is important where it is important to avoid presenting files to testers which are not defective. In SDP defective files are called positive instances and non-defective files are called negative instances.

$$\textbf{\textit{Precision}} = (\text{True Positive})/(\text{True Positive} + \text{False Positive}) \tag{1}$$

Recall—It is the proportion of defective files correctly predicted out of all the files that actually defective. Concisely, it is the probability that defective files are correctly classified. This measure is effective for defect prediction in risk averse mission critical software projects where it is important to maximize detection of defective files to avoid life threatening risks [14].

$$\textbf{\textit{Recall}} = (\text{True Positive})/(\text{True Positive} + \text{False Negative}) \tag{2}$$

F-measure. Gives the inverse of harmonic mean of precision and recall and its important where it is needed to evaluate the quality of a classifier and evaluate the composite performance of a SDP model with respect to Precision and Recall.

$$\boldsymbol{F\text{-}measure} = (2 * \text{Precision} * \text{Recall})/(\text{Precision} + \text{Recall}) \tag{3}$$

Any ML, AIS, EC classifier used to build a SDP model is valuable only if it performs better in F-measure than a dummy classifier [6]. A dummy classifier is one which makes a random prediction regarding a file being defective or non defective with probability 0.5. The formula for F-measure of a dummy classifier is

$$F\text{-measure dummy classifier} = (2 * (0.5 * \text{ percent\_defective\_files})/(0.5 + \text{ percent\_defective\_files})) \tag{3}$$

Thus, the dummy classifiers F-measure values for etc., Mylyn and team sub-project data sets are 0.1, 0.22 and 0.42.

4 Experimental Results

The initial investigation in this paper started with 10 ML, 8 AIRS and 15 EC algorithms, that is a total of 33 algorithms. The following observations were made:

(i) Evolutionary hierarchical rule learning algorithm HIDER labeled all files as non-defective on Mylyn subproject. Clonal Selection algorithm CLONALG, CLONALG.CSCA, Immunos2 and Particle swarm optimization algorithms (PSO) CPSO and REPSO also could not detect defective files and labeled all files as non-defective in all three sub-projects and were eliminated from further analysis.

(ii) Evolutionary fuzzy rule learning algorithms GFSAdaboost, GFSLogitboost, GFSMaxLogitboost, GFS-SP labeled all files as defective and could distinguish between defective and non-defective files were also removed from further analysis. GANN an evolutionary neural network was also removed for the same reason

(iii) XCS algorithm labeled several files in all three subprojects as unclassified and was also removed from further analysis.

Thus out of 33 algorithms only those 22 algorithms were retained for further comparison and analysis that could distinguish between defective and non-defective files. The experimental results in Table 3 and Figs. 1, 2, 3 and 4 show that all the retained 22 classifiers perform better than a dummy classifier. However, the results of F-measure, Recall and Precision are divergent across three sub-projects and it is difficult to find a single best classifier that performs the best on all accuracy measures across all data sets. Previous research has highlighted that there are no silver-bullet solutions to prediction and learning problems in software engineering

Table 3 Accuracy measures with ML, EC and AIRS classifiers

Dataset	Etc			Mylyn			Team		
Name of technique	F-measure	Recall	Precision	F-measure	Recall	Precision	F-measure	Recall	Precision
Random Forest	**0.62**	0.51	0.78	**0.32**	**0.25**	0.45	**0.80**	**0.75**	0.85
Rotation Forest	**0.70**	**0.56**	0.91	**0.28**	0.19	0.54	**0.75**	0.68	0.83
Bagging	**0.65**	0.51	0.91	0.18	0.11	0.57	**0.78**	0.67	0.92
AdaBoost	0.60	0.46	0.87	0.04	0.02	0.50	**0.79**	**0.71**	0.91
BayesNet	0.60	0.46	0.87	**0.33**	**0.41**	0.28	0.73	0.60	0.94
Logistic	**0.70**	**0.74**	0.67	**0.34**	0.24	0.56	0.74	**0.80**	0.69
Kstar	0.22	**0.74**	0.13	**0.27**	**0.59**	0.17	0.54	**0.80**	0.41
REPTree	**0.63**	**0.56**	0.71	0.18	0.11	0.50	0.73	0.64	0.86
J48	**0.63**	**0.56**	0.71	**0.31**	**0.29**	0.34	**0.75**	**0.74**	0.77
Naïve Bayes	0.29	**0.75**	0.18	**0.29**	**0.84**	0.17	**0.75**	0.66	0.86
AIRS1	0.43	0.37	0.51	0.23	**0.24**	0.22	0.69	**0.72**	0.67
AIRS2	0.52	0.39	0.79	0.19	0.14	0.26	0.68	0.58	0.82
AIRSParallel	0.53	0.49	0.58	0.25	**0.23**	0.27	0.60	0.61	0.60
Immunos1	0.12	**0.77**	0.07	**0.26**	**0.60**	0.17	0.48	**0.76**	0.35
Immunos99	0.14	**0.65**	0.08	0.25	**0.57**	0.16	0.45	0.64	0.35
DT-GA-C	**0.72**	**0.81**	0.64	**0.27**	0.21	0.36	0.73	**0.71**	0.75
GGA	0.53	0.40	0.79	0.20	0.12	0.56	**0.76**	0.67	0.88
MPLCS	**0.67**	**0.56**	0.82	0.06	0.03	0.38	**0.75**	0.68	0.83
NNEP	**0.68**	**0.60**	0.79	0.10	0.06	0.31	**0.75**	**0.71**	0.81
GAS-I	0.56	0.46	0.72	0.04	0.02	0.27	0.71	0.64	0.79
GAS-ADI	0.56	0.42	0.83	0.04	0.02	0.19	0.71	0.64	0.81
UCS	**0.62**	0.53	0.75	**0.31**	**0.24**	0.43	**0.82**	**0.74**	0.91

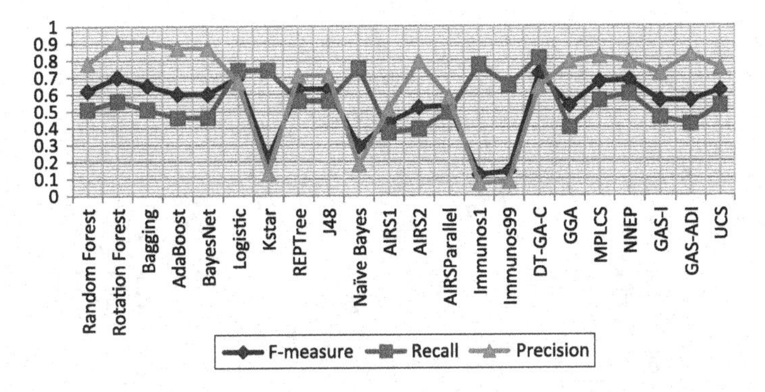

Fig. 1 F-measure, Recall and Precision of different classifiers on etc. sub-project

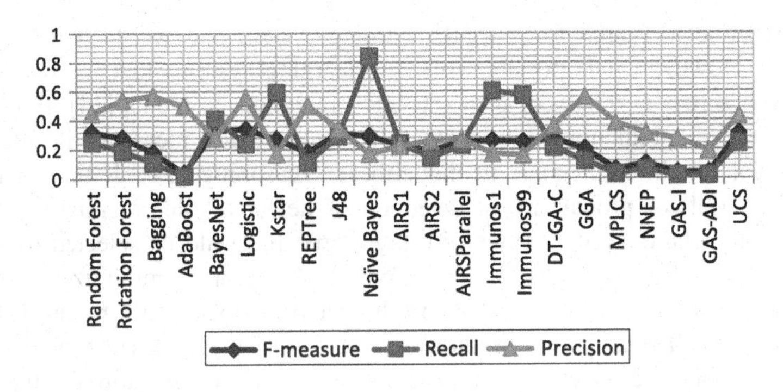

Fig. 2 F-measure, Recall and Precision of different classifiers on Mylyn sub-project

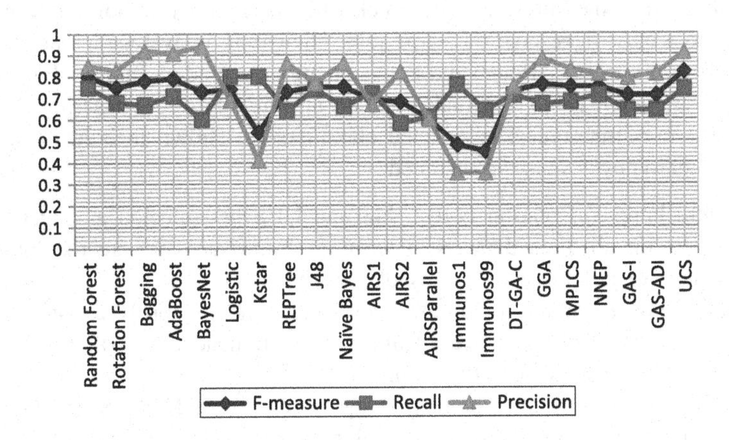

Fig. 3 F-measure, Recall and Precision of different classifiers on Team sub-project

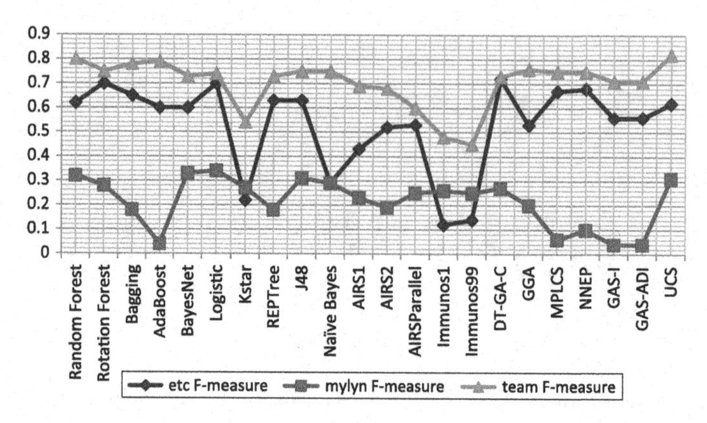

Fig. 4 F-measure comparison of classifiers on etc., Mylyn and Team subprojects

and it is futile to search for a single learning system that would handle all the problems. However, it is possible to select learning systems that frequently performs well across a variety of data sets and prediction problems [14]. Further, in software defect prediction, optimizing different accuracy measures is related to achievement of different goals in model-based reasoning [14]. If the goal is to maximize recall or probability of detection of defective files in mission critical software projects, then a classifier with higher recall should be selected for defect prediction, even if it has lower precision. If the goal is minimize testing of non-defective files, then a classifier with higher precision should be used. If balancing of overall quality of classification is the goal the a classifier with higher F-measure should be selected. In light of above discussion, we adopt the following strategy to recommend classifiers for different goals.

We select top 10 out of 22 classifiers on each sub-project's dataset by ranking them on accuracy measures. The top 10 classifiers are shown in bold in Table 3. If two or more classifiers have the same rank, their rank is averaged. The ranks of classifiers on F-measure, Recall and Precision are presented in Table 4. Further, we select a classifier if it has $1 \leq \text{rank} \leq 10$ on all three data sets in terms of a particular accuracy measure. The selected classifiers are shown in bold in Table 4. Thus the following results are obtained:

- Random Forests, Rotation Forest, J48 and UCS get selected amongst top 10 classifiers on all three datasets in terms of F-measure and are recommended classifiers for good overall quality of classification.
- Logistic, J48, Kstar and Immunos1 get selected amongst top 10 classifiers on all three datasets in terms of Recall and are recommended classifiers for defect prediction in risk averse software projects
- Random Forests, Rotation Forest, Bagging, Adaboost MPLCS and GGA get selected amongst top 10 classifiers on all three datasets in terms of Precision and are recommended classifiers for cost averse projects

Table 4 Ranks of ML, EC and AIRS classifiers on accuracy measures

Dataset	Etc			Mylyn			Team		
Name of technique	Rank on F-measure	Rank on Recall	Rank on Precision	Rank on F-measure	Rank on Recall	Rank on Precision	Rank on F-measure	Rank on Recall	Rank on Precision
Random Forest	**9.5**	13.5	**10**	**3**	7	**7**	**2**	4	**8**
Rotation Forest	**2.5**	9.5	**1.5**	7	13	**4**	**8**	11.5	**9.5**
Bagging	6	13.5	**1.5**	16.5	16.5	**1**	4	13.5	**2**
AdaBoost	11.5	17	**3.5**	21	21	**5.5**	3	9	**3.5**
BayesNet	11.5	17	3.5	2	5	13	13	21	1
Logistic	2.5	**4.5**	15	1	9	**2.5**	11	**1.5**	17
Kstar	20	**4.5**	19.5	8.5	**3**	20	20	**1.5**	20
REPTree	7.5	9.5	13.5	16.5	16.5	5.5	13	17.5	6.5
J48	**7.5**	**9.5**	13.5	**4.5**	**6**	11	**8**	**5.5**	15
Naïve Bayes	19	3	19.5	6	1	20	8	15	6.5
AIRS1	18	22	18	13	9	17	17	7	18
AIRS2	17	21	8	15	14	16	18	22	11
AIRSParallel	15.5	15	17	11.5	11	14.5	19	20	19
Immunos1	22	**2**	21	10	**2**	20	21	**3**	21.5
Immunos99	21	6	22	11.5	4	22	22	17.5	21.5
DT-GA-C	1	1	16	8.5	12	10	13	9	16
GGA	15.5	20	8	14	15	2.5	5	13.5	5
MPLCS	5	9.5	6	19	19	9	8	11.5	9.5
NNEP	4	7	8	18	18	12	8	9	12.5
GAS-I	13.5	17	12	21	21	14.5	15.5	17.5	14
GAS-ADI	13.5	19	5	21	21	18	15.5	17.5	12.5
UCS	**9.5**	12	11	**4.5**	9	8	**1**	5.5	**3.5**

5 Conclusion

In this study we evaluated ten Machine Leaning (ML), fifteen Evolutionary Computation (EC) and eight Artificial Immune recognition system (AIRS) classifiers to predict defective Java files of three sub-projects of Java project eclipse-etc., mylyn and team with micro-interaction metrics as predictors. Our experimental results indicate that there is no single best classifier that performs best in terms of all accuracy measures on all data sets. However it is possible to select suitable classifiers for different goals of software defect prediction. If a software company is doing a mission critical software project, then a classifier with higher recall should be selected for defect prediction, even if it has lower precision. Because the goal is to identify maximum number of defective files as even small number of defects remaining in post release software can be life threatening. Logistic, J48, Kstar and Immunos1 classifiers are recommended in this scenario. If a software company wants to minimize the testing of non-defective files and thereby reduce the cost of testing, then a classifier with higher precision should be selected for defect prediction. Random Forests, Rotation Forest, Bagging, Adaboost, MPLCS and GGA classifiers are recommended in this scenario. If balancing of overall quality of classification with respect to Recall and Precision is the goal, then a classifier with higher F-measure should be selected. Random Forests, Rotation Forest, J48 and UCS classifiers are recommended in this scenario.

References

1. Radjenovic, D., et al.: Software fault prediction metrics: a systematic literature review. Inf. Softw. Technol. **55**, 1397–1418 (2013)
2. Menzies, T., Greenwald, J., Frank, A.: Data mining static code attributes to learn defect predictors. IEEE Trans. Softw. Eng. **3**, 2–13 (2007)
3. Catal, C., Diri, B.: A systematic review of software fault prediction studies. Expert Syst. Appl. **36**(4), 7346–7354 (2009)
4. Lessmann, S., Baesens, B., Mues, C., Pietsch, S.: Benchmarking classification models for software defect prediction: a proposed framework and novel findings. IEEE Trans. Softw. Eng. **34**(4), 485–496 (2008)
5. Abaei, G., Selamat, A.: A survey on software fault detection based on different prediction approaches. Vietnam J. Comput. Sci. **1**, 79–95 (2014). doi:10.1007/s40595-013-0008-z
6. Lee, T., Nam, J., Han, D., Kim, S., In, H.: Micro interaction metrics for defect prediction. In: ESEC/FSE '11: Proceedings of the 19th ACM SIGSOFT Symposium and the 13th European Conference on Foundations of Software Engineering, pp. 311–321 (2011)
7. Shepperd, M., Bowes, D., Hall, T.: Researcher bias: the use of machine learning in software defect prediction. IEEE Trans. Softw. Eng. **40**(6), 603–616 (2014)
8. Malhotra, R., Pritam, N., Singh, Y.: On the applicability of evolutionary computation for software defect prediction. In: Proceedings of ICACCI, pp. 2249–2257 (2014)
9. Catal, C., Diri, B., Ozumut, B.: An artificial immune system approach for fault prediction in object-oriented software. In: Proceedings of Second International Conference on Dependability of Computer Systems, pp. 238–245 (2007)

10. Catal, C., Diri, B.: Software fault prediction with object-oriented metrics based artificial immune recognition system. In: Proceedings of PROFES, pp. 300–314 (2007)
11. Malhotra, R., Khanna, M.: Analyzing software change in open source projects using artificial immune system algorithms. In: Proceedings of ICACCI, pp. 2674–2680 (2014)
12. Malhotra, R., Gupta, V., Khanna, M.: Applicability of inter project validation for determination of change prone classes. Int. J. Comput. Appl. (0975–8887) **97**(8) (2014)
13. Hall, T.: A systematic literature review on fault prediction performance in software engineering. IEEE Trans. Softw. Eng. **38**(6), 1276–1304 (2012)
14. Menzies, T., Kocagunelli, E,. Minku, L., Peters, F., Turhan, B.: Sharing Data and Models in Software Engineering. Morgan Kauffmann, Los Altos (2015)
15. Kaur, A., Kaur, K.: An empirical study of robustness and stability of machine learning classifiers in software defect prediction. Adv. Intell. Inf. 383–397 (2014)
16. Kaur, A., Kaur, K.: Performance analysis of ensemble learning for predicting defects in open source software. In: Proceedings of ICACCI, pp. 219–225 (2014)
17. Alcalá-Fdez, J., et al.: KEEL: a software tool to assess evolutionary algorithms to data mining problems. Soft. Comput. **13**(3), 307–318 (2009). doi:10.1007/s00500-008-0323-y
18. Hall, M., Frank, E., Holmes, G., Pfahringer, B., Reutemannthe, P., Witten, I.H.: WEKA data mining software: an update. SIGKDD Explor. **11**(1), 10–18 (2009)
19. Brownlee, J.: Artificial immune recognition system: a review and analysis. Technical Report 1–02, Swinburne University of Technology (2005)
20. Brownlee, J.: Learning classifier systems. Technical Report, No. 070514A, Swinburne University of Technology (2005)
21. Weka Plugin for AIS algorithms in Java. Available online at http://wekaclassalgossourceforge.net

Soil Moisture Forecasting Using Ensembles of Classifiers

N. Rajathi and L.S. Jayashree

Abstract In the field of agriculture, accurate and timely forecast of soil moisture has great influence on crop growth and cultivation. The soil water status of an irrigated crop needs to be monitored regularly to make effective irrigation decisions. The challenge is to develop a feasible method to collect and examine large volume of soil moisture data on continuous base. The developments in wireless technologies have made practical deployment of reliable sensor nodes possible for various agricultural monitoring operations, which facilitate to meet the goal. The historical status of the soil moisture needs to be known advance in order to predict future readings. This work introduces a Soil Moisture Forecasting Ensemble Model (SMFEM) by combining the features of various machine learning approaches. The experimental results confirm that the prediction accuracy of the proposed approach is better when compared to the individual classifiers.

Keywords Ensembles of classifiers · Machine learning · Soil moisture wireless sensor network · Prediction accuracy

1 Introduction

Many scientific applications requires soil moisture, including land surface models, weather prediction models, water and energy balance models [1]. Agriculture is the main stay in India with most of the population depending on it. Due to increase in population, large quantity of water is preserved for drinking purposes. While lower

N. Rajathi (✉)
Department of Information Technology, Kumaraguru College of Technology, Coimbatore, India
e-mail: rajathi.n.it@kct.ac.in

L.S. Jayashree
Department of Computer Science and Engineering, RVS College of Engineering and Technology, Coimbatore, India
e-mail: jayashreecls@gmail.com

S.C. Satapathy and S. Das (eds.), *Proceedings of First International Conference on Information and Communication Technology for Intelligent Systems: Volume 1*, Smart Innovation, Systems and Technologies 50, DOI 10.1007/978-3-319-30933-0_25

quantity is recommended for irrigation. This increased demand on water supply will lead to drought problems in near future. Insufficient soil moisture content causes drought problem during the growing session of the crop. Monitoring of soil moisture variations is more important for efficient water management in areas where water resources are scarce. It is also helps to determine effective irrigation pattern in crop lands in order to decrease water usage for profitable crop production.

Recent evolution in wireless networking technology has lead to deployment of low cost reliable sensors. This make sensing is a viable solution [2, 3]. Usually soil moisture sensor nodes are deployed densely over a event of interest to produce real time and fine grained measurements at low cost. The purpose is to have the network function without human intervention for long time. A Wireless Sensor Network (WSN) consisting of thousands of multifunctional heterogonous sensor nodes with one or more sink nodes. These nodes are densely installed in the monitoring region to continuously observe physical phenomenon of interest. Wireless sensor networks with self organizing, self diagnosing, self configuring and self healing ability have been developed to solve problems that traditional technologies could not solve [4].

In case of data gathering applications each sensor node has to continuously collect local measurement of interest, such as pressure, temperature, humidity etc., and report the sampled data to the sink node. The sink node can estimate or reconstruct the phenomenon of interest in the sensing region by using the reported measurements. The sensor nodes in the wireless sensor network rely upon battery for their operation. Therefore minimizing energy consumption becomes a major problem for any data collection applications using resource constraint sensor networks. The sink node maintains the historical status of the data collected from the network over a period of time. The sink node predicts the future observations from the past history in order to decrease the volume of unnecessary communications in the sensor networks.

Machine learning methods have gained popularity compared to statistical methods [5–7]. It learns automatically from experience. Normally ANNs are adaptive in, and adapt to changing environment over time. Artificial Neural Networks (ANN) is one of the widely used promising machine learning approach for solving many types of non-linear problems that are hard to solve with the help of traditional techniques.

Normally ANNs are self-adaptive in nature and does not require any specific assumptions about the underlying model. With the help of ANNs, it is possible to mine the association between the inputs and outputs of a process [8]. Thus, these properties suits well to the problem of soil moisture prediction under consideration. Depending only on the predictive power of single method will not be realistic always. It was noted that the generalization ability of neural networks are improved by ensembling technique.

Combining the predictive power of appropriate methods will provide better results and is proved in large number of successful applications in diversified areas such as face recognition [9, 10], optical character recognition [11–13], analysis of scientific images [14], medical analysis [8], classification of seismic signals [15] etc. Hence, a two level ensemble model for soil moisture forecasting is proposed for water saving irrigation in agriculture.

The content of this paper is organized as follows. Overview of various ensemble approaches and related literatures is presented in Sect. 2. The proposed ensemble forecasting model is presented in detail in Sect. 3. In Sect. 4, experimental results are shown. Finally, Sect. 5 presents the concluding remarks.

2 Overview of Ensemble Learning and Related Works

In ensemble learning, finite set of learners are trained to solve the problem. This approach was introduced by Hansen and Salamon's [16]. Combining multiple learners has been intensively studied and widely recognized as a successful technique [11, 12] to resolve the limitations of traditional learning algorithms [6].

The learners or classifiers used in an ensemble are normally called as base learners. Normally, ensemble learning is s capable of boosting weak learners and allow us to create powerful learner from a set of base learners. Dietrich gave three reasons for the generalization ability of an ensemble is stronger when compared to a single learning algorithm [6]. The most important representatives of ensembling techniques are Bagging, Boosting, Stacking, Voting and Random Forest [17–19].

Combining the predictive power of appropriate methods will provide better results and is proved by researchers through large number of successful applications. Kim et al. developed SVM ensemble for classification of IRIS data, fraud detection and hand written digits [20]. They proved that the performance of SVM ensemble is better than a single SVM with respect to classification accuracy. A two level neural network ensemble for time series prediction was proposed and proved that the proposed approach provides better accuracy than single classifier models for Mackey glass time series data and sunspot data set [21]. Kaur et al. [22] and Maqsood et al. [23] describe a model that predicts various environmental values on hourly basis.

Hayati et al. [24] described short term temperature forecasting system using a three layer MLP network. A fully connected, 3 layer MLP network for temperature prediction is also presented by Santhosh Baboo and Kadar Shereef [25]. The training is done with the help of back propagation algorithm. Sergio makes use of evolutionary neural networks in combination with generic algorithms to predict the maximum temperature daily [26]. The works in [27] study the applicability of ANN approach to forecast daily maximum temperature for the year.

Weather forecasting models have been developed by Mathur et al. [28]. The performances of resulting ensembles are compared with a classical statistical method. In addition, a RBFN, which is is also applied [29]. Application of Time Delay Neural Network for modelling and forecasting of rainfall for the forthcoming month in the region of Tizi-Ouzou (Algeria) is investigated in [30]. The authors used two learning algorithms: the first order algorithm (recursive gradient with the constraint of shared weights) and the second order algorithm (Lavenberg-Marquardt algorithm). They proved that the Lavenberg-Marquardt algorithm is more accurate and produced better results for forecasting of rain fall for the coming month.

Different regression models were developed for long term electricity forecasting [31]. Neural network ensemble based electricity load forecasting was discussed in [32]. A soil moisture forecast system is designed using neural network based annealing algorithm and proved the outcome of the proposed one is superior when compared to single BPN and RBF networks [33]. Tabari et al. [34] developed network model for one day ahead forecast on soil temperature. Both Soil and air temperature were used as input to the model. All these existing literatures created an interest to apply ensemble classifier for the problem of study. Hence a two level neural network ensemble method has been proposed to forecast the soil moisture content more precisely.

3 Soil Moisture Forecasting Ensemble Model (SMFEM)

A wireless soil moisture sensor network deployed in the phenomenon of interest is useful to collect observations of surface level soil moisture content Each sensor node gathers enough samples and sends the time series data to a remote base station called sink node. The sink node receives and stores the time series data from all the sensor nodes in the network. A number of communication efficient routing algorithms have been proposed by researchers. They may be chain based, tree based or clustering based algorithms with single hop or multi hop communication to the sink node [35, 36]. Once collection of data gets completed, the base station executes the proposed ensemble model. The proposed study aims to predict the future communications more precisely in the sink node. The proposed SMFEM model and its algorithm is diagrammatically shown in Figs. 1 and 2. The model includes three types of individual learner's namely Radial Basis Function neural network (RBF), Multi Layer Perceptron (MLP) and Support Vector Machines (SVM).

The algorithm of the proposed approach is given below.

The various statistical metrics used to assess the performance of the model are Mean Absolute Error (MAE), Root Mean Square Error (RMSE) and Mean Absolute Percentage Error (MAPE) These are defined below in the Eqs. 1–3.

$$\mathrm{MAE} = \frac{\sum_{i=1}^{n} |x_i' - x_i|}{n} \tag{1}$$

$$\mathrm{RMSE} = \sqrt{\frac{\sum_{i=1}^{n} (x_i' - x_i)^2}{n}} \tag{2}$$

$$\mathrm{MAPE} = \frac{1}{n}\sum_{i=1}^{n} \left|\frac{(x_i' - x_i)}{xi}\right| * 100 \tag{3}$$

where

x_i represents the measured value

x_i' represents the predicted value

n represents the total number of measurements

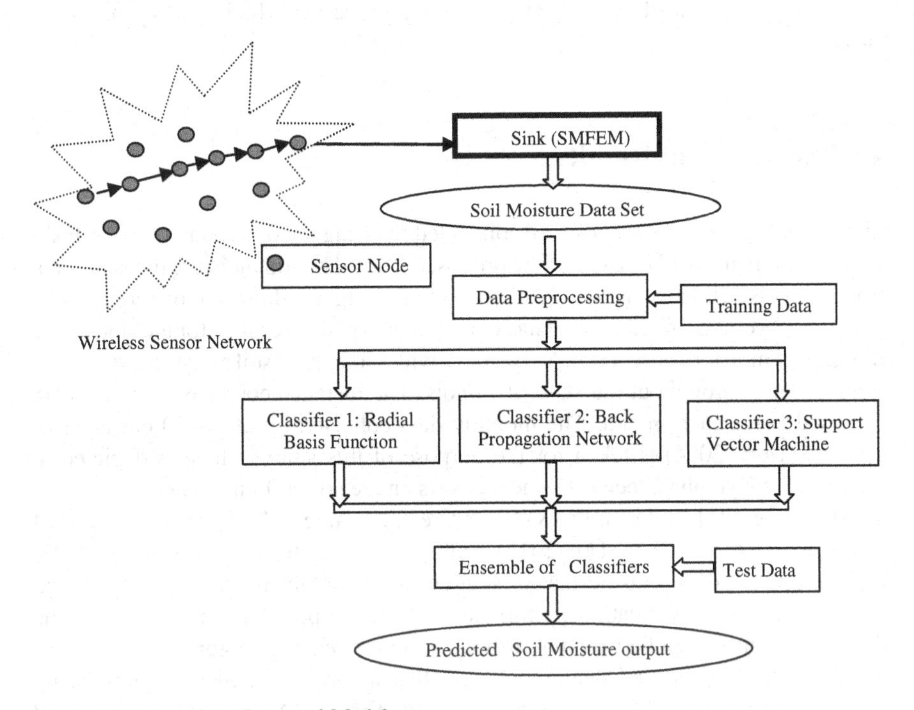

Fig. 1 Working of the Proposed Model

Input: Original data set D_d. Preprocessed Dataset D_d'. Set of heterogeneous level1 learning classifiers *Lf* ={RBF, MLP, SVM }, Level2 learning algorithm called *Ls*.

Step 1 : Start

Step 2 : Prepare the dataset for ensemble training and testing.

Training dataset - Dataset used for ensemble training (TR'). This dataset contains soil moisture data where TR'€D_d'.

Testing dataset-Dataset used for ensemble testing (TE'). This dataset contains soil moisture data where TE'€D_d'

Step 3 : for i = 1 to n // where n represents the various heterogeneous classifiers

hti = Lf_i(TR') % Train each individual level1 learner LF_i by using the % corresponding algorithm for learning and data set

end

Step 4 : Combine the best rules obtained in Step3 and apply to second level learning algorithm Ls.

Step 5 : Train the second level learning algorithm *Ls* by and applying TE'.

Step 6 : Output the predicted soil moisture value

Step 5 : Stop

Output: Predicted value of soil moisture

Fig. 2 Algorithm of the proposed approach

The objective function is to reduce MAE, which is the difference between the actual and predicted measurements. Testing were conducted using BPN, RBF, SVM, single level BPN ensemble, single level RBF ensemble, single level SVM ensemble and the proposed SMFEM model for Illinois and Kyemba soil moisture datasets. Single level ensembles such as RBF ensemble, MLP ensemble and SVM ensemble are obtained by applying bagging process to RBF, BPN and SVM classifiers.

4 Experimental Results

Simulation based experiments were conducted by using the open source tool WEKA 7.0 to measure the performance of the proposed work. The soil moisture measurement from Illinois climate network (climate.envsci.rutgers.edu/soil_moisture/Illinois.html) and Kyemba soil moisture data (www.oznet.org.au) are taken for the study. The Illinois soil moisture observation network provides long term soil moisture time series from 18 sites throughout the state of Illinois. The instantaneous measurements are collected twice each month. The monthly data collected by of 10–30 cm layer of station 1(1995–2004) is taken for the purpose of this study, which is depicted in Fig. 3a. The Kyemba Creek catchment covers an area of 600 km^2. There are 14 soil moisture monitoring sites in the Kyemba Creek catchment. Daily measurements of station K4 (01/03/2012 to 31/08/2012) is taken for the study and is shown in Fig. 3b. One third of the data set is normally used for testing and the remaining for training.

To claim the classification performance of the proposed ensemble model, the datasets taken are applied for the various machine learning algorithms. The performance of the proposed one among the other algorithms used was tested using Prediction Accuracy, Mean Absolute Percentage Error (MAE), Root Mean Squared Error (RMSE) and Mean Square Error (MSE).The results obtained for Illinois soil

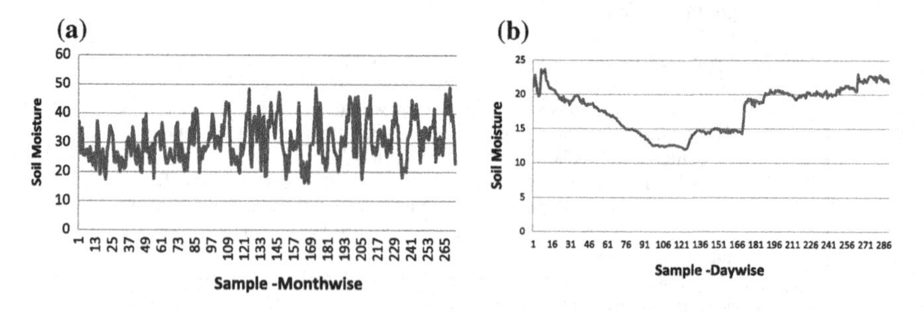

Fig. 3 **a** Illinois soil moisture time series, **b** Kyemba soil moisture time series

Table 1 Prediction accuracy of various classifiers

Classifier/result	Prediction accuracy (%)	
	Illinois data	Kyemba data
SVM	76.85	79.06977
MLP	73.21	70.93023
RBF	74.32	76.74419
RBF-Bagging	76.34	77.90698
MLP-Bagging	75.37	73.25581
SVM-Bagging	78.81	82.55814
SMFEM	83.52	89.53488

moisture time series dataset is given in Table 1 and its graphical representation is in Fig. 4a. Performance of the various classifiers with the proposed one for Kyemba soil moisture data is depicted in Fig. 4b.

The measurement prediction accuracy that is defined as the percentage of the perfectly predicted over the entire testing set of various classifiers with the proposed one is depicted in given in Table 1 and its graphical representation is depicted in Fig. 5a, b for both the data sets.

The results of the experiment have proved that the proposed ensemble model performs superior when compared to other single level ensemble classifiers.

The result of this study shows that the application of the ensemble neural networks through the proposed method can improve the prediction accuracy. This model would be helpful to agricultural practitioners, farmers and policy planners to build future situations. In addition, water saving irrigation schedule can be planned for farm lands.

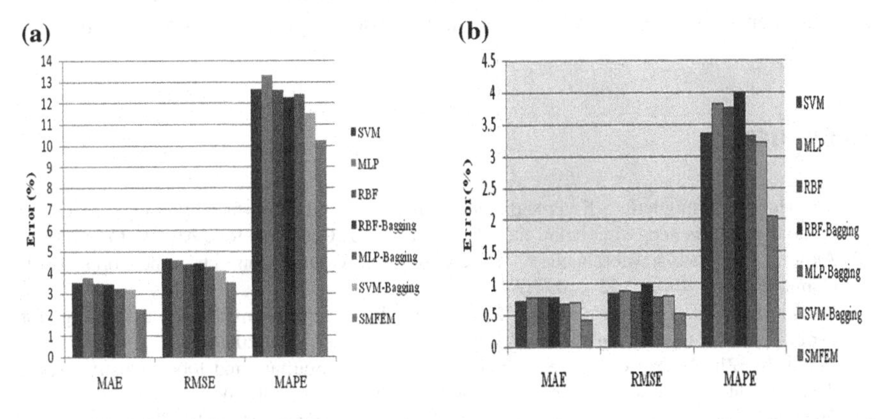

Fig. 4 **a** Performance of the classifiers with proposed model for Illinois Dataset, **b** performance of the various with proposed model for Kyemba Dataset

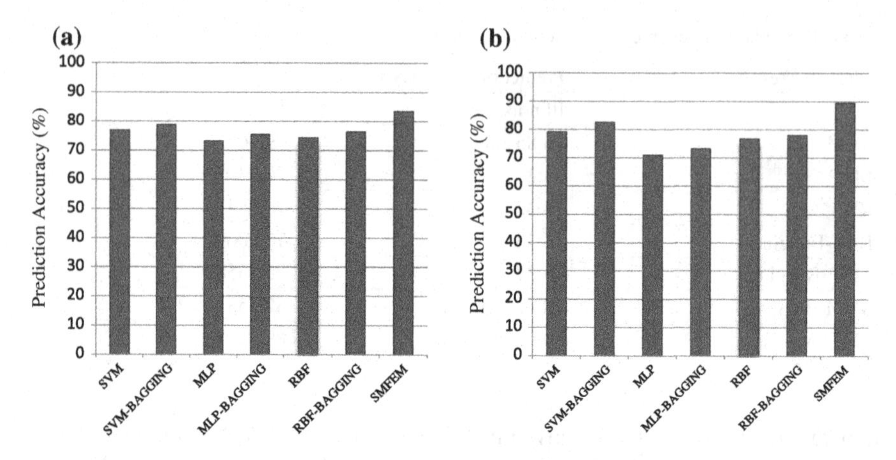

Fig. 5 **a** Prediction accuracy of different classifiers for Illinois soil moisture data, **b** prediction accuracy of different classifiers for Kyemba soil moisture data

5 Conclusion and Future Work

This paper proposes a soil moisture forecasting system for long term data gathering scenario. The ensemble model proposed combines the best rules of classifiers RBF, MLP and SVM. The experimental result shows that combining multiple models in an effective way, improves prediction performance of soil moisture content compared to individual learners. This system is helpful to the farmers by providing realistic information about the soil moisture statistics in future and the changes to be incorporated in the cropping pattern. In future, this work can be extended to the real deployment of soil moisture sensor nodes in the field of interest to collect the measurements and the performance of the proposed approach is to be investigated.

References

1. Min-Hui, L., Famiglietti, S.F.: Precipitation response to land subsurface hydrologic processes in atmospheric general circulation model simulations. J. Geo-phys Res. Atmos. **116** (2002)
2. Akildiz, I.F., Sankarasubramaniam, Y., Cayirci, E.: A survey on sensor networks. IEEE Commun. Mag. **40**(8), 102–114 (2002)
3. Ramanathan, N., Schoellhammer, T., Kohler, E.: Suelo: human-assisted sensing for exploratory soil monitoring studies. In: Sensys'09, pp. 197–210 (2009)
4. Wang, N., Zhang, N., Wang, M.: Wireless sensors in agriculture and food industry-Recent development and future perspective. Comput. Electron. Agric. (2006)
5. Freund, Y., Schapire, R.E.: A decision-theoretic generalization of on-line learning and an application to boosting. J. Comput. Syst. Sci. **55**(1), 119–139 (1997)
6. Dietterich, T.: Ensemble methods in machine learning. In: Proceedings of International Workshop on Multiple Classifier Systems, pp. 1–15 (2000)

7. Kuncheva, L.I., Whitaker, C.J.: Measures of diversity in classifier ensembles and their relationship ith the ensemble accuracy. Mach. Learn. **51**(2), 181–207 (2003)
8. Cunningham, P., Carney, J., Jacob, S.: Stability problems with artificial neural networks and the ensemble solution. Artif. Intell. Med. **20**(3), 217–225 (2000)
9. Gutta, S., Wechsler, H.: Face recognition using hybrid classifier systems. In: Proceedings of ICNN-96, Washington, DC, IEEE Computer Society Press, Los Alamitos, CA, pp. 1017–1022 (1996)
10. Huang, F.J., Zhou, Z.H., Zhang, H.J., Chen, T.H.: Pose invariant face recognition. In: Proceedings of 4th IEEE International Conference on Automatic Face and Gesture Recognition, Grenoble, France, IEEE Computer Society Press, Los Alamitos, CA, pp. 245–250 (2000)
11. Hansen, L.K., Liisberg, L., Salamon, P.: Ensemble methods for handwritten digit recognition. In: Proceedings of IEEE Workshop on Neural Networks for Signal Processing, pp. 333–342 (1992)
12. Drucker, H., Schapire, R., Simard, P.: Improving performance in neural networks using a boosting algorithm. Adv. Neural Inf. Process. Syst. (1993)
13. Mao, J.: A case study on bagging, boosting and basic ensembles of neural networks for OCR. In: Proceedings of IJCNN-98, vol. 3, Anchorage, AK, IEEE Computer Society Press, Los Alamitos, CA, pp. 1828–1833 (1988)
14. Cherkauer, K.J.: Human expert level performance on a scientific image analysis task by a system using combined artificial neural networks. In: Chan, P., Stolfo, S., Wolpert, D. (eds.) Proceedings of AAAI-96 Workshop on Integrating Multiple Learned Models for Improving and Scaling Machine Learning Algorithms, Portland, pp. 15–21 (1996)
15. Shimshoni, Y., Intrator, N.: Classification of seismic signals by integrating ensembles of neural networks. IEEE Trans. Signal Process. **46**(5), 1194–1201 (1998)
16. Hansen, L.K., Salamon, P.: Neural network ensembles. IEEE Trans. Pattern Anal. Mach. Intell. **12**(10), 993–1001 (1990)
17. Wolpert, D.H.: Stacked generalization. Neural Netw **5**(2), 241–259 (1992)
18. Drucker, H., Cortes, C., Jackel, L., LeCun, Y., Vapnik, V.: Boosting and other ensemble methods. Neural Comput. **6**, 1289–1301 (1994)
19. Breiman, L.: Bagging predictors. Mach. Learn. **24**(2), 123–140 (1996)
20. Kim, H.C., Pang, S., Je, H.M., Kim, D., Bang, S.Y.: Constructing support vector machine ensemble. Pattern Recogn. **36** (2003)
21. Chitra, A., Uma, S.: An ensemble model of multiple classifiers for time series prediction. Int. J. Comput. Theory Eng. **2**(3) (2010)
22. Kaur, A., Sharma, J.K., Agrawal, S.: Artificial neural networks in forecasting maximum and minimum relative humidity. Int. J. Comput. Sci. Netw. Secur. **11**(5), 197–199 (2011)
23. Maqsood, I., Khan, M.R., Abraham, A.: An ensemble of neural networks for weather forecasting. Neural Comput. Appl. **13**, 112–122 (2004)
24. Hayati, M., Mohebi, Z.: Application of artificial neural networks for temperature forecasting. World Acad. Sci. Eng. Technol. (2007)
25. Baboo, S., Kadar Shereef, S.: An efficient weather forecasting system using artificial neural network. Int. J. Environ. Sci. Dev. **1**(4), 321–326(2010)
26. Caltagirone, S.: Air temperature prediction using evolutionary artificial neural networks. Master's thesis, University of Portland College of Engineering, Portland, OR 97207, 12 (2001)
27. Maqsood, I., Abraham, A.: Weather analysis using ensemble of Connectionist learning paradigms. Appl. Soft Comput. (2007)
28. Mathur, S., Kumar, A., Chandra, M: A feature based neural network model for weather forecasting. World Acad. Sci. Eng. Technol. **34** (2007)
29. Park, J., Sandberg, I.W.: Universal approximation using radial basis function. Neural Comput. **3**(2), 246–257 (1991)
30. Benmahdjouba, K., Ameura, Z., Boulifaa, M.: Forecasting of rainfall using time delay neural network in Tizi-Ouzou (Algeria). Energy Proc. **36**, 1138–1146 (2013)

31. Bianco, V., Manca, O., Nardini, S.: Electricity consumption forecasting in Italy using linear regression models. Energy **34**, 1413–1421 (2009)
32. Jayashree, L.S.: Forecasting of electricity demands using ensemble of classifiers. Int. J. Appl. Eng. Res. (2015)
33. Liang, R., Ding, Y., Zhang, X., Zhang, W.: A real–time prediction system of soil moisture content using genetic neural network based on annealing algorithm. In: Proceedings of International Conference on Automation and Logistics, China (2008)
34. Tabari, H., Hosseinzadeh Talaee, P., Willems, P: Short term forecasting of soil temperature using artificial neural network. Metrolog. Appl. (2015)
35. Shamshirband, S., Maghsoudlou, H., Altameem, T.A., Gani, A.: A clustering model based on an evolutionary algorithm for better energy use in crop production. Stoch. Environ. Res. Risk. Assess. (2015)
36. Goyal, D., Tripathy, M.R.: Routing protocols in wireless sensor networks: A survey. Second International Conference on Advanced Computing and Communication Technologies, pp. 474–480 (2012)

Design of Smart and Intelligent Power Saving System for Indian Universities

Monika Lakra, Kappala Vinod Kiran and Suchismita Chinara

Abstract Now-a-days power management plays a vital role in reducing the consumption and efficient utilization of the resource. In traditional system, manual operation of electrical devices in university gets unnoticed, that leads to maximum wastage of power i.e., with different device operating even when the classroom is abandoned. These extended hours of operation leads to maximum power wastage. In order to overcome this problem, we have designed a "smart and intelligent power saving system for Indian Universities", where every classroom is equipped with passive infrared sensor (PIR) which responds to occupancy, and corresponding devices are switched ON/OFF automatically. The entire system is monitored and controlled by the central base station.

Keywords Wireless sensor network · Passive infrared sensor · ATMEGA

1 Introduction

The economy of any nation mostly depends on energy utilization in different fields of science and technology. Therefore, efficient energy utilization and consumption in every field is a burning research issue in the present day scenario. In a developing country like India, the gap between electricity supply and its requirement in terms of both capacity and energy has been steadily grown. The formidable challenge of meeting the energy needs and providing adequate and varied energy of desired quality to user in a sustainable manner and at reasonable cost, improving efficiency has become critical component of energy policy [1]. In India, almost all the places

M. Lakra · S. Chinara (✉)
Department of Computer Science and Engineering,
National Institute of Technology, Rourkela, India
e-mail: suchismita@nitrkl.ac.in

K.V. Kiran
Department of Electronics and Communication,
National Institute of Technology, Rourkela, India

S.C. Satapathy and S. Das (eds.), *Proceedings of First International Conference on Information and Communication Technology for Intelligent Systems: Volume 1*,
Smart Innovation, Systems and Technologies 50, DOI 10.1007/978-3-319-30933-0_26

adopt manual switching system for the electrical appliances like lights and fans, which often goes unattended in public places, class rooms, seminal halls etc. This results in considerable energy wastage as well as under-utilization of resources. Similar situations demand for a smart and automatic energy utilization system to reduce the gap. This paper, aims for the efficient energy utilization in a university campus. In most of the Indian schools/colleges/Universities, the class rooms, seminar halls, Auditoriums, toilets are left with lights and fans in switched ON mode even when there is no occupancy in the same. This situation has motivated for the design of a wireless sensor network based smart energy monitoring system, that considers the human occupancy in a particular room and controls the electrical devices like lights and fans automatically without any human intervention. The state of the electrical lights, fans and the sensors in various class rooms are monitored centrally through the base station.

2 Related Work

Electricity is wasted in an enormous amount in any university of India. The authors of [2] have tried to analyze the electricity profile of a University of the Aligarh Muslim University, India. This is also a major hub of research and innovations. The analysis says that 66 % of the total electric supply is being consumed by the offices, departments and street lights whereas rest 34 % is being consumed by the staff quarters, halls of residences and line losses. Similar situation exists in most of the universities in India. Almost all the Universities consume more electricity than their sanctioned limits while meeting the technological demands and facilities. The increased consumption also imposes high penalty on the Universities. Such an alarming situation demands for the smart utilization of the electricity and energy conservation. In the International scenario also the authors of [3] have made a study on the wastage of electricity in the classrooms of Shandong University. They have calculated that by saving 30 % of energy, an amount of 150 to 200 k Yuan of electricity fee will be saved. The authors have suggested to install sensors on every lamp of the class room, so that the light goes ON/OFF as per the occupancy of anybody in the sensor zone. However, deploying sensors for each lamp and coordinating the communication of several sensors is a tedious task. Recently various works have been done in designing the general overview of the remote access approaches to control devices in which switches of the different lamps and fans are controlled automatically according to the need and requirement, so as to avoid energy wastage, and save expenditure. Controlling the electricity consumption in the university according to the occupancy, as well as required light intensity in any particular area according to the need and requirement [3, 4], direction of movement as well as the speed of the movement of human being are monitored using the concept of dual sensors [5].

Various lights can be monitored and controlled all around using the client-server architecture [6]. Doppler sensor is used for entity detection and user interface that enable the remote monitoring and control of the street lights and run through the computerized unit located in the computer center consisting of the computerized unit that has TCP/IP connection.

3 Proposed System Model

We have proposed a smart energy monitoring system for a university campus by deploying a wireless sensor network around it. The sensors are planned to be deployed at the entrance doors of every class room, halls, office and toilets of the building. Similarly the sensors are planned to be deployed at the corridor lamps as well. The sensors at the entry of every room would detect the occupancy inside it and would turn ON/OFF the electrical devices like the lights, fans, and other switch points to run the overhead projectors, hand dryers etc. The block diagram of the proposed monitoring system is shown in Fig. 1. The sensor SN(1) … SN(n) are deployed at the different rooms entrance. The corridor are deployed with sensor CD (1) … CD(m) around the lamp.

Here, the sensors are used to detect the occupancy of any room and the corridor. An RTC provides a real time to the ATMEGA micro-controller. The LCD displays the current occupancy in any room and the corresponding devices are controlled using driver circuit. Corridor lights are controlled with regulator circuit optimizing the power utilization.

The PIR sensors installed at the entrance of the class room detects the presence of any person entering and leaving the class room giving the total number of occupancy in that room. Two sensors SN1 and SN2 are placed at the door of any room determining the direction of movement of the person as shown in the Fig. 2.

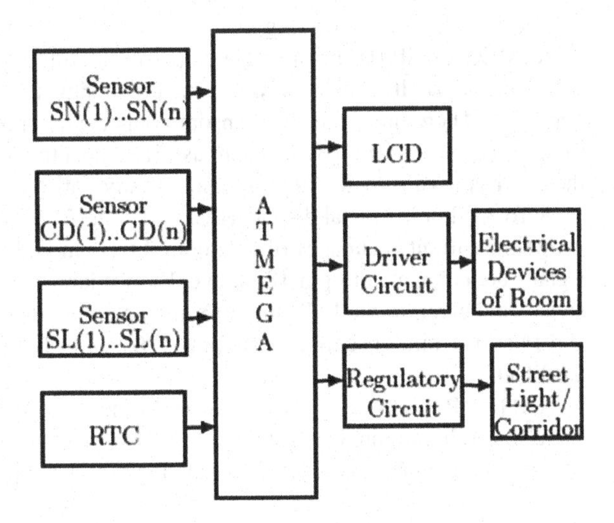

Fig. 1 Block diagram of the system

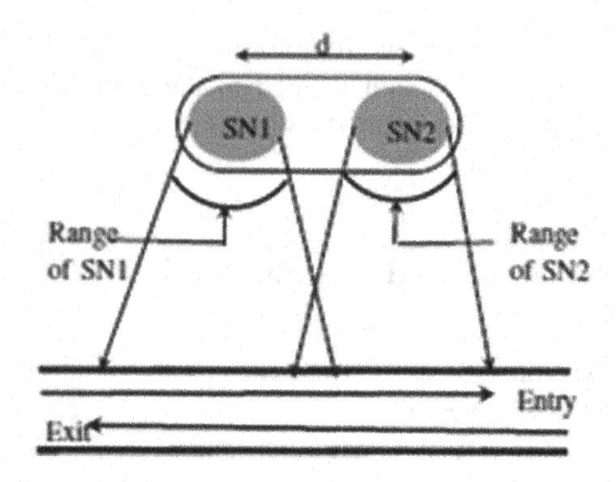

Fig. 2 Direction of student movement

For the entry of any person node SN1 detects first followed by node SN2, incrementing the occupancy by 1. Similarly, for an exit case, the node SN2 detects first followed by node SN1, decrementing the occupancy counter by 1. The proposed monitoring system is designed in such a way that, the electrical devices in the room get switched ON as soon as the occupancy counter gets a non-zero value, and they remain in the same state till the value of the occupancy counter becomes zero again. All the electrical devices are switched OFF as the counter becomes zero. A room having multiple doors can also be designed in the similar way. The data collected from all the sensor nodes are communicated to an ATMEGA microcontroller that controls the entire operation of the electrical devices.

4 System Hardware Description

The circuit diagram of our proposed model is shown in Fig. 3. It consists of ATMEGA microcontroller (ATMEGA328P), PIR sensor, LCD display, a RTC (RTCDS1302), and the driving circuit for controlling the loads of the classroom.

ATMEGA328P (U1) is a single chip with inbuilt flash memory of 32bytes and RAM of 2 KB. It is used as a central processing unit, which control the entire operation. There are three PIR sensors connected to pin 12, 15, 16 of microcontroller (U1) of which two sensors are used to detect the occupancy of classroom and the other is used to detect the presence of any entity in the corridor.

A 16 × 2 liquid crystal display (LCD1) is interfaced to microcontroller (U1) that shows the current occupancy of any room. LCD1 modules with pin 4, 5, 6 as control signals are connected to pin 18, 8, and 17 of micro-controller (U1). The command signals are given to the LCD to perform some predefined tasks and pin 11, 12, 13, and 14 as data pins are connected to pin 11, 6, 5, and 4 of U1 that displays the data on the LCD.

A RTC (RTCDS1302) contains a real-time clock/calendar and of 31 bytes of static RAM. It communicates with micro controller U1 via a simple serial interface, where serial data (SDA) at pin 5, serial clock (SCL) at pin 6 is connected to pin 27

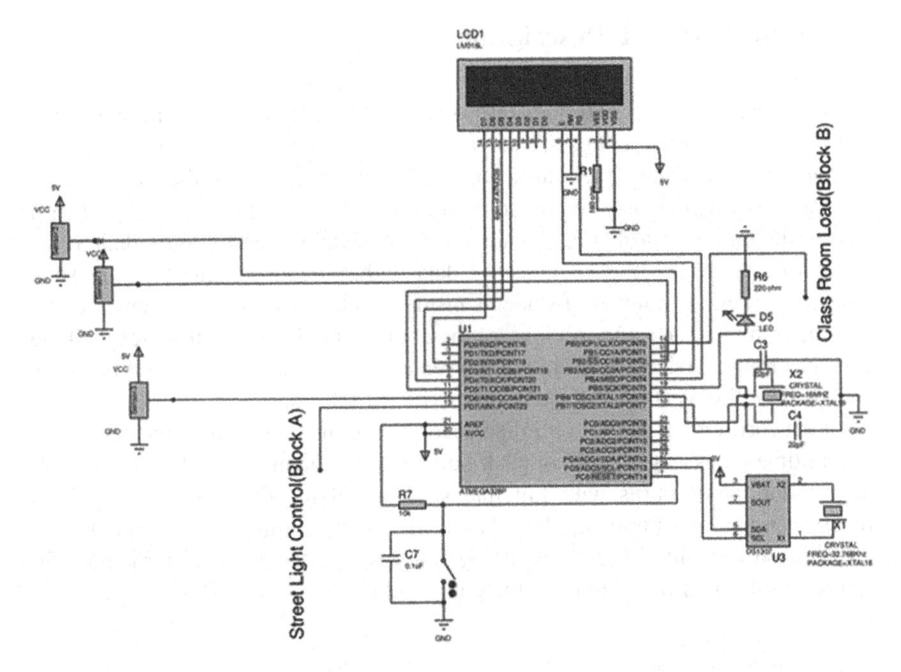

Fig. 3 Circuit diagram of the system model

and pin 28 of micro-controller. Data can be transferred to and from the RAM of 1 byte at a time or in a burst of up to 31 bytes. In addition, it has the features of dual power pins for primary and backup power supplies providing the updated time even if the circuit resumes from power failure.

The Driver Circuit is shown in Fig. 4. It is used to check the load in the class room, which is controlled by a 5 V relay (RL3) driven by a transistor where switching takes place according to the instruction given by the micro-controller (U1).

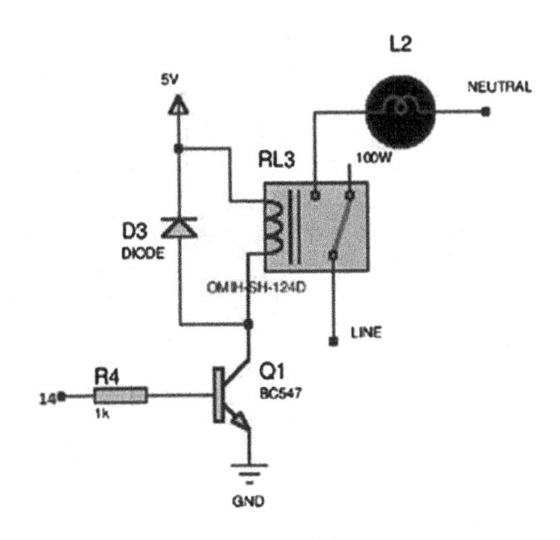

Fig. 4 Driver circuit (Block B)

5 System Software Description

The flow chart in Fig. 5 describes the working principle of controlling any room electronic devices such as lights, fans, and other points for dryer, projector etc. in different conditions and situations according to the occupancy in the rooms in the university. In which, if any sensor node from SN(1) … SN(n) gets the data then it sends to the micro controller, and waits for its consecutive sensor nodes data, if the consecutive sensor node also sends the data within the fixed time limit then the count value (i.e., the number of students inside the class room) gets incremented by one other remains the same. As long as the count value is greater than zero, all the of the electronic device of the class room remains switch ON, and once it becomes zero, then all the electronic devices gets switched OFF automatically.

All the corridor lamps are controlled according to the time. If time received from the real time clock is greater than peak hours and less than peak-off hours then all the corridor lamps burns with full intensity, and if the received time becomes greater then peak-off hours and less than peak hours then the corridor lamps burn with reduced intensity. Mean while if any entity is detected, the corridor lamps will burn with full intensity as long as the entity get clears.

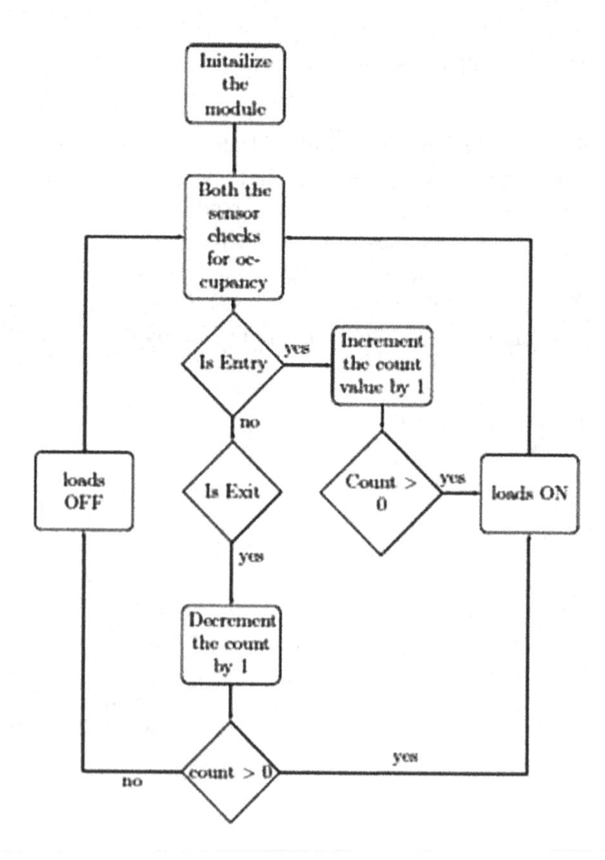

Fig. 5 Flow chart of smart class room

6 Observation

Figure 6 shows the different cases of the corridor burning under different condition. In case 1, corridor lamps are OFF. In case 2, the corridor lights are ON automatically burning with the full intensity i.e., during the peak hours, and in case 3, the corridor lamps are in minimum intensity level i.e., during the off-peak hours. Also, during the off-peak hours if any traffic gets detected then the street lights restores its full intensity until the traffic gets clear.

As calculated in Table 1 we can save up to 81,180 kWh energy per year from university room by controlling various electrical devices and nearly 15,552 kWh energy per year can be saved from street light/corridor of the university by reducing the intensity of the light during the off-peak hours.

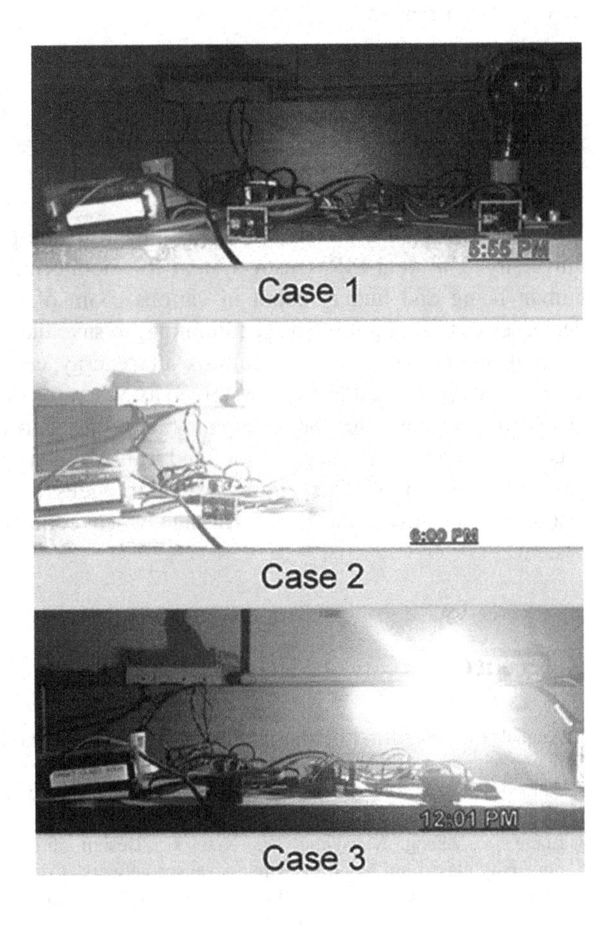

Fig. 6 Different cases for corridor lamp

Table 1 Case study of energy consumption in NIT Rourkela, India

Average time of lectures in any classroom per day	5 h
Average number of classes running simultaneously	60
Time for which the electricity supply remains ON in each room/hall	8 h
Average load per class	14.4 kWh
Average electricity loss per day per class (for 3 lectures per day)	6.15 kWh
Loss for 60 class per day	369 kWh
Loss per month	8118 kWh
Loss per year (220 working days)	81180 kWh
Total Street light/corridor	180
Average power consumed per day	324 kWh
Average power consumed per month	9720 kWh
Average loss per month	1296 kWh
Average loss per year	15552 kWh

7 Conclusion

This paper presents the hardware development of the power saving system in the university. The dual PIR sensor system for monitoring direction of movement of human being and human count in various room of university. The system also decreases corridor light intensity (dimming) to save the energy consumption during off-peak hours. This system minimize the energy consumption and provides an autonomous power control in the university where classroom devices are switched ON/OFF according to the occupancy and the corridor lamps are controlled according to real time, date, season, and traffic by varying the intensity during the peak/off peak hours. This entire system provides an efficient utilization of the resources and saves the power wastage when compared with the conventional system.

References

1. Annual Report 2013–14, Ministry of power, Government of India, Available at http://powermin.nic.in
2. Ahmad, F., Iqbal, S.: Reducing Electricity Consumption in Educational Institutes: A Case Study on Aligarh Muslim University's Electricity Usage Scenario
3. Liang, Y., Zhang, R., Wang, W., Xiao, C.: Design of energy saving lighting system in university classroom based on wireless sensor network. Commun. Netw. **5**, 117–120 (2013)
4. Radhakrishnan, A., Anand, V.: Design of an intelligent and efficient light control system. Int. J. Comput. Appl. Technol. Res. **2**, 117–120

5. Hung, P., Tahir, M., Farrell, R., McLoone, S., McCarthy, T.: Wireless sensor networks for activity monitoring using multi-sensor multi-modal node architecture 2009, IET
6. Kumaar, A.A., Sudarshan, T.S.B., et al.: Intelligent lighting system using wireless sensor networks. arXiv preprint arXiv:1101.0203 (2010)

Multiview Image Registration

Mehfuza Holia

Abstract Human brain mosaics the split images of a very large object which have been captured through eyes and each eye functions as a camera lens. But, it is not possible to cover very large area with the help of single eye than a pair of eyes. Similarly, multi view registration is essential because it may not be possible to capture a large object with a given camera in a single exposure. The field of view (FOV) of the commercial camera is much smaller than that of humans. Multiview image registration is an extremely challenging problem because of large degree of variability of the input data such as the images that are to be registered may contain visual information belonging to very different domains and can undergo many geometric distortions such as scaling, rotations, projective transformations, non rigid perturbations of the scene structure, temporal variations, and photometric changes due to different acquisition modalities and lighting conditions [1]. In proposed algorithm, transformation parameters are estimated using affine warp for corresponding matching points for the images to be registered. Here LM (Levenberg-Marquardt) optimization algorithm is used to optimize transformation matrix, which gives the minimum of a multivariate function that is specified as the summation of squares of nonlinear and real-valued functions, so some error can be tolerated in selection for control points. Final registered image is formed using backward mapping for sampling and distance based image blending. This proposed algorithm is compared with Euclidean warp algorithm and limitation of Euclidean warp is overcome and results are compared.

Keywords Image registration · Multi view · Image resampling · Image blending

M. Holia (✉)
BVM Engineering College, VV Nagar, Anand, India
e-mail: msholia@bvmengineering.ac.in

S.C. Satapathy and S. Das (eds.), *Proceedings of First International Conference on Information and Communication Technology for Intelligent Systems: Volume 1*, Smart Innovation, Systems and Technologies 50, DOI 10.1007/978-3-319-30933-0_27

1 Introduction

Image registration is the technique to determine a geometrical transformation which converts the various sets of data into one coordinate system, which is required for comparing or combining the data obtained from different measurement such as different view angle, different times, different imaging modalities for acquisition etc. In multiview image registration, images of the same scene or object are acquired from different viewpoints. The most traditional application of multi view registration is obtain large aerial and satellite photographs of high resolution. Multiview registration is process of generating a single integrated image by combining the visual information from multiple images. Here the aim is to attain high field of view without compromising the image quality. The problem of multiview image registration is mainly a combination of the process of estimating a transformation that maps the points from one image to the other and the process of combining the registered images into a composite image, in such a way that no seams are visible in mosaic, at the image boundaries. Image when joined leaves the some sort of distorted region behind, these regions must be dealt with. Some of the region obtained the pure black part, this part indicate either the part of the region which is not there in the image or the part that is distorted due to applied algorithms. All these things are eradicated by image blending. This feature is the most important in the implementation of registration for multi view images. The other important feature is to detect the overlapped region in the series of given images. After detection of overlap region synthesis is performed on the extracted overlapped region. The series of images may or may not be in the same alignment as per desired output image, thus mainly scaling, rotation and image transformation must be done to the series of images to get the proper aligned registered image.

2 Image Warping

Image warping is a mapping function which gives correspondance of all positions in one image to another image. It is a one kind of transformation that changes the spatial configuration of an image [2].

2.1 *Affine Warp*

The Affine warp is also called Affine transform involving four parameters P = [s θ t_x t_y s_1 s_2], which is scaling, rotation and translation(in x and y direction) and shearing (in x and y direction) respectively.

In Euclidean transformation, lines transform to lines, planes to planes, circles to circles. All that changes are the position and orientation, hence it preserve length and orientation. With affine transformations, lines transform to lines but circle become ellipses, affine transformation includes combined effect of translation, rotation, scaling and shear.

Here assume that an image $f(x', y', w')$ and due to geometric distortion produce an image g *(x,y,w)*. So it can be expressed as $(x', y', w') = T\{(x, y, w)$

Using homogeneous coordinates, we can describe the transformation matrix

$$\begin{bmatrix} x' \\ y' \\ w' \end{bmatrix} = \begin{bmatrix} m_0 & m_1 & m_2 \\ m_3 & m_4 & m_5 \\ m_6 & m_7 & m_8 \end{bmatrix} \begin{bmatrix} x \\ y \\ w \end{bmatrix} \quad (1)$$

Where

$$x' = \frac{m_0 x + m_1 y + m_2}{D} \quad y' = \frac{m_3 x + m_4 y + m_5}{D} \quad (2)$$

$$D = m_6 x + m_7 y + 1 \quad (3)$$

Here An image $f(x', y'')$, due to geometric distortion produce an image g *(x,y)*. then this transformation may be expressed as $(x', y') = T\{(x, y,)$. Where T is (x,y) rotated by θ, scaled by s and translated by t_x in x–direction, t_y in y direction, sheared by s_1 in x direction and s_2 in y direction.

$$\begin{bmatrix} x' \\ y' \\ 1 \end{bmatrix} = \begin{bmatrix} s\cos\theta & -ss_1 \sin\theta & t_x \\ ss_2 \sin\theta & s\cos\theta & t_y \\ 0 & 0 & 1 \end{bmatrix} \begin{bmatrix} x \\ y \\ 1 \end{bmatrix} \quad (4)$$

3 Levenberg-Marquardt (LM) Optimization

Direct search methods require many function evaluations to converge to the minimum point. Gradient based methods uses derivative information and are fast searching methods [9]. Different Gradient based methods are available in which Cauchy 's method works well when the initial point is far away from the minimum point and Newton's method works well when the initial point is near the minimum point. So to find optimization solution, it is not known whether the selected initial point is away from the minimum or close to the minimum, but wherever be the minimum point, a method can be devised to take advantage of both these methods. Marquardt method, follows Cauchy's method initially and after Newton's method is followed. The transition from Cauchy's method to Newton‘s method is adaptive and depends on the history of the obtained intermediate solutions [9, 10]. In the algorithm of registration, this optimization method is used to minimize the intensity difference in overlapped region of the images to be registered and according to that transformation matrix of Eq. (3) is updated.

Suppose I(x,y) and $I'(x', y')$ are images to be registered,for all corresponding pair of pixels (overlapping region) in I(x,y) and $I'(x', y')$,the sum of the squared intensity error is given by

$$E = \sum [I'(x', y') - I(x, y)]^2 \sum e^2 \tag{5}$$

To perform minimization, LM algorithm calculated the partial derivatives of e_i with respect to the unknown motion parameters $\{m_0 \ldots\ldots m_7\}$, also known as Jacobian matrix J, presents all first-order partial derivatives of a vector with respect to another vector.

$$J = \begin{bmatrix} \frac{\partial e_1}{\partial m_0} & \frac{\partial e_1}{\partial m_1} & \frac{\partial e_1}{\partial m_2} & \frac{\partial e_1}{\partial m_3} & \frac{\partial e_1}{\partial m_4} & \frac{\partial e_1}{\partial m_5} & \frac{\partial e_1}{\partial m_6} & \frac{\partial e_1}{\partial m_7} \\ \frac{\partial e_2}{\partial m_0} & \frac{\partial e_2}{\partial m_1} & \frac{\partial e_2}{\partial m_2} & \frac{\partial e_2}{\partial m_3} & \frac{\partial e_2}{\partial m_4} & \frac{\partial e_2}{\partial m_5} & \frac{\partial e_2}{\partial m_6} & \frac{\partial e_2}{\partial m_7} \\ \vdots & \vdots & \vdots & \vdots & \vdots & \vdots & \vdots & \vdots \\ \vdots & \vdots & \vdots & \vdots & \vdots & \vdots & \vdots & \vdots \\ \frac{\partial e_n}{\partial m_0} & \frac{\partial e_n}{\partial m_1} & \frac{\partial e_n}{\partial m_2} & \frac{\partial e_n}{\partial m_3} & \frac{\partial e_n}{\partial m_4} & \frac{\partial e_n}{\partial m_5} & \frac{\partial e_n}{\partial m_6} & \frac{\partial e_n}{\partial m_7} \end{bmatrix} \tag{6}$$

Where e_n is the intensity error (Eq. 5) at corresponding pairs of pixels *n* inside both images I(x,y) and $I'(x', y')$. So partial derivative of e_n with respect to m_k is given by

$$\frac{\partial e}{\partial m_k} = \frac{\partial I'}{\partial x'}\frac{\partial x'}{\partial m_k} + \frac{\partial I'}{\partial y'}\frac{\partial y'}{\partial m_k} \tag{7}$$

With respect to motion parameter m_0 to m_7, finding partial derivative of e

$$\frac{\partial e}{\partial m_0} = \frac{x}{D}\frac{\partial I'}{\partial x'} \tag{8}$$

$$\frac{\partial e}{\partial m_1} = \frac{y}{D}\frac{\partial I'}{\partial x'} \tag{9}$$

$$\frac{\partial e}{\partial m_2} = \frac{1}{D}\frac{\partial I'}{\partial x'} \tag{10}$$

$$\frac{\partial e}{\partial m_3} = \frac{x}{D}\frac{\partial I'}{\partial y'} \tag{11}$$

$$\frac{\partial e}{\partial m_4} = \frac{y}{D}\frac{\partial I'}{\partial y'} \tag{12}$$

$$\frac{\partial e}{\partial m_5} = \frac{1}{D}\frac{\partial I'}{\partial y'} \tag{13}$$

$$\frac{\partial e}{\partial m_6} = \frac{x}{D}\left(x'\frac{\partial I'}{\partial x'} + y'\frac{\partial I'}{\partial y'}\right) \tag{14}$$

$$\frac{\partial e}{\partial m_7} = \frac{y}{D}\left(x'\frac{\partial I'}{\partial x'} + y'\frac{\partial I'}{\partial y'}\right) \tag{15}$$

where D is the denominator in Eq. (2) and $\left(\frac{\partial I'}{\partial x'}, \frac{\partial I'}{\partial y'}\right)$ is the image intensity gradient of I' From these partial derivatives, the LM algorithm computes an approximate Hessian matrix A and the weighted gradient vector B. Hessian matrix is derivative of jacobian matrix (Eq. 6). Hessian matrix A is given by

$$A_{kl} = \sum \frac{\partial e}{\partial m_k}\frac{\partial e}{\partial m_l} \tag{16}$$

So

$$A = \begin{bmatrix} \sum_n \frac{\partial e_n^2}{\partial m_0^2} & \sum_n \frac{\partial e_n^2}{\partial m_0 \partial m_1} & \sum_n \frac{\partial e_n^2}{\partial m_0 \partial m_2} & \sum_n \frac{\partial e_n^2}{\partial m_0 \partial m_3} & \sum_n \frac{\partial e_n^2}{\partial m_0 \partial m_4} & \sum_n \frac{\partial e_n^2}{\partial m_0 \partial m_5} & \sum_n \frac{\partial e_n^2}{\partial m_0 \partial m_6} & \sum_n \frac{\partial e_n^2}{\partial m_0 \partial m_7} \\ \sum_n \frac{\partial e_n^2}{\partial m_0 \partial m_1} & \sum_n \frac{\partial e_n^2}{\partial m_1^2} & \sum_n \frac{\partial e_n^2}{\partial m_1 \partial m_2} & \sum_n \frac{\partial e_n^2}{\partial m_1 \partial m_3} & \sum_n \frac{\partial e_n^2}{\partial m_1 \partial m_4} & \sum_n \frac{\partial e_n^2}{\partial m_1 \partial m_5} & \sum_n \frac{\partial e_n^2}{\partial m_1 \partial m_6} & \sum_n \frac{\partial e_n^2}{\partial m_1 \partial m_7} \\ \vdots & \vdots & \vdots & \vdots & \vdots & \vdots & \vdots & \vdots \\ \sum_n \frac{\partial e_n^2}{\partial m_0 \partial m_7} & \sum_n \frac{\partial e_n^2}{\partial m_1 \partial m_7} & \sum_n \frac{\partial e_n^2}{\partial m_2 \partial m_7} & \sum_n \frac{\partial e_n^2}{\partial m_3 \partial m_7} & \sum_n \frac{\partial e_n^2}{\partial m_4 \partial m_7} & \sum_n \frac{\partial e_n^2}{\partial m_5 \partial m_7} & \sum_n \frac{\partial e_n^2}{\partial m_6 \partial m_7} & \sum_n \frac{\partial e_n^2}{\partial m_7 \partial m_7} \end{bmatrix} \tag{17}$$

And gradient vector B is given by

$$B_k = -\sum e \frac{\partial e}{\partial m_k} \tag{18}$$

So

$$B = -\left[\sum_n e_n \frac{\partial e_n}{\partial m_0} \quad \sum_n e_n \frac{\partial e_n}{\partial m_1} \quad \sum_n e_n \frac{\partial e_n}{\partial m_2} \quad \sum_n e_n \frac{\partial e_n}{\partial m_3} \quad \sum_v e_n \frac{\partial e_n}{\partial m_4} \quad \sum_n e_n \frac{\partial e_n}{\partial m_5} \quad \sum_n e_n \frac{\partial e_n}{\partial m_6} \quad \sum_n e_n \frac{\partial v_n}{\partial m_7}\right]' \tag{19}$$

Amount of update in motion parameter estimate **m** is given by,

$$\Delta m = (A + \lambda I_d)^{-1} B \tag{20}$$

Where λ is a stabilization parameter which is time varying.
Now solve the system by

$$(A + \lambda I_d)\Delta m = B \tag{21}$$

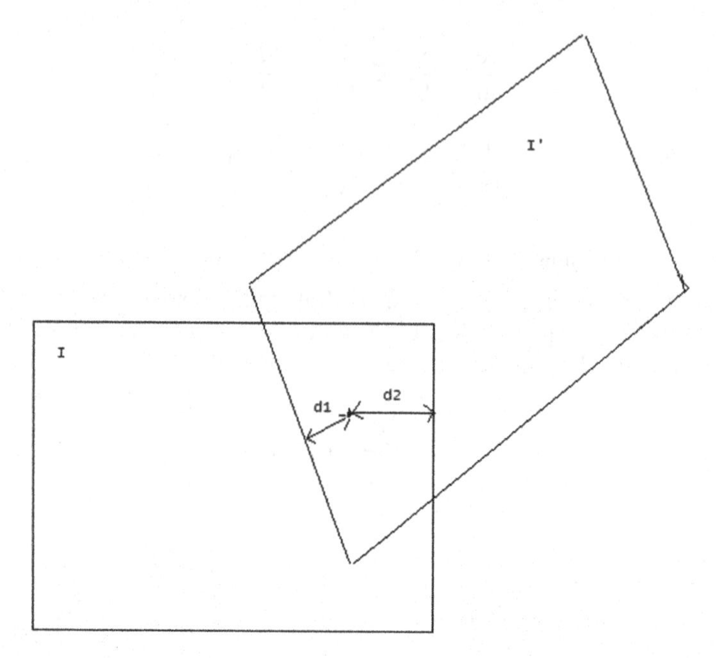

Fig. 1 Distance based image blending

The motion parameters estimation is given by

$$m(t+1) = m(t) + \Delta m \tag{22}$$

The error using Eq. (5) is checked, if it is incremented,then compute new Δm by changing the value of λ and Continue this iteration until the error is less than threshold or a given number of steps has been completed Fig. 1.

4 Image Blending Using Distance Weight Factor

Image blending is used for providing smooth transition between images and remove small residues of misalignments resulting from mis-registrations. In the overlapped portion the image blending procedure compute the contribution of the images to be registered at every pixel. Image blending method is advantageous to minimize the intensity variations and making the edges invisible so as a result visual quality of the registered image is improved. In our implementation, weight based blending is used, in which least distance from the pixel to the boundary of each image is used as shown in Fig. 2.

Fig. 2 3 X 3 region

I'(a-1,b-1)	I'(a-1,b)	I'(a-1,b+1)
I'(a,b-1)	I'(a,b)	I'(a,b+1)
I'(a+1,b-1)	I'(a+1,b)	I'(a+1,b+1)

The resultant image after blending consists of pixels,

$$N(x,y) = \alpha * I(x,y) + (I - \alpha) * I'(x',y') \tag{23}$$

Where,

I(x, y) and $I'(x', y')$ are image pixel in overlapping region

N(x,y) is the composite image pixel, α is the weight factor which is distance from image edge.

$$\alpha = \frac{d_2}{d_1 + d_2} \tag{24}$$

Thus blending function refines the overlapping region by using the least distance from the pixel to the boundary of each image.

5 Multiview Image Registration Using Affine Warp and LM Optimization

Inputs: Multiview Images (Two or more)
Output: Registered Image

1. Load Input images to be registered.
2. Define matching points in both images for correspondance.
3. Estimate transformation parameters (scaling, rotation, translation in x direction (tx), translation in y direction(ty),shearing in x direction and shearing in y direction) using Euclidean Warping and construct the transformation matrix using Eq. (4).
4. Apply Levenberg Marquardt algorithm to minimize the intensity difference in overlapped region of the images to be registered and according to that transformation matrix of Eq. (4) is updated.

LM algorithm along with sampling are stated below.

(i) For each pixel *i* at location (*xi*, *yi*), Calculate the corresponding position in the second image using Eq. (2).
(ii) Calculate the error in intensity between the images in overlapped region.

$$e = I'(x',y') - I'(x',y') \tag{25}$$

Compute the sum of the squared intensity error using Eq. (5).

Compute the intensity gradient $\frac{\partial I'}{\partial x'}, \frac{\partial I'}{\partial y'}$. The gradient of an image measures how it is changing. Bilinear interpolation is used when we need to know values at random position on a regular 2D grid as an image. In this approach the four nearest neighborhood is used to estimate the intensity at a given location [15].

Suppose (i,j) is center point in 3 X 3 region of an overlapping area of an image I′. Then intensity gradient in x direction is given by

$$\frac{\partial I'}{\partial x'} = \frac{(i_1 - i_2) + (i_3 - i_4)}{2} \tag{26}$$

Intensity gradient in y direction is given by

$$\frac{\partial I'}{\partial y'} = \frac{(i_3 - i_1) + (i_4 - i_2)}{2} \tag{27}$$

Where

$$\text{where}\begin{array}{ll} i_1 = I'(a-1, b+1) & i_3 = I'(a+1, b+1) \\ i_2 = I'(a-1, b-1) & i_4 = I'(a+1, b-1) \end{array}$$

(iv) Compute $\frac{\partial e}{\partial m_k}$ where k = 1 to 8 using Eqs. (8–15)
(v) Compute a Hessian matrix A using Eq. (17) and the weighted gradient vector B using Eq. (18)
(vi) Using Eq. (22), motion estimate is updated
(vii) Apply new transformation matrix to find correspondence between images to be registered Using Eq. (2)
(viii) If error in Eq. (5) has not reduced, then increment λ, and a new Δm is calculated.
(ix) Iterations will continue until the error is less than a threshold or a given number of steps have been completed.

5. Apply backward mapping as stated in Sect. 2 and blend two images (or more than two).
6. To reduce visible artifacts blend overlapped region by finding weight-distance parameter using Eq. (24) and find new intensity value using Eq. (23).
7. Show the result.
8. End

6 Results and Discussion

Proposed algorithm based on affine warp and LM optimization are implemented on different multiview images and results are compared with conventional multiview registration algorithm which uses euclidean warp (which includes only

Rotation-Scaling –Translation) without optimization for transformation matrix which is derived from selection points in input images Fig. 3.

Without LM optimization, seems are visible because of some minor mistakes in selection points results in misregistration. Here transformation matrix is used directly to find warp for reference image. So here matching points must be selected very carefully with great accuracy, any minor change in the selection of those points will lead to distortion in the output image that we can see from the above results and second limitation is that the reference points chosen are only static objects, those whose position remain stationary with respect to images to be registered. Euclidean warp takes only translation; rotation and scaling only, shearing of the correspondence points are not considered Fig. 4.

From Figs. 5, 6c show the results using LM optimization, Here after finding transformation matrix for correspondence, Levenberg-Marquardt optimization algorithm is applied in which transformation matrix using affine parameters is optimized to see the minimum error by using all nearest neighbour pixel, so here though minor changes are there in selection of reference points will be neglected, which is also shown in results. Input images in Fig. 6 includes Rotation-Scaling-Translation-shearing in all direction. Due to affine warp in transformation matrix all the affine parameters are obtained.

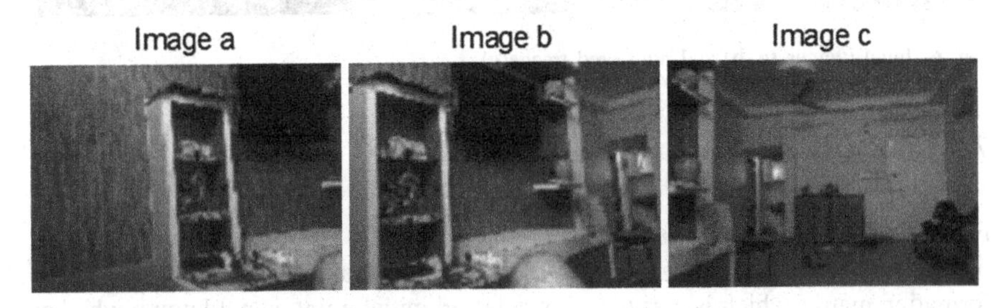

Fig. 3 Input images to be registered

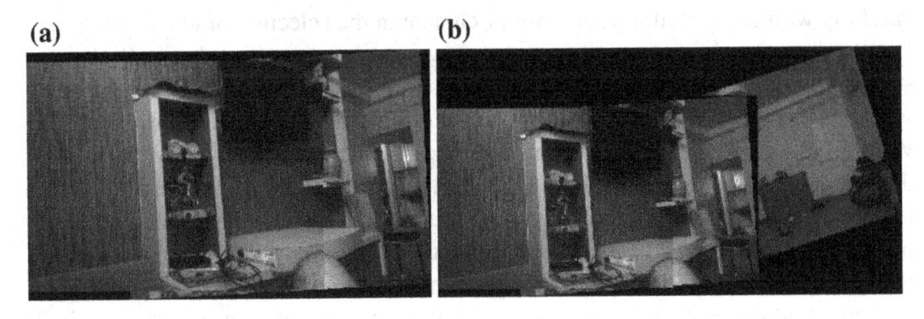

Fig. 4 Registration of two images (**a**) and three images (**b**) using Euclidean warp and without optimization

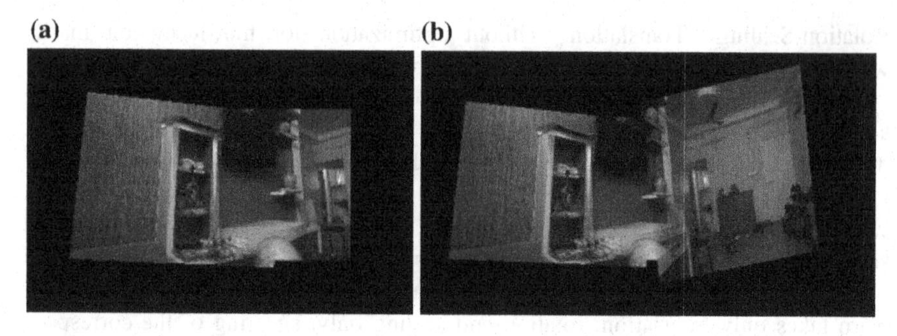

Fig. 5 Registration of two (**a**) and three (**b**) images using affine warp and LM optimization

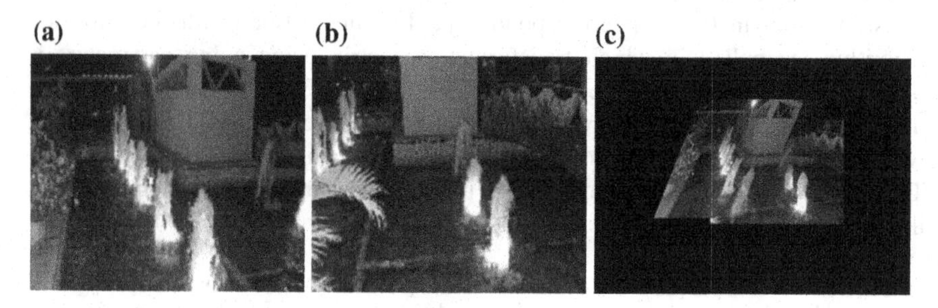

Fig. 6 Input images (**a**–**b**) and registered image(**c**)

7 Conclusion

From the above result, we can see that affine warp with LM optimization gives the best result. In euclidean warp based registration without optimization, the transformation matrix which is derived from selected initial point, is not known whether it is away or close to the minimum. This transformation matrix is used directly to find warp for reference image. So here matching points must be selected very carefully with great accuracy, any minor change in the selection of those points will lead to distortion in the output image that we can see from the above results and second limitation is that the reference points chosen are only static objects, those whose position remain stationary with respect to images to be registered. Euclidean warp takes only translation, rotation and scaling only, shearing of the correspondence points are not considered but in second algorithm using affine warp also takes the effect of shearing along with rotation, translation and scaling. Transformation matrix using affine parameters is optimized to see the minimum error by using all nearest neighbor pixel, so here though minor changes are there in selection of reference points will be neglected. Another major difference between both approaches is in computational efficiency. Euclidean warp is taking less time as directly formed transformation matrix is used (In proposed algorithm, it is taking average of 35 s, derived from different size images) but LM optimization algorithm is taking more time compared to previous(average of 70 s). Our experimental results suggest that second approach are competitive in quality with the best currently available technique but main limitation of this algorithm is, it is not automatic so we have to give control point selection and there must be some overlapping region in images to be registered.

References

1. Zitova, B., Flusser, J.: Image registration methods: a survey. J. Image vis. comput. **21,** 977–1000
2. Glasbey, C.A., Mardia, K.V.: A review of image-warping methods. J. Appl. Stat. **25**(2), 155–172 (1998)
3. Cyganek, B., Siebert, J.P.: An Introduction to 3D Computer Vision Techniques and Algorithms. 2nd edn. John Wiley & Sons Ltd. (2007) (Top of Form Bottom of Form)
4. Fischler, A., Bolles, R.C.: Random sample consensus: a paradigm for model fitting with apphcatlons to image analysis and automated cartography. Martin SRI Int. **24**(6), 381–396 (1981)
5. Cho, S., Chung, Y., Lee, J.: Automatic image mosaic system using image feature detection and taylor series. In: Proceeding VII th Digital Image Computing Techniques and Applications, pp. 549–557,10–12 Dec 2003
6. Li, N., Wu, L., Zhang, S.: An algorithm of fast corner match for image mo-saic. Int. J. Res. Rev. Soft Intell. Comput. (IJRRSIC) **1**(2), 17–24 (2011)
7. Ke, Y., Sukthankar, R.: PCA-SIFT: a more distinctive representation for local image descriptors. Technical Report at Intel Research Pittsburgh, pp. 1–12 (2003)
8. Wolberg, G.: Digital Image Warping. 1st edn. IEEE Computer Society Press (1998)
9. Deb, K.: Optimization for Engineering Design—Algorithms and Examples. 2nd edn. PHI Publication (2009)
10. Joshi, M., Moudgalya, K.M.: Optimization-Theory and Practice. 1st edn. Narosa Publishing House (2005)
11. Ranganathan, A.: Notes on the levenberg-marquardt algorithm (June 2004)
12. Szeliski, Richard: Video mosaics for virtual environments. IEEE Comput. Graphics Appl. **16**, 22–30 (1996)
13. Brown, M., Szeliski, R.: Multi-image matching using multi-scale oriented patches. In: IEEE Computer Society Conference on Computer Vision and Pattern Recognition, vol. I, pp. 510–517, June 2005
14. Manassah, J.T.: Elementary Mathematical and Computational Tools for Electrical and Computer Engineers. CRC Press, New York Washington, D.C (2008)
15. Venaz, P., Unser, M.: An efficient mutual information optimizer for multi resolution image registration. In: Proceeding of IEEE International Conference on Image Processing (ICIP-98), Chicago, pp. 833–837, 4–7 Oct 1998
16. Bentoutou, Y., Taleb, N., Kpalma, K., Ronsin, J.: An automatic image registration for applications in remote sensing. IEEE Trans. Geosci. remote sens. **43**(9), 3137–3137 (2005)
17. Yetik, I.S., Nehorai, A.: Performance bounds on image registration. IEEE Trans. Sig. Process. **54**(5), 1737–1750 (2006)
18. Pluim, J.P.W., Fitzpatrick, J.M.: Image registration. IEEE Trans. Med. Imaging **22**(11), 1341–1344 (2003)
19. Nelder, J.A., Mead, R.: A simplex method for function minimization. Comput. J. **7**, 308–313 (1965)

20. Lagariasy, J.C., Reeds, J.A., Wright, M.H., Wright, P.E.: Convergence properties of the Nelder-Mead simplex method in low dimensions. SIAM J optim. **9,** 112–1247 (1999)
21. Holia, M.S., Thakar, V.K.: Image registration for multi focus and multi modal images using windowed PCA. In: 2014 IEEE International Advance Computing Conference (IACC)

Part III
Applications for Intelligent Systems

Optimization Techniques for High Performance 9T SRAM Cell Design

Pramod Kumar Patel, M.M. Malik and Tarun Gupta

Abstract The new era begin to understand these beyond semiconductor CMOS devices, and circuit capabilities, to test and analysis the performance of CNTFET based structures compared to conventional silicon processes. These new devices can replace silicon in logic, analog, memory and data converters applications. The most close to the design of CNTFET based include the carbon-based options of graphene and carbon nanotube technologies, and also compound semiconductor-based Carbon nanotubes FET (CNTFETs). The study of high performance nine transistor static random access memory arrays and its optimization in 9-nm CNTFET technology are presented and comparative study done with the conventional six-transistors (6T) and previously issued eight-transistor (8T) Static RAM cell. The 9T CNTFET Static RAM cell provides same read speed comparatively 6T and 8T but has read data stability is enhanced by 1.56×. The proposed new memory cell consumes 53.40–19.17 % low leakage power comparatively the 6T and 8T Static RAM cell, respectively.

Keywords Optimization techniques · CNTFETS devices · High performance · High electron and hole mobility · Low leakage current · Static RAM cell · Static noise margin · Read and write voltage margin

P.K. Patel (✉) · M.M. Malik
Nano Science and Engineering Center, Maulana Azad National Institute of Technology, Bhopal, India
e-mail: pk_patel05@yahoo.com

M.M. Malik
e-mail: manzar.malik@gmail.com

T. Gupta
Electronics and Communication Engineering, Maulana Azad National Institute of Technology, Bhopal, India
e-mail: taruniet@rediffmail.com

S.C. Satapathy and S. Das (eds.), *Proceedings of First International Conference on Information and Communication Technology for Intelligent Systems: Volume 1*, Smart Innovation, Systems and Technologies 50, DOI 10.1007/978-3-319-30933-0_28

1 Introduction

Here a novel design technique proposed for digital circuits and system such as memory circuits implemented with 9 nm CNTFETs and compared with the conventional CMOS technology. As CMOS approaches physical and technological limits, new CNTFET based devices have been proposed. A CNTFET device has high mobility of charge carriers and subthreshold operation due to its gate geometry over silicon-based CMOS. The proposed high performance static random access memory (Static RAM) array have improved data stability, low leakage power dissipation and stronger write ability characteristics. The Fig. 1 shows the three-dimensional structure of CNFETs device.

First step to fabricate carbon nanotube transistors that are compatible with current CMOS processes. Further fabrication in is based on implanting the aligned carbon nanotubes transister from quartz substrates to Si/SiO2 substrates followed by electrode patterning. Novel chirality vector alignment techniques have also been developed to alleviate nonidealities in CNTFET fabrication, such as the presence of metallic CNTs and semiconducting CNTs; a stochastic modeling and analysis and a metallic-CNT tolerant SRAM architecture The CNTFET based memory circuits is available as another possible to the based on CMOS memory cell and digital circuit design. Since it is possible to achieve simplicity and highly power efficient with high performance in recently used memory cells design due to the minimization in circuit overhead such as reduce gate size of transistor and chip area.

2 9-nm Carbon Nanotube FET Technology

The design of Static RAM cell based on CNTFET devices implemented with 9 nm technology provides optimization for power and chip area. Therefore, CNTFETs based Static RAM cell design having significance of high integration density than its conventional CMOS counterpart. The designs of high performance memory circuits have the highest condition and ballistic transport phenomena make it very

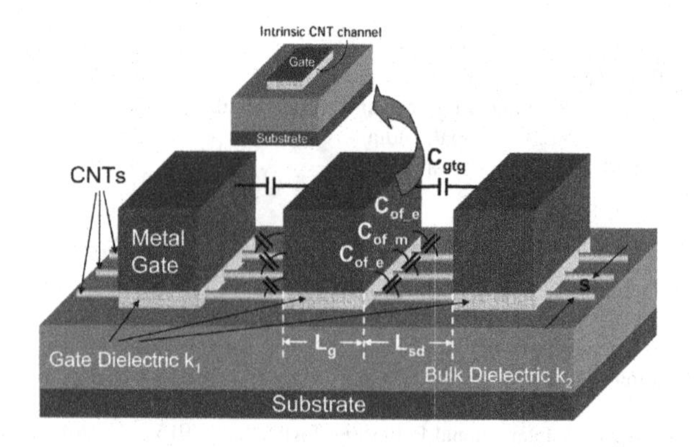

Fig. 1 The CNFETs device [1]

low ON/OFF current. The CNTFET a useful device for high speed operation and have high integration density of Static RAM cell.

HSPICE modeling and simulations accomplished using CNTFETs parameter by Stanford University's Nanoelectronics Group [1]. The gate channel lengths L_g of all the carbon nanotube based transistor that are assumed to be 9-nm(L_g = 9 nm).Here the drain power supply voltage (V_{dd}) is 0.6 V considered for the CNTFETs based Static RAM structure. The CNTFETs utilize semiconducting single-wall CNTs to assemble electronic devices and circuits. A single-wall carbon nanotube(or say SWCNT) consists of only one cylinder, and the SWCNT have simple to fabricate makes it very useful for alternative to new devices designing of MOSFET. The 3-D view of CNTFET device is shown in Fig. 1. The complete design parameters for 9-nm CNFETs are introduced in [1]. An single-wall CNT have unusual property that can act as alternative specified as a conductor or a semiconductor, depends on the angle of the atom arrangement along the tube referred to known as the chirality vector and can be represents as the integer pair (j,k). To determine if a CNT has the property of the metallic or semiconducting by considering the indexes (j,k), considered the nanotube is metallic if j = k or the index may be j−k = 3i, where i is an integer, moreover tube will behave like semiconducting. The diameter of the CNT can be determined as:

$$D_{cnt} = \frac{\sqrt{d_0}}{\pi}\sqrt{j^2 + k^2 + jk}$$

where d_0 = 0.142 nm represents the interatomic distance between each carbon atom and its neighbor. The similar carbon nanotube diameter and nano array pitch are considered 0.74–4.7 nm respectively, to accomplish high performance and high integration density.

3 The 9T CNTFET SRAM Cell

The high performance 9T CNTFET based technology Static RAM memory array is presented in this section to improve the quality of read data stability, strong write ability and low leakage power dissipation characteristic in Sect. 3.1. To find improved read data stability in the 9T SRAM cell in conventional silicon CMOS technology is review in Sect. 3.2.

3.1 Proposed High Performance 9T-CNTFET SRAM Cell with 9 nm

In the proposed paper presented a new technique 9T Static RAM cell in 9 nm CNTFETs technology have better efficacy for the read and write operations in the same time span. Here discuss on the optimization techniques for improved write and read operations employed in the 9T SRAM cell.

For the read and write enhancement techniques for the proposed Static RAM cell used in additional single NMOS and PMOS transistor are used to become SRAM cell OFF duration of write operation. A separate wordline (RWL) is used to control the gate of additional transistor. The additional transistors and RWLs increase the layout area, but there is no need have add these in every section.

To design the SRAM cell similar column are used for common rail power line in the all cells. The similar column will increase write power because there will exist of leakage current. For this purpose here considered the effect in power dissipation due to leakage current. Here one selected and other 32 unselected SRAM cells with similar column are use to design this architecture with write power to write 0 was executed. For this two cases were assumed.

(1) *Case 1: R*ight internal nodes voltage in unselected cells is 0 so there is a existence of sneaky current.
(2) *Case 2: R*ight internal node voltage in unselected cells is V_{DD} so, for this path there will not exist a leakage current.

For the proposed work structure 2, 3c (Fig. 2c) is selected for implementation of the CNTFETs SRAM. This structure has the state of half selection will be valid if each bit select by separately.

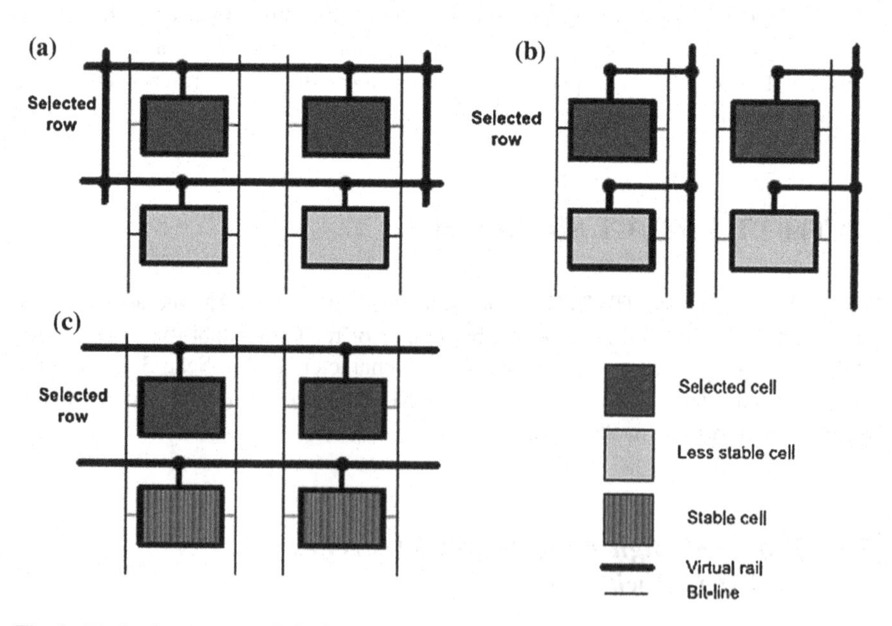

Fig. 2 The basic structures of sharing techniques for power and ground **a** Use of virtual rails for all cells **b** Common column and **c** Common row have separate virtual rail and all cells in similar column [6]

However, the leakage current for a selected Static RAM cell that is same to store a 0 on the right internal node; because in this case, in the read section two ON and one OFF transistors are used. The threshold voltage can expressed as

$$V_T = V_{TO} + \gamma(\sqrt{|-2\phi_F + V_{SB}|} - \sqrt{|-2\phi_F|} \quad (1)$$

and I_{Dsub} can expressed as

$$I_{Dsub} = I_s 10^{\frac{V_{GS} - V_T + \lambda V_{DS}}{S}}(1 - 10^{\eta V_{DS}}) \quad (2)$$

For the optimization of power and area in SRAM cells proposed a new technique for 9T CNTFET SRAM cell. Figure 3 shows a proposed optimized 9T SRAM cell in 9 nm technology. The proposed SRAM cells have the read path of this design at the left part of section. The structure has added a PMOS transistor M9 is ON for unselected cells so the voltage of node Qc will reach to V_{DD}. Increasing V_T will decrease the subthreshold leakage current I_{Dsub}. Due to decreasing leakage current by the increase in voltage of source of M6 and due to reduction of subthreshold current I_{Dsub} the V_{GS} also decreases. Add more number of cells sharing same bitline in this design comparatively others techniques, results a save in the chip area as well as power of circuits used for each column in SRAM cell design. Table 1 shows the optimization of high performance SRAM cells on the basis of write and read delay, static power dissipation, energy and Static Noise Margin(SNM) and layout area between 6T, 8T and 9T.

3.2 *Existing 9-CMOS Static RAM Cell*

As before alternative 9T Static RAM cell based on *Si* CMOS technology is presented and published in [2]. The schematic diagram of this existing 9-CMOS Static RAM array is draw in Fig. 4. The sub half circuit of 9-Si-MOSFET

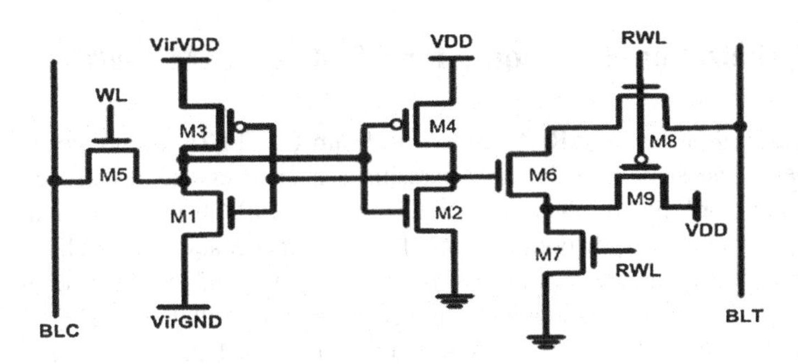

Fig. 3 Proposed high performance 9T CNTFET SRAM cell

Table 1 The optimization of SRAM Cells based on CMOS and CNTFET technology

Parameters	CMOS	CNTFET based			Quality metrics ratio[a]	
	6T	6T	8T	9T	6T	8T
Write delay(ps)	4.71	3.06	2.46	1.02	2.75	2.22
Read delay(ps)	28.43	29.23	8.96	3.94	0.97	1.78
Static power(nW)	4.58	0.185	0.11	0.04	24.7	41.63
Energy(fJ)	4.5	1.62	1.83	2.85	2.78	2.00
SNM(mV)	145.8	290.8	327.4	510.7	0.5	0.89
Area(μm^2)	0.438	0.015	0.02	0.024	2.89	2.89

[a]The quality metrics ratio = CMOS/CNTFET

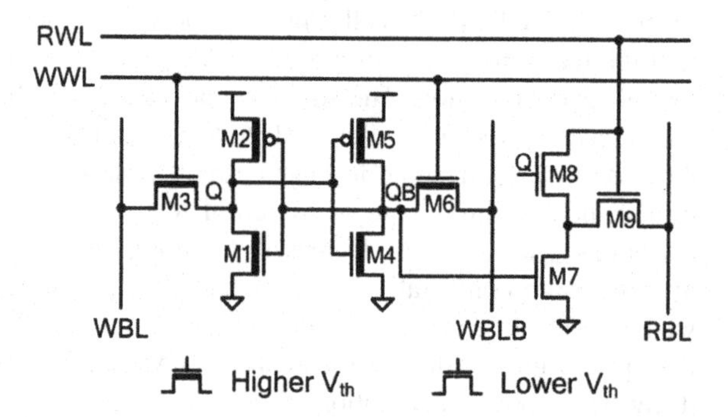

Fig. 4 The schematic diagram of this previously published 9-CMOS static RAM cell

SRAM array is implemented using a conventional 6T SRAM cell. The sub half circuit of 9-CMOS memory circuit is a read access point. To store and move the binary information into the existing 9T Static RAM cell [3] have big problem due to very high hole mobility pCNTFETs (*P*1 and *P*2) with a CNTFET based technology. The new Static RAM cell is introduced in Sect. 3.1 is consequently more suitable 9T Static RAM cell to accomplish stable write ability with improved read data stability in the structure of memory cells implementing CNTFETs based technology.

4 Optimization Techniques for 9T Static RAM Cells

The optimization for 9T SRAM cells with 9-nm CNTFETs technology are introduce here. The static and dynamic parasitic capacitances and resistances for the bitline and wordline are obtained with Clever [4]. HSPICE modeling and simulation [5] accomplished for optimization of the different Static RAM cells at 80 °C. A substrate bias voltage V_{SB} of 0.3 V is selected and a uniform diameter and pitch array of 0.83–5.3 nm, respectively are considered for all the CNTFETs based devices. The most favorable guidelines of Static RAM cells are considered in subparts 4.1.1–4.1.3.

4.1 Optimization for Proposed 9T Static RAM Cell

The optimization for the proposed 9T Static RAM cell is considered in Sect. 4.1 and shown in Fig. 3. The number of tubes of $N5$ and $N6$ are both 2. To accomplished high integration and stable write ability uses the number of tubes is one in the cross-coupled inverters of transistors $P1$, $P2$, $N1$, $N2$, and $N4$. The number of tubes from 1 to 16 is considered in this to achieve high performance. The minimum write voltage margin of the 9T Static RAM cell is consequently obtain while storing 0 onto Q. Write delay for the 9T Static RAM cell has the highest time duration for the half of transition from low to high of the write wordline at the event of Q_bar is discharged to $V_{DD}/2$ (to write 1 onto Q) or say Q_bar is charged to $V_{DD}/2$ (to write 0 onto Q). The design of the optimized 9T Static RAM cell is shown in Fig. 7 [7].

4.1.1 Active Power Dissipation

Figure 5 shows the write and read power which is dissipated by separate memory circuits. To measure the write power dissipated by the CNTFETs SRAM circuits for binary data received during every clock cycle. The rate of write power dissipation calculated for 3–10 f_s (Fig. 6).

For data stability or to save data in not selected memory 16- bit data, distinct write wordlines are make to use for different 16- bit data in the 9T memory arrays. This removes the waste of switching action on the write wordlines of not selected cells.

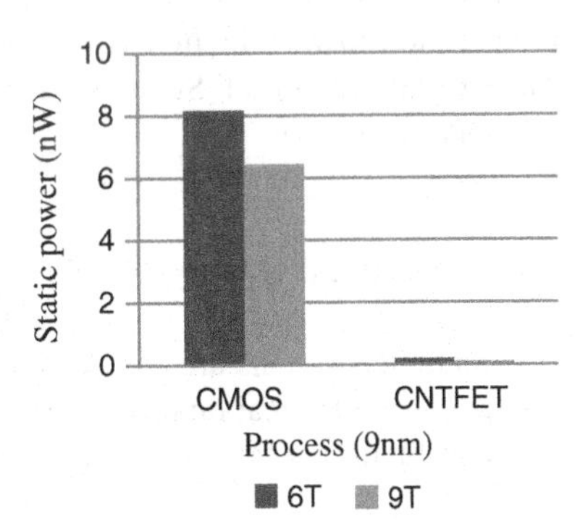

Fig. 5 Static power dissipation of SRAM cell

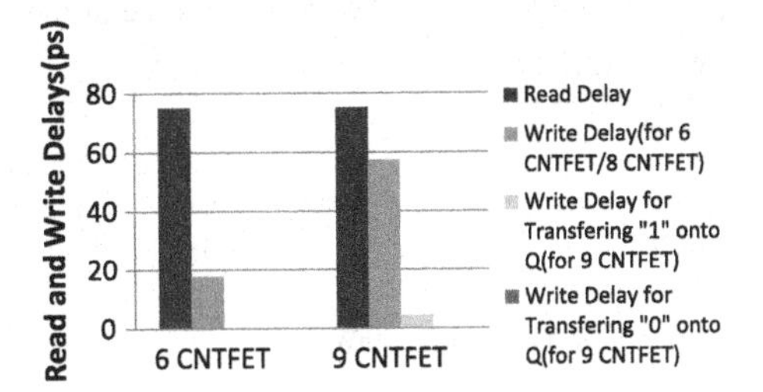

Fig. 6 Delays of SRAM cells using different technology, at T = 80 °C

Hence the write power dissipation of the 9T memory arrays is reduced by 38.52 %, respectively, as compared to the conventional 6T CMOS SRAM cell. The bitlines charged to V_{DD} in the conventional 6T CMOS memory cell. The write power dissipation of the 9T Static RAM cells is listed in Table 2 with the two different voltage case: First is write bitlines are precharged from 0 to V_{DD} for every f_{clk} (case I, as presented in Sect. 4.1.1) and write bitlines are write operation with the reverse type of incoming data (considered as case II). The write power dissipation of a memory cells with 6T and 9T SRAM cells comparative reduced with case II, as listed in Table 2. In case write bitlines is not occurring at interval precharged, the write power dissipation of the 9T memory cells are reduced by 72.70 %, as comparatively conventional 6T Static RAM cell. The read bitline is discharged with some condition that depends on the data stored in a 9T Static RAM cell. The condition that for a stable 0 on node *Q*_bar, the read bitline, the voltage is maintained at $V_{DD.}$ The next condition is when 1 is accumulating on node *Q*_bar, the read bitline is discharged in 9T Static RAM cells. As shown in Fig. 5 the switching activity of the read bitline cells read power dissipation is increased by 20.62 %, because of the higher bitline voltage change as comparatively to the 6T memory array [8].

4.1.2 The Chip Area

The chip areas optimizations for 9T Static RAM cells are discussed here and shown in Fig. 7. The layout design and area of memory cells is considered for 32 × 32-bit memory circuits. The chip area of new 9T Static RAM cell is increase by 53.1 % comparatively the 6T SRAM cells. For the 9T SRAM arrays, 33 nm more length is required to separate the first 16-bit line to from the second 16-bit line in every row. The overall area of 9T Static RAM cell is increase by 59.49 % comparatively 6T memory arrays [9].

4.1.3 Electrical Quality Metric

The Static RAM cells are categorized for different design parameters are shown in Table 2. An electrical quality metric (EQM) is calculated here for the comparative study of Static RAM cells is shown here. The QM can be expressed as

Table 2 Leakage power dissipation for the memory circuits

Write power dissipation of 6T SRAM cell(µW)	9 CNTFETs SRAM cell				6 CNTFETs	9 CNTFETs
	Case-I		Case-II			
	Power (µW)	Saving (%)	Power (µW)	Saving (%)		
16.20	13.04	38.52	10.31	72.70		
SRAM cell						
Leakage power dissipation (nW)					3.00	1.97

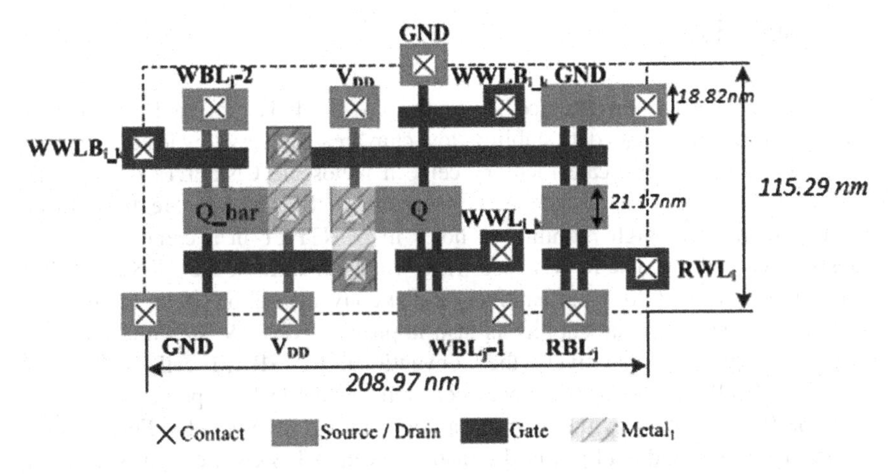

Fig. 7 The optimized layout of 9T static RAM cell. Dimensions are 115.29 nm × 208.97 nm = 24092.15 nm^2

$$QM = \frac{\text{Read\_SNM} \times \text{WVM}}{\text{Cell\_Max\_Delay} \times \text{Cell\_}P\text{leakage} \times \text{Cell\_Area}} \tag{3}$$

From the Eq. 3 the WVM define as the write voltage margin. Cell_Max_Delay is the highest read and write delays of the Static RAM cell including peripheral devices. Cell_*P*leakage is the minimum leakage power dissipation of the memory arrays including peripheral devices [10].

The 9T Static RAM circuits have leakage power dissipation measured with these two data storage cases: For storing 0 in all the cells and also for storing 1 in all the cells. Cell_*P*leakage can define as average leakage power dissipation of these two data storage cases. Cell_*P*read and Cell_*P*write shows the read and write power dissipation of the memory cells including peripheral devices. For the 9T Static RAM cell, the write bitlines are assumed to be periodically recharged to V_{DD} for every f_{clk}. The calculated write power dissipation (Table 2). The new 9T memory circuit have highest overall EQM due to the improved the quality of read data ability, more write voltage margin, and have lower leakage power dissipation parameters (comparatively 6T Static RAM arrays) the overall electrical quality is improved by 15.33× for proposed 9T SRAM circuit as compared to the 6T memory circuits. The new 9T CNTFETs based Static RAM cell have the most suitable technology which provides best overall electrical quality with improved stronger write ability, read data stability and very low leakage power dissipation properties with high performance in nanoscale [11].

5 Conclusion

In this paper a high performance 9T Static RAM cell is proposed which have stronger write ability, read data stability, low chip area and very low leakage current phenomena in new gigascale memory cells in nanoscale CNTFETs devices. To store and move the binary data into the proposed SRAM cell are have to be considered the very high mobility of holes in pCNTFET in a carbon nanotube transistor technology. The lowest case write voltage margin of the 9T Static RAM cell is improved by 3.90× comparatively to the conventional 6T. With the new 9T SRAM cell, the read data stability is also improved by 98.89 %, while have the same read speed comparatively the conventional 6T SRAM cell. A 1 K 9T Static RAM cell with the new memory cells reduces the leakage power dissipation by 34.68 % comparatively the memory arrays with 6T SRAM cells. Furthermore, the average writes and read power dissipations of the 9T SRAM circuit is decrease by 31.42 %, comparatively the conventional 6T Static RAM circuit. The new 9T SRAM circuit suffers from more worst-case write delay comparatively the 6T Static RAM circuits. The longest delay in write process not because a lowering the grade of clock frequency with the 9T Static RAM circuit but the write delay is shorter comparatively the read delay in other Static RAM arrays. The 9T Static RAM cells also have more 59.49 % area overheads comparatively the 6T memory cells. With the consideration of the different design parameters, the proposed 9T memory array improve the electrical quality metric with 15.33× comparatively the 6T memory circuits, respectively. Further leakage current can be reduce by using multithreshold CNTFETs technology to reduce the static power dissipation. The gate delay can be improved by reduction in contact resistance and metal-CNT contact and also the optimized device structure to minimize the tunneling current.

References

1. Stanford CNFET Model—HSPICE: [Online]. Available http://nano.stanford.edu/stanford-cnfet-model-hspice (2014)
2. Kureshi, A.K., Hasan, M.: Performance comparison of CNFET- and CMOS-based 6T SRAM cell in deep submicron. Microelectron. J. **40**(6), 979–982 (2009)
3. Hamzaoglu, F., et al.: Bit cell optimizations and circuit techniques for nanoscale SRAM design. IEEE Des. Test Comput. **28**(1), 22–31 (2011)
4. Clever User's Manual: [Online]. Available http://www.silvaco.com. (2014)
5. Wang, Y., et al.: A 1.1 GHz 12 μA/Mb-leakage SRAM design in 65 nm ultra-low-power CMOS with integrated leakage reduction for mobile applications. IEEE J. Solid-State Circ. **43** (1), 172–179 (2008)
6. Chang, L., et al.: An 8T-SRAM for variability tolerance and low-voltage operation in high-performance caches. IEEE J. Solid-State Circ. **43**(4), 956–963 (2008)
7. Sun, Y., Kursun, V.: Carbon nanotubes blowing new life into NP dynamic CMOS circuits. IEEE Trans. Circuits Syst. I, Reg. Papers **61**(2), 420–428 (2014)

8. Sun, Y., Kursun, V.: Uniform diameter and pitch co-design of 16 nm n-type carbon nanotube channel arrays for VLSI. In: Proceedings of IEEE 3rd Asia Symposium on Quality Electronic Design, pp. 211–216, July 2011
9. Sun, Y., Kursun, V.: Uniform carbon nanotube diameter and nanoarray pitch for VLSI of 16 nm P-channel MOSFETs. In: Proceedings IEEE 19th International Conference on VLSI System-on-Chip (VLSI-SoC), pp. 226–231, Oct 2011
10. Takeda, K., et al.: A read-static-noise-margin-free SRAM cell for low-VDD and high-speed applications. IEEE J. Solid-State Circ. **41**(1), 113–121 (2006)
11. Amlani, I., Lee, K.F., Deng, J., Wong, H.-S.P.: Measuring frequency response of a single-walled carbon nanotube common-source amplifier. IEEE Trans. Nanotechnol. **8**(2), 226–233 (2009)

Comparative Analysis of Data Gathering Protocols with Multiple Mobile Elements for Wireless Sensor Network

Bhat Geetalaxmi Jayram and D.V. Ashoka

Abstract WSNs has verity of applications as it has feature of collecting sensed data from different sources and redirect it to the sink. Through this paper authors have made an effort to find the best data gathering protocol among latest two protocols EEDG and IAR by comparing performance metrics of both in the perspective of improvising existing protocol in future. Both protocols performance is compared by taking statistics based on performance metrics. Based on this Authors have came up with the conclusion that EEDG is efficient than IAR protocol but at the same time this protocol has few drawbacks like Redundancy removal and Idle listening of sensor nodes issues are not taken care. At the same time Mobile Sink's does not contain feature like pause/wait state at node when node has not sensed any data. In future work authors have mentioned their focus towards minimizing mentioned drawbacks so that authors may come up with the new energy efficient protocol whose moto will be on improving life time of the WSN.

Keywords Aggregation · Consumption · Data gathering · Efficiency

1 Introduction

Wireless sensor networks has many applications in different fields as it is emerging technique [1]. WSNs are typically self organizing ad-hoc systems that consist of many small, low cost devices. Each node is equipped with sensor, microprocessors, memory, wireless transceivers, and battery. Once deployed, sensor nodes form a network through short-range wireless communication. They monitor the physical environment and subsequently gather and relay information to one or more sink

B.G. Jayram (✉)
The National Institute of Engineering, Mysuru, Karnataka, India
e-mail: g.nettar@rediffmail.com

D.V. Ashoka
JSS Academy of Technical Education, Bengaluru, Karnataka, India
e-mail: ashok_d_v@rediffmail.com

S.C. Satapathy and S. Das (eds.), *Proceedings of First International Conference on Information and Communication Technology for Intelligent Systems: Volume 1*, Smart Innovation, Systems and Technologies 50, DOI 10.1007/978-3-319-30933-0_29

nodes. The sensed data are relayed towards the sink in hop by hop manner. During data collection energy is consumed in WSN. This Energy consumption is the bottleneck of WSN. Reduction in energy consumption to increase the life time of network is a challenging issue. Consumption of energy of the entire network can be reduced by reducing number of transmissions. WSNs is more and more used for delay sensitive applications such as battle field monitoring. In these applications delay reduction between data generated and data processing becomes mandatory.

Most of the existing work on the data gathering techniques concentrates on energy, reliability and network lifetime for the efficient data gathering. Most of the mobile element based data gathering considers energy, data collection latency, Network life time, reliability, cost, load balancing etc, for sufficient data gathering.

Data gathering is one of the basic distributed data processing procedures in WSNs for conserving energy and reducing MAC contention [2]. This mechanism does network aggregation of data essential for energy-efficient data flow [3]. This protocols can reduce the communication cost and improves the lifetime of WSN. The redundant data sensed from the sensors can be eliminated by data gathering [4]. Data gathering increase data accuracy, reduces the number of redundant packets transmitted intern increasing data collection efficiency.

2 Related Work

Saad et al. [5] proposed a mobile sink with moving feature in large-scale networks. The mobile sink starts gathering data from a fixed depot, moves around the network, gathers data and then returns to depot. Cluster heads sends sensed data to mobile sink when the mobile sink comes near its transmission range. This approach avoids multi-hop relays for packets from all cluster heads to reach the sink. This reduces energy consumption due to multihop routing protocols.

Bi et al. [6] has presented a mobile sink based data-gathering scheme for WSN. In this scheme, authors have explained three key tasks of a data-gathering, like mobility of the sink, data collection along with intimation of position of the sink to the sensors. Using this scheme, authors have proposed an autonomous moving strategy due to sink mobility using which energy consumed by sensor nodes can be balanced by improving lifetime of the network.

Zhao et al. [7] discussed the data gathering issue in wireless sensor network using mobility and space-division multiple access (SDMA) technique to improve network performance. Authors have used mobile data collector by name SenCar which is deployed in sensor network. SenCar's function is to collect data from each sensor node in its transmission range without any relay this in turn improves the life time of the entire network.

Jayaraman et al. [8] have proposed the use of context aware mobile devices as it carriers data from sensor. The mobile nodes have enough spare capacity to form a distributed shared access network that can be used for sensor data collection and delivery. This is made possible with available powerful and context aware mobile devices that can work in pervasive environments like VPAN. This paper has proposed the architectural framework of the components on the mobile node that act as the data carrier.

Xiang et al. [9] proposed energy-efficient intra-cluster data gathering. This technique is designed for preserving the energy as much as possible and increasing life of the network. Here every cluster head works as local control center and will be substituted by the candidate cluster head only when its operation times reach to its optimum value. This will increase the energy consumption ratio of data collected to broadcased message. But this techniques has not consider the buffer overflow of the cluster head.

Puthal et al. [10] has proposed Mobile Sink Wireless Sensor Network (MSWSN) model which uses a mobile sink to gather the data. In this mobility model the sink moves in the network and covers the whole network area. But this technique has not addressed the visiting schedule of the mobile sink.

Apart from this authors have done extensive survey on the existing protocols and came up with the conclusion that in the existing data gathering techniques metrics like visiting schedule and the buffer overflow are not much considered [11]. To address these issues it is important to know which is the efficient protocol, based on which in future work authors can concentrate on eliminating the above issues. In connection with this here authors have made an attempt to prove the efficiency of the protocol by comparing latest energy efficient protocols. Here concepts are implemented which are common to both (EEDG and IAR) like deployment of mobile sinks, generating visiting schedule, collecting data from cluster member.

In the following section authors have compared two latest data gathering protocols, one is "Energy-Efficient Data Gathering with Tour Length-Constrained Mobile Elements in Wireless Sensor Networks (EEDG) [12]" and another is "An Intelligent Agent-based Routing Structure for Mobile Sinks in WSNs (IAR) [13]", to find out which is efficient protocol and what are the limitation of that protocol. Authors have decided to take the existing limitations as further research topic in the perspective of improvising the existing protocol efficiency.

3 Comparison Perspective of EEDG and IAR Protocols for WSN

This paper contains two data gathering techniques using multiple mobile sinks/elements meant for collecting data from sensor nodes. One is Mobile Element based Energy-Efficient Data Gathering with Tour Length-Constrained in WSNs (EEDG), in which mobile sinks uses vehicle routing protocol to collect data from sensor nodes. Another is an Intelligent Agent-based Routing Structure for Mobile Sinks in WSNs (IAR), in which mobile Agents uses prim's algorithm to collect data from sensor nodes.

3.1 Algorithmic Approach of EEDG [12]

The main objective of this algorithm is to improve life time of the WSN network by finding the most efficient Mobile Element (ME) Tour. Here authors propose a Cluster–Based (CB) algorithm that iteratively finds the ME tour. This protocol finds visiting schedule of the Mobile Element (ME) based on Vehicle Routing Protocol (VRP) with the objective to find Minimum Shortest Tour (MST). This MST determines the path with the shortest length using which ME can visits nodes based on the Earliest Deadline First (EDF) exactly once in each cluster. This algorithm is designed such that large number of clusters should satisfy tour length constraints. This algorithm start with the depot node. In every iteration this algorithm finds a next node in the cluster with EDF and added to the Tour which will be assigned to the ME as its visiting schedule. Summary of the algorithm is given in Table 1.

3.2 Algorithmic Approach of IAR [13]

This protocol is composed of Mobile Agent Based Data Gathering. To select the Mobile Agent, sink broadcasts Hello request message (HREQ). The nodes which receive a HREQ reply with a hello message (HELLO). The HELLO packet unicast to sink. Hello packet includes the sender's address and coordinates, so the sink node can easily determine which node is the closest node. This node is called Mobile Agent. This protocol finds visiting schedule of the Mobile Agent (MA) based on prims algorithm. Using prims algorithm this protocol finds the Minimum Spanning Tree (MST). This MST determines the path with the shortest length using which MA can visits nodes based on the Earliest Deadline First (EDF) exactly once in each cluster. Summary of the algorithm is given in Table 2.

Table 1 ME's visiting schedule

Input output	G (network topology graph), T (tour), c clusters, L (tour length constraint) T′ (tour to be assigned to the ME)
1	T < −T'
2	While T′ < L
3	Do T < −T'
4	Start with depot (d), add 'd' to tour
5	Find next node 'm' from cluster which is having Earliest Dead Line Fist (EDF)
6	Add 'm' to the tour
7	Go back to step 5 until all node has been added
8	Update edges, connect first and last node to form complete tour
9	Assign final tour to ME
10	ME follows the tour, collects data during visiting schedule and finally transmits collected data to sink

Table 2 Pseudo code for ME's visiting schedule

1	Agent selection by the sink
2	Start with depot as first node of visiting schedule
3	From the depot find minimum spanning tree based on the Earliest Dead Line First (EDF) using Prims algorithm
4	Assign minimum spanning tree to MA as visiting schedule
5	MA visits each node in cluster, based on visiting schedule, collects data and finally transmits collected data to sink

4 Simulation Experiments

4.1 Simulation Scenario Along with Metrics for EEDG and IAR

To evaluate performance of the Mobile Elements based protocols in WSN, authors have implemented EEDG and IAR protocols. To implement EEDG and IAR protocols, authors have used NS-2.32 All-in-One tool. Implementation is divided in two parts:

- Front end design using TCL scripting: Front end includes topology design and connection procedure using TCL scripting.
- Back end design using C++: Back end includes implementation of EEDG and IAR protocol.

Wireless sensor environment is created using flat grid topology over area 500 × 500 m. Energy model is included to obtain energy available in each scenario. For Performance Analysis, authors have taken 5 different scenarios i.e. for 20 nodes with 3MS and 1BS, 40 with 4MS and 1BS, 60 with 4MS and 1BS, 80 with 5MS and 1BS,100 with 5MS and 1BS where (BS) is Base Station and (MS) is Mobile Element/Agent. For each scenario authors have taken the reading for 5 performance metrics (Delay, Packet Drop, Packet Delivery Ratio, Energy Available, Control Overhead). Table 3 gives summary of the simulation parameters.

4.2 Simulation Results and Performance Analysis of EEDG and IAR

For performance Analysis purpose author have taken results based on different number of nodes. Analysis of EEDG and IAR is done based on following performance metrics:

- *Energy*: It is calculated by extracting energy field from trace file and taking its average.

Table 3 Summary of the simulation parameters

Parameters	Values
Number of nodes (N)	20, 40, 60, 80, 100
Simulation area (A)	500 × 500 m
Transmission range (R)	250 m
Simulation time	50 s
Traffic source	CBR
Packet size	512 byte
Rate	50, 100, 150, 200, 250 kb
Initial energy	20.1 J
Transmission power	0.660 W
Receiving power	0.395 W

- *Overhead*: It is calculated by taking ratio of sum of control information to the actual data received.
- *Delay*: It is calculated by using formula ∑ (receive time-sent time)/∑(count)
- *Packet Delivery ratio*: It is calculated by using formula (∑ No. of packet received/∑ No. of packet sent)
- *Packet Drop*: It is calculated by extracting dropped packet from trace file.

In this part different topologies are set based on the number of nodes (20, 40, 60, 80, 100). These topologies are simulated over the simulation time for both protocols and various performance metrics are calculated. Different data files are generated for Delay, Drop, Delivery Ratio, Energy Available, Overhead. After collecting the statistics of performance metrics respective graphs are plotted.

Figure 1 shows the overhead of EEDG and IAR techniques for different number of nodes. Statistics says that the overhead of EEDG approach has 49 % of less than EEDG approach.

Figure 2 shows the available energy of EEDG and IAR techniques for different number of nodes. Statistics says that the energy consumption of EEDG approach has 3 % of less than IAR approach.

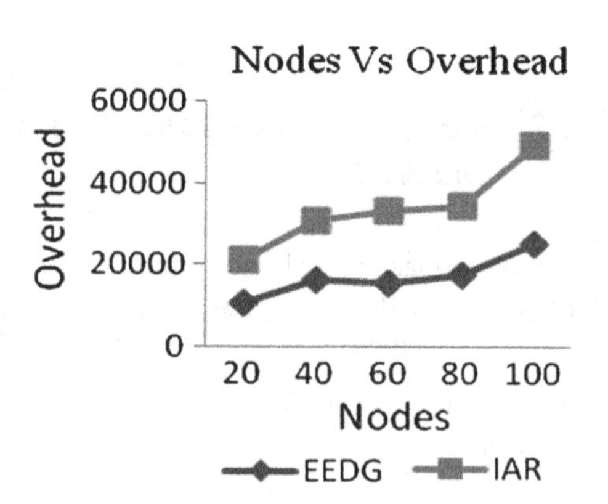

Fig. 1 Nodes versus overhead

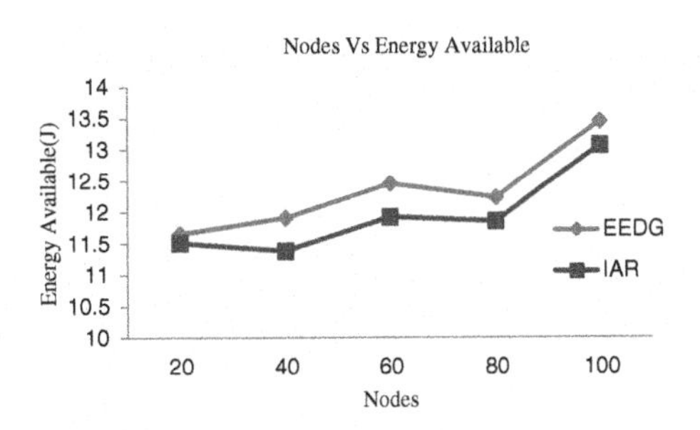

Fig. 2 Nodes versus energy available

Figure 3 shows the delivery ratio of EEDG and IAR techniques for different number of nodes. Statistics says that the delivery ratio of EEDG approach has 28 % of higher than IAR approach.

Figure 4 shows the delay of EEDG and IAR techniques for different number of nodes. Statistics says that the delay of EEDG approach has 13 % of less than IAR approach.

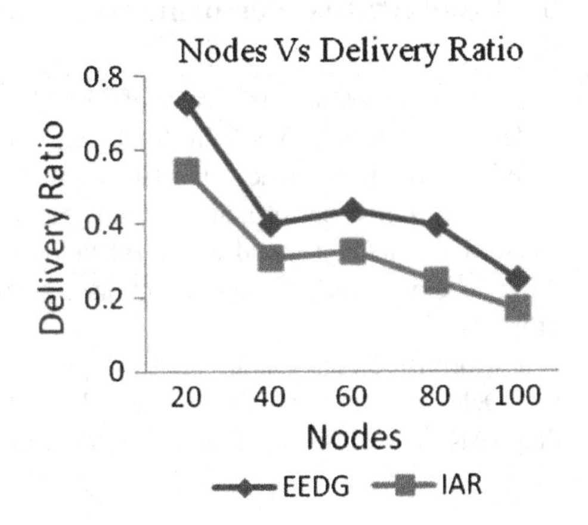

Fig. 3 Nodes versus delivery ratio

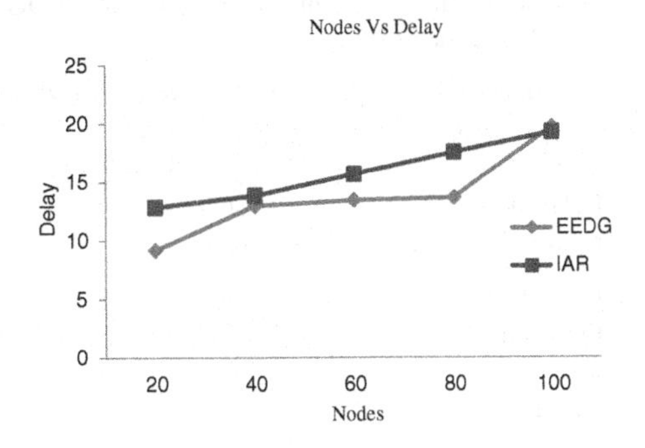

Fig. 4 Nodes versus delay

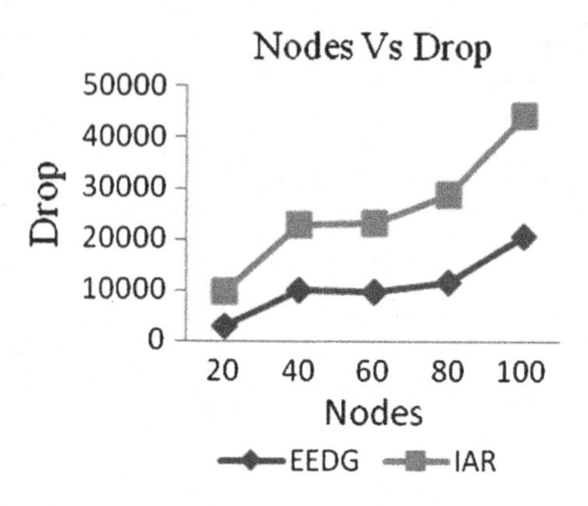

Fig. 5 Nodes versus drop

Figure 5 shows the drop of EEDG and IAR techniques for different number of nodes. Statistics says that the drop EEDG approach has 57 % of less than IAR approach.

5 Comparative Performance Analysis

A summary of simulation results of EEDG and IAR protocols based on number of nodes is shown in Table 4. In this table, first field represents performance metrics based on which efficiency of EEDG and IAR is decided. For efficient protocol, above mentioned performance parameters should be less or more is clearly indicated in this field. Second and third fields represents average values obtained for different topologies of above mentioned performance metrics for EEDG and IAR protocols.

Performance Analysis of both the protocols based on number of nodes shows that Delay incurred in EEDG is less than IAR, Delivery ration of EEDG is more than IAR, Packet drop is less in EEDG compared to IAR, Energy consumption in

Table 4 Summary of simulation results of EEDG and IAR based on nodes against different performance metrics

Performance metrics	EEDG (Average of values obtained for different topologies)	IAR (Average of values obtained for different topologies)
Delay (less)	13.78676	15.83537
Delivery ratio (more)	0.440678	0.317602
Drop (less)	11125.6	25772.8
Energy available (more)	12.33875	11.94521
Overhead (less)	17115	33817.4

EEDG is less than IAR and Control Overhead incurred is less in EEDG compared to IAR. This quantitative analysis clearly says that EEDG protocol is more efficient than IAR protocol.

6 Conclusion and Future Enhancement

Authors have compared performances of two latest Data Gathering Protocols EEDG [12] and IAR [13], which uses Mobile Element/Agent for collecting data in wireless sensor networks, based on nodes. Authors have made an attempt to show that the performance analysis of both protocols is the evidence to prove the efficiency of EEDG protocol is best than IAR protocol.

Even though EEDG is efficient, Authors have identified few limitations of this protocol. These limitations are Redundancy removal aspect and Idle listening concept which are not taken care while implementing EEDG protocol. Also this protocol will not allow Mobile Element/Agent to pause or wait at node, which will affect the life time of the entire network. In forth coming days authors are going to concentrate more on overcoming limitations of EEDG protocol to make it more energy efficient which will contribute towards improving lifetime of Wireless Sensor Network.

Acknowledgments This work is sponsored and supported by grant from VGST, Govt. of Karnataka (GRD-128). The authors wish to thank Dr. S. Ananth Raj, Consultant, VGST, and Prof. G.L. Shekar, Principal, NIE., for their encouragement in pursuing this research work.

References

1. Tumaishat, M., Madria, S.: Sensor networks: an overview. IEEE Potentials **22**, 20–23 (2003)
2. Ye, Z., Abouzeid, A.A., Ai, J.: Optimal policies for distributed data aggregation in wireless sensor networks. In: Proceedings of 26th International Conference on Computer Communication, IEEE, Anchorage, AK, pp. 1676–1684, May 2007
3. Krishnamachari, B., Estrin, D., Wicker, S.: The impact of data aggregation in wireless sensor networks. In: Proceedings of 22nd International Conference on Distributed Computing Systems Workshops. pp. 575–578 (2002)
4. Fan, K.W., Liu, S., Sinha, P.: Structure-free data aggregation in sensor networks. IEEE Trans. Mobile Comput. **6**(8), 929–941 (2007)
5. Saad, E.M., Awadalla, M.H., Saleh, M.A., Keshk, H., Darwish, R.R.: A data gathering algorithm for a mobile sink in large scale sensor networks. In: Proceedings of 10th International Conference on Mathematical Methods and Computational Techniques in Electrical Engineering, Bulgaria, pp. 288–294 (2008)
6. Bi, Y., Sun, L., Ma, J., Li, N., Khan, I.A., Chen, C.: HUMS: an autonomous moving strategy for mobile sinks in data gathering sensor networks. EURASIP J. Wireless Commun. Netw. **7**, 1–15 (2007)

7. Zhao, M., Ma, M., Yang, Y.: Mobile data gathering with space division multiple access in wireless sensor networks. In: Proceedings of 27th International Conference on Computer Communications, IEEE INFOCOM, pp. 1958–1965 (2008)
8. Jayaraman, P.P., Zaslavsky, A., Delsing, J.: Sensor data collection using heterogeneous mobile devices. In: International Confernce on Pervasive Services, IEEE, pp. 161–164 (2007)
9. Xiang, M., et al.: Energy-efficient intra-cluster data gathering of wireless sensor networks. Int. J. Hybrid Inf. Technol. J. Netw. **5**(3), 383–390 (2010)
10. Puthal, D., Sahoo, B., Sharma, S.: Dynamic model of efficient data collection in wireless sensor networks with mobile sink. IJCST **3**(1) (2012)
11. Geetalaxmi, B.J., Ashoka, D.V.: Merits and demerits of existing energy efficient data gathering techniques for wireless sensor networks. Int. J. Comput. Appl. (0975–8887) **66**(9) (2013)
12. Almi'Ani, K., et al.: Energy-efficient data gathering with tour length-constrained mobile elements in wireless sensor networks. In: Proceedings of 35th International Conference on Local Computer Networks (LCN), pp. 582–589 (2010)
13. Kim, J.W., In, J.S., Hur, K., Kim, J.W., Eom, D.S.: An intelligent agent-based routing structure for mobile sinks in WSNs. IEEE Trans. Consum. Electron. **56**(4), 2310–2316 (2010)

Hybrid Decision Model for Weather Dependent Farm Irrigation Using Resilient Backpropagation Based Neural Network Pattern Classification and Fuzzy Logic

Saroj Kumar Lenka and Ambarish G. Mohapatra

Abstract Irrigation in agricultural lands plays a crucial role in water and soil conservation. Real-time prediction of soil moisture content using wireless sensor network (WSN) based soil and environmental parameters sensing may provide an efficient platform to meet the irrigation requirement of agriculture land. In this research article, we have proposed Resilient Back-propagation optimization technique to train neural network pattern classification algorithm for the prediction of soil moisture content. Finally, the predicted soil moisture content is used by fuzzy weather model for generating adequate suggestions regarding irrigation requirement. The fuzzy model is developed by considering different weather parameters like sun light intensity, wind speed, environment humidity and environment temperature. Different weather conditions like cloudy situation, low pressure, cyclone and storm conditions are simulated in the fuzzy model. The soil moisture content prediction algorithm is tested with soil moisture content in each 1 h advance by considering eleven different soil and environmental parameters collected during a field test. The prediction errors are analysed using MSE (Mean Square Error), RMSE (Root Mean Square Error), and R-squared error.

Keywords Wireless sensor network · Neural network · Resilient Back-propagation · Fuzzy logic weather model · Soil moisture content

S.K. Lenka
Department of Information Technology, Mody University, Lakshmangarh, Rajasthan 332311, India
e-mail: lenka.sarojkumar@gmail.com

A.G. Mohapatra (✉)
Department of Applied Electronics and Instrumentation, Silicon Institute of Technology, Bhubaneswar, Odisha 751024, India
e-mail: ambarish.mohapatra@gmail.com

S.C. Satapathy and S. Das (eds.), *Proceedings of First International Conference on Information and Communication Technology for Intelligent Systems: Volume 1*, Smart Innovation, Systems and Technologies 50, DOI 10.1007/978-3-319-30933-0_30

1 Introduction

Irrigation is a crucial practice in several agricultural cropping systems in semiarid and arid areas, and also useful water applications and management are key concerns. The efficiency and uniformity of irrigation could be maintained from the complex and diverse information based systems by considering weather, soil, water, and crop data. Irrigation water management forms a major part of precision agriculture. It involves better assessment of need and availability of soil water level for crop cultivation. The statistical data from the United Nations indicate worldwide, agricultural accounts for 70 % of all water consumption, compared to 20 % for industry and 10 % for domestic use [1]. There are models developed that use the remotely sensed data to profile the soil moisture. However, remote sensing can be used to directly infer soil moisture. Microwave emissivities and infrared data were proven to be highly correlated with the soil moisture [2]. A lot of research has been concentrated in this area during the last two decades. Recently many researchers have been reported for soil moisture level control in an agriculture land by collecting soil and environmental information using wireless sensor network technique. Liang et al. [3] have proposed a design and implement of a real-time soil moisture prediction system based on GPRS and wireless sensor network. The front-end of the system utilizes wireless sensor network (WSN) environment to collect soil moisture data and a GPRS network to transmit the acquired data. The back-end of the system utilizes genetic backpropagation neural network to analyze and process the acquired data, simulate using annealing algorithm to optimum result, and to gives a real-time prediction output. The experimental results concludes that this system has the major advantages like low cost, high degree of accuracy and convenient maintenance [3].

The method proposed by Liang et al. [3] is a direct sensing of soil moisture content using WSN environment but the proposed method does not emphasizes on the irrigation mechanism. In the similar context, the other land related parameters and environmental parameters may affect the irrigation requirements. Timely prediction of soil moisture content will provide better irrigation management in a land by considering almost all the required parameters which affects the soil water evaporation. This prediction requires efficient algorithms which will produce future irrigation requirements in an agriculture land. The algorithm is tested for the prediction of soil moisture content in 1 h advance by considering eleven different soil and environmental parameters. The performance of Resilient Back-propagation is compared using MSE (Mean Square Error), RMSE (Root Mean Square Error), and R-squared error. In the similar context, Goumopoulos et al. [4] describe the design of an adaptable decision support system and its integration with a wireless sensor/actuator network (WSAN) to implement autonomous closed-loop zone-specific irrigation. Using ontology for defining the application logic emphasizes system flexibility and adaptability and supports the application of automatic inferential and validation mechanisms. Furthermore, a machine learning process has been applied for inducing new rules by analyzing logged datasets for extracting new

Table 1 Summarized literature review

Sl. no.	Review article	Comparison statement
1.	Liang et al. [3]	Experiment was conducted using Bluetooth based WSN environment and only few parameters like soil moisture content and environmental temperature data were used for the linear irrigation sprayer control
2.	Chai et al. [5]	Experiment was conducted for the prediction of soil moisture content using Resilient Back-propagation (Rprop). The data were collected by utilizing airborne microwave measurements. Here only H- and V-polarized brightness temperature data were considered
3.	Goumopoulos et al. [4]	Experiment conducted only in a greenhouse environment. The timely prediction of soil moisture content is also not included

knowledge and extending the system ontology in order to cope, for example, with a sensor type failure or to improve the accuracy of a plant state diagnosis. A deployment of the system is presented in zone specific irrigation control in a greenhouse setting. The method proposed by Goumopoulos et al. [4], consider only on the soil moisture content data from a greenhouse environment. This technique does not include other soil and environmental data like soil temperature, environmental humidity, environmental temperature and sun light intensity which are most required parameters for soil evapotranspiration. The timely prediction of soil moisture content is also not included in the method proposed by Goumopoulos et al. [4]. Our proposed method emphasises on the timely prediction of soil moisture content in any farming area like a greenhouse, polyhouse and large farmland by considering different soil and environmental parameters. The summarized literature review of the articles used in this work is listed with some of the comparison statements in Table 1.

2 Proposed Methodology

Agriculture system models are complex nonlinear systems which can be solved using powerful estimation methodologies like neural network algorithms [6]. In this work neural network pattern classification technique is proposed for the prediction of soil moisture content by considering soil and environmental parameters. A feedforward neural network is trained using Resilient Back-propagation (Rprop) which update weight and bias values. In the similar context, Chai et al. [5] have proposed Resilient Back-propagation (Rprop) for the prediction of soil moisture content during a field test during National Airborne Field Experiment 2005 (NAFE'05). According to the experiment, the RMSE obtained is about 4.93 %v/v [5].

In our proposed technique, the collection of soil and environmental data is performed using wireless sensor network (WSN) environment for the real-time monitoring and analysis of soil moisture content (MC) of the test site located in the eastern region of India. The test site of (50 m × 100 m) is planted with bermuda grass (Cynodon dactylon) which is used for stomach ulcer, colitis, and stomach infections as ayurvedic medicine. The test site contains nine WSN nodes with nine individual irrigation valves. The WSN nodes are used to collect data from soil moisture sensor, soil temperature sensor, environmental temperature sensor, environmental humidity sensor, CO_2 sensor, and sun light intensity sensor for multi-point measurement of land data in agricultural production. The proposed single node fabricated during this work is shown in Fig. 1a. The collected data were stored in a gateway node consisting of Raspberry Pi, Zigbee series 2 and wifi connectivity as shown in Fig. 1b. The proposed data collection architecture is shown in Fig. 1c. The complete scheme is utilized to collect twenty four numbers of

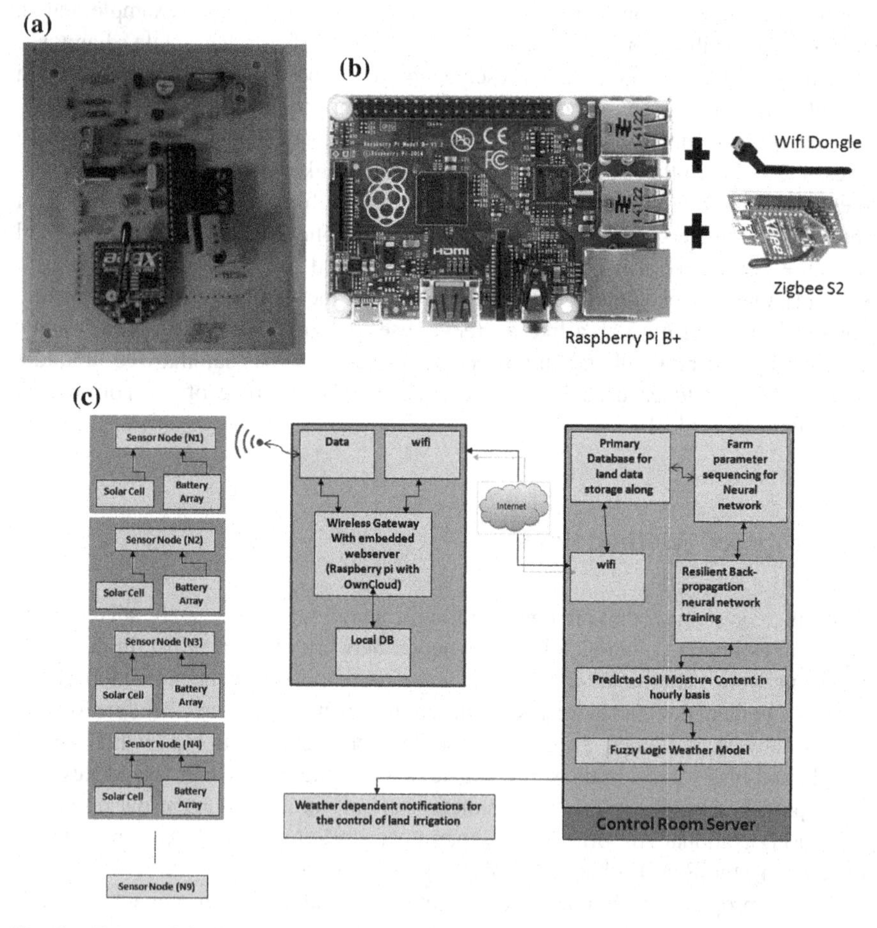

Fig. 1 **a** Proposed single node fabricated for acquiring farm data. **b** Proposed methodology for the design of embedded webserver with Zigbee and wifi connection. **c** Proposed architecture of the complete real-time soil moisture content prediction methodology

data sets consisting of all the above mentioned parameters. These twenty four numbers of datasets signifies the collection of all the data in 24 h in a day.

2.1 Resilient Back-Propagation (Rprop)

Resilient Back-propagation is an efficient learning approach for supervised learning mechanism in feedforward artificial neural networks architectures [7–9]. It is a first order optimization technique. This technique is also an efficient new learning scheme focusing on a direct adaptation of the weight steps depending on local gradient information. To achieve this, individual update-value Δ_{ij} is introduced for each weight, which solely determines the size of the weight-update [10–12]. This adaptive update-value evolves during the learning process based on its local sight on the error function E, according to the following learning-rule:

The learning rule is represented as:

$$\Delta_{ij}(t) = \begin{cases} \eta^{+}.\Delta_{ij}(t-1) \ , \ if \ s_{ij} > 0 \\ \eta^{-}.\Delta_{ij}(t-1) \ , \ if \ s_{ij} < 0 \\ (t-1) \ , \ otherwise \end{cases}$$

Where $s_{ij} = \frac{\partial E}{\partial w_{ij}}(t-1).\frac{\partial E}{\partial w_{ij}}(t)$

and $0 < \eta^{-} < 1 < \eta^{+}$

The weight step is represented as:

$$\Delta w_{ij}(t) = \begin{cases} +\Delta_{ij}(t) \ , \ if \ \frac{\partial E}{\partial w_{ij}}(t) > 0 \\ -\Delta_{ij}(t) \ , \ if \ \frac{\partial E}{\partial w_{ij}}(t) < 0 \\ 0 \ , \ otherwise \end{cases}$$

Exception: $\Delta w_{ij}(t) = -\Delta w_{ij}(t-1)$,

$$if \ \frac{\partial E}{\partial w_{ij}}(t-1).\frac{\partial E}{\partial w_{ij}}(t) < 0$$

2.2 Fuzzification and Defuzzification

Fuzzification is the mechanism of constructing a crisp fuzzy set. It is performed by recognizing that several of the quantities is thought about to be crisp and settled which are actually not deterministic at all [13]. If the shape of uncertainty exists to arise as a result of impreciseness, ambiguity, or unclearness, then the variable may be fuzzy and portrayed by a membership operate. Defuzzification is that the conversion of a fuzzy amount to an explicit quantity, just as fuzzification is the conversion of an explicit amount to a fuzzy quantity. In this work, the fuzzy membership functions are considered as sunlight intensity, wind speed, environmental humidity, environmental temperature and predicted soil moisture content which was obtained using neural network pattern classification. The membership function selected for the fuzzy weather model is a triangular shape function. The fuzzy logic input and output membership functions are listed in Figs. 6, 7, 8, 9, 10, and 11.

3 Results and Discussions

The overall performance of the optimization techniques for the prediction of soil moisture content is examined by considering eleven soil and environmental parameters. The input and output data are listed in Table 2. The predicted soil moisture content for 1 h advance is compared with ideal moisture content using time, MSE (Mean Square Error), RMSE (Root Mean Square Error), and R-squared error during ideal weather condition.

Resilient Back-propagation based neural network pattern classification was used to predict soil moisture content from soil and environmental data. The soil moisture content prediction response and error graph is shown in Fig. 2.

The Resilient Back-propagation based neural network pattern classification training performance is shown in Fig. 3. The best training performance is obtained at 2000 epochs. The confusion matrix of Resilient Back-propagation based neural network pattern classification is shown in Fig. 4. It is observed that twenty four datasets are correctly predicted with 100 % performance. Also from the neural network training state as shown in Fig. 5, the best MSE is obtained at epoch 2000 with gradient 0.00059969.

Table 2 Input and outputs used for soil moisture content prediction

Inputs					Output
Current time	Soil moisture content	Sunlight intensity	Air flow	Air CO_2	Soil moisture content after 1 h duration
Environment temperature	Soil temperature	Sunrise time	Rate	UV index	
Environment humidity	soil type				

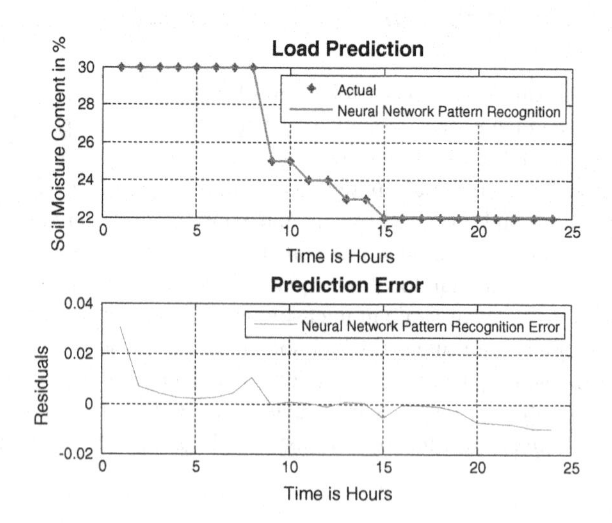

Fig. 2 Actual soil moisture content and predicted soil moisture content versus time in hours. Error graph plotted between residuals and time in hours (resilient back-propagation based neural network pattern classification)

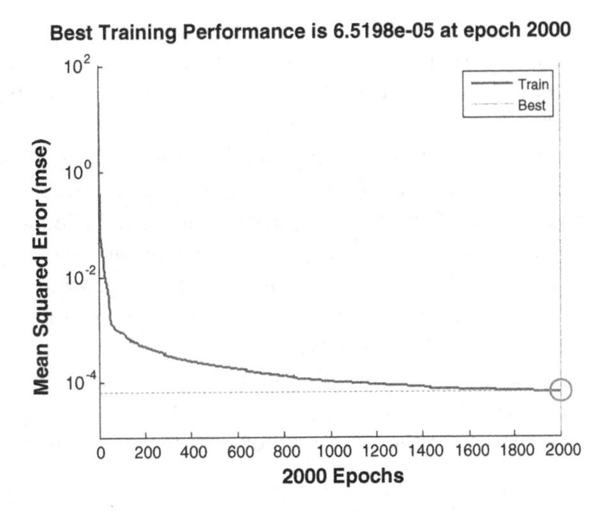

Fig. 3 Resilient back-propagation based neural network pattern classification training performance

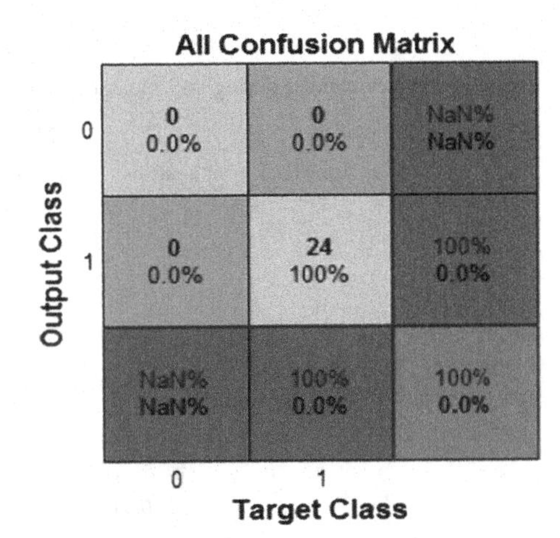

Fig. 4 The confusion matrix of the resilient back-propagation neural network pattern classification training

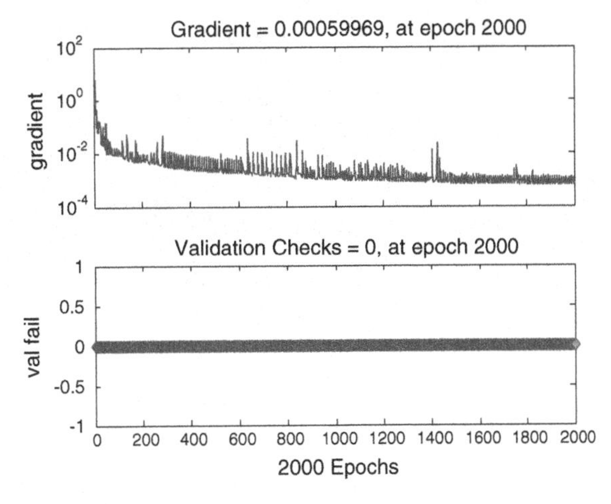

Fig. 5 The training state of the resilient back-propagation neural network pattern classification training

The performance of all the four optimization techniques for neural network pattern classification is also found out using MSE (Mean Square Error), RMSE (Root Mean Square Error), and R-squared error. All the above errors are listed in Table 3.

It is observed from the errors listed in Table 3 that Resilient Back-propagation based neural network pattern classification produces better result during soil moisture content prediction by considering the inputs as mentioned in Table 2. After successfully prediction of soil moisture content, the predicted soil MC is supplied to fuzzy logic weather model for generating appropriate irrigation suggestions. The fuzzy logic input and output membership fuctions are shown in Figs. 6, 7, 8, 9, 10 and 11.

The proposed hybrid irrigation model is tested with different test inputs and it is observed that the neural network and fuzzy logic algorithms are performing well good. Some of the test conditions are listed in Table 4.

Table 3 Errors obtained during soil moisture content prediction

Sl. no.	Algorithm type	MSE	RMSE	R-squared
1.	Resilient back-propagation	6.5198e-05	0.008074	0.9999998

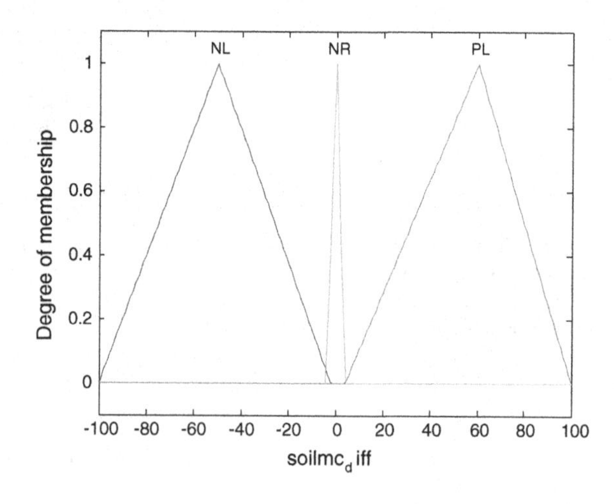

Fig. 6 Fuzzy logic membership function for soil moisture content difference

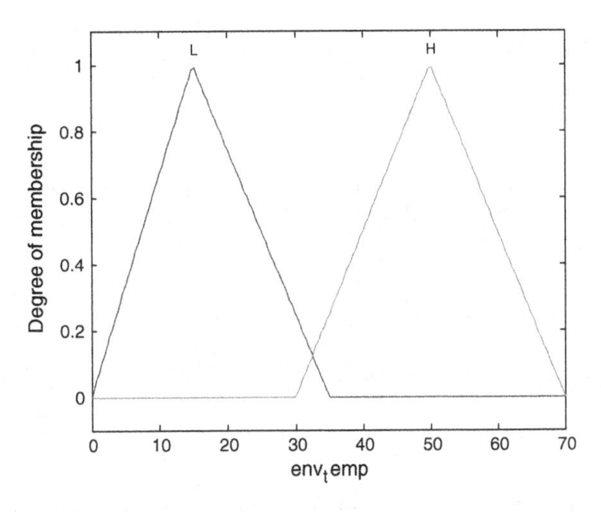

Fig. 7 Fuzzy logic membership function for environmental temperature

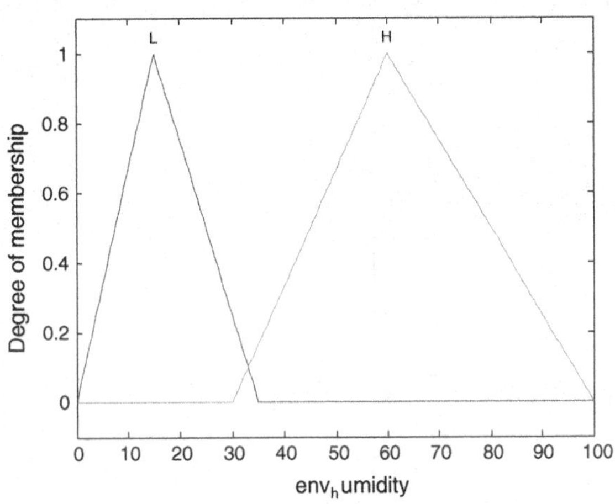

Fig. 8 Fuzzy logic membership function for environmental humidity

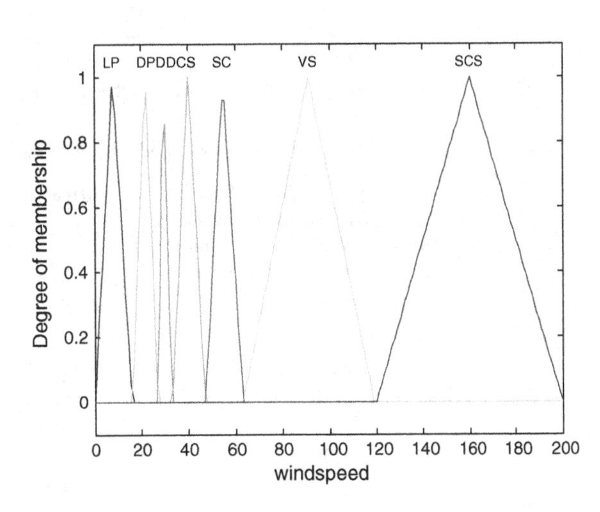

Fig. 9 Fuzzy logic membership function for wind speed

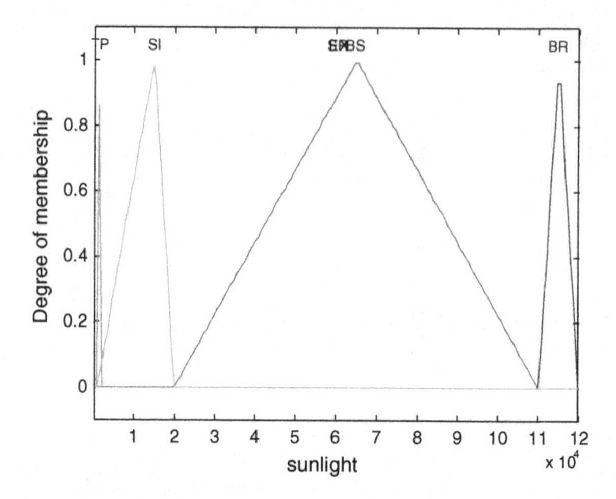

Fig. 10 Fuzzy logic membership function for sunlight intensity

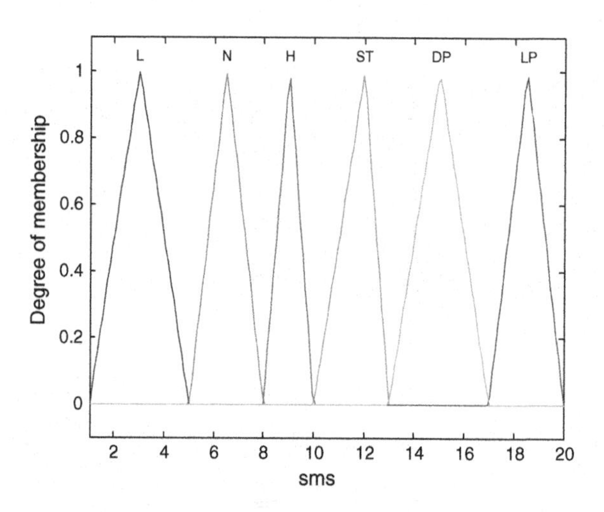

Fig. 11 Fuzzy logic membership function for weather dependent irrigation suggestions

Table 4 Test inputs and outputs of the proposed irrigation model

Sl. no.	Input parameters [soil MC difference (%), environment temperature (centigrade), environment humidity (RH), wind speed (Knots), sunlight intensity (Lux)]	Irrigation suggestions
1.	[7.5872, 27, 40, 40, 1500]	Cyclonic storm. Irrigation valve closed
2.	[5, 40, 30, 10, 110400]	Required water level is less and irrigation is required
3.	[−2, 29, 38, 20, 1800]	Required water level is maintained and irrigation is not required
4.	[−5, 29, 38, 20, 2300]	Required water level is high and water extraction pump should be activated

4 Conclusion

The soil moisture content in 1 h advance is efficiently predicted by taking various soil and environmental parameters. The performance analysis of Resilient Back-propagation optimization technique for neural network pattern classification is done by calculating MSE, RMSE and R-squared error. From the errors obtained during prediction shows that the performance of Resilient Back-propagation based neural network pattern classification gives better result in soil moisture content prediction. The MSE and RMSE obtained by Resilient Back-propagation based neural network pattern classification are 6.5198e-05 and 0.008074. The predicted soil moisture content is successfully used in the fuzzy logic weather model for generating adequate irrigation suggestions. Hence it concluded that the proposed weather condition dependent hybrid irrigation model will help to predict soil MC and to generate necessary irrigation suggestions using real-time sensor data from the agriculture locations. The complete work is successfully tested in the test land (Latitude: 20.351323, Longitude: 85.806547). The proposed model needs more rigorous multiple farm testing to obtain complete performance of the soil moisture control mechanism.

Acknowledgements The authors thank to All India Council for Technical Education (AICTE), New Delhi, India to provide support under Career Award for Young Teachers (CAYT) scheme 2013–2014 and help for continuing research work in the area of precision agriculture mechanism.

References

1. Worldometers about world statistics updates in real time (2011).: Water consumed this year, http://www.worldometers.info/water/
2. Zhang, M., Li, M., Wang, W., Liu, C., Gao, H.: Temporal and spatial variability of soil moisture based on WSN. Math. Comput. Model. **58**(3–4), 826–833 (2013)
3. Liang, R., Ding, Y., Zhang, X., Zhang, W.: A real time prediction system of soil mc using genetic neural-network based on annealing algorithm. In: IEEE International Conference on Automation and Logistics (ICAL-2008), pp. 2781–2785, 1–3 Sept 2008 (Computer & Inf. Eng. College, Hohai University, Changzhou)
4. Goumopoulos, C., O'Flynn, B., Kameas, A.: Automated zone-specific irrigation with wireless sensor/actuator network and adaptable decision support. Comput. Electron. Agric. **105**, 20–33 (2014) (Computer Technology Institute, DAISy Research Unit, 26500 Rio Patras, Greece, Information & Communication Systems Engineering Department, Aegean University, Greece, Tyndall National Institute, Lee Maltings, Prospect Row, Cork, Ireland)
5. Chai, S.-S., Veenendaal, B., West, G., Walker, J.P.: Back-propagation neural network for soil moisture retrieval using nafe'05 data: a comparison of different training algorithms. In: The International Archives of the Photogrammetry, Remote Sensing and Spatial Information Sciences, vol. XXXVII, Part B4. Beijing (2008)
6. Phillips, A.J., Newlands, N.K., Liang, S.H.L., Ellert, Benjamin H.: Integrated sensing of soil moisture at the field-scale: measuring, modeling and sharing for improved agricultural decision support. Comput. Electron. Agric. **104**, 73–88 (2014)

7. Rehman, M.Z., Nawi, N.M.: The effect of adaptive momentum in improving the accuracy of gradient-descent back-propagation algorithm on classification problems. In: Software Engineering and Computer Systems Communications in Computer and Information Science, vol. 179, pp. 380–390. Springer, Berlin (2011)
8. Andrei, N.: Scaled conjugate gradient algorithms for unconstrained optimization. Comput. Optimization Appl. **38**(3), 401–416 (2007)
9. Chel, H., Majumder, A., Nandi, D.: Scaled conjugate gradient algorithm in neural network based approach for handwritten text recognition. In: Trends in Computer Science, Engineering and Information Technology Communications in Computer and Information Science, vol. 204, pp. 196–210. Springer, Berlin (2011)
10. Riedmiller, M., Braun, H.: A direct adaptive method for faster back-propagation learning, The RPROP algorithm. In: IEEE International Conference on Neural Networks, vol. 1, pp. 586–591, 28 Mar 1993–01 Apr 1993 (Institut fur Logik, Komplexitat und Deduktionssyteme, University of Karlsruhe)
11. Dai, Y.-H.: A perfect example for the BFGS method. Math. Program. **138**(1–2), 501–530 (2013)
12. Nocedal, J., Yuan, Y.-X.: Analysis of a self-scaling quasi-Newton method. Math. Program. **61**(1–3), 19–37 (1993)
13. Cetişli, B., Barkana, A.: Speeding up the scaled-conjugate-gradient algorithm and its application in neural network fuzzy logic classifier training. Soft Comput. **14**(4), 365–378 (2010)

Energy-Hole Minimization in WSN Using Active Bonding and Separating Coverage

Chinmaya Kumar Nayak, Subhashree Rath, Manoranjan Pradhan and Satyabrata Das

Abstract A wireless sensor network (WSN) is typically comprised of a large number of nodes spread over a large area sensors. Sensor nodes are small, low battery, limited storage and processing power. Each node is usually equipped with a wireless radio transceiver, a small micro-controller, a source of power sensors and multiple types such as temperature, light, pressure, sound, vibration nodes etc. These ways to communicate directly with network among themselves or through other nodes. The first objective is to collect data on WSN sensor nodes. If the data on the web passes through the one or more sensor nodes, power consumption in the first node differs from the second node. Therefore, the loaded nodes too quickly lose battery power and shut down. Imagine this situation happens to a group of neighboring nodes and fall of premature death, leading energy gap in the network. Energy gap affects other nodes in the network nodes that share the power load port has a load to the other nodes in the network. As a result, life of a network will end soon. Our research is aimed at maximizing the coverage area, increasing the detection range of remaining sensor nodes when a node fall in premature death, to prevent the formation of holes in the supply network of wireless sensors.

Keywords Separating coverage · Active bonding · Energy holes · Sensing range

C.K. Nayak (✉) · S. Das
Department of CSE, VSSUT, Burla, Sambalpur, Odisha, India
e-mail: cknayak85@gmail.com

S. Das
e-mail: satya.das73@gmail.com

S. Rath · M. Pradhan
Department of CSE, GITA, Bhubaneswar, Odisha, India
e-mail: sratha@gietbbsr.com

M. Pradhan
e-mail: manoranjanpradhan72@gmail.com

S.C. Satapathy and S. Das (eds.), *Proceedings of First International Conference on Information and Communication Technology for Intelligent Systems: Volume 1*, Smart Innovation, Systems and Technologies 50, DOI 10.1007/978-3-319-30933-0_31

1 Introduction

Wireless sensor networks deployed in harsh environments and left unattended for a relatively long time. In some worst conditions, a group of sensor nodes do not perform network operations [1]. The sensor node fails its assigned function and can not communicate with other nodes is considered a destroyed node. In sensor networks, you can come across a type of node called as faulty node. A node is considered defective when giving results that are significantly different from the results of the work of defective sensor closest nodes. But produces abnormal results. Defective nodes can be seen as special, the only point in the sensor itself consists open. Thus, destroyed node completely stops working and do not participate in the socket [2, 3]. On the other hand, faulty nodes to participate in networking activities, but produces significantly abnormal results are very different from their neighbors.

We highlight the main reasons leading to the destruction of the sensor nodes that lead to the formation of holes in the network. The node destruction takes place, either through some kind of external entity destroy physical nodes or may happen as a result of the deployment of the network. Some of the main reasons for the destruction and hole creating problems in this section.

We formally define different types of holes that can happen in a wireless sensor network and discuss their features.

1.1 Coverage Holes

Literature review, covering, among other problems, though a series of sensors and a hole through the target area is a target site, to the extent required k, where k is the end, all points are covered with at least one sensor coverage for the application of this definition to cover a hole problem depending on application requirements, not to mention that the appropriate (see Fig. 1). Some programs fault tolerance/discharge or tenderness to a more accurate location information can be a cover for a target

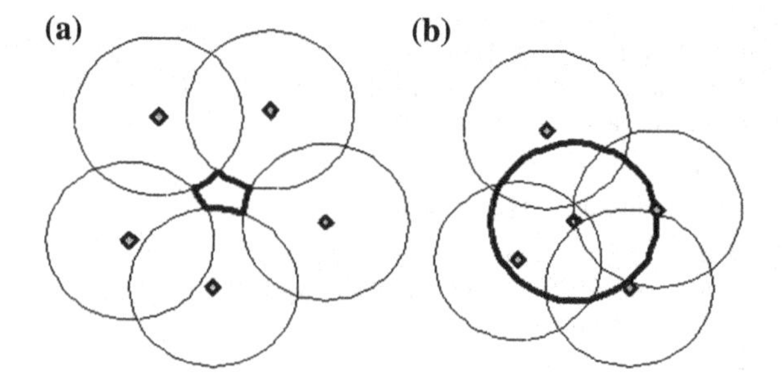

Fig. 1 **a** Single hole surrounded by a disk sensing model. **b** If the required coverage rate of 2 *dark gray* with a touch sensing area needs

based on the protocols required to use Trilateration positioning or location based triangulation [4–6].

In summary, and a single device in any direction (Fig. 1). Disk is represented by the model such as covering a sensor node is usually found. But on the basis of a realistic assessment idealized model covers all the sensors in a circle and at the same time [7, 8]. In addition, the ability to include detection sensor, as well as military exercises, the nature and the characteristics of the case will depend on the sensitivity of the detection sensors detect targets as military tanks, compared to only depends on the characteristics.

Nodes that are able to communicate with each other, a network of sensors constantly blocked. As mentioned above, different coverage requirements, as well as a switch protect the net working capital of link or node failure would be shared. Individual coverage requirements, Wang et al. They not only give coverage to the protocols and restrictions will ensure communication range of sensors to work at least twice as likely to have a detection range of the show.

1.2 Routing Holes

The nodes for various reasons, cannot participate in the actual data—A network routing holes are either not available or working sensor nodules [4, 5, 9, 10], is a field. Due to the structure of an accident, such as a reset or discharge of the battery or an external event, such as inadequate for various reasons, either because of flaws in the implementation of a lack of sensor or sensor nodes can create holes in the destruction of the physical nodes.

Routing hole greedy expedition often face geographical phenomenon is attributed to a minimum. Here, based on the location of the destination. Figure 2 neighbors who are trying to send traffic to a destination node in a node hop is geographically close. X 21 cannot 1-hop destination in and of itself is no closer to solving the only way to package requires temporarily moving away when the transmission destination and not Abo. Local routing hole to offset a minimum phenomenon is more likely to occur in this special event, as it is called.

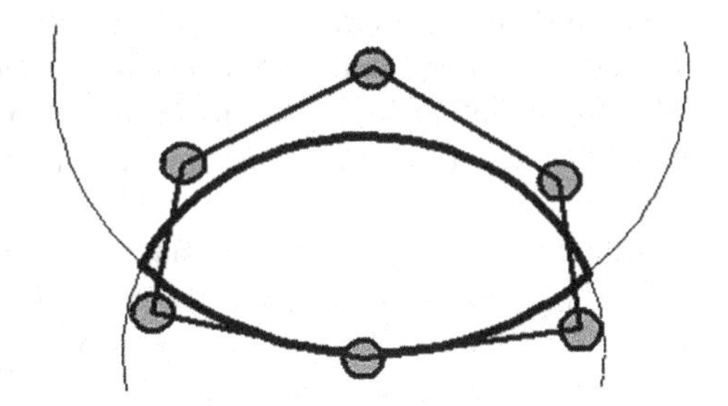

Fig. 2 Greedy forwarding the local minimum phenomenon

1.3 Jamming Holes

When the object to be tracked by radio frequency jammers capable of jamming an interesting scenario is used for communication between the sensor nodes tracking applications may occur [5, 11, 12]. These communication nodes are still able to detect the presence of objects in the field of communication jamming head back to the sink, but will be able to. Jammer central effect of the zone, it is referred to as paper jamming hole.

1.4 Sink Holes/Black Holes/Worm Holes

That's release on the basis of the mechanism or selected from littering or relaying some malicious content before you can change the message. The malicious node thus established himself at the center of a sink hole.

Sink holes and channel access to limited bandwidth [8, 13] harmful to the resource node between neighboring nodes is characterized by controversy. This result in congestion due to the failure of the routing nodes, the nodes involved in forming holes can accelerate energy consumption. Sink hole forming in a sensor network, the service is available after attacks on several other types of refuse.

Worm hole of service attack [11, 14, 15] is a kind of denial. Malicious nodes located in different parts of the sensor network, which creates a tunnel between them. They are using different radio channels of communication; the other end of the tunnel is a part of the sensor network and start forwarding packets [16]. Visiting a malicious node, then the other part of the network replays the message. This is wrong as a result of the convergence of routing; neighbors believe that there are nodes located in different parts of the networks. Section 6 of this denial of service attacks against some of the details of the measures.

2 Proposed Scope of Work to Cut Holes

After working for a different kind of hole we just focus on the holes in the cover. The aim of our research is to reduce the wireless sensor network to cover the hole. We found a number of WSN nodes in sensor coverage of the entire area taken at periodic intervals of time to change the setting to reduce the number of sensor nodes are found.

Life coverage hole detection range of energy reduces the life of the hard drive [13]. Compared to the stage as the network increases the detection range of variables. We cover the hole by increasing the detection range of the sensor nodes in order to reduce the four algorithms. Here is the time to increase the detection range of the various algorithms. Logic and sensors into a small area of the detection range of the scope of the amendment is to discover the logic cover.

Compared to the fixed sensing range sensing range of variable height increases. Network [13] as a life time helps to reduce the increases to cover the energy hole. We cover the hole by increasing the number of sensor nodes detection algorithms to minimize the four listed. Increasing the number of times different sensing algorithms. We are a small, fragmented logistics sector and the scope of their sensing ranges of sensors, adjusting the sensing range to cover the field of logistics fall. The algorithms are as follows:

(i) Covers an area of fixed time interval logic
(ii) Active bonding with Neighbor node
(iii) Active bonding with sink node
(iv) Touching the Active target point.

2.1 Covers an Area of Fixed Time Interval Logic

The total area of the distribution of some of the logic in this article. Logic in each region such as sensor nodes that will be distributed equally divided. In a time interval T, the area covered by the logic of the zone sensors and sensors for detecting any logical node covers an area of dead time you increase the detection range for the number of nodes increases, the survey covers the field of logic. Kill all the nodes in a logical partition of the logic that if the sensor nodes, each time you get active again for the remainder of the interval T (Figs. 3 and 4).

2.2 Active Bonding with Neighbor Node

This is the logic of the whole area is divided into a small area. It borders on all nodes in the field of logistics. This dynamic means that a bond formed between nodes. Any node touched the dead link detection range of the rest of the nodes, install the most logical area. When the bond increases the number of dead found in a single node. The obvious and logical area is growing again after a time interval. The logic of the detection range and the rest of the nodes is a link between the node is dead (Figs. 5 and 6).

2.3 Active Bonding with Sink Node

Here logical area is separated according to the receiving node. Receiving node is a connection or association with other nodes. If any node that is dead then the

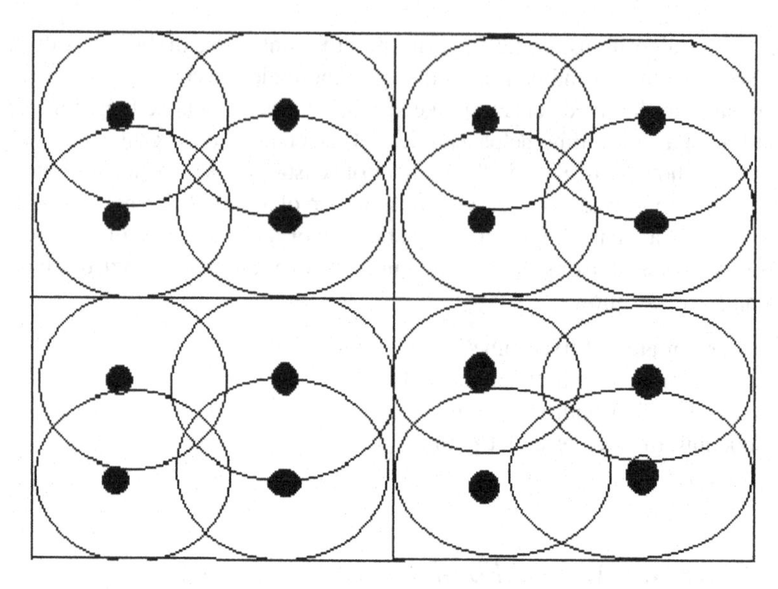

Fig. 3 Sensor nodes having initial sensing range point in logical Separated area

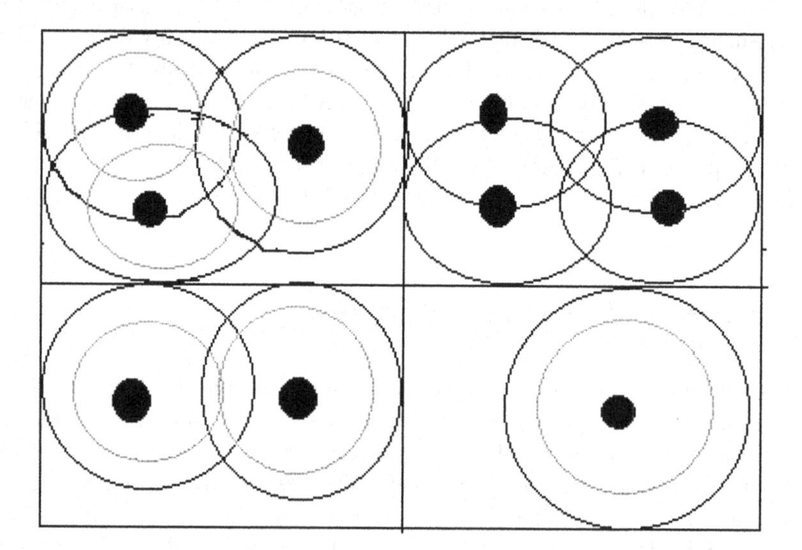

Fig. 4 Sensor nodes having increased sensing range point in logical separated area

relationship with the neighboring node on the way to the sink node. The neighboring nodes increase its detection range to cover the area logical partition (Figs. 7 and 8).

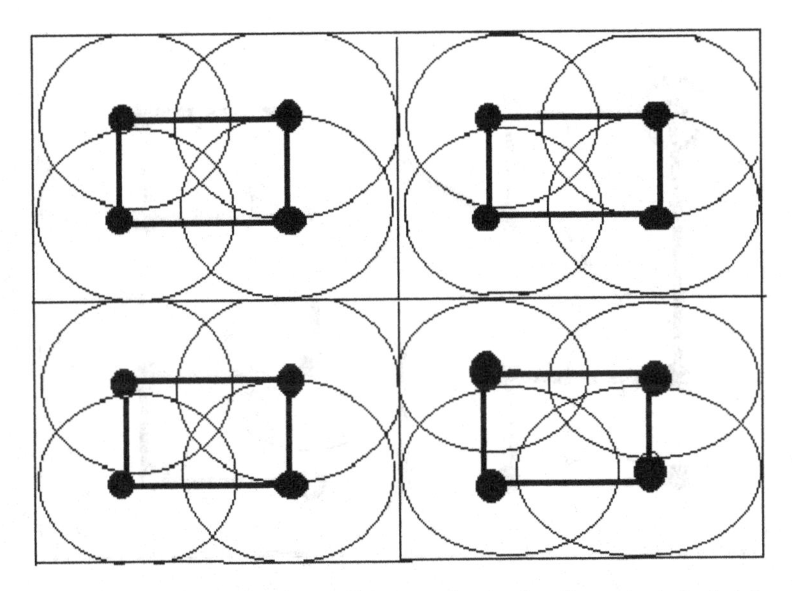

Fig. 5 Sensor nodes having initial sensing range and active bonding point in logical separated area

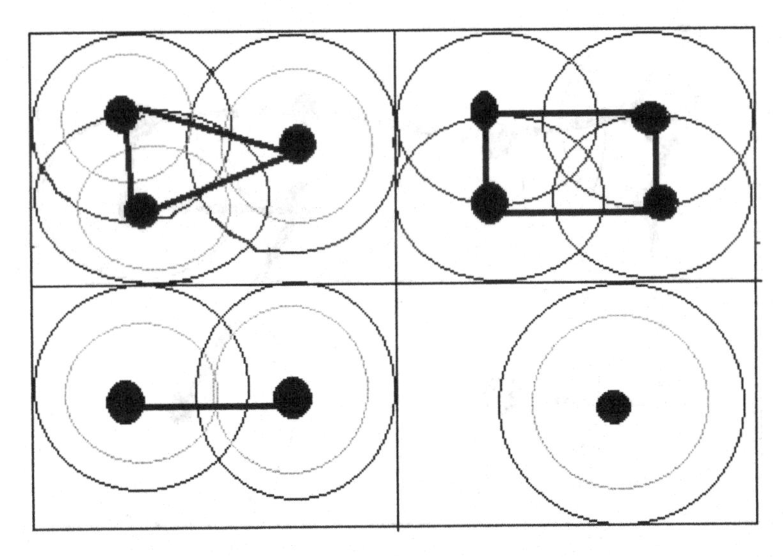

Fig. 6 Sensor nodes having increased sensing range and active bonding point in logical separated area

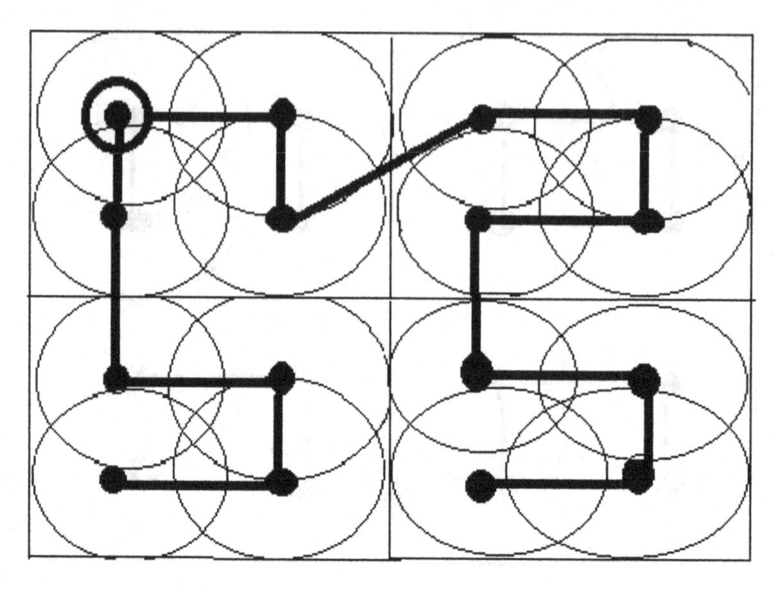

Fig. 7 Sensor nodes having initial sensing range and active bonding with the sink node point in logical separated area

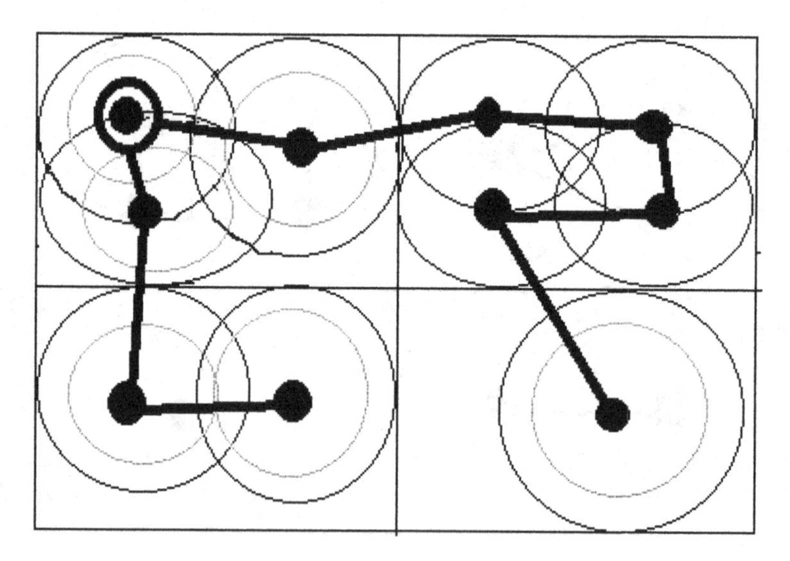

Fig. 8 Sensor nodes having increases sensing range and active bonding with the sink node point in logical separated area

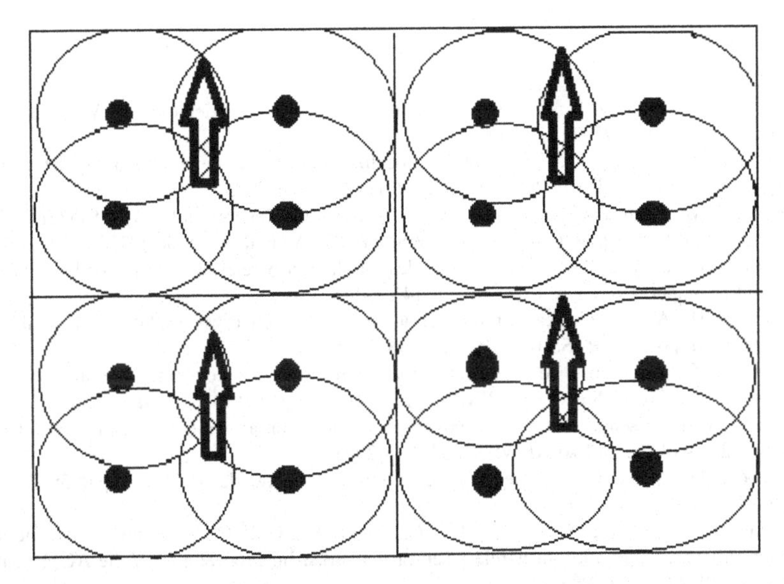

Fig. 9 Sensor nodes having initial sensing range and touch the active target point in logical separated area

2.4 *Touching the Active Target Point*

Destination point is a coordinate point in the area of the logical partition dynamically created according coverage hole. The neighbour sensor nodes must become the point of destination, with its detection range. If any sensor node is dead, then contact point to change the cover hole in the logic area. When the destination neighbouring nodes is changed to the target to increase its detection range to become the destination point (Fig. 9).

3 Conclusion

We have four different ways to increase the detection range of the sensor nodes for wireless sensor network of four methods to cut holes in the cover offered. Here we creating a sensor node wireless sensor network to cover the gap in the series to detect when the time is damaged.

Acknowledgement We have four different ways to increase the detection range of the sensor nodes for wireless sensor network of four methods to minimize holes in the cover offered. Here we creating a wireless sensor network to cover the hole in the series to detect when a node is damaged.

References

1. Bojkovic, Z., Bakmaz, B.: A survey on wireless sensor networks deployment. WSEAS Trans. Commun. **7**(12), 1172–1181 (2008)
2. Ding, M., Chen, D., Xing, K., Cheng, X.: Localized fault-tolerant event boundary detection in sensor networks. In: IEEE INFOCOM, pp. 902–913, Mar 2005
3. Tripathy, M.R., Nayak, C.K., Pradhan, M.: Minimization of energy-holes By DBPC (Dynamic Bonding and Partitioning Coverage) in a WSN. Int. J. Comput. Appl. **5**(3), 2250–1797 (2015)
4. Ahmed, N., Kanhere, S.S., Jha, S.: The holes problem in wireless sensor networks: a survey. Mobile Comput. Commun. Rev. **1**(2), 1–14 (2009)
5. Son, J.A.M.: A jammed-area mapping service for sensor networks. In: 24th IEEE Real Time System Symposium (RTSS'03), pp. 286, 298, Dec 2003
6. Adlakha, S., Srivastava, M.: Critical density thresholds for coverage in wireless sensor networks. In: Proceedings of the IEEE WCNC '03, 3, pp. 1615–1620, March 2003
7. Huang, C.-F., Tseng, Y.-C.: The coverage problem in a wireless sensor network. In: Proceedings of the 2nd ACM WSNA'03, Sep 2003
8. Wood, A.D., Stankovic, J.A.: Denial of service in sensor networks. IEEE Comput. **35**(10), 48–56 (2002)
9. Wang, X., Xing, G., Zhang, Y., Lu, C., Pless, R., Gill, C.: Integrated coverage and connectivity configuration in wireless sensor networks. In: Proceedings of the ACM SenSys '03, pp. 28–39, Nov 2003
10. Youssef, M.W.: Securing computer networks communication by modifying computer network communication protocols. In: An IEEE Communications Society Security and Applications Workshop, the 11th International Conference on Telecommunications for Intelligent Transport Systems "ITST-2011", St. Petersburg, Russia, Aug 2011
11. Hu, Y.-C., Perrig, A., Johnson, D.B.: Wormhole detection in wireless adhoc networks. Technical Report TR01-384, Department of Computer Science, Rice University, June 2002
12. Curiac, D., Volosencu, C., Doboli, A., Dranga, O., Bednarz, T.: Neural network based approach for malicious node detection in wireless sensor networks. In: Proceedings of WSEAS International Conference on Circuits, Systems, Signal and Telecommunications, pp. 17–19, Jan 2007
13. Cardei, M., Wu, J., Lu, M.: Department of Computer Science and Engineering Florida Atlantic University, Boca Raton, FL 33431 (2005)
14. Akyildiz, I.F., Su, W., Sankarasubramaniam, Y., Cyirci, E.: Wireless sensor networks: a survey. Comput. Netw. **38**(4), 393–422 (2002)
15. Pires, W.R., Jr., Figueiredo, T.H., Wong, H.C., Loureiro, A.A.: Malicious node detection in wireless sensor networks. In: IEEE 18th International Parallel and Distributed Processing Symposium (2004)
16. Curiac, D., Banias, O., Dranga, O.: Malicious node detection in wireless sensor networks using an autoregression technique. In: Third International Conference Networking and Services (2007)

"Mahoshadha", the Sinhala Tagged Corpus Based Question Answering System

J.A.T.K. Jayakody, T.S.K. Gamlath, W.A.N. Lasantha, K.M.K.P. Premachandra, A. Nugaliyadde and Y. Mallawarachchi

Abstract "Mahoshadha" the Sinhala Question Answering Systems aims at retrieving precise information from a large Sinhala tagged corpus. This paper describes a novel architecture for a Question Answering System which summarizes a tagged corpus and uses the summarization to generate the answers for a query. The summarized corpuses are categorized according to a set of topics enabling fast search for information. K-Nearest Neighbor Algorithms is used in order to cluster the summarized corpuses. The query will be tagged, the tagged query will be used to get more accurate results. Through the tagged query the question will be identified clearly with the category of the query. Support Vector Machine is used in order to both automate the summarization and question understanding. This will enable "Mahoshadha" to answer any type of query as well as summarize any type of Sinhala corpus. This enables the Question Answering System to be more useable through many applications.

Keywords Question answering · Document summarization · Document categorization · SVM algorithm · k-NN classification

---

J.A.T.K. Jayakody · T.S.K. Gamlath · W.A.N. Lasantha (✉) · K.M.K.P. Premachandra · A. Nugaliyadde · Y. Mallawarachchi
Sri Lanka Institute of Information Technology, Malabe, Sri Lanka
e-mail: it12520640@my.sliit.lk; nlasantha22@gmail.com

J.A.T.K. Jayakody
e-mail: it11189640@my.sliit.lk

T.S.K. Gamlath
e-mail: it12096176@my.sliit.lk

K.M.K.P. Premachandra
e-mail: it12065936@my.sliit.lk

A. Nugaliyadde
e-mail: anupiya.n@sliit.lk

Y. Mallawarachchi
e-mail: yashas.m@sliit.lk

S.C. Satapathy and S. Das (eds.), *Proceedings of First International Conference on Information and Communication Technology for Intelligent Systems: Volume 1*, Smart Innovation, Systems and Technologies 50, DOI 10.1007/978-3-319-30933-0_32

1 Introduction

People are always in a quest for information. With the rapid growth of the available information, Question Answering (QA) systems are mandatory for areas ranging from medical science to personal assistants.

QA differs from Information Retrieval (IR) or Information Extraction (IE). IR systems provide a set of documents related to the query, but do not exactly indicate the correct answer for the query. In IR, the relevant documents are obtained by matching the keywords from the query with a set of index terms from the set of documents. QA is the process of extracting the most precise answer to natural language question asked by the user.

"Mahoshadha" is a fully automated free and open source QA system for Sinhala Language, which helps to answer any question within the provided content. Implications of "Mahoshadha" are immense as it can be adopted by using a tagged corpus according to the application of use. Any annotated Sinhala text document is allowed to input to the system. Some applications where this system can be used are as a medical instructor, artificial teacher, self-learning tool and also can be used in call centres. Therefore it has a high business value as well as its technological service.

"Mahoshadha" is a high research contribution project as it opens a new research area, Question Answering in Sinhala Natural Language Processing.

2 Methodology

The system consists of four components. Document Summarizing, Document categorizing, Question processing and Answer processing.

2.1 Document Summarization

"Mahoshadha" has used summarization to summarize the multiple Sinhala documents enter by the user to increase the efficiency by reducing number of terms using Support Vector Machine (SVM) Algorithm. SVM is a Classification method which can be used for Classification and Regression as it supports text mining and pattern recognition.

This uses set of sample documents that have summarized manually as the training documents [1]. As the first step those documents and their respective summaries are provided for the learning process of the algorithm.

After the learning process Classifier (A classifier is a supervised function where the learned attribute is categorical) is used to classify new records (new documents to summarize) by giving them the best target attribute (predicted summary) using text mining and recognizing text patterns as shown in Fig. 1.

Fig. 1 Classification

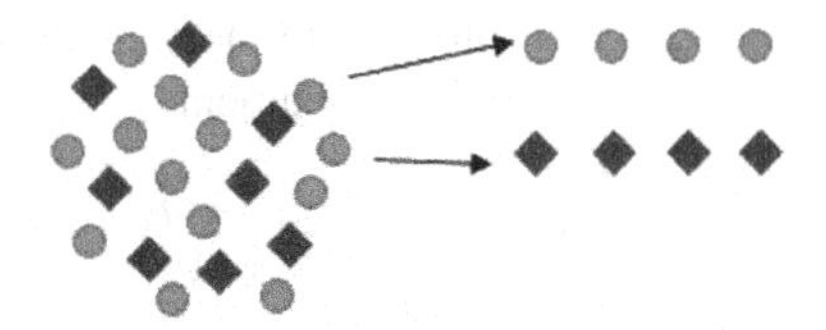

2.2 *Document Categorization*

In a QA system, organizing documents in a convenient way is important in order to increase the efficiency of retrieving the answer. In "Mahoshadha" organizing the documents has done by categorizing them to predefined categories considering its content using k-Nearest Neighbor (k-NN) Classification [2]. The method makes use of training documents, which have known categories, and finds the closest neighbors of the new sample document among all [3]. The solution involves a similarity function in finding the confidence of a document to a previously known category.

While categorizing documents, terms that do not have any importance and effect in categorizing documents should be eliminated [4]. Further accuracy can be provided to the categorization as terms added dynamically while adding training documents.

Term Space Model is an important concept in text categorization. First calculate weightages of each terms using the given formulas to use in similarity function [5].

$$wtf(t) = tf * idf(t) \text{ for term t,} \tag{1}$$

$$idf = \log 2(N/n) \tag{2}$$

N Number of all documents
n Number of documents where that term appears

Similarity function takes one training document and the new document as parameter [6]. It returns a value that corresponds to the amount of similarity between these documents.

$$Sim(X, Dj) = [\Sigma ti€(X \cap Dj)xi * dij]/[||X||*||Dj||] \tag{3}$$

X New document
Dj Training document *j*
ti term in both vectors
xi *wtf*(*i*)
dij *wtf*(*ij*)

$$||X|| = \sqrt{x1^2 + x2^2 + x3^3 + \cdots} \tag{4}$$

where all x's are wtf of all terms in X. ||Dj|| same as ||X|| for the terms in training documents (D)

Find the category of new document using k-NN algorithm taking k as the number of training documents.

$$Conf(c, d) = [\Sigma ki€K|(Class(ki') = c)Sim(ki', d)]/[\Sigma ki€KSim(k, d)] \quad (5)$$

(Conf is confidence in long terms.)

c Any category
d New document (X in the formula above)
K Neighborhood of size k for the document X

All similarities between the new document and the documents that belong to class c are added. Then they are divided to all similarities between the new document and training documents belonging to the neighborhood of X in size k [7] (k value used in k-NN algorithm is chosen as the number of training documents). Finally the confidences (conf) are compared and the category, for which the greatest confidence is calculated, is chosen as the category for the new document d (or X).

Using the same method "Mahoshadha" identify the category of the query asked by the user and it will search for the answer in the document belong to that particular category. If the answer couldn't find in that category "Mahoshadha" can search it in the category with next greatest confidence. Likewise the system is capable of finding the answers category wise as a solution for increasing the efficiency.

2.3 *Question Processing*

This module is the entry point for the project. Under this module discuss about how process on the question which entered to the system and how this module help for the project success. Question processing contain two key components.

- Analyze the Question

  POS tagging
  Language model

- Identify Answer Type (Fig. 2)

Analyze the Question. Inserted question need to be analyzed. It is handled by this component. When analyzing a sentence or set of words first need to come up with a good knowledge about the language. Appearance of the language, behavior of the language, grammar rules are key areas when analyzing a sentence.

POS Tagging. Part of Speech (POS) Tagging is an important process of Natural Language Processing (NLP) and a prerequisite to many other NLP activities. Automatic POS tagging process identifies the syntactic category of each word in a given sentence according to the context where the word of that sentence. POS

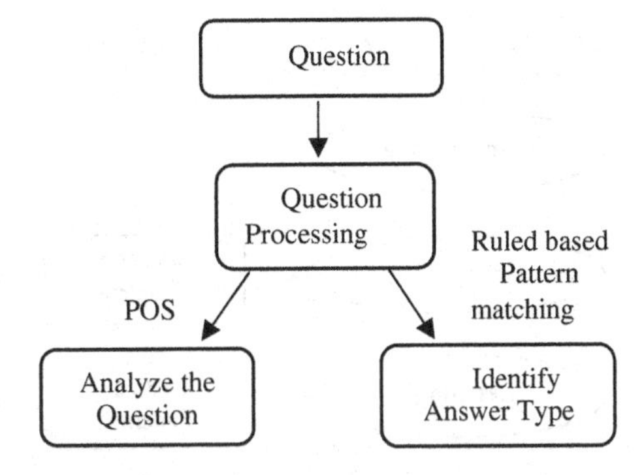

Fig. 2 Question processing

tagger assign the weightages to the each word in dynamically changing context. Tagger cannot process by its own. Need to consider the behavior of the language. (In Sinhala) in this research the POS tagger has been developed using tagged Sinhala corpus of university of Colombo school of computing in Sri Lanka (Fig. 3).

Language model. Language model has responsible to inform about the language to the POS tagger. Language model describe Appearance of the language, behavior of the language, grammar rules of the Sinhala language.

Identify Answer Type. Answer type is a one of a key input to answer processing component. Using answer type, extract answer form selected line in summarized document while answer processing [8]. Many approaches available for answer type identification. For "Mahoshada" project rule based with pattern matching is used to identify the question type.

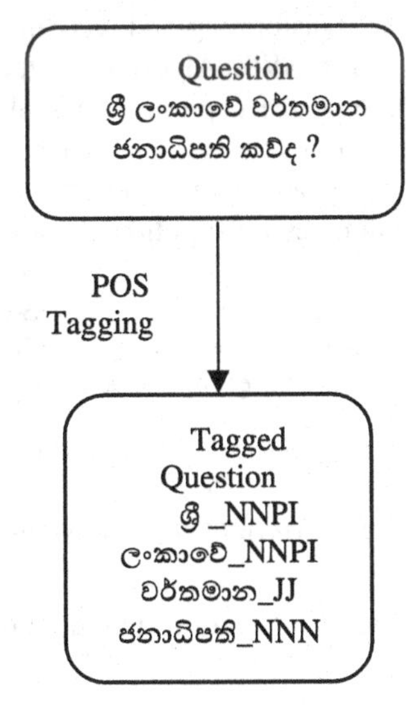

Fig. 3 POS-tagging

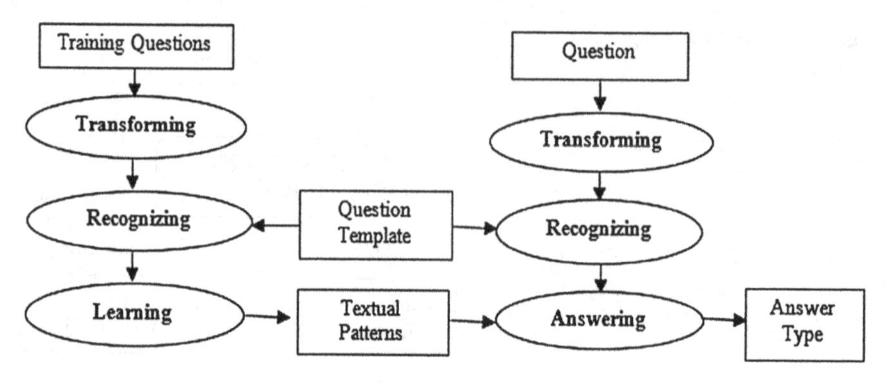

Fig. 4 Architecture of answer type identification using SVM algorithm

Example.
ශ්‍රී ලංකාවේ වර්තමාන ජනාධිපති කව්ද ?
Answer type = Person Type
සමාධි පිළිමය පිහිටා ඇත්තේ කොහේද?
Answer type = Location Type

Question types thus derived are used to extract and filter answers in order to improve the overall accuracy of question answering system. To enhance this question classification, we use the question informer feature with Support Vector Machines (SVM). In machine learning approach feature selection is an optimization problem that involves choosing an appropriate feature subset.

As shown in Fig. 4, the system entails two main functions, one is learning and the other is answering. For both functions, the question has to be pre-processed by the transforming and recognizing module.

The SVM Algorithm determines the class of the question. This was done by finding the appropriate template of the question. Currently, [9] we have defined question classes, and each class has several templates formulated as regular expressions which indicate the possible appearance of this type questions. Some examples for question classes are as the follow

කවුද? (who) කුමක්ද? (what)
කොහෙද? (where) කීයද ? (how much)
කොපමනද? (how much) කුමන ?(what)

2.4 Answer Processing

The architecture of this module is shown in Fig. 5.

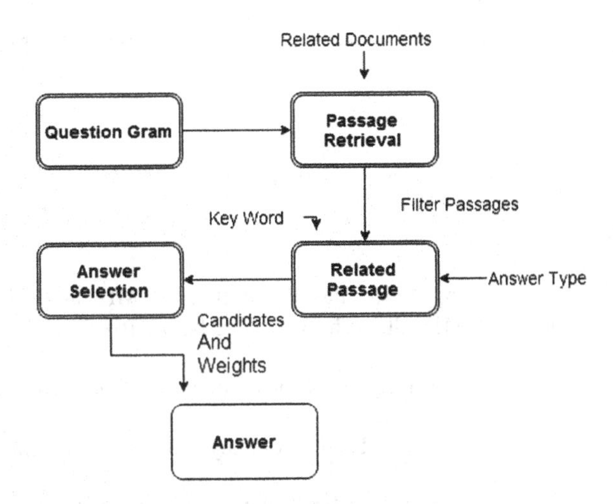

Fig. 5 Architecture of answer processing

The input of this module is constituted by the relevant summarized documents returned by the Question Categorization module, the answer type of the question obtained through the Question Classification module. An n-gram module [10], which we named Question-gram since it creates grams by tokenizing the question, is instantiated for each and every retrieved documents looking for one or several matching passages. This gram starts with a 2-gram and ends with an n (question_lenght-1)-gram. More details about how question-gram works shown in the below (Fig. 6).

There may be more than one passage for a particular question, it may include even unrelated passages also. The process of identifying the most relevant passage is the most crucial process, ability to provide correct answers defend on this process. After retrieving matching passages for a question, those passages goes through a kind of a filter process, end of this process output will be filtered most relevant passage for a particular question. Filter process identifies the passage which has the highest number of question gram.

Example. Suppose for a question there are three passages and highest gram is 4-gram. Then,

Passage 1 contains only 2-gram,
Passage 2 contains only 3-gram and 4-gram,
Passage 3 contains only 2-gram, 3-gram and 4-gram. According to this most relevant passage is passage 3.

This AP module has the relevant passage, keyword identification process starts. This process is to identify an important word for question which helps to identify the most suitable answer. The identification of the answer is done using Positive Point wise Mutual Information (PPMI). Using PPMI equation get the values for each and every question words, simply this tells how many times each word of the question occurred in the retrieved passages, as the key word it generates the

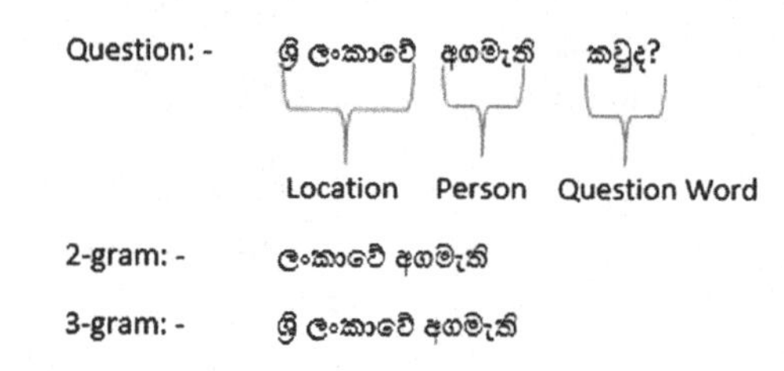

Fig. 6 Example for question gram

question word which has the highest PPMI value [11–13]. Equations used to generate PPMI values shown in the below (6).

$$\text{Word1}_{ppmi} = (\text{Line Frequency} * \text{Question Frequency}) * 100 \tag{6}$$

$$\text{Line Frequency} = \text{Line Count}/\text{Line Size} \tag{7}$$

$$\text{Question Frequency} = \text{Question Count}/\text{Question Size} \tag{8}$$

Finally to get the correct answer keyword and answer type goes through the answer selection process. In this process it retrieves the candidate answers by going through the relevant passage.

Example. If answer type is Person type then it retrieves all the persons in that passage. Then using distance calculation methods it gets the shortest distance between candidate answers and the keyword. As the correct answer it outputs the candidate answer which has the shortest distance.

3 Research Findings/Results and Evidence

"Mahoshadha" is the only attempt of Sinhala QA based on NLP this has become an encouragement for other researchers who are interested in this area. A result of this research, new researches has started for the first time in Sinhala. Best example is a research for QA system for mathematical operations has started as the second part of "Mahoshadha". We have encouraged new researchers to come up with an advanced POS tagger using NLP.

Project "Mahoshadha" is the core application of the research which is based on a Sinhala News corpus. With the success of the core application we have come up with other applications of "Mahoshadha" like "Self Learning Tool for School Children", "Call Assistant Application" for call centres, and "Patient Assistant System" for hospitals that can be very useful in real world.

We provided the self-learning tool to school teachers and students as the first step to get feedback. The product received high positive feedback on finding solution for subject matters themselves, fast and easy to operate the application. Initial request of all of them to provide this application around all the schools island wide. Therefore we are planning to provide the learning tool for government school free of charge as a social service. Then we will consider about the other applications.

4 Conclusions

The goal of "Mahoshadha" was to retrieving the most accurate answers to the questions ask by users in Sinhala Language. According to the test results the goal has accomplished with 93 % of accuracy and with a high efficiency of generating the answer. "Mahoshadha" opens new paths in Sinhala Natural Language Processing for the researchers as it kept first steps.

Accuracy and efficiency of "Mahoshadha" is increased automatically through machine learning used in summarization, categorization and also in retrieving the answers. "Mahoshadha" can be applied for different types of real world applications.

5 Future Works

We are currently working on delivering the self-learning tool to government schools free of charge as a social service. Using the feedback we hope to improve it further as an answering system for Multiple Choice Questions. This could also be developed as a call assistant system for companies. This could also be used as a Medical Instructor, User Guidance, and Patient Assistant for hospitals and other applications that can be used in mundane activities in order to make our lives easier.

Our next step is to apply "Mahoshadha" for mobiles as mobile application in Android and IOS that can be adapted to any user requirement.

The Application will focus on a voice model which is capable of getting inputs and generating outputs through voice.

Acknowledgment "Mahoshada" team would like to thank Dr. A.R. Weerasingha and Mr. Viraj Welgama of the University of Colombo Language Center for providing a tagged Sinhala News corpus to successfully complete our research.

References

1. Hovy, E.H.: Automated Text Summarization. The Oxford Handbook of Computational Linguistics, pp. 583–598. Oxford University Press, Oxford (2005)
2. Danesh, A., Moshiri, B., Fatemi, O.: Improve text classification accuracy based on classifier fusion methods. In: 10th International Conference on Information Fusion, pp. 1–6 (2007)

3. Guo, G., Wang, H., Bell, D., Bi, Y., Greer, K.: KNN model-based approach in classification. In: Proc. ODBASE, pp. 986–996 (2003)
4. Shin, K., Abraham, A., Han, S.Y.: Improving kNN Text Categorization by Removing Outliers from Training Set, in Computational Linguistics and Intelligent Text Processing, vol. 3878. Series Lecture Notes in Computer Science, pp. 563–566 (2006)
5. Rahal, I., Najadat, H., Perrizo, W.: A P-tree Based K-Nearest Neighbor Text Classifier Using Intervalization. Computer Science Department, North Dakota State University
6. Soucy, P., Mineau, G.W.: A Simple k-NN Algorithm for Text Categorization. Department of Computer Science, Universite Laval, Quebec, Canada, pp. 647–648
7. Miah, M.: Improved k-NN Algorithm for Text Classification. Department of Computer Science and Engineering University of Texas at Arlington, TX, USA, vol. 3, pp. 80–84
8. Tong, S., Koller, D.: Support Vector Machine Active Learning with Applications to Text Classification. Computer Science Department, Stanford University, pp. 287–295 (1998)
9. Cristianini, C., Taylor, J.S.: An Introduction to Support Vector Machine, pp. 206–240. Cambridge University Press, Cambridge (2000)
10. Buscaldi, D., Rosso, P., Gómez-Soriano, J.M., Sanchis, E.: Answering questions with an n-gram based passage retrieval engine. J. Intell. Inf. Sys. **34**, 113–134 (2010)
11. Church, K.W., Hanks, P.: Word association norms, mutual information, and lexicography. Comput. Linguist. **16**(1), 22–29 (1990)
12. Turney, P.D., Pantel, P.: From frequency to meaning: vector space models of semantics. J. Artif. Intell. Res. **37**(1), 141–188 (2010)
13. Dagan, I., Pereira, F., Lee, L.: Similarity-based estimation of word co-occurrence probabilities. Mach. Learn. **34**(1), 43–69 (1999)

Trust Appraisal Based Neighbour Defense Secure Routing to Mitigate Various Attacks in Most Vulnerable Wireless Ad hoc Network

Tada Naren, Patalia Tejas and Patel Chirag

Abstract The Mobile Ad hoc network is the most growing field as it is dynamic in nature and requires no central authority to manage it. In this paper we have worked on routing layer of MANET. We have worked upon AODV (Ad hoc on-demand distance vector) routing protocol. In this paper we have developed a very vulnerable network and a number of attackers and also modified AODV with our trust-based approach. To the best of our knowledge no one has implemented strong attackers and prevented them. In our approach each node monitors its neighbours' activities and develops a trust table which in turn is very difficult to maintain when there is no cooperation between nodes. Based on the trust table it decides whether to eliminate the attacker. We have also optimized HELLO packet of AODV in such a way that it propagates information about malicious node among other available nodes and isolates the attacker from the entire network. At the end we present the results which clearly show that proposed protocols perform better than AODV in vulnerable situation. To evaluate the performance, we have carried out an extensive simulation study of AODV and modified AODV using NS2.35 with and without attacker showing significant improvement in Packet Delivery Fraction (PDF), Packet Drop Ratio (PDR), Average Throughput (AT), Normalized Routing Load (NRL) and E2E (Average End-2-End delay).

Keywords AODV · MANET · HELLO · HAT

T. Naren (✉) · P. Tejas · P. Chirag
VVP Engineering College, Gujarat Technological University, Ahmedabad, India
e-mail: Narentada@gmail.com

P. Tejas
e-mail: pataliatejas@rediffmail.com

P. Chirag
e-mail: cecrp@vvpedulink.ac.in

S.C. Satapathy and S. Das (eds.), *Proceedings of First International Conference on Information and Communication Technology for Intelligent Systems: Volume 1*, Smart Innovation, Systems and Technologies 50, DOI 10.1007/978-3-319-30933-0_33

1 Introduction

Wireless Ad hoc network is collection of autonomous nodes which behave like a router. There is no governance in network as all nodes assume that there is cooperation between them. Due to mobility, our traditional Internet routing algorithms are not suitable in Ad hoc network [1]. As Ad hoc network requires cooperation in each node there is a chance that some intruders may be dangerous to network. AODV is the best suitable protocol for wireless Ad hoc network. AODV performs better with respect to PDF (Packet delivery ratio), E2E (End-2-End delay) and throughput of network with comparison to DSR, DSDV and ZRP routing protocols [2]. AODV adopts quick change to mobility in wireless network. To find the path, it implements on-demand approach. As of now many researchers are working to implement security of routing in mobile Ad hoc network. We have implemented trust framework in which we prevent flood attack, gray-hole attack, drop attack and black hole attack. Many researchers worked upon cryptographic approaches. Paper contains related work, proposed algorithm followed by simulation environment, results, conclusion and future work.

1.1 Theory of AODV (Ad hoc on Demand Distance Vector) Protocol

AODV is a reactive routing protocol in MANET. Each node maintains its neighbour's information in its routing table. It has four control messages namely Route Request, Route Reply, Route Error and HELLO packets to perform and maintain routing operation. Request is broadcast in nature [3]. The source node sends it to find the destination node. Each node receives and forwards it until the packet reaches the destination node. Each node maintains a reverse route in their routing table. Destination node will send the route-reply packet towards the source. Each node maintains forward route towards destination as reply relay from destination to source. Route error message is generated by nodes in certain conditions when active route is broken due to mobility, node is crashed or down, or some node goes out of range. Route error packet is broadcast in nature and it will flood into the network. Error packet has unreachable destination count field which denotes the number of destination for which this error has occurred. HELLO packet is periodic in nature and it is generated by each neighbour of the node. It is very useful to maintain liveliness of network. Each node maintains neighbour table for taking routing decision [4]. AODV is plain protocol, there is no security provision added to it. In next session we will talk about security in MANET (Fig. 1).

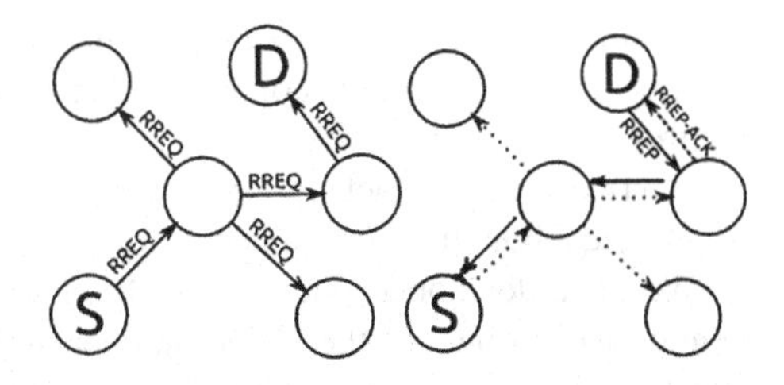

Fig. 1 AODV route discovery process

2 Security Implementation Method in Protocol

As AODV is very poor in matter of security, many researchers have shown security method either by implementing it as cryptographic approach or by calculating trust value of a node. Protocols like SAODV [5], SEAD, ARAN and SAR provide security implementation in AODV. When we apply security primitives, there is requirement to maintain key distribution centre which works as a central entity in network. Pirazada and all, applied trust method which calculates trust value in the neighbouring nodes [6]. In above protocols no one has provided complete framework of security and none of them gives complete isolation of the attacker from the network after it is detected. We have extended it with our extensive simulation setup and optimized HELLO packet to impose complete isolation of attacker from network which is the strength of our protocol design.

2.1 Different Attacks Possible in AODV

Anybody can intercept the AODV packet as there is no security provision in it. We have implemented four different strong attacks with various numbers of malicious nodes in network. Topology Table 3.2 depicts entire simulation parameters of network.

Black-Hole attacks: A malicious node sends fake route reply packet, claiming that it has the best path and causes other good nodes to route data packets through the malicious node. In reply packet, attacker inputs the highest sequence number. After becoming part of active route, attacker drops each data packet passing through it [7].

Gray-Hole attacks: Contrary to black-hole attacker, a gray hole attacker only drops selective data packet passing though it. Comparing to black-hole attack, it is a challenging task to detect it [7][8].

Drop Attacks: It blindly drops each type of packet including control packet as well as data packet. This attack is easy to detect.

Flood attacks: It is a denial-of-service attack. Using this attack, attacker floods the network by injecting control packet or data packet. We have taken request control packet as flood attacker.

Delay attacks: Attacker delays each control and data packet while forwarding it to its neighbour. It incurs unnecessary delay in transmission.

We have developed our security framework in such a way that it covers prevention techniques for the above mentioned attacks. Our proposed protocol incurs less burden in presence of trust appraisal method and gives the best performance while dealing with more number of attackers compared to standard protocols.

3 Proposed Protocols and Its Implementation

Proposed protocol maintains two counters and two timers in making smooth operation that detects, prevents and isolates bad nodes. The entire operation is divided into three categories mentioned earlier.

3.1 Counter and Timer

1. **Retardous_counter**:
   It maintains a list of malicious nodes detected by their neighbour. This list is propagated via HELLO message using **H**ELLO **A**larm **T**echnique (HAT) (Table 1).
2. **CP_Counter**:
   It maintains frequency of control packets coming from neighbour within a second. This structure is specially used to detect flood attacks. We apply threshold value 10[4] on each packet and detect whether it is coming from genuine node or from malicious node. By applying following formula we can count delay for packet forwarding (Table 2).

Table 1 Retardous_counter

Attacker ID 1
Attacker ID 2
Attacker ID 3

Table 2 CP_counter

Sender_ID	Packet_type	Arrival_frequency	Time_stamp	Arrival_time
1	RREQ	23	0.3456	0.3537
4	RERR	12	0.3446	0.3498
8	RREP	5	0.1245	0.1268

$$\textbf{Forwarding\_delay} = |\textbf{Time\_stamp} - \textbf{Arrival\_time}|$$

3. **Timer**:
   It is used to check CP_Counter entry for expiry time as well as count for the request to check whether it exceeds peak value or not. It is also use to flush the entry which does not exceed from peak value for forwarding_delay and arrival_frequency.

3.2 Algorithm

```
If (CacheTimer out)
     Then
     Erase CP_counter table entries
If (Sender is found in Retardous_counter)
     Then
     Drop the packet
If ((Sender is found in neighbor_table) && (There is no
entry found in CP_Counter table))
     Then
     Add (Sender_ID and Packet_type for in CP_Counter
  table)
If (Arrival_friquency >PeakValue) || (Forwarding_delay >
PeakDelay)
     Then
     Put the Sender is in Retardous_Counter
     Else
     Flush the entry from CP_Counter
  Activate HAT
     Put malicious node list in HELLO packet
```

3.3 Simulation Environment and Parameter

No.	Simulation parameter	Value
1	Simulator	NS-2.35
2	Simulation area	1000 m × 1000 m
4	Mobile nodes	25
5	Antenna type	Omni antenna
6	Propagation model	Two ray ground
7	Number of connections	5
8	Packet size	512 byte
9	Routing protocols	AODV,NAODV
10	Data traffic	CBR (UDP)
11	Simulation time	100 s
12	Percentage of malicious node	0, 10, 20, 30 and 40 %
13	Pause time	0, 5, 10, 15 and 20 ms
14	Flood interval	0.067 s
15	Cache interval	1 s
16	Peak value for flood prevention	11(RREQ,RERR,RREP)

4 Result Analysis

(See Figs. 2, 3 and 4.)

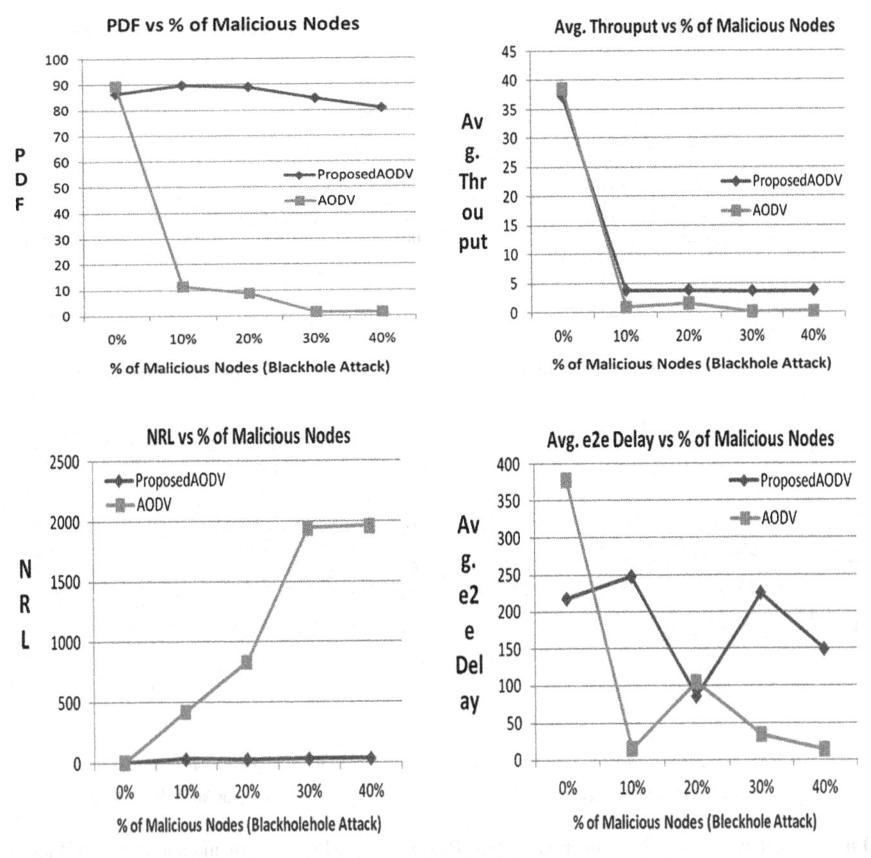

Fig. 2 CBR with 5 CBR connection pattern Pause time 0 (Presence of malicious node), Blackhole attacker

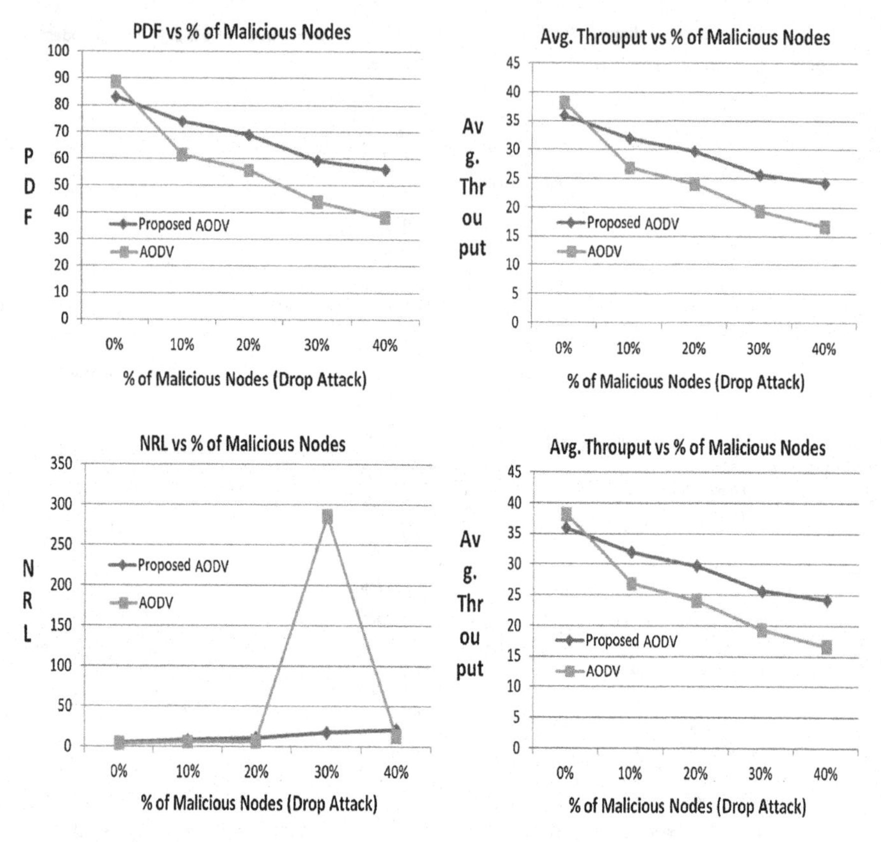

Fig. 3 CBR with 5 CBR connection pattern Pause time 0 (Presence of malicious node), Flood attacker

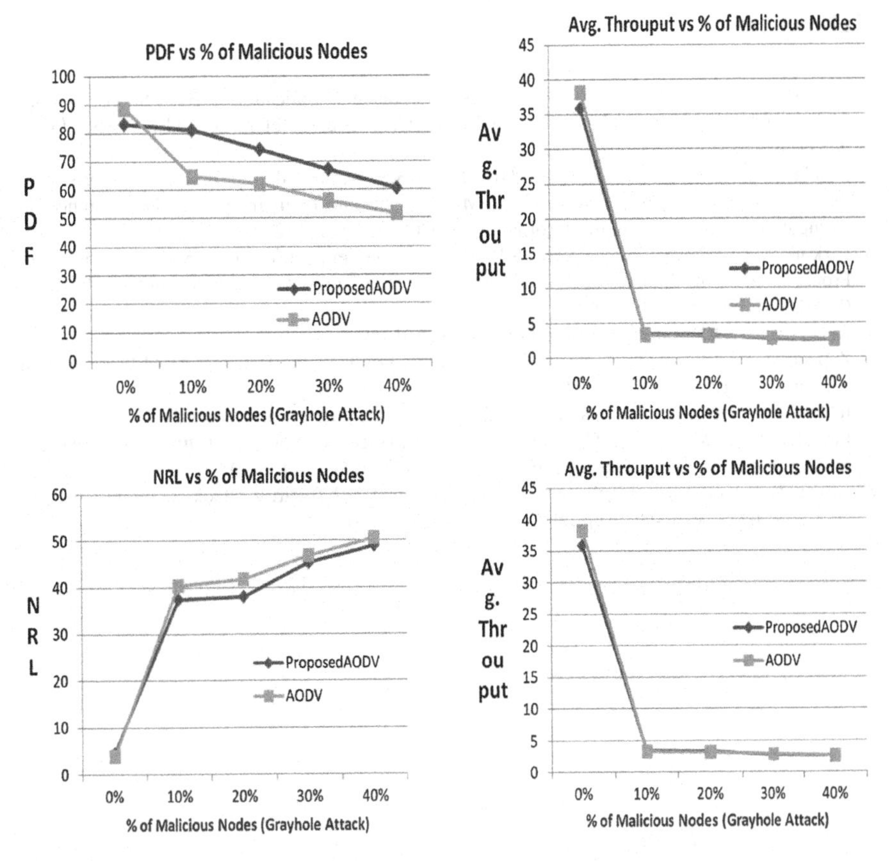

Fig. 4 CBR with 5 CBR connection pattern Pause time 0 (Presence of malicious node), Grayhole attacker

5 Conclusion and Future Work

The above graph has depicted that in the presence of malicious node our proposed AODV performs better than AODV. We have checked that without malicious node present in the network our proposed AODV incurs less burden compared to traditional AODV. It is also concluded from above graph that as the number of malicious nodes increases the efficiency of proposed AODV also increases as compared to AODV. In said approach we have assumed that source and destination nodes must not be considered as malicious node. Future work can include more number of attacks which in turn can be prevented with the same approach.

References

1. Ad-hoc-mobile-wireless-networks-principles-protocols-and-applications—By Subir Kumar Sarkar and Basavraju. February 5, 2013 by CRC Press Reference—349 Pages—ISBN 9781466514461
2. Gandhi, S., Chaubey, S., Tada, N., Trivedi, S.: Scenario-based performance comparison of reactive, proactive & hybrid protocols in MANET. In: 2nd IEEE International Conference on Computer Communication and Informatics (ICCCI-2012), 10–12, January 2012
3. Perkins, C.E., Royer, E.M., Das, S.R.: Ad-hoc on-demand distance vector routing. In: Proceedings of the International Workshop on Mobile Computing Systems and Applications (WMCSA), pp. 90–100 (1999)
4. RFC 3561 (AODV); http://www.ietf.org/rfc/rfc3561.txt
5. Zapata, M.G., Asokan, N.: Securing Ad hoc routing protocols, In: Proc. 1st ACM Workshop on Wireless Security, pp. 1–10 (2002)
6. Internet draft of SAODV https://tools.ietf.org/html/draft-guerrero-manet-saodv-03
7. Pirzada, A.A., McDonald, C., Datta, A.: Performance comparison of trust-based reactive routing protocols. IEEE Trans. Mobile Comput. **5**(6), 695–710 (2006)
8. Goyal, Priyanka: S.B., Singh, A.: A literature review of security attack in mobile Ad-hoc networks. Int. J. Comput. Appl. **9**(12), 0975–8887 (2010)

Behavior of Adhoc Routing Protocols for H-MANETs

Mitul Yadav, Vinay Rishiwal, Omkar Singh and Mano Yadav

Abstract Finite battery power (energy) of the nodes in Mobile Adhoc Networks (MANETs) leads to frequent route failures if not handled properly by a routing protocol. Conditions become worst when a pre assumption is made that all the nodes of the network have equal battery power to communicate. Network scenarios are usually heterogeneous when they are used and realized in real life practical situations. Therefore, considering only homogeneous scenarios restrict the designer to benchmark the performance of a routing protocol to actual practical aspects. Existing protocols and network designs for MANETs not only consider homogeneous scenarios but they are also tested for small size of networks with few numbers of nodes. The throughput achieved by MANETs has been extended its applications to commercial acceptability also. Therefore, to render MANETs services in terms of internet or large number of users, existing protocols should be tested with both large size homogeneous and heterogeneous networks. Thus, scalability becomes another big issue of concern for the better acceptability of a routing protocol for large networks. These limitations make the task of routing very difficult in MANETs. Therefore, routing protocols especially in heterogeneous MANETs (H-MANETs) must be energy efficient and scalable. In this paper, the behavior of some well accepted communication protocols namely AODV, DSR and AOMDV is analyzed for H-MANETs. These protocols have been comprehensively analyzed considering different performance parameters (both traditional and energy

M. Yadav
IFTM University, Moradabad, UP, India
e-mail: mitul_yadav@yahoo.com

V. Rishiwal (✉) · O. Singh
MJP Rohilkhand University, Bareilly, UP, India
e-mail: rishi4u100@gmail.com

O. Singh
e-mail: omkar650@gmail.com

M. Yadav
Galgotias University, Greater Noida, UP, India
e-mail: mano425@gmail.com

S.C. Satapathy and S. Das (eds.), *Proceedings of First International Conference on Information and Communication Technology for Intelligent Systems: Volume 1*,
Smart Innovation, Systems and Technologies 50, DOI 10.1007/978-3-319-30933-0_34

related). This paper is aimed to judge the competence of considered protocols in an energy constrained environment with varying network size.

Keywords Mobile Adhoc networks · Energy · Routing protocols · Heterogeneity · H-MANETs

1 Introduction

MANETs [1] are only networks which can be cheaply and rapidly deployed in any hostile territory like military and search-rescue operations. Due to regular movement of nodes, it is very difficult to enable communication among them. This job is usually done by efficient and dynamic routing strategies [2]. Routing protocols are used to find a feasible path between a pair of nodes. This job in MANETs is done in different phases: firstly, the nodes need to discover the route information in the network, since they do not have a prior knowledge of topology. Then, the routes often get disconnected even after discovery due to frequent node movements. So it is the routing protocol which controls the searching the recovery of the routes between computing devices in a MANET.

Routing protocols in MANETs can be broadly categorized as reactive (AODV [2], DSR [3], TORA etc.), Proactive (DSDV, OLSR, FSR etc.) and hybrid [4–6] (ZRP, SHARP and DST etc.). These variants have their own strengths and weaknesses. Reactive protocols are often best suited for MANETs environment as they setup a routing path on need basis hence incurs minimum routing overheads. Most of the existing MANET routing protocols utilize the single route for communication between a given pair of nodes. At the time of route break, finding alternative routes introduces additional delay. Therefore, multipath routing like AOMDV [7] can be used to reduce latency and overheads.

In this paper, the performance of AODV and DSR (single path) is compared with one of the well accepted multipath routing protocol i.e. AOMDV over the different performance parameters. Though many such comparative studies [8–16] are available in the existing literature, but none of them has covered energy related metrics for performance evaluation with heterogeneous node configurations. The objective of this work is to observe the behavior and energy efficiency of single as well as multipath routing protocols for heterogeneous network environment.

Heterogeneous MANETs (H-MANETs) [17] are composed of nodes with different value of initial energy, from where node starts their transmission. Practically, networks are heterogeneous in nature. Communication is relatively easier in homogeneous networks than in heterogeneous networks. Heterogeneity imposes many additional requirements and challenges in the network [6, 10, 17–19].

This paper is organized as follows. Section 2, discusses the considered routing strategies. Section 3 provides the details of the parameters used in the simulation and energy model. Considered metrics and performance results are shown and discussed in Sect. 4. Section 5 presents the concluding remarks.

2 AODV, DSR & AOMDV Protocols

2.1 Ad Hoc on Demand Distance Vector (AODV)

AODV [2] is a reactive, distributed, deterministic, single path and state dependent protocol. In AODV, a route is setup by the destination on first come basis and route breakages can be repaired locally. AODV uses the concept of sequence numbers to maintain the freshness of the routes. AODV incurs less overheads and connection setup delay. But, the regular hello message in AODV consumes bandwidth.

2.2 Dynamic Source Routing (DSR)

DSR [3] is also a reactive protocol that performs both route search and recovery as in AODV. It maintains a route cache, which limits the excessive request propagations and provide alternative routes. Unlike AODV, DSR supports asymmetric links. In DSR, no routing table is maintained as such. But DSR is not suitable for the large network and it require more processing resources compare to AODV.

2.3 Ad-hoc On-Demand Multipath Distance Vector Routing (AOMDV)

AOMDV [7] protocol is different from AODV in different aspects as it computes multiple link disjoint paths. It can easily opt an alternative path al any link failure. For efficiency and load balancing only link disjoint paths are computed. Link disjoint paths, on the other hand, may have common nodes [16]. Unlike the single path case, different routes with different hop counts are offered by AOMDV.

Route maintenance in AOMDV is similar to AODV except that a destination is declared unreachable only when all routes to the destination are broken.

3 Heterogeneity in MANETs

It is assumed in the existing literature that nodes in MANETs are homogeneous in their capabilities. However in real life applications, the networks are practically heterogeneous [6, 17]. The devices may differ in various aspects from the communication point of view, such as transmission power, antenna type, mode of communication, initial energy etc. [18].

Efficient communication is relatively easier in homogeneous networks than in heterogeneous networks. Heterogeneity always brought additional challenges in the network. Some fundamental issues raised by heterogeneity of the nodes are as following [6, 17–19]:

- An unambiguous classification of devices is needed.
- An appropriate device identification scheme and addressing mechanism is needed.
- Need to define different mobility patterns to support heterogeneous networks.
- Energy constraints and power management issues.
- Issue of handling real time traffic over such networks.
- Quality of service.
- Security issues due to heterogeneous nodes in the network.
- Nodes' cooperation for the network functioning.
- Support for different proactive/reactive/hybrid routing protocols.
- Inter-operation with other types of wireless networks.

In this work, the initial energy among the nodes is distributed randomly, to create an instance of heterogeneity. This is done to understand the behavior of considered protocols in energy constrained heterogeneous environment.

4 Simulation Methodology

To compute and compare the routing protocols' performance in heterogeneous environment, we have considered varying number of nodes. To generate heterogeneous network scenario, each node has been given different initial energy value.

The performance analysis required for this work has been done through its simulation on network Simulator with inbuilt energy model [20, 21, 23]. The various parameter used in forming different scenario patterns are shown in Table 1.

Table 1 Simulation parameters

Parameters	Values
Simulation area	1000×1000 m^2
Simulation time	500 sec
Routing protocol(s)	AODV, DSR and AOMDV
Mobility model	Random waypoint
MAC	IEEE 802.11
Number of nodes	10, 15, 20, 25, 30, 35, 40, 45
Pause time	100 s
Traffic type	CBR
Number of Connections	50 % of the network size
Initial energy	Randomly initialized: 50 to 100 Joules
Data rate	2 Mbps

5 Performance Metrics & Result

Five performance metrics have been used in this paper to evaluate the efficacy of the considered protocols. The different scenarios which have been simulated for different performance parameters are discussed below:

Scenario 1: Packet Delivery Ratio (PDR) defines the fraction of successfully delivered packets. In this scenario, PDR for AODV, DSR and AOMDV has been compared against the varying network size. The performance is better when PDR is high [22].

As shown in Fig. 1, the PDR with AODV and DSR is almost similar. The PDR is comparatively less in case of AOMDV routing protocol. This shows the effective packet delivery of AODV and DSR over AOMDV protocol.

Scenario 2: Throughput defines the number of bytes delivered to destination in unit time. This scenario is having comparison of AODV, DSR and AOMDV w.r.t. throughput, while heterogeneous nodes are there in the network.

According to Fig. 2, throughput is quite high with DSR as compare to other protocols, as the network size increases, the throughput of AOMDV and AODV are almost same.

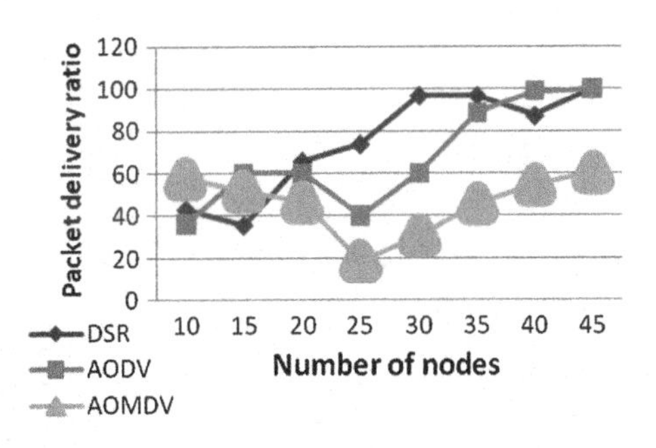

Fig. 1 Nodes versus Packet Delivery Ratio

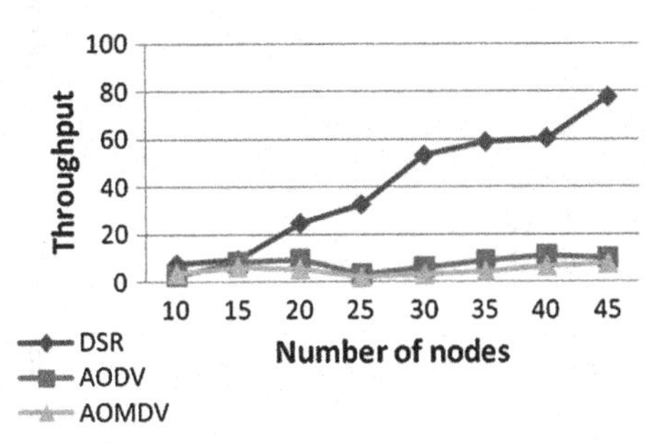

Fig. 2 Nodes versus Throughput

Scenario 3: Average End to End Delay (AEED) is the another parameter for which, the considered protocols have been evaluated for a heterogeneous network. The performance is better when AEED is low [22]. In Fig. 3, AEED is quite high in case of DSR and it is decreasing with the increasing size of network. For both, AODV and AOMDV, the AEED is quite low and consistently range bound.

Scenario 4: In this scenario the performance of AODV, DSR and AOMDV has been compared in terms of total energy consumed (TEC) by alive nodes, during packets transfer.

As shown in Fig. 4, the TEC with AODV and AOMDV is almost similar. The TEC is comparatively less in case of DSR. This shows the energy efficiency of DSR with heterogeneous nodes.

Scenario 5: This scenario is formed for measuring average energy consumption (AEC) for AODV, DSR and AOMDV.

It is evident from Fig. 5 that the AEC of AODV and AOMDV is almost similar and quite high, as compared to DSR. The AEC is minimum for DSR.

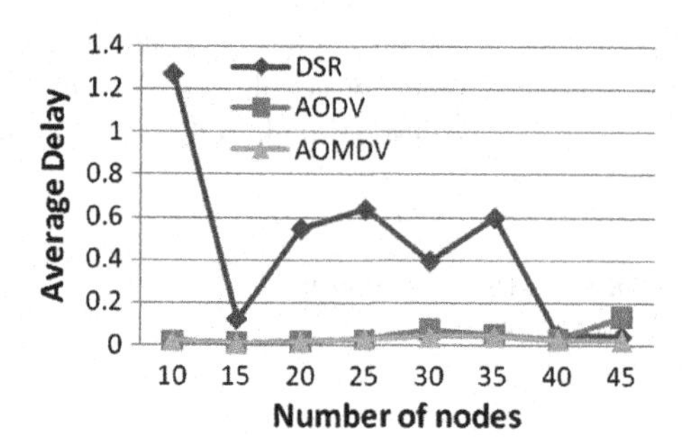

Fig. 3 Nodes versus Average End to End Delay

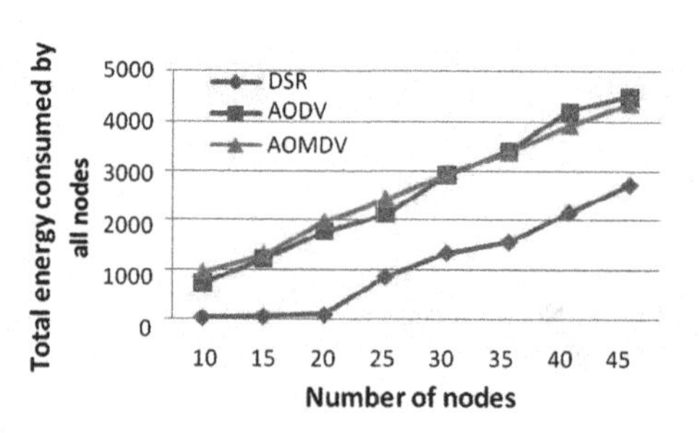

Fig. 4 Total Energy Consumption versus Nodes

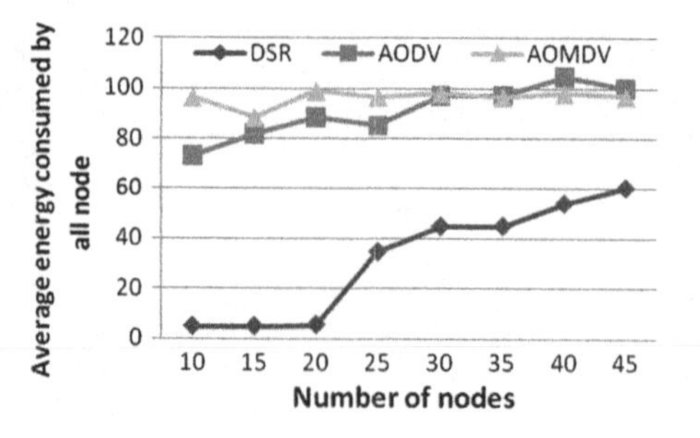

Fig. 5 Average Energy Consumed versus Nodes

6 Result Analysis

For varying network size, the AEED remains range bound in case of heterogeneous networks. The value of throughput and AEED for DSR is better as compare to AODV and AOMDV for heterogeneous environment. The AODV and DSR outperform the AOMDV protocol, when PDR is considered.

TEC is also increasing with the increase network size for H-MANETs for all three protocols. But in case of AEC, DSR outperforms AODV and AOMDV due to its source routing concept.

7 Conclusion and Future Work

In this paper, the behavior of AODV, DSR and AOMDV is analyzed for different heterogeneous network scenarios, with respect to varying network size from 10 to 45 nodes. The results clearly indicate the difference in the behavior of different reactive routing protocols.

It is found that none of the protocol is a clear winner in all considered network scenarios. In some cases DSR and AODV is better while for others AOMDV is better. The relative performance of the protocols also depends on the mobility and traffic conditions. This work will probably develop interest to analyze the performance of other protocols with H-MANETs. This work is restricted only for 10 to 45 nodes. In future, work will be done for large scale networks with more energy related parameters.

References

1. Perkins, C.E.: Ad hoc networking. Addison-Wesley, Boston (2001)
2. Perkins, C. E., Royer, E. M.: Ad hoc on-demand distance vector routing. In: Proceedings of IEEE Workshop on Mobile Computing Systems and Applications, pp. 90–100 (1999)
3. Johnson, D.B., Maltz, D.A.: Dynamic source routing in Ad hoc wireless networks. In: Mobile Computing, Chapter 5, pp. 153–181. Kluwer Academic Publishers, Dordrecht (1996)
4. Royer, E.M., Toh, C.K.: A review of current routing protocols for Ad hoc mobile wireless networks. IEEE Pers. Commun. **6**(2), 46–55 (1999)
5. Boukerche, A., Turgut, B., Aydin, N., Ahmad, M.Z., Bölöni, L., Turgut, D.: Routing protocols in Ad hoc networks: a survey. Elsevier Comput. Netw. **55**, 3032–3080 (2011)
6. Vijaya Kumar, G., et al.: Current research work on routing protocols for MANET: a literature survey. Int. J. Comput. Sci. Eng. **02**(03), 706–713 (2010)
7. Marina, M., Das, S.: On-demand multipath distance vector routing in ad hoc networks. In: Proceedings of the International Conference on Network Protocols (ICNP), pp. 14–23 (2001)
8. Trung, H.D., Benjapolakul, W., Duc, P.M.: Performance evaluation and comparison of different ad hoc routing protocols. Department of Electrical Engineering, Chulalongkorn University, Bangkok, Thailand (2007)

9. Ehsan, H., Uzmi, Z.A.: Performance comparison of Ad hoc wireless network routing protocols. In: Proceedings of the IEEE 8th International Multi Topic Conference, pp. 457–465 (2004)
10. Kaosar, M.G., Asif, H.M., Sheltami, T.R., Hasan Mahmoud, A.S.: Simulation-based comparative study of on demand routing protocols for MANET. In: Proceedings of the Internaional Conference on Wireless Networking and Mobile Computing, vol. 1, pp. 201–206 (2005)
11. Biradar, S.R., et al.: Performance evaluation and comparison of AODV and AOMDV. Int. J. Comput. Sci. Eng. **02**(02), 373–377 (2010)
12. Kumawat, R., Somani, V.: Comparative study of on-demand routing protocols for mobile Ad-hoc network. Int. J. Comput. Appl. **27**(10), 6–11 (2011)
13. Wadhwa, D.S., et al.: Performance comparison of single and multipath routing protocols in Adhoc networks. Int. J. Comp. Tech. Appl. **2**(5), 1486–1496 (2011)
14. Singh, P., Kaur, H., Ahuja, S.P.: Brief Description of routing protocols in MANETS and performance and analysis (AODV, AOMDV, TORA). Int. J. Adv. Res. Comput. Sci. Soft. Eng. **2**(1) (2012)
15. Chadha, M.S., Joon, R.: Simulation and comparison of AODV, DSR and AOMDV routing protocols in MANETs. Int. J. Soft. Comput. Eng. **2**(3), 375–381 (2012)
16. Khiavi, M.V., Jamali, S.: Performance comparison of AODV and AOMDV routing protocols in mobile Ad hoc networks. Int. Res. J. Appl. Basic Sci. **4**(11), 3277–3285 (2013)
17. Zuhairi, M.F., Harle, D.A.: AODV Routing Protocol in Heterogeneous Networks, PGnet (2010). ISBN: 978-1-902560-24-3
18. Chekkiralla, S., et al.: Routing in heterogeneous wireless Ad-hoc networks. Thesis submitted at MIT, USA (2008)
19. Wahi, C., Sonbhadra, S.K., Chakraverty, S., Bhattacharjee, V.: Effect of scalability and mobility on on-demand routing protocols in a mobile Ad-hoc network. Lecture Notes Soft. Eng. **1**(1), 12–16 (2013)
20. Rishiwal, V., Verma, S.: Analysis of energy consumption pattern for on demand Adhoc routing protocols. Int. J. Comput. Sci. Netw. Secur. **8**(11), 429–435 (2008)
21. Agarwal, S.K., Rishiwal, V., Arya, K.V.: Fallout of different routing structures for different mobility patterns in large Ad-hoc networks. IEEE Confluence **2013**, 242–249 (2013)
22. Gupta, A.K., Sadawarti, H., Verma, A.K.: Performance analysis of AODV, DSR & TORA routing protocols. IACSIT, pp. 226–231 (2010)
23. Network Simulator NS2 and Network Animator NAM. [Online]. Available: http://www.isi.edu/nsnam

Mathematical Treatment of ABC Framework for Requirements Prioritization

Sita Devulapalli, O.R.S. Rao and Akhil Khare

Abstract The conceptual ABC framework for requirements prioritization for software products development is analyzed mathematically adapting sets for each layer of the framework. Weights are associated with classes of each set and requirements priorities are determined based on contribution from membership in each set. A unique number sequencing scheme is proposed to visually interpret the priorities associated with requirements.

Keywords ABC framework · Requirements prioritization · Multi-level framework · Software product development

1 Introduction

ABC Framework proposed [1] for the requirements prioritization takes in to account different aspects encountered in the product development flow in a structured and in a sequence of layers. Requirements prioritization is invariably linked to cost of development and benefit to be achieved in most of the methods proposed for prioritization. In general the cost factor is considered to the extent of time taken to develop or resources cost. Business value is normally understood to the extent of immediate revenue. Considering the "other than software world" projects and cost and benefit analysis done for taking up projects—Business value encompasses present value of future returns, indirect benefits, return on investment periods. The costs involve not just development costs, but also opportunity costs and impact costs.

S. Devulapalli (✉) · O.R.S. Rao
ICFAI University, Jharkhand, India
e-mail: sitadpalli@yahoo.co.in

O.R.S. Rao
e-mail: orsrao.icfai@gmail.com

A. Khare
MVSR Engineering College, Hyderabad, India
e-mail: khareakhil@gmail.com

S.C. Satapathy and S. Das (eds.), *Proceedings of First International Conference on Information and Communication Technology for Intelligent Systems: Volume 1*, Smart Innovation, Systems and Technologies 50, DOI 10.1007/978-3-319-30933-0_35

Typically Software requirements prioritization does not start or stop at one time or in one step. The prioritization of what will finally get into the product release goes through levels of decision making considering different aspects. Trying to club all the aspects into one or two parameters or trying to prioritize at one time considering all aspects generally results in suboptimal or not so well understood prioritization. Don Reinertsen [2] proposed Weighted Shortest Job First, which talks of the economic value in the product development process flow. This model takes into consideration business value, time criticality, risk reduction, future value for determining cost of delay. The model considers job size or time to develop and proposes a ratio of cost of delay to job size as the single weight for prioritization of requirements. Nevertheless, this model comes closest to the ABC Framework in considering Business value aspects for requirements prioritization.

In order to understand—What is to be made available in the next release, How to manage the requirements under expanding client needs, cost and time implications, what set of requirements implementation will increase revenues—a layered approach using ABC Framework paves way. It helps in prioritization of requirements and planning releases, streamlining the project deliveries to client's satisfaction without overworking the teams or missing time to market deadlines.

The framework proposed is conceptualized based on enterprise products' development experience. It enables practical use through its tabular format. This paper provides mathematical modeling through formation of sets for representing each layer of the framework and association of weights at each layer based on the class, the requirement belongs to, for arriving at a combination of weights for each requirement.

A Unique numbering scheme is proposed for easy interpretation and visualization of the parameters of the prioritization and basis of prioritization in the paper.

This paper describes the ABC framework in Sect. 2. Interpretation through sets is presented in Sect. 3. Unique weights numbering scheme is presented in Sect. 4.

2 ABC Framework for Prioritization of Requirements

The purpose of getting a set of requirements implemented for the next release (time bound) is to maximize the business value of the release for the most valued customers. A strict ordering of requirements may not be the need. Need is more for a near optimal sets of requirements. Since a release is always timed to meet customers expected needs, the following constraints are considered for prioritization of requirements:

1. Time/duration—minimum time required for development
2. Nature of development needed for the requirements
3. Resources—knowledgeable in domain/technology/skill
4. Uncertainties—changes due to expanded/extended scope
5. Impacts on existing customers and existing product modules

Based on the above considerations, ABC Framework enables simple and effective prioritization at multiple levels enabling implicit weights application for

Table 1 Framework—sets, classes

Sets	Classes/bins—A, B, C
S1. Business Value(BV) in conjunction with Customer Base (CB)	A: 20 % of CB with 70 % BV
	B: 30 % of CB with 25 % BV
	C: 50 % of CB with 5 % BV
S2. Requirements Applicability with respect to product, where UW: User Interface, BI: Business Logic, CP: Core	A: 70 % UW, 30 % BI, 0 % CP
	B: 50 % UW, 40 % BI, 10 % CP
	C: 30 % UW, 50 % BI, 20 % CP
S3. Implementation Cost, where MI: Marginal Implementation, NI: New Implementation, IR: Impact Recovery.	A: 70 % MI, 25 % NI, 5 % IR
	B: 50 % MI, 40 % NI, 10 % IR
	C: 30 % MI, 50 % NI, 20 %IR
S4. Time Requirement, where L: 8 to 16 person weeks, M: 4 to 8 person weeks, S: 2 to 4 person weeks	A: 10 % L, 20 % M, 70 % S
	B: 15 % L, 25 %M, 60 %S
	C: 20 % L, 30 % M, 50 % S
S5. Resource Requirement, where RC: Core aware, RI: Industry aware, RT: Technology aware	A: 10 %RC, 20 %RI, 70 %RT
	B: 15 %RC, 25 %RI, 60 %RT
	C:20 %RC, 30 %RI, 50 %RT

relevant parameters for the requirements, which enables flexible planning through the development cycle. The framework provides visualization for the changes in requirements during the release cycle and acts as an easy communicator to the involved stakeholders including testing team members.

The Framework is defined as 5 sets based on most used parameters in the sequence of priority determination. Each set is defined by three classes/bins defined by % value of the respective set parameters. Requirements are grouped into the classes in the sets in the process of prioritization. The % bands may vary from industry to industry and organization to organization to some extent.

Prioritization sets—S1 to S5 and classes/bins—A, B, C within are described in Table 1.

The Framework is applied in a layered approach through the sets. The order of preference emerges for the requirements Set through the filtering process. Not all sets may be required to be used. When all sets are used for classification, we will arrive at 243 bins of requirements. Based on the constraints and release theme, the bins can be selected in the order of preference for the releases.

3 Interpretation Through Sets for the ABC Framework

Business Value encompasses Value to customer now and repeat value to customer, value to other customers and value possible through being used as a platform component in other products. The assumption is—there is at least one customer for each requirement and all requirements have equal business value of unit 1. R1 to R3

Table 2 Set 1

Requirements	Normalized CB ($CB/\sum CB$)	Normalized BV in descending order ($BV/\sum BV$)	Class A	Class B	Class C
			CB (0–0.2)	CB (0.2–0.5)	CB (0.5–1)
			BV (1–0.7)	BV(0.05–0.25)	BV (0–0.05)
R1	0.1	0.7	R1		
R2	0.3	0.2		R2	
R3	0.6	0.1			R3

are feature level requirements. Table 2 provides sample classification in to classes A, B, C for set 1. The table has CB and BV normalized to Total CB and BV, for ease of classification.

Classifying R1, R2 into class A, class B respectively maps to the default framework suggestions of class boundaries for Customer base and Business Value. R3 is classified in to class C with slightly adjusted class boundaries to suit the requirements at hand.

With the possibility of requirements being incremental on top of an existing product or with development utilizing some of the proprietary frameworks or open-source frameworks or the development involving totally new product from scratch, we can consider each feature will have partly User Interface or data input and output forms, partly business logic implementation for processing the data, industry/vertical/domain specific and partly core data model/architecture development. Table 3 below details the membership association for set 2. Normalizing UW, BI, CP to Total Effort (TE) allows classification into A, B, C classes based on the normalized values ranges given in the table. Here the effort can be considered in terms of code base to be developed or person weeks or story points required to complete the activities related to UI, BI or CP.

In the sample classification above, R1, R2 are classified as per default class boundaries, whereas R3 required a slight adjustment to the class boundaries. The class boundaries can be tuned to the nature of projects and type of development.

Table 4 below details the membership association for set 3, which considers whether the feature requires entirely new implementation or marginal implementation is sufficient and if there is going to be impact on existing features and customers due to new requirements. Normalizing MI, NI, IR to Total Effort

Table 3 Set 2

Requirements	UW	BI	CP	Class A	Class B	Class C
				UW (0.7–1)	UW (0.5–0.7)	UW (0.3–0.5)
				BI (0–0.3)	BI (0.3–0.4)	BI (0.4–0.5)
				CP (0)	CP (0–0.1)	CP (0.1–0.2)
R1	0.8	0.2	0	R1		
R2	0.5	0.5	0		R2	
R3	0.2	0.5	0.3			R3

Table 4 Set 3

Requirements	MI	NI	IR	Class A	Class B	Class C
				MI (0.7–1)	MI (0.5–0.7)	MI (0.3–0.5)
				NI (0–0.25)	NI (0.25–0.4)	NI (0.4–0.5)
				IR (0–0.05)	IR (0.05–0.1)	IR (0.1–0.2)
R1	0.6	0.3	0.1		R1	
R2	0.2	0.5	0.3			R2
R3	0.4	0.5	0.1			R3

Table 5 Set 4

Requirements	L	M	S	Class A	Class B	Class C
				L (0–0.1)	L (0.1–0.15)	L (0.15–0.2)
				M (0–0.20)	M (0.20–0.25)	M (0.25–0.3)
				S (0.7–1)	S (0.6–0.7)	S (0.5–0.6)
R1	0	0.3	0.7	R1		
R2	0.2	0.3	0.5			R2
R3	0.1	0.2	0.6		R3	

(TE) allows classification into A, B, C classes based on the normalized values ranges given in the table. Here the effort can be considered in terms of code base to be developed or person weeks or story points required to complete the activities related to UI, BI or CP.

R2 is classified into Class C with Impact Recovery beyond 20 % and marginal implementation being less than 30 %.

Set 4 considers the overall effort required to develop the feature or requirement and decides on classes based on the duration—large, Medium or Small as defined in the framework or as practical for a particular organization—required to complete a requirement. In order for a requirement to be completed, some parts of the requirement would need long duration—large and some parts can be completed in shorter duration, while others may take medium durations. The extent of each of these durations influences the classes association. Table 5 below describes the associations for set 4. The durations are normalized to Total duration in the table for the classes association.

R2, R3 map to framework default boundaries and R1 is put in class A with Medium class boundary adjustment, based on effort required under Large.

Set 5, the last layer focuses on right resources requirement in order to develop the requirement within the constraints. Here the knowledge needs of resources are emphasized. Table 6 provides the classes association for set 5 with resource needs normalized to total resource requirements.

Set 5 classification of R1, R2, R3 fits in to default boundaries. Now looking at the classification across the sets S1 to S5 for the three requirements—R1, R2, R3. Assuming weights of 3/3, 2/3, 1/3 for classes A, B, C respectively, macro level

Table 6 Set 5

Requirements	RC	RI	RT	Class A	Class B	Class C
				RC (0–0.1)	RC (0.1–0.15)	RC (0.15–0.2)
				RI (0–0.20)	RI (0.20–0.25)	RI (0.25–0.3)
				RT (0.7–1)	RT (0.6–1.0)	RT (0.5–1.0)
R1	0.1	0.2	0.7	R1		
R2	0.15	0.25	0.6		R2	
R3	0.2	0.3	0.5			R3

Table 7 Priority values

Requirements	S1	S2	S3	S4	S5	Pm
R1	A	A	B	A	A	34*2/35(or 0.66)
R2	B	B	C	C	B	23/35(or 0.0109)
R3	C	C	C	B	C	14*2/35(or 0.00274)

priority—Pm can be arrived at for each requirement by multiplying the class weights across sets. The priority can vary from AAAAA resulting in 1 to CCCCC resulting in 0.001372, providing a range of priorities for each of the requirements. The priorities need not necessarily be unique. Same priority requirements can be grouped together for simultaneous development. Table 7 indicates Pm calculations for R1, R2, R3.

4 Unique Numbering Scheme for the Framework

As we have seen in Sect. 2 the framework has 5 sets—ranging from S1 to S5 with S1 being the first level and determining the Business Value for the requirement. S2 looks at the existing capabilities in terms of components, products and effort required broadly for the new requirement. S3 goes deeper with effort understanding along with impact insights. S4 attempts to get at time requirements for the job at hand for the requirement, whereas S5 assesses the capabilities in terms of resources.

Section 3 has the discussion on how the requirements can be assigned into classes. Distinct Priority is arrived at by multiplying across the sets the class weights. While a single number may be useful to look at relatively at the requirements, the intelligence of classification into classes is lost from visibility. In order to retain the class information and yet arrive at a weighted priority scheme, the following number sequence is proposed.

Assigning a five digit sequence with each set holding the positional value from S1 to S5 in that order, the sequence will be a number -

S1 S2 S3 S4 S5—With S1 to S5 holding

S1 (10000ths place)
S2 (1000ths place)

Table 8 Priority sequences

Requirement	S1	S2	S3	S4	S5
R1	0	0	0	0	0
R2	1	0	0	1	2
R3	2	1	0	2	1
R4	1	2	1	1	1

S3 (100ths place)
S4 (10 s place)
S5 (unit place)

Where each of S1, S2, S3, S4, S5 can have values—0 or 1 or 2 based on which class of A, B, C, a requirement falls into.

Each position having three values—0, 1, 2 and with 5 positions of value, the number of sequences equals to 35, that is 243 sequences.

A requirement falling into class A across sets S1 to S5 will have a sequence 00000.

A requirement falling into class B across sets S1 to S5 will have a sequence 11111.

A requirement falling into class C across sets S1 to S5 will have a sequence 22222.

All the 243 values of sequence will range from 00000 to 22222, with each value in each position representing the class and set the requirement belongs to. This enables immediate interpretation of the priority with respect the requirements associated Business value, resources availability, time requirements, cost implications.

While the requirements are being prioritized not all sequences need be used or get used. When new requirements come in to picture during the development cycle, it is easy to insert the requirements once the sequence is determined for the requirement. More than one requirement can have the same sequence and it becomes easy to group the requirements instantly. Table 8 shows sample sequences for requirements R1 to R4.

A new requirement with a priority sequence of 20100 can be placed above R3 and below R2 instantly. A requirement with a priority sequence 10012 can be placed along R2 forming a group, indicating same priority sequence.

5 Conclusion

The framework provides a unique representation for prioritization of the requirements. The framework enables understanding and interpreting prioritization in a visual and instant way. In order to understand and utilize the framework in practice, two mathematical representations are adopted. One based on sets and associations of requirements into the classes within sets and the other based on unique number sequence representation. The Framework and both the methods adapted enable simple and effective methodology for Requirements Prioritization for successive releases under dynamic changes and lead to better understanding and planning of releases. Both Methods help build traceability and visualize effects of plan changes and help in informed quality planning.

References

1. Sita, D., Khare, A.: A framework for requirement prioritization for software products, IUJ J. Manage. **2**(1) (2014)
2. Reinertsen, D.: Principles of Product Development Flow: Second Generation Product Development. Celeritas Publishing (2009)
3. Moisiadis, F.: The fundamentals of prioritising requirements. In: Systems Engineering, Test & Evaluation Conference, Sydney, Australia (2002)
4. Karlsson, J., Ryan, K.: A cost-value approach for prioritizing requirements. IEEE Softw. **14**(5), 67–74 (1997)
5. Wiegers K.E.: Software Requirements. MicrosoftPress, Redmont (1999)
6. Karlsson, J., Wohlin, C., Regnell, B.: An evaluation of methods for prioritizing software requirements. Inform. Softw. Technol. **39**(14–15), 939–947 (1998)
7. Davis, A.M., Yourdon, E., Zweig, A.S.: Requirements Management Made Easy, 39-947. www.omni-vista.com
8. Regnell, B., Höst, M., och Dag, J.N., Beremark, P., Hjelm, T.: An industrial case study on distributed prioritization in market-driven requirements engineering for packaged software. Requirements Eng. **6**(1), 51–62 (2001)
9. Lehtola, L., Kauppinen, M.: Suitability of requirements prioritization methods for market-driven software product development. Softw. Process Improv. Pract. **11**, 7–19 (2006)

Reduction of Micro Aneurysms in Retinal Identification Based on Hierarchical Clustering in Terms of Improved Circular Gabo Filter

G. Srinivasa Rao and Y. Srinivasa Rao

Abstract Programmed recognition of miniaturized scale aneurysms (MA) in shading retinal pictures is proposed in this paper. At present days acknowledgment of MA is a pivotal stride in conclusion and evaluating of diabetic retinopathy. Customarily directional cross segment profile acknowledges MA location on nearby most extreme pixels of pre-prepared retinal picture. Top acknowledgment is connected on every pixel, and an arrangement of traits like measurement, size (length), tallness and state of every pixel computed precisely and accordingly. However, cross-segment profile examination is not material for location of MA as for dimensional in retinal pictures. Customarily propose to create CPHC (Classification by Pattern-Based Hierarchical Clustering) a semi managed order calculation that uses example based group chain of importance as the immediate which means, of agreement. In any case, a few insufficiencies like the effort of emphasize removal and complicated are available in SIFT (Scale Invariant Function Transformation)-based ID. To take care of these problems, a novel preprocessing technique with CPHC in perspective of the Enhanced Round Gabor Convert suggested. After planning by the iterated spatial an isotropic sleek technique, the quality of the uninformative SIFT key concentrates is reduced considerably. Tried on the VARIA and eight duplicated retina data source collaborate rotate and climbing, designed technique provides appealing outcomes and reveals heartiness to radical changes and range changes.

Keywords Retinal scanning · Scale steady feature extraction · Improve circular Gabor filter · Micro aneurysms · Color retinal images · Pattern based classification

G. Srinivasa Rao (✉)
Instrument Technology Department, Andhra University College of Engineering, Visakhapatnam, AP, India
e-mail: gollapudi_srinivas15@yahoo.com

Y. Srinivasa Rao
Andhra University College of Engineering, Visakhapatnam, AP, India
e-mail: srinniwasarau@gmail.com

S.C. Satapathy and S. Das (eds.), *Proceedings of First International Conference on Information and Communication Technology for Intelligent Systems: Volume 1*,
Smart Innovation, Systems and Technologies 50, DOI 10.1007/978-3-319-30933-0_36

1 Introduction

The World Health Organization (WHO) indicators that are at present 347 million individuals experiencing diabetes and undertakings that this sickness will be the seventh driving reason for death worldwide in 2030. Throughout the years, patients with diabetes tend to show variations from the norm in the retina, building up an intricacy called Diabetic Retinopathy (DR) [1]. DR is a standout amongst the most genuine infections influencing the eye, and it is viewed as the most widely recognized reason for visual impairment in grown-up somewhere around 20 and 60 years old. In Fig. 1 is demonstrated a trust us picture caught non-invasive, containing smaller scale aneurysms and hemorrhages.

Small sized range aneurysms can bring about reducing and at times obstruction of blood vessels of the retina, other than devastating of the blood vessels divider panel. These small-scale aneurysms may break, providing on hemorrhages [2]. As indicated by an international order of severity levels of DR, the more compact range aneurysms are normally present in the beginning levels of this condition and the hemorrhages usually show in the later levels. Early finding and moreover the sporadic testing of DR possibly allows in reducing the activity of this disease and in maintaining the following loss of visible capability. The recognition and testing of RD are conducted by research of finance us pictures in Fig. 2.

As proven in above determine more compact range aneurysms (MA) are the most prompt medically limited regular for DR. Their place can be used to evaluate the DR level. Various techniques for MA place have been allocated. T. Spencer, M.J. Cree and A. Summarize. Recommend the medical geomorphology program to

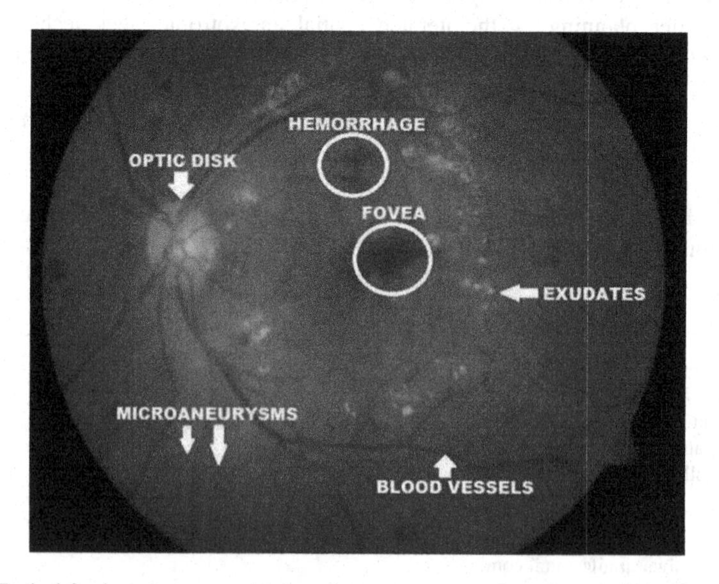

Fig. 1 Retinal funds us images containing micro aneurysms and exudates and blood vessels

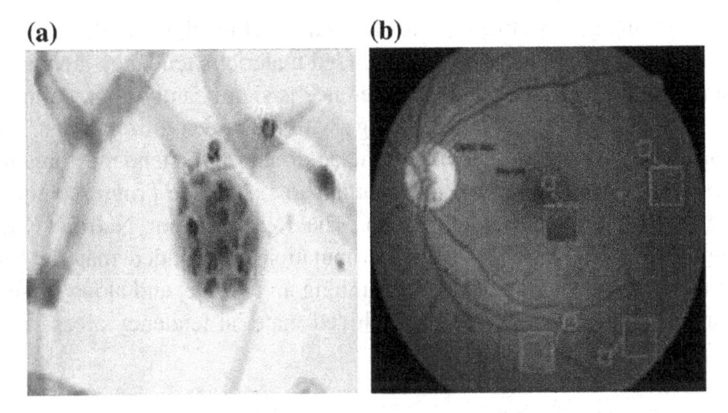

Fig. 2 (**a**) Photography of micro aneurysm; (**b**) A fundus showing micro aneurysms which are zoomed out

fragment MA inside fluoresce in X-rays. Most techniques say before processor away at fluoresce in angiographies or covering images handled sufferers with increased learners in which the MA and other retinal elements are simply apparent [1]. The evaluation efforts and effectively on the affected person could be reduced if the place structure could be successful on images introduced from sufferers with non-enlarged students In any case, the characteristics of these images will be more terrible and it extremely impacts the performance of those said computations. Designed MA finding on images obtained without university student growth is analyzed in this perform with the point of giving choice support despite reducing the amount of work of eye specialists. In [3] our past performs, we have shown techniques for programmed recognition using a medical morphological system, a FCM collection process, a mix of an FCM and mathematical morphology, a guileless Bayesian classifier, a SVMs classifier and a nearest next door neighbor classifier.

The suggested cross-segment discovery centered factor removal and potential set could be used in the other therapeutic image looking after appropriate projects, particularly in circumstances of issue ID that integrate the ID of about circular or a smidgen indicated section in a retinal image. By considering these details in suffering from diabetes retinopathy of retinal images we recommend to develop up a CPHC (Classification by Design centered Hierarchal Clustering) a novel semi-administered depiction requirements that uses an example centered team framework as a immediate path for agreement. Be that as it may, a few weak points like the effort of emphasize removal and jumbling are available in SIFT-based identifiable evidence. Be that as it may, the retinal images dependably encounter the ill results of low dim stage complexes and factor varies which can impact key factor removal and managing and immediate inadequate performance in SIFT-based identifying evidence. In this work, another technique in mild of ICGF with enhancement of CPHC is suggested for improving the simple components of retinal images, and the base is officially clothed all the while. We first entitle an ICGF structure of the image extremes the area around a picture element, then following to being complex with the structure, and included material's tendency area is obtained from the first picture [4]. The picture difference is significantly upgrade in a waken of losing tendency area from the first image with the unrecognizable hair like veins being elucidated. The by and large recognized doubt on the tendency area in that it is progressively changing, and it is important to secure the progressively moving residence of the prepared tendency area. The Round Gabor Narrow (CGF) has promisingly great performance in the computation of included material tendency areas for its residence of an isotropy in climbing and online, and along these lines is a reasonable probability for handling included material tendency areas.

2 Related Work

A few scientists have proposed techniques for programmed discovery of smaller scale aneurysms and hemorrhages, displaying diverse answers for the issue. Hatanaka et al. present a classifier of hopeful locales for red sores that performs counts utilizing the three channels of the HSV (Hue Saturation Value) shading space. Ismailia et al. separate the hopeful locales for red injuries utilizing the curvet change. Mariño et al. extricate the hopeful areas by joining distinctive channels and afterward a district creating process is used to disregard the places whose dimension does not fit in the red damage design [4]. Shahin et al. use administrators of numerical morphology to extricate the diverse objects of the retina and a classifier of locales taking into account neural systems is therefore connected. Balasubramanian et al. remove the applicant areas for red sores with a system called Automatic Seed Generation (ASG), in view of the power of the pixels and its similitude with the neighbors. Badea et al. identify all objects of the retina with a proposed strategy, called Expanding Gradient Method (EGM), and afterward the competitor districts for red injuries are separated. Kande et al. use administrators of numerical morphology to concentrate applicant locales, proposing a classifier in light of Support Vector Machines (SVM) [5]. Jaafar et al., Niemeyer et al. furthermore, Ravi Shankar et al. likewise utilized methods of numerical morphology for location of red sores, getting agreeable results. Programmed identification of small scale aneurysms and hemorrhages is not a simple errand. Some red sores are exceptionally close to the veins, making it hard to recognize them. The bigger size hemorrhages have the same shade of the vessels, varying just in geometry, yet littler small scale aneurysms and hemorrhages have shading, composition and geometry fundamentally the same to the structure of veins. Moreover, smaller scale aneurysms typically have a breadth under 125 μm, and this little size and power varieties out of sight make the recognition of miniaturized scale aneurysms a muddled errand. Reserve us pictures regularly experience the ill effects of non uniform enlightenment, poor difference and clamor, consequently these pictures need to experience pre handling stages. As indicated by, the greatest difficulties for recognition of red injuries are the division of miniaturized scale aneurysms in zones of low differentiation and the vicinity of white sores in the retina (exudates). It might likewise be said that smaller scale aneurysms and hemorrhages have different sizes and shapes, making it hard to perform location of every one of them in the picture, principally in light of the fact that the ordinary structures of the retina, as the veins, can be mistaken for injurious.

The utilization of open databases of fundus pictures is vital to demonstrate the precision and unwavering quality of each proposed system. Moreover, a few techniques have a high run time, taking numerous minutes to investigate every trust us pictures [5, 6]. This may block the utilization of these systems in handling of a major measure of trust us pictures. Moreover, numerous works concentrate on accomplishing elevated amounts of affectability without organizing the outcomes' specificity. This makes these routines delicate to numerous structures in the picture, getting a high number of false positives. The affectability and specificity measures will be clarified in subtle elements.

The MA applicants are eliminated by restricting the acquired strategy, and organized by procurement back slide allotment. The process proposed review comprehends the recommendation of MA's through the quality's evaluation requirements along distinct range places of unique guidelines placed between candidate pixel. These information are the most aspect known as cross-segment or energy information. In a previous execute we have confirmed how this convention can be used to without guidance MA recommendation, which program got to be intense with several best in education ones in a start on the web competitors. The strategy suggested in this document recognizes an absolutely different strategy by part the identification strategy into the actions of applicant evacuation, emphasize removal, category, and placement dedication. The aftereffects of the suggested system in the same on the web competitors confirmed that it is prepared to a great level surpass its harbinger, as well as it transformed into the best non- troupe based MA locator among all people. Techniques with indistinguishable concept of the retinal boat department have furthermore been suggested by a few students. In [7], the students recommend a relative cross-sectional process for the pontoon boat department, and the 2D "tramlines" restricted suggested by Lowell. In [8] furthermore said, explanation that it is flexible for both concentrate line recommendation.

3 Hierarchal Clustering for MA Detection

We now prescribe in this report CPHC (i.e., Classification by Pattern-based Ordered Clustering), a novel semi-regulated classification criteria that uses design lengths as a method for creating gathering (i.e., hub) loads. CPHC first is appropriate a without supervision example driven example based progressive bunching criteria to the entire data set to create a gathering structure [9]. Contrasted and current semi-regulated class systems, CPHC straight use the bringing about gathering structure to arrange break down circumstances and thus evacuates the extra

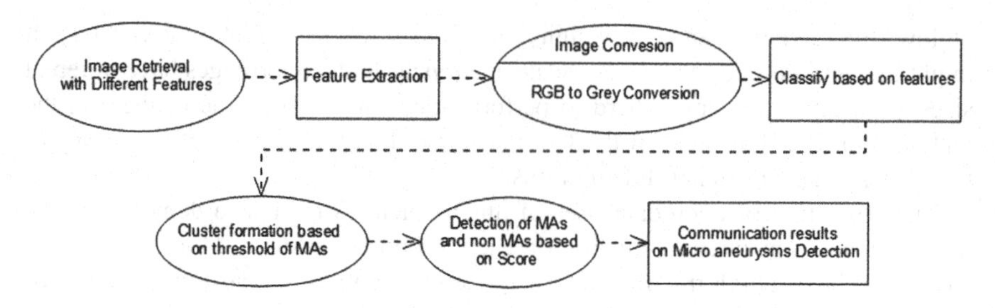

Fig. 3 Procedure for hierarchal clustering in detection of micro aneurysms

instructing stage. To arrange an investigate sample, CPHC first uses the various leveled system to perceive hubs that contain the break down illustration, and afterward utilizes appearance of coinciding honing circumstances, with a weight of them by hub design lengths (i.e., by developing the hub design interestingness esteem with the example length) to get classification label(s) for the examiner case [10]. These permits CPHC to arrange unlabeled dissect circumstances without making any assumptions about their conveyance in the data set. Technique of the proposed calculation as takes after:

Procedure was shown in Fig. 3 will give better improvement in handling micro aneurysms. The data picture is changed over in different perspectives to distinguish the best smaller scale aneurysms over different subtle elements areas. The advantages of spatial pivoting are to figure out the closeness exercises for finding small scale aneurysms in the retina pictures [8]. The unverifiable viewpoints, for example, record work; reason works and our proposed design order id used to discover a group in a viable approach to get the considerable bunching sum. The points of interest's scope areas are recognized in view of the vague perspectives i.e., the reach evaluate over the data angles are computed. The classes are built up in view of the spatial questionable angles furthermore as opposed to regular container with same spot of turning [11]. On the off chance that two group points of interest over spatial unverifiable are same then the eye is not experiencing the smaller scale aneurysms generally the eye is experiencing the miniaturized micro aneurysms.

4 Implementation

The in this proposed acknowledgment method incorporates two essential stages: preprocessing and CPHC-Based Identification as is reliably portrayed by the running counteract outline in Fig. 4.

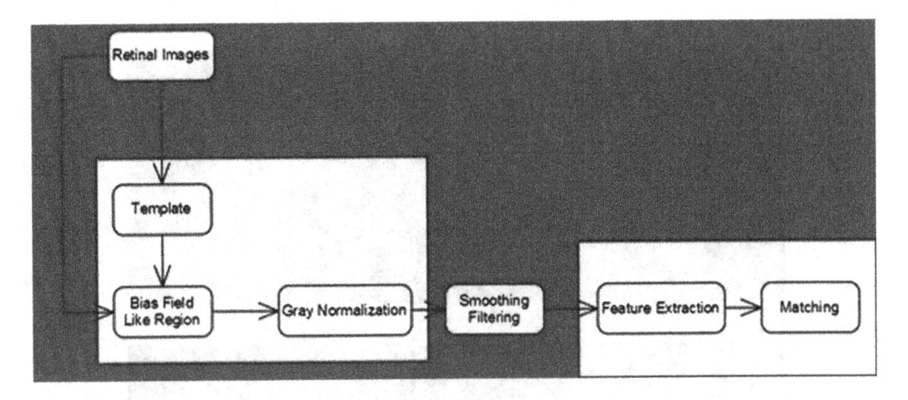

Fig. 4 Procedure for CPHC with circular Gabo filter

4.1 Preprocessing

(1) Qualifications information. This level standardizes the no continually furnished base by eliminating the prejudice area like area, which is obtained by convoy the one of a type image with the ICGF summarize, from the one of a type retinal images [3]. At that point the centralization of the new picture is resolved by taking requirements between 0 and 255. (2) Eliminating. Decrease the frustration in similar areas utilized the spatial anisotropic sleek process. Besides, sections of retinal blood vessels are making progress. Acknowledgment: (1) Function evacuation. Stable key concentrates are provided by using the SIFT criteria; the key concentrates can define the creativeness of every category. (2) Related. Identify the quality of printed sheets in two retinal images. Every level is particular and showed up in the associated with sections.

4.2 Image Preprocessing

Round Gabor Purification (CGF) were presented and used into steady framework department by Zhang. Schedule Gabor programs are recognized as agreement receptors, yet being used of getting agreement invariant capabilities, their important benefits get to be important. Circular Gabor programs are appropriate for getting switching invariant capabilities on the reasons that there is no believed of course in them. The Gobar Filter is settled as requires after:

$$G(a,b) = g(a,b)\exp\left(2\pi \mathrm{jF}\sqrt{x^2+y^2}\right)$$

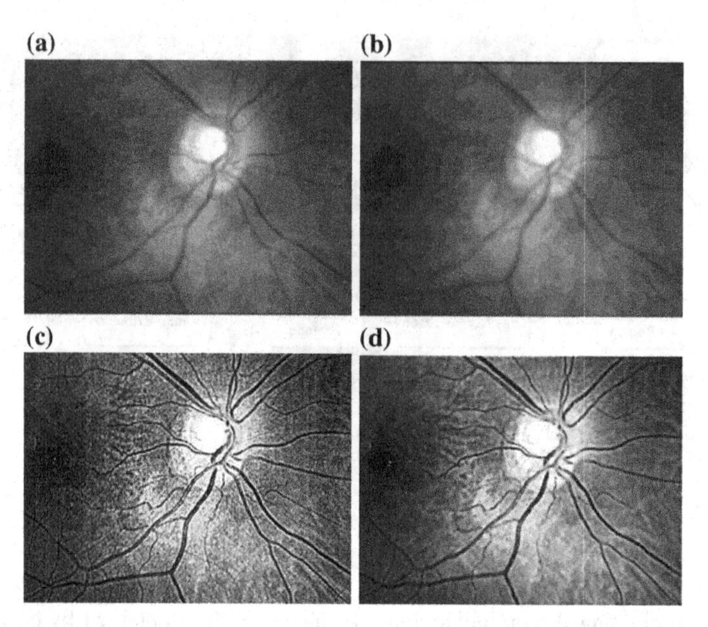

Fig. 5 (**a**) The distinct retina picture; (**b**) the partiality like picture which is formed by the ICGF; (**c, a**) subtract (**b, d**) Result picture of preprocessing which is after changing anisotropic distribution

Here F represents the first frequency of a round Gabor filter, and g(a, b) represents 2-D Gaussian program considered to be isotropic, which is identified as:

$$g(a,b) = \frac{1}{2\pi\sigma^2}\exp\left(-\frac{x^2+y^2}{2\sigma^2}\right)$$

The Fourier representation of filtering process as follows:

$$F(a,b) = \frac{\sqrt{2\pi}}{2}a\exp\left(-\frac{\left(\sqrt{x^2+y^2}-F\right)^2}{2a^2}\right)$$

Where $a = \frac{1}{2\pi\sigma}$ Reliability area can better illustrate you will of a Round Gabor slimmer. The 2 Dimensional and 3-Dimensional perspective of Round Gabor purification appears a group in the repeat section it is the same as undulating common water browse in spatial part in Fig. 5.

$$R = I - P$$

The last retinal Image handling

$$R = \frac{R' - low(R')}{high(R') - low(R')} \times 255$$

R is a described retinal image cause from the suggested ICGF. This potential is resembled as getting out the preservative choice from the outstanding retinal image. R′ is the temporary modifying, its requirements, combined bag range between −255 to 255 and is healthy out of [0, 255] further managing of small aneurysms.

4.3 SIFT Based Detection

Obtained retina picture has unusual shadow near the images restrict which may mediate with the key concentrates end, we basically give up the restrict variety of the retinal image. The portioned area, R, is taken is required after:

$$R = I(\mathrm{h1} : \mathrm{h2}, \mathrm{W1} : \mathrm{W2})$$

In our research, we set $h_1 = 40$, $h_2 = 520$, $w_1 = 25$, $w_2 = 730$

Range Invariant Function Modification (SIFT) strategy was suggested for getting special invariant capabilities from pictures [2, 4]. The SIFT requirements features four large phases of calculation:

(1) Ranking—space excessive location; (2) Untrustworthy key point's evacuation; (3) Positioning task; (4) key point descriptor. Taking after a quick appearance of the every stage. The major phase chooses the areas of upcoming concern elements in the picture by discovering the high and low of an agreement of Difference of Gaussian (DoG) pictures acquired at dissimilar devices everywhere throughout pictures.

4.4 Matching

At this phase, potential caption provided by two retinal pictures are printed. As per the designed potential description, the combined bag of relevant places is used to evaluate the likeness of these two retinal images. At that factor as far as possible T (the combination of relevant sites) is selected in the awaken of looking at the whole retinal details resource. The two retinal pictures will be requested into the identical category if the varieties of relevant places are larger than T, usually, these two retinal pixels will be organized as unique groups.

5 Experimental Evaluation

The analysis is actualized in JAVA, and load with a PC with a 2.4 GHz speed of CPU and 2.0 Giga Byte storage. The regular pre planning time is 0.28 s when removing is furnished with, which is more speeder than the suggested strategy, however managing duration of emphasize remove and managing are considerably extended. The total regular planning duration of the suggested strategy is 7.87 s, which is extremely tedious. Distinct with the PBO program is 2.31 is completed, SIFT-based technique is more slowly however both are in a few moments. Very few methods to decrease the computation time. The size retina pictures used as a part of our research are usually extensive and the planning time can reduce reasonably as indicated by the range changes [1]. The most far attaining program for the performance evaluation of different from the standard recognition plans in healing pictures is the use of Free reaction ROC (FROC) turns. A FROC fold plots affectability against the regular number of incorrect advantages per picture. Since the FROC fold does not stand upon the quality of authentic negative concentrates on the image, it is appropriate to think about the presentation of recognition techniques that give an hidden variety of hopefuls, obviously evaluating the modified on a picture [10]. After all there is no highest possible highest of the right half of a FROC fold, that is the normal diversity of incorrect advantages per picture can be hypothetically endless, no generally recognized file variable that represents a whole FROC fold prevails.

The understanding and uniqueness actions were used to assess the results achieved by the recognition program. These actions were selected considering the relevant work, where most shows implement these actions. For an unique understanding of these actions, it is very important contemplate a few categories used as a part of analysis service assessments for finding of diseases. At the factor when a analyze result is sure, the person can display the problem, which is known as “Genuine Positive” (TP) or can’t display it, which is known as “False Positive” (FP). Then again, when the result is adverse, the person can’t have the illness, which is known as “Genuine Negative” (TN) or can have it, what is known as “False Negative” (FN).

The sensitivity is characterized as the capacity of a test to recognize effectively individuals with a malady or condition. The affectability is figured concurring after mathematical statement:

$$Sensitivity = \frac{TP}{TP + FN}$$

The specificity is characterized as the capacity of a test to prohibit legitimately individuals without a malady or condition. The specificity is figured by mathematical statement:

$$Specificity = \frac{TN}{TN + FP}$$

It is essential to have some harmony between the estimations of affectability and specificity. Get high values for affectability and low values for specificity implies that the technique recognized most sores (genuine positives), however different structures were wrongly delegated injuries (false positives). Then again, get high values for specificity and low values for affectability implies that the strategy erased accurately locales without sores (genuine negative) however was not ready to recognize all injuries (genuine positives) [12].

The strategy's execution was tried on pictures with occurrence of sores gave by the DIARETDB1. The pixels' bookkeeping in TP, FN, FP and TN was performed for every subsequent picture of the strategy [13, 14]. Table 1 demonstrates the normal estimations of TP, FN, FP and TN for every fundus picture.

Every subsequent picture of the strategy was contrasted and its ground-truth, where it was performed an examination taking into account the zone of every area [3]. The affect ability and specificity measures were figured by picture, and after it was performed the general normal of the qualities acquired for every picture, where the system accomplished an affect ability of 87.69 % and a specificity of 92.44 %. Table 1 thinks about the outcomes got by the technique with different works of writing [13, 14].

The greatest position of a technique is calculated as the regular knowing at seven wrong valuable prices (1/8, 1/4, 1/2, 1, 2, 4, and 8 incorrect advantages per image) in Fig. 6.

The recommended technique is appropriate a two way order, so it needs two locations of readying cases. Building the precious (MA) set is pretty easy. Despite, selection of the contrary (non-MA) set is most complex, since its elements have to be choosing individually. To create the teaching operate set for the classifier of the recommended technique, we took the authentic labeling of the MAs on the teaching set as a base, and we classified out the unclear one's individually. The non MA set comprised boat fragments.

Our proposed technique achieves efficient classification results in micro aneurysms detection in retinal images. However, there are a few problems that ought to

Table 1 Comparative results in terms of sensitivity and specificity with previous and our proposed technique

Author	Publicly Available Images	Sensitivity (%)	Specificity (%)
G. Srinivas	Yes: DIARETDB0	98.6	87.9
Jaafar	DIARETDB1	97.6	86
Ravishankar	Not specified	87	90.5
Balasubramamian	Yes: STARE, DIARETDB0 and DIARETDB1	100	92

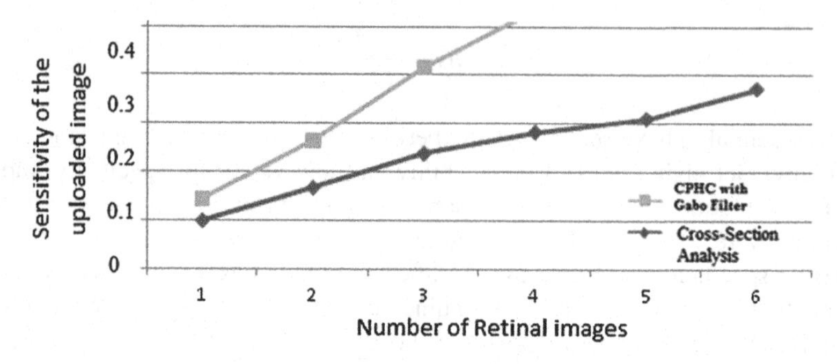

Fig. 6 Sensitivity comparison with respect to Rotational cross section analysis and CPHC Gabo Filter

be said. The technique furthermore got attractive outcomes with regard to its playback, on the reasons that it's low multi-dimensional characteristics. Every image was analyzed with threshold pixel values with respect micro aneurysms in retinal images. This may be a strategy's differential for reasons of wide range planning of fundus images.

6 Conclusion

We had showed a strategy for identification of MA's on retinal pictures, in light of standard of splitting down online trail area information concentrated on the optimistic p of the pre- prepared picture. The quality of p is to be managed completely decreased by just taking the encompassing extreme of the pre ready picture. We try to identification of every information; furthermore, determine an agreement of features that represent the size, prominence, and state of the central top. The actual actions of these principles as the cross' release section changes represent the list of abilities utilized as a part of a depiction project to take out incorrect candidates. Our work concentrates on micro aneurysm finding from suffering from diabetes retinopathy persistent non-enlarged student innovative images. It is an expansion to our already proposed robotized DR screening framework. The framework is means to offer assistance the oculist in the diabetic retinopathy screening procedure to distinguish manifestations speedier and all the most effortlessly. The calculation could recognize MAs on exceptionally low quality pictures. Albeit further advancement of this calculation is still required, the outcomes are fulfilling. The result is entirely fruitful with affectability and specificity of 81.61 and 99.99 %, separately. The framework additionally furnished optometrist of the quantity of MAs for evaluating the DR stage.

References

1. Lazar, I., Hajdu, A.: Retinal micro aneurysm recognition through regional spinning cross-section information analysis. In: Proceedings of IEEE Transactions on Medical Imaging, vol. 32, no. 2 (2013)
2. You, J., Mayonnaise, A., Li, Q., Hatanaka, Y., Cochener, B., Roux, C., Karray, F., Garcia, M., Fujita, H., Abramoff, M.: Retinopathy online challenge: automatic detection of retinal images. In: Proceedings of the IEEE Annual International Conference on EMBC, pp. 5939–5942 (2011)
3. Meng, X., Yin, Y., Yang, G., Xi, X.: Retinal recognition depending on an enhanced round Gabor narrow and range invariant function transform. In: Proceedings of Receptors, vol. 13, pp. 9248–9266 (2013). doi:10.3390/s130709248
4. Antal, B., Hajdu, A.: An ensemble-based system for micro aneurysm detection and diabetic retinopathy grading. IEEE Trans. Biomed. Eng. **59**(6), 1720–1726 (2012)
5. Structure, A.J., Undrill, P.E., Cree, M.J., Olson, J.A., McHardy, K.C., Distinct, P.F., Forrester, J.: A comparison of internet centered category techniques used to the recognition of micro aneurysms in ophthalmic fluoresce in angiograms. Comput. Biol. Med. **28**, 225–238 (1998)
6. Mendonca, A., Campilho, A., Nunes, J.: Automatic segmentation of micro aneurysms in retinal angiograms of diabetic patients. In: Proceedings of International Conference on Image Analysis and Processing, pp. 728–733 (1999)
7. Mizutani, A., Muramatsu, C., Hatanaka, Y., Suemori, S., Hara, T., Fujita, H.: Automated micro aneurysm recognition technique depending on dual band narrow in retinal fondues pictures. In: Proceedings of SPIE Medical Imaging (2009)
8. Ram, K., Joshi, G.D., Sivaswamy, J.: A successive clutter-rejection- based approach for early detection of diabetic retinopathy. IEEE Trans. Biomed. Eng. **58**(3), 664–673 (2011)
9. Ricci, E., Perfetti, R.: Retinal blood vessel segmentation using line operators and support vector classification. IEEE Trans. Med. Imag. **26**(10), 1357–1365 (2007)
10. Giancardo, L., Meriaudeau, F., Karnowski, T.P., Li, Y., Tobin, K.W., Chaum, E.: Micro aneurysm detection with radon transform-based classification on retina images. In: Proceedings of the IEEE Annual International Conference on EMBC, pp. 5939–5942 (2011)
11. Zhu, T.: Fourier cross-sectional profile for vessel detection on retinal images. Comput. Med. Image. Grap. **34**, 203–212 (2010)
12. Júnior, S.B., Welfer, D.: Automatic detection of micro aneurysms and hemorrhages in color eye fundus images. In: Proceedings of Worldwide Publication of Pc Technological innovation & Details Technological innovation (IJCSIT), vol. 5, no 5 (2013)
13. Peters, S., Vivó-Truyols, G., Marriott, P.J., Shoemakers, P.J.: Development of an algorithm for optimum recognition in extensive two-dimensional chromatography. J. Chromatogr. A **1156**, 14–24 (2007)
14. Zhang, B., Wu, X., You, J., Li, Q., Karray, F.: Detection of micro aneurysms using multi-scale correlation coefficients. Pattern Recognit. **43**(6), 2237–2248 (2010)

Part IV
Applications of ICT in Rural and Urban Areas

A Modified Representation of IFSS and Its Usage in GDM

B.K. Tripathy, R.K. Mohanty, T.R. Sooraj and A. Tripathy

Abstract Soft set is a new mathematical approach to solve the uncertainty problems. It is a tool which has the prospects of parameterization. Maji et al. defined intuitionistic fuzzy soft set (IFSS). However, using the approach of provided by Tripathy et al. (2015) we re-define IFSS and use it in deriving group decision making (GDM). An application is used for illustration of the process.

Keywords Soft sets · Fuzzy sets · Fuzzy soft sets · Intuitionistic fuzzy soft sets · Decision making · Group decision making

1 Introduction

Fuzzy sets (FS) [1] and soft set (SS) are well-known tools to handle uncertainty in data. Hybrid model of FSS was handled in [2–4]. Some applications of soft sets are discussed in [5] and an application to decision making (DM) is proposed in [3]. However, there are some problems in this approach identified in [6].

Intuitionistic fuzzy set introduced in [7] is an extension of fuzzy set. Extending their own work IFSS was used by Maji et al. and applied to GDM problems.

In [8] characteristic function approach was used to define SS. Here we re-define IFSS extending our work on FSS in [6] by using the approach in [8] and propose an algorithm for DM using IFSS.

B.K. Tripathy (✉) · R.K. Mohanty · T.R. Sooraj · A. Tripathy
School of Computing Science and Engineering, VIT, Vellore, Tamilnadu, India
e-mail: tripathybk@vit.ac.in

R.K. Mohanty
e-mail: rknmohanty@gmail.com

T.R. Sooraj
e-mail: soorajtr19@gmail.com

A. Tripathy
e-mail: anurag7642@gmail.com

S.C. Satapathy and S. Das (eds.), *Proceedings of First International Conference on Information and Communication Technology for Intelligent Systems: Volume 1*, Smart Innovation, Systems and Technologies 50, DOI 10.1007/978-3-319-30933-0_37

2 Preliminaries

Let U and E be the Universe and parameter set respectively. By P(U) and I(U) we denote the set of all subsets of U and the set of all intuitionistic fuzzy subsets of U respectively.

Definition 2.1 (*Soft Set*): A soft set over (U, E) denoted by (F, E), where

$$F : E \rightarrow P(U) \tag{1}$$

Definition 2.2 (*Fuzzy soft set*): We denote a FSS over (U, E) by (F, E) where

$$F : E \rightarrow I(U) \tag{2}$$

Definition 2.3 (*Intuitionistic Fuzzy soft set*): An intuitionistic fuzzy set (IFS) in *E*, a fixed set, is an object of the form $A = \langle x, \mu_A(x), \nu_A(x)\rangle | x \in E$ where, the function $\mu_A : E \rightarrow [0,1]$ and $\nu_A : E \rightarrow [0,1]$ define the degree of membership and degree of non-membership respectively of the element $x \in E$ to the set *A*. Also, we can say for any $x \in E$, $0 \leq \mu_A(x) + \nu_A(x) \leq 1$. The remaining part $\pi_A(x) = 1 - \mu_A(x) - \nu_A(x)$ is called the indeterministic part of x. In this section, we describe the main notions used and the operations of intuitionistic fuzzy soft sets.

Definition 2.4 Let *U* is the universal set and *E* be a set of parameters. Let *IF* (*U*) denotes the set of all intuitionistic fuzzy subsets of *U*. The pair (*F, E*) is called as intuitionistic fuzzy soft set over *U*, where *F* is a mapping given by

$$F : E \rightarrow IF(U) \tag{3}$$

3 Application of IFSS in GDM

In [5] many applications of soft set are provided. In [9] the decision making example given is depend on the decision of a single person and in [10] one example of GDM using FSS. Sometimes it's not appropriate to handle a problem by a single decision maker. So we need a combined decision of multiple persons. Here we discuss an example of GDM in IFSS.

In [6], the concept of negative parameters was introduced.

Following formula is used to get a score from an IF value to make computation similar to the fuzzy values.

$$Score = (Membership\,Score - Non\text{-}Membership\,Score + Hesitation\,Score + 1)/2 \tag{4}$$

Equation (4) reduces to only membership score in case of a fuzzy soft set.

3.1 Algorithm

1. Input the priority given for each parameter by the panel (J_1, J_2, J_3 …, J_n), where 'n' is the number of judges.
2. Construct the parameter rank table by ranking according to the absolute value of parameters. If the priority for any parameter has not given than take the value as 0 and that column can be opt out from further, unless there is more than one has the highest score in the decision table. The boundary condition for a positive parameter is [0, 1] and for a negative parameter is [−1, 0].
3. For every judge J_i (i = 1, 2, 3,…, n) repeat the following steps.
    a. Input the parameter data table.
    b. Input the intuitionistic fuzzy soft set (*U*, *E*) provided by Judge J_i in tabular form.
    c. Multiply the priority values with the corresponding parameter values to get the priority table.
    d. Compute the sum of membership, non-membership and hesitation values separately for each row in the priority table.
    e. Construct the comparison table by finding the entries as differences of each row sum in priority table with those of all other rows taking membership, non-membership and hesitation separately.
    f. Compute the sum of membership values, non-membership values and hesitation values separately for each row in the comparison table.
    g. Construct the decision table by taking the sums of membership values, non-membership values and hesitation values separately for each row in the comparison table. Compute the score for each candidate using the formula (4). Assign rankings to each candidate based upon the score obtained.
        i. If more than one have same score than who has more score in a higher ranked parameter will get higher rank.
4. Construct the rank table by computing the sum of rankings given by all judges to particular candidate. Lowest rank sum will give the highest rank and so on. In case of same rank sum conflict, resolve by the process used in the step 3.g.i.

The following illustration is provided:

Let U be a set of selected candidates for an interview given by $U = \{c_1, c_2, c_3, c_4, c_5, c_6\}$ and *E* be the parameter set given by $E = \{e_1, e_2, e_3, e_4, e_5, e_6\}$ for the parameters Subject Knowledge, Communication, Reaction, Presentation, Experience, Other Activities respectively. There are three judges to analyze the performance of the candidates.

Table 1 Parameter data table

U	e_1	e_2	e_3	e_4	e_5	e_6
Parameters	Subject knowledge	Communication	Reaction	Presentation	Experience	Other activities
Priority	0.4	0.3	−0.15	0.05	0.1	0
Parameter rank	1	2	3	5	4	6

The panel of judges assigns priority values for each parameter. The parameters ranked as per the priority values. The parameter having highest absolute value as priority will get the highest rank and so on. If more than one parameter having same priority value then the panel of judge can decide the rank among those parameters. All the data about parameters are listed in the Table 1.

The performance of all candidates according to each judge can be represented as IFSS in the tabular form. Tables 2, 3 and 4 represent the performance of a candidate as per judges (J1, J2, J3) respectively. Each candidate will get a rank from every judge. It can be noted that due to zero priority, column e_6 is not present in further computation tables.

The priority tables for each judge can be obtained by multiplying the values in the Tables 1, 2 and 3 with respective parameter priority values fixed by the panel. In this paper priority Table 5 and comparison Table 6 constructed only for judge J1. In the same way priority table and comparison table can also be constructed for other judges. Note that, parameter 'Reaction' is having negative priority, shows that, it is a negative parameter.

The respective comparison tables are constructed as per the procedure given in the step 3.e of the Algorithm 3.1. Table 6 is the comparison table for the judge j1. In the same way comparison tables can be constructed for other judges.

By using the formula (4), the decision table can be formulated and a rank will be given to each candidate. If multiple same score conflict occurs, than the higher score in higher ranked priority will have the upper hand and so on. By this way all the decision tables for all judges can be obtained (Tables 7, 8 and 9).

The rank table can be constructed by adding the ranks given to the candidates by each of the judge as shown in Table 10. If same rank sum obtained by more than one candidate, then the conflict can be resolved by the same way as in decision table formation.

Decision Making: The person having highest rank is the best performer of the interview. If the vacancy is more than one than the next subsequent rank holders can be chosen.

Table 2 Tabular representation of the intuitionistic fuzzy soft set (U, E) for J1

U	e_1			e_2			e_3			e_4			e_5		
c_1	0.30	0.50	0.20	0.10	0.80	0.10	0.20	0.70	0.10	0.30	0.40	0.30	0.40	0.60	0.00
c_2	0.90	0.00	0.10	0.60	0.40	0.00	0.80	0.20	0.00	0.80	0.10	0.10	0.30	0.60	0.10
c_3	0.30	0.40	0.30	0.40	0.50	0.10	0.40	0.50	0.10	0.50	0.20	0.30	0.10	0.90	0.00
c_4	0.70	0.10	0.20	0.70	0.20	0.10	0.60	0.40	0.00	0.60	0.20	0.20	0.60	0.30	0.10
c_5	0.10	0.70	0.20	0.00	0.90	0.10	0.20	0.70	0.10	0.10	0.80	0.10	0.80	0.20	0.00
c_6	0.90	0.10	0.00	0.50	0.40	0.10	0.60	0.30	0.10	0.60	0.20	0.20	0.50	0.40	0.10

Table 3 Tabular representation of the intuitionistic fuzzy soft set (U, E) for J2

U	e_1			e_2			e_3			e_4			e_5		
c_1	0.00	0.50	0.50	0.00	0.70	0.30	0.00	0.50	0.50	0.00	0.60	0.40	0.60	0.00	0.40
c_2	0.70	0.00	0.30	0.40	0.20	0.40	0.60	0.00	0.40	0.60	0.00	0.40	0.10	0.40	0.50
c_3	0.10	0.30	0.60	0.00	0.60	0.40	0.00	0.50	0.50	0.10	0.20	0.70	0.20	0.40	0.40
c_4	0.50	0.00	0.50	0.50	0.00	0.50	0.40	0.20	0.40	0.40	0.00	0.60	0.40	0.10	0.50
c_5	0.10	0.20	0.70	0.20	0.30	0.50	0.20	0.30	0.50	0.30	0.00	0.70	−0.10	0.70	0.40
c_6	0.70	0.00	0.30	0.30	0.20	0.50	0.40	0.10	0.50	0.40	0.00	0.60	0.30	0.20	0.50

Table 4 Tabular representation of the intuitionistic fuzzy soft set (U, E) for J3

U	e_1			e_2			e_3			e_4			e_5		
c_1	0.20	0.30	0.50	0.30	0.40	0.30	0.30	0.40	0.30	0.40	0.10	0.50	0.00	0.80	0.20
c_2	0.80	0.10	0.10	0.50	0.30	0.20	0.70	0.10	0.20	0.70	0.10	0.20	0.20	0.50	0.30
c_3	0.10	0.60	0.30	0.10	0.80	0.10	0.10	0.60	0.30	0.10	0.70	0.20	0.70	0.10	0.20
c_4	0.60	0.10	0.30	0.60	0.10	0.30	0.50	0.30	0.20	0.50	0.10	0.40	0.50	0.20	0.30
c_5	0.20	0.40	0.40	0.10	0.70	0.20	0.10	0.60	0.30	0.20	0.30	0.50	0.30	0.50	0.20
c_6	0.80	0.10	0.10	0.40	0.30	0.30	0.50	0.20	0.30	0.50	0.10	0.40	0.40	0.30	0.30

Table 5 Priority table for the judge J1

U	μ_{e1}	v_{e1}	h_{e1}	μ_{e2}	v_{e2}	h_{e2}	μ_{e3}	v_{e3}	h_{e3}	μ_{e4}	v_{e4}	h_{e4}	μ_{e5}	v_{e5}	h_{e5}	$\sum\mu$	$\sum v$	$\sum h$
c_1	0.12	0.2	0.08	0.03	0.24	0.03	−0.03	−0.105	−0.015	0.015	0.02	0.015	0.04	0.06	0	0.175	0.415	0.11
c_2	0.36	0	0.04	0.18	0.12	0	−0.12	−0.03	0	0.04	0.005	0.005	0.03	0.06	0.01	0.49	0.155	0.055
c_3	0.12	0.16	0.12	0.12	0.15	0.03	−0.06	−0.075	−0.015	0.025	0.01	0.015	0.01	0.09	0	0.215	0.335	0.15
c_4	0.28	0.04	0.08	0.21	0.06	0.03	−0.09	−0.06	0	0.03	0.01	0.01	0.06	0.03	0.01	0.49	0.08	0.13
c_5	0.04	0.28	0.08	0	0.27	0.03	−0.03	−0.105	−0.015	0.005	0.04	0.005	0.08	0.02	0	0.095	0.505	0.1
c_6	0.36	0.04	0	0.15	0.12	0.03	−0.09	−0.045	−0.015	0.03	0.01	0.01	0.05	0.04	0.01	0.5	0.165	0.035

Table 6 Comparison table for judge J1

U	c_1			c_2			c_3			c_4			c_5			c_6			M	N	H
	m	n	h	m	n	h	m	n	h	m	n	h	m	n	h	m	n	h			
c_1	0	0	0	−0.315	0.26	0.055	−0.04	0.08	−0.04	−0.315	0.335	−0.02	0.08	−0.09	0.01	−0.325	0.25	0.075	−0.915	0.835	0.08
c_2	0.315	−0.26	−0.055	0	0	0	0.275	−0.18	−0.095	0	0.075	−0.075	0.395	−0.35	−0.045	−0.01	−0.01	0.02	0.975	−0.725	−0.25
c_3	0.04	−0.08	0.04	−0.275	0.18	0.095	0	0	0	−0.275	0.255	0.02	0.12	−0.17	0.05	−0.285	0.17	0.115	−0.675	0.355	0.32
c_4	0.315	−0.335	0.02	0	−0.075	0.075	0.275	−0.255	−0.02	0	0	0	0.395	−0.425	0.03	−0.01	−0.085	0.095	0.975	−1.175	0.2
c_5	−0.08	0.09	−0.01	−0.395	0.35	0.045	−0.12	0.17	−0.05	−0.395	0.425	−0.03	0	0	0	−0.405	0.34	0.065	−1.395	1.375	0.02
c_6	0.325	−0.25	−0.075	0.01	0.01	−0.02	0.285	−0.17	−0.115	0.01	0.085	−0.095	0.405	−0.34	−0.065	0	0	0	1.035	−0.665	−0.37

Table 7 Decision table for Judge J1

	Membership	Non-membership	Hesitation	Score	Rank
c_1	−0.915	0.835	0.08	−0.335	5
c_2	0.975	−0.725	−0.25	1.225	2
c_3	−0.675	0.355	0.32	0.145	4
c_4	0.975	−1.175	0.2	1.675	1
c_5	−1.395	1.375	0.02	−0.875	6
c_6	1.035	−0.665	−0.37	1.165	3

Table 8 Decision table for Judge J2

	Membership	Non-membership	Hesitation	Score	Rank
c_1	−0.9	1.21	−0.31	−0.71	6
c_2	0.84	−0.38	−0.46	0.88	3
c_3	−0.87	0.67	0.2	−0.17	5
c_4	0.84	−1.1	0.26	1.6	1
c_5	−0.81	0.19	0.62	0.31	4
c_6	0.9	−0.59	−0.31	1.09	2

Table 9 Decision table for Judge J3

	Membership	Non-membership	Hesitation	Score	Rank
c_1	−0.81	0.19	0.62	0.31	4
c_2	0.84	−0.38	−0.46	0.88	3
c_3	−0.9	1.21	−0.31	−0.71	6
c_4	0.84	−1.1	0.26	1.6	1
c_5	−0.87	0.67	0.2	−0.17	5
c_6	0.9	−0.59	−0.31	1.09	2

Table 10 Rank table

	J1	J2	J3	Rank-sum	Final-rank
c_1	5	6	4	15	5
c_2	2	3	3	8	3
c_3	4	5	6	15	6
c_4	1	1	1	3	1
c_5	6	4	5	15	4
c_6	3	2	2	7	2

4 Conclusions

Here, we proposed an algorithm GDM using IFSS. We use the definition proposed in [8] to re-define IFSS. Earlier FSS and IFSS were defined in [3] and [4] respectively. Flaws in [3] were rectified in [6] by us. Here we rectify the flaws in [4] and propose an algorithm using the new approach. In fact our approach works on group DM.

References

1. Zadeh, L.A.: Fuzzy sets. Inf. Control **8**, 338–353 (1965)
2. Maji, P.K., Biswas, R., Roy, A.R.: Fuzzy soft sets. J. Fuzzy Math. **9**(3), 589–602 (2001)
3. Maji, P.K., Biswas, R., Roy, A.R.: An application of soft sets in a decision making problem. Comput. Math Appl. **44**, 1007–1083 (2002)
4. Maji, P.K., Biswas, R., Roy, A.R.: Soft set theory. Comput. Math Appl. **45**, 555–562 (2003)
5. Molodtsov, D.: Soft set theory—first results. Comput. Math Appl. **37**, 19–31 (1999)
6. Tripathy, B.K., Sooraj, T.R., Mohanty, R.K.: A new approach to fuzzy soft set theory and its application in decision making. Comput. Intell. Data Min. **2**, 305–313 (2016)
7. Atanassov, K.: Intuitionistic fuzzy sets. Fuzzy Set Syst. **20**, 87–96 (1986)
8. Tripathy, B.K., Arun, K.R.: A new approach to soft sets, soft multisets and their properties. Int. J. Reasoning-Based Intell. Syst. **7**(3/4), 244–253 (2015)
9. Tripathy, B.K., Sooraj, T.R., Mohanty, R.K., Arun, K.R.: A new approach to intuitionistic fuzzy soft set theory and its application in decision making. In: Proceeding of ICICT 2015, 9–10th Oct, Udaipur (2015)
10. Sooraj, T.R., Mohanty, R.K., Tripathy, B.K.: Fuzzy soft set theory and its application in group decision making. In: Proceedings of ICACCT2015

A State of Art Survey on Shilling Attack in Collaborative Filtering Based Recommendation System

Krupa Patel, Amit Thakkar, Chandni Shah and Kamlesh Makvana

Abstract Recommendation system is a special type of information filtering system that attempts to present information/objects that are likely to the interest of user. Any organization, provides correct recommendation is necessary for maintain the trust of their customers. Collaborative filtering based algorithms are most widely used algorithms for recommendation system. However, recommender systems supported collaborative filtering are known to be extremely prone to attacks. Attackers will insert biased profile information or fake profile to have a big impact on the recommendations made. This paper provide survey on effect of shilling attack in recommendation systems, types of attack, knowledge required and existing shilling attack detection methods.

Keywords Recommendation system · Collaborative filtering shilling attack · Detection and evaluation parameters · Information filtering

1 Introduction

Recommendation systems (RS) provide information or item that is interest of the user by analysing rating pattern and stable information of user. The huge growths of information on the web as well as variety of guests to websites add some key challenges to recommender systems technology; these are producing accurate recommendation and handling several recommendations with efficiency [1].

K. Patel (✉) · A. Thakkar · C. Shah · K. Makvana
Department of Information Technology, CSPIT, CHARUSAT, Anand, India
e-mail: 14pgit010@charusat.edu.in

A. Thakkar
e-mail: amitthakkar.it@charusat.ac.in

C. Shah
e-mail: chandnishah.it@charusat.ac.in

K. Makvana
e-mail: kamleshmakvana.it@charusat.ac.in

S.C. Satapathy and S. Das (eds.), *Proceedings of First International Conference on Information and Communication Technology for Intelligent Systems: Volume 1*,
Smart Innovation, Systems and Technologies 50, DOI 10.1007/978-3-319-30933-0_38

Therefore, new recommender system technologies are required which will quickly turn out prime quality recommendations even for immense information sets.

Content based and collaborative filtering (CF) based are two approaches for developing recommendation systems. In content based system items are recommended based on users past rating history and content of items. Collaborative filtering recommendation system is based on U-I rating matrix. In a typical Collaborative filtering system, an $n \times m$ user-item matrix is created, where n users' preferences about m products are represented as ratings, either numeric or binary. To obtain a prediction for a target item i or a sorted list of items that might be liked, an active user u sends her known ratings and a query to the system. CF system estimates similarities between u and each user in the database, forms a neighbourhood by selecting the best similar users, and estimate a prediction (p_{ui}) or a recommendation list (top-N recommendation) using a CF algorithm [2, 3]. Profile injection attacks degrade the quality and accuracy of a CF based recommender system over inflicting frustration for its users and probably resulting in high user defection [4]. CF based recommendation systems are extremely prone for shilling attacks then content based recommendation systems [2]. New technology are needed that cannot biased to the various fake profile, and generate recommendation with high precision.

Overall success of CF based recommendation system is depends on how it handle shilling attack and how effectively detect shilling attacks [5]. In this paper we provide survey of various types of shilling attack in CF based RS. Also classification of shilling attacks, detection attributes detection techniques and some evaluation parameters of recommendation systems.

The paper is designed, as follows: In Sect. 2 we briefly discuss theoretical background after that in Sect. 3, contain related work then in Sect. 4 contain various detection attributes of shilling attack detection after that in Sect. 5 evaluation matrix and parameters and then in Sect. 6 we discuss some future work and open issues. Finally, we conclude our paper in Sect. 7.

2 Theoretical Background

Section 1, provides basic introduction about recommendation system so now, we focus on shilling attack in collaborative filtering based recommendation system.

2.1 Shilling Attacks

Recommendation schemes are successful in e-commerce sites; they are prone to shilling or profile injection attacks. Shilling attack or profile injection attacks is outlined as,

Table 1 U-I matrix without Shilling attack

User	Items						Similarity with user 1
	I1	I2	I3	I4	I5	I6	
U1	5	2	3	3		?	1.00
U2	2		4		4	1	−1.00
U3	3	1	3		1	2	0.76
U4	4	2	3	1		1	0.72
U5	4	3		3	3	2	0.94

Table 2 U-I matrix with Shilling attack

User	Items						Similarity with user 1
	I1	I2	I3	I4	I5	I6	
U1	5	2	3	3		?	1.00
U2	2		4		4	1	−1.00
U3	3	1	3		1	2	0.76
U4	4	2	3	1		1	0.72
U5	4	3		3	3	2	0.94
U6	4	2		3	3	5	0.98

Malicious users and/or competitive vendors may attempt to insert fake profiles into the user-item matrix in such a way so they will have an effect on the predicted ratings on behalf of their benefits [2].

To understand how shilling attack works, consider Table 1. Contain 6 users and 6 items and we want to predict rating on item 6 by user 1 which is our target user.

Without shilling attack similarity with user 1 is given in Table 1. This similarity is calculated using Pearson correlation coefficient (PCC). If we chose k = 1 then most similar user with user 1 is user 5 and rating by user 1 on item 5 is 2, which is our correct answer.

Now, if attacker enters shilling attack profile which is user 6 then similarity with target user 1 is shown in Table 2. Here with shilling attack profile user 6 is most similar user with target user 1 with similarity 0.98, and rating for item 6 by user 1 is 5 instead of 2 (original rating without shilling attack).

Hence, shilling attacks is reducing quality of data and hence reduce accuracy of recommendation system.

2.2 *Classification of Shilling Attacks*

Shilling attacks are classified based on intent and based on amount of knowledge required to build shilling attack profiles.

Based on intent. Based on intent shilling attacks are classified as push and nuke attacks. Push attack try to increase popularity of target item by giving high rating to

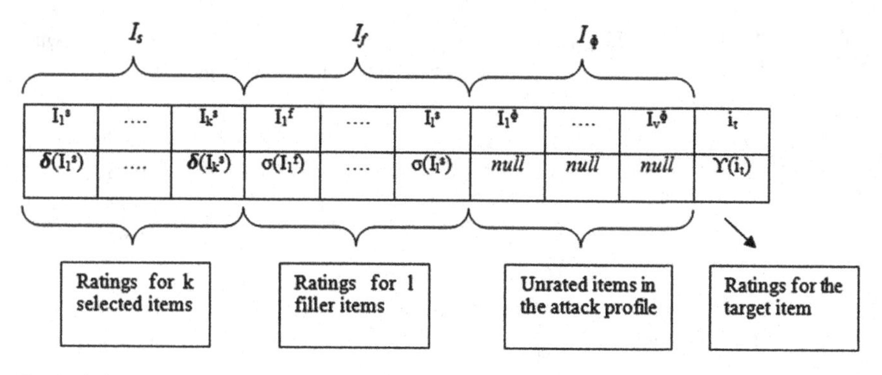

Fig. 1 Shilling attack profile

target item. Nuke attack tries to reduce popularity of target item by giving low rating to target item [6].

Based on knowledge required. Based on amount of knowledge required there are different shilling attack models like *average attack, random attack, bandwagon attack, reverse bandwagon, segment attack* etc.

Attack profile for shilling attack is shown in Fig. 1.

The attack profile is an m-dimensional vector of ratings as per Fig. 1, were m is the total range of items within the system. The profile is partitioned into four components. The empty partition, I_Φ is those items with no ratings in the profile. The only target item i_t will be given a rating as determined by the function ϒ, usually this will be either the highest or lowest possible rating, looking on the attack type (push/nuke). As described below, some attacks need distinguishing a group of items for special care during the attack. This special set I_s receives ratings as given by the function δ. Finally, there is a group of filler items I_f whose ratings are given as specified by the function σ. It is the strategy for choosing items in I_s and I_f and the functions ϒ, σ, and δ that outline an attack model and provides it its character [6].

For different attack models different strategies are used for creating attacker profiles and how to provide rating for items to create attacker profiles are shown in Table 3.

Table 3 Attack profile summary [2]

Attack model	I_s		I_f		I_Φ	I_t (push/nuke)
	Items	Ratings	Items	Ratings		
Random	Null	–	Randomly chosen	System mean	Null	r_{max}/r_{min}
Average	Null	–	Randomly chosen	Item mean	Null	r_{max}/r_{min}
Bandwagon	Popular items	r_{max}/r_{min}	Randomly chosen	System mean/item mean	Null	r_{max}/r_{min}
Reverse bandwagon	Unpopular items	r_{max}/r_{min}	Randomly chosen	System mean	Null	r_{max}/r_{min}
Segment	Segmented items	r_{max}/r_{min}	Randomly chosen	r_{max}/r_{min}	Null	r_{max}/r_{min}

3 Related Work

In this section we have a tendency to represent some related works in field of shilling attacks in recommendation systems. Since shilling profiles look like authentic profiles, it is very tough to spot them. To discover shilling profile various statistical, classification, clustering techniques are used. Bryan et al. [7] Suggest new algorithm known as "Unsupervised Retrieval of Attack Profiles" (UnRAP). They recommend new measure known as Hv-score measure to find shilling profile from genuine profile. They said that Hv-score value of attacker profile is higher than genuine profile. Based on this assumption they identify attacker profile. Lu [8] Extends work of [7] to find group of attacker instead of individual attackers. With help of various detection matrices and analysing raring pattern of attacker [9] Propose unsupervised learning method for detection of fake profile using target item analysis. Algorithm find potential attack profiles using digsim and rdma (Rating Deviation from Mean Agreement) and then refine set of potential profile using target item analysis. Supervised learning is another approach to detect shilling attack in memory based CF. Zhang and Zhou [10] suggest Ensemble learning concept for shilling attack detection using back propagation neural network classifier, finally output is combined using voting strategy. Semi-supervised learning also helpful to detect shilling profiles. Bilge et al. [11] Use bisecting k-means algorithm to generate binary decision tree. Intra cluster correlation (ICC) is used to find correlation within cluster between the profiles. This method assumes that attacker profiles in cluster have high ICC between them. And cluster with high value for ICC is considered as attacker cluster. But Performance of this scheme is slightly worse with increasing filler size in segment attack. Zhang et al. [12] Detect shilling attacks using clustering social trust information between the users. They propose two algorithms, CluTr and WCluTr, to mix clustering with "trust" among users. According to them user with no incoming trust is considered as attacker profiles. Cao et al. [13] Use Semi-supervised learning method semi-SAD. Combination of EM-λ and naïve-bayes is used for detection of shilling attacks.

4 Detection of Shilling Attack

CF based recommendation systems are vulnerable to shilling attacks. We begin this section with a review of some of the statistical measures that have been designed to detect shilling attack in recommendation system. Some of the Standard shilling attack detection metrics are explains below:

Rating Deviation from Mean Agreement (*RDMA*). This measures a user's rating disagreement with other genuine users in the system, weighted by the inverse number of item that user rated. It is defined as,

$$RDMA_u = \frac{\sum_{i=0}^{N_u} \frac{|r_{u,i} - Avg_i|}{NR_i}}{N_u} \quad (1)$$

Weighted Deviation from Mean Agreement (*WDMA*). This measure is strongly based on RDMA; however it places higher weight on rating deviations for sparse items [6]. It is defined as,

$$WDMA_u = \frac{\sum_{i=0}^{N_u} \frac{|r_{u,i} - Avg_i|}{NR_i^2}}{N_u} \quad (2)$$

Where, N_u is the range of items user u rated, $r_{u,i}$ is the rating given by user u to item i, NR_i is the overall range of ratings in the system given to item i. Avg_i is average rating of item i.

Degree of similarity (*DegSim*). Which based on hypothesis that is attacker profiles is highly similar with each other because of theirs characteristics and they are generated with same process [7]. But this profile has low similarity value with genuine profiles. It can be defined as,

$$DigSim_u = \frac{\sum_{v \epsilon\, neighbors(u)} W_{u,v}}{k} \quad (3)$$

Where, $W_{u,v}$ similarity between u and k-nearest neighbours v. and k is number of nearest neighbours of user u.

Equation for similarity between u and v using Pearson correlation coefficient is given as below [9],

$$W_{u,v} = \frac{\sum_{i \epsilon I} \left(r_{u,i} - \overline{r_u}\right)\left(r_{v,i} - \overline{r_v}\right)}{\sqrt{\sum_{i \epsilon I} \left(r_{u,i} - \overline{r_u}\right)^2 \left(r_{v,i} - \overline{r_v}\right)^2}} \quad (4)$$

Where, I is the set of items that users u and v both rated, r_{ui} is the rating user u gave to item i, and $\overline{r_u}$ is the average rating of user u.

Length Variance (*LengthVar*). This attribute relies on the length of user profile. Most of the attacker enters shilling profile that contains large number of rated items [6]. Thus shilling profile has high value for this attribute. Length Variance (LengthVar) that is a measure of what proportion the length of a given profile varies from the average length within the database. It is defined as,

$$LengthVar_u = \frac{|n_u - \overline{n_u}|}{\sum_{u \epsilon U} \left(n_u - \overline{n_u}\right)^2} \quad (5)$$

Where, nu is the average length of a profile in the system.

5 Evaluation Parameters and Matrices

Various evaluation matrices are used for evaluation of effect of shilling attack in recommendation system and evaluating detection algorithms and measure accuracy of recommendation system. These measures are shown in Table 4.

6 Discussion and Open Issues

The internet has been increasing attention now a days. Due to the continuously increasing popularity of internet large amount of data are available. This large data create information overload problem. Recommendation system is system that predict users interest and recommend items to the user. Therefore, the research about recommender systems seems to remain popular. Similarly, shilling attacks against such systems will be in place, as well. Hence, number of researchers are doing research in this area and they are try to find various solutions but there are still missing gaps in this area that is need to be filled.

From all above surveys in Sect. 3 some interesting points are found these are almost all detection methods are unsupervised learning methods. Using supervised learning high accuracy is possible to achieve then unsupervised method. Detection

Table 4 Evaluation parameters [2]

Parameter	Significance	Equation
Precision	Precision (also referred to as positive predictive value) is that the fraction of retrieved instances (attacker) that are really attacker	$precision = \frac{TP}{TP+FP}$
Recall	Recall (also called sensitivity) is that the fraction of relevant instances (attacker) that area unit retrieved as attacker	$recall = \frac{TP}{TP+FN}$
F1 measure	Combination of precision and recall Use for Accuracy of detection algorithm	$F1 = \frac{2*precision*recall}{precision+recall}$
Prediction shift	Prediction shift is that the average change within the predicted rating for the attacked item before and when the attack. This measure is employed for assessing impact of shilling attack	$prediction\ shift = r_{u,i} - r'_{u,i}$ $r_{u,i}$ is rating before shilling attack $r_{u,i}'$ is rating after shilling attack
MAE (mean absolute error)	MAE measures however close the estimated predictions to their discovered ones	$MAE = \frac{1}{N}\sum_{t=1}^{N}\frac{\lvert A_t - F_t \rvert}{A_t}$ A_t = actual value F_t = predicted value

of fake profile in model based recommendation is also one interesting point. Sparse database are vulnerable to shilling attack hence this direction is also need to be investigated. Design a various methods that effectively improves recommendation accuracy in presence of sparsity and shilling attack profile are one of the good idea of research. Using social relationship between users we can also find fake profile that help to improve recommendation accuracy. Using content and social information of user analyze their search history to determine that given user is attacker or not this is also one of the new topic of research.

7 Conclusion and Future Work

In this survey we discuss about effect of shilling attack in recommendation system, their types, detection parameters, evaluation parameters and related works that was done in this In future we are planning to conduct detail survey in field of fake review detection in model based and hybrid recommendation system. We are also planning to propose method for detection of shilling attack using supervised learning.

References

1. Almazro, D., Shahatah, G., Albdulkarim, L., Kherees, M., Martinez, R., Nzoukou, W.: A Survey Paper on Recommender Systems (2010)
2. Gunes, I., Kaleli, C., Bilge, A., Polat, H.: Shilling attacks against recommender systems: a comprehensive survey. Artif. Intell. Rev., 1–33 (2012)
3. Adomavicius, G., Tuzhilin, A.: Toward the next generation of recommender systems: a survey of the state-of-the-art and possible extensions. IEEE Trans. Knowl. Data Eng. **17**(6), 734–749 (2005)
4. Sandvig, J., Mobasher, B., Burke, R.: A survey of collaborative recommendation and the robustness of model-based algorithms. IEEE Data Eng. Bull., 1–11 (2008)
5. Asanov, D.: Algorithms and Methods in Recommender Systems (2011)
6. Burke, R., Mobasher, B. Williams, C., Bhaumik, R.: Classification features for attack detection in collaborative recommender systems. In: Proceedings of the 12th ACM SIGKDD International Conference Knowledge Discovry and Data Mining KDD 06, p. 542 (2006)
7. Bryan, K., O'Mahony, M., Cunningham, P.: Unsupervised retrieval of attack profiles in collaborative recommender systems. In: Recsys'08 Proceedings of the 2008 ACM Conference on Recommender Systems, pp. 155–162 (2008)
8. Lu, G.: Engineering, "a group attack detector for collaborative filtering recommendation." IEEE Internet Comput., 2–5 (2014)
9. Zhou, W., Wen, J., Koh, Y.S., Xiong, Q., Gao, M., Dobbie, G., Alam, S.: Shilling attacks detection in recommender systems based on target item analysis. PLoS ONE **10**(7), e0130968 (2015)
10. Zhang, F., Zhou, Q.: Ensemble Detection Model for Profile Injection Attacks in Collaborative Recommender Systems Based on BP Neural Network, vol. 9, pp. 24–31 (2015)

11. Bilge, A., Ozdemir, Z., Polat, H.: A novel shilling attack detection method, Procedia Comput. Sci. **31**, 165–174 (2014)
12. Zhang, X.L., Lee, T.M.D., Pitsilis, G.: Securing recommender systems against shilling attacks using social-based clustering. J. Comput. Sci. Technol. **28**, 616–624 (2013)
13. Cao, J., Wu, Z., Mao, B., Zhang, Y.: Shilling attack detection utilizing semi-supervised learning method for collaborative recommender system. World Wide Web **16**, 729–748 (2013)

Labeling of Quadratic Residue Digraphs Over Finite Field

R. Parameswari and R. Rajeswari

Abstract The study of digraphs provides a proving ground where mathematicians' ability to bind together multiple disciplines of mathematics becomes evident. The new class of graph called Arithmetic graph was introduced on the basis of Number theory, particularly the Theory of Congruence. Graham and Spencer brought forth the idea of using quadratic residues to construct a tournament with p vertices where $p \equiv 3 \pmod 4$ is a prime. These tournaments were appropriately named Paley digraphs in honor of the late Raymond Paley, who used quadratic residues 38 years earlier to construct Hadamard matrices. In this paper we prove that quadratic residue digraphs over a finite field called Paley tournament allows Edge product cordial labeling, K edge graceful labeling and H_n cordial labeling.

Keywords Quadratic residue digraph · Graph labeling · Cordial labeling · Edge product cordial labeling · K edge graceful labeling · H_n cordial labeling

AMS Subject Classification 05C78

1 Introduction

This article consists of directed and strong regular graph. Prior to the introduction of Paley digraphs, there was much interest in the class of undirected graphs known as Paley graphs [1], given their ability to draw together number theory and graph theory. Paley digraph, the name was given to the tournaments to honor late Raymond Paley, who used quadratic residues 38 years earlier to construct Hadamard matrices [2]. Sachs [3] noticed the achievements of Paley with quadratic residues. He defined Paley graph as a graph whose vertices are finite field elements, and an edge ab exist if and

R. Parameswari (✉) · R. Rajeswari
Sathyabama University, Chennai, Tamil Nadu, India
e-mail: paramsumesh2011@gmail.com

R. Rajeswari
e-mail: rajeswarivel@yahoo.in

S.C. Satapathy and S. Das (eds.), *Proceedings of First International Conference on Information and Communication Technology for Intelligent Systems: Volume 1*, Smart Innovation, Systems and Technologies 50, DOI 10.1007/978-3-319-30933-0_39

only if a − b is a quadratic residue. Let F_q be a finite field with q elements such that either q ≡ 1 (mod 4) or q ≡ 3(mod 4). Let ω be a primal element in F_q and S the set of nonzero squares in F_q, so $S = \{\omega^2, \omega^4,\ldots,\omega^{q-1} = 1\}$. The standard examples of pseudo-random graphs are Paley graphs. It is also a self-complementary and symmetric graph [4].

Graham and Spencer [5] introduced Paley digraphs as a tournaments based on a property of random tournaments. In this graph every undersized split of vertices is conquered by some other vertex.

Under a mapping a series of graph elements send to a series of real values (generally positive or non-negative integers) is called labeling or valuation of a graph. It is called as vertex labeling if the field of the function is the vertex set and is called edge labeling if it is edge set. A collection of values for allocation of vertices, a rule that disperse a label to each edge and some restriction(s) that this labeling is to satisfy are the three common description of labeling.

Several labeled graphs has been used as an model for an extensive reach of application in the field of coding theory, X-ray crystallography, radar, astronomy, circuit design, cryptography, communication network addressing and data base management. The quadratic residue graphs is used in asymmetric cryptographic schemes such as Rabin cryptosystem and the oblivious transfer.

1.1 *Related Survey*

In the last 50 years more than 1500 papers were published in graph labeling. It was discussed as a survey by Gallian [6].

In 1987, Cahit [7] introduced a labeling called cordial labeling and in 1996 he investigated the newly introduced concept H_n cordial labeling for undirected graphs [8]. In 2004, the product cordial labeling was introduced by Sundaram et al. [9]. In 2012, Vaidya and Barasara [10] introduced the edge analogue of product cordial labeling.

Vaidhya and Barasara study that C_n is edge product cordial labeling only for n odd and proved that it is not true for n even. Trees are edge product cordial if n > 2, unicyclic graphs of odd order; $C_n^{(t)}$, for t multiples of 2 or t and n = 1, 3, 5, 7,… but it is not for t odd and n even. $C_n^{(t)}\ \Theta K_1$ called armed crowns, helms, closed helms, webs, flowers, gears are all edge product cordial graphs. Shells Sn for odd n is edge product cordial and it is not for even n. Tadpoles $C_n@P_m$ for m + n even or m + n odd and m > n while not for m + n odd and m < n. For all n triangular snakes are edge product cordial while double triangular snake is only for n odd. Quadrilateral snake, double quadrilateral snake P_n^2 are all edge product cordial for odd n values and the same graph is not for n even. Middle of path $M(P_n)$, Total graph of path and Split of path are all edge product cordial for even n. Shadow graph, middle graph, total graph and split graph of cycle also split graph of path are not for odd n. Tensor product of P_m and P_n and C_m and C_n is edge product cordial if m and n are even. Now we verify it for a strong regular graph called Paley digraph.

Cahit [8] introduced H-cordial and H_n cordial labeling. He investigated H_n cordial labeling for some graphs and generalize the conception of H-cordial labeling in the same paper. In [11] Ghebleh and Khoeilar have given away that complete graph is H-cordial if and only if n = 4n or n = 4n + 3. Wheels are H cordial only when n is odd, but they are H_2 cordial for all n. K_n is H_2-cordial if n = 4n or n = 4n + 3, and Kn is not H_2-cordial if n = 4n + 1. In [12], Freeda and Chellathurai prove that the link of two paths, the link of two cycles, ladders, and the tensor product $P_n \otimes P_2$ are H_2-cordial: They also prove that the join of W_n and W_m where $n + m \equiv 0 \pmod 4$ is H-cordial.

The edge graceful labeling was introduced in the year 1985 by Lo [13]. Lee and Wang [14] described the k-edge graceful graphs for trees. The k-edge graceful trees were described for k = 0, 2, 3, 4, and 5. Gayathri and Sarada Devi [15] obtained some necessary conditions and characterizations for k-edge-gracefulness of trees. They also proved that specific families of trees are edge graceful and k-edge-graceful and speculate that the entire odd trees are k-edge-graceful. In 2013, Rajeswari and Tamizharasi [16] introduced H_n cordial and Edge graceful labeling for directed graphs. We extend the same for the quadratic residue digraph which is strong regular.

2 Preliminaries

Let G = G (V, E) be a finite, strong regular directed graph with V set of vertices and E set of edges. Here the domain of the labeling is whichever, the set of all vertices or the set of all edges or the set of all vertices and edges. These are called vertex labeling or edge labeling or total labeling respectively.

Definition 2.1 Let q be a prime number such that $q \equiv 3 \pmod 4$. The graph P = (V, E) with $V(P) = F_q$ and $E(P) = \{(x, y): x, y \in F_q, x - y \in (F_q^*)^2\}$ is called Paley digraph of order q.

Definition 2.2 For a graph G, a mapping is defined as f: E(G) $\rightarrow$ {0, 1} and an induced mapping f^*: V(G) $\rightarrow$ {0, 1}, if $e_1, e_2, \ldots, e_n$ are the edges incident to vertex v then $f^*(v) = f(e_1)f(e_2) \ldots f(e_n)$. The mapping is known as edge product cordial labeling of a graph G if $|v_f{*}(0) - v_f{*}(1)| \leq 1$ and $|e_f(0) - e_f(1)| \leq 1$.

Definition 2.3 A digraph G (V, E) is supposed to have k edge graceful labeling if its leaving arcs can be labeled with k, k + 1,..., k + q − 1 such that the sum of the leaving arcs of every vertex are distinct modulo p where p stand for the total vertices and q stand for the edges.

Definition 2.4 A digraph is H_n cordial if there is a mapping f: E $\rightarrow$ {±1, ±2,..., ±n}, at every vertex v the addition of all leaving edges of v is in the set {±1, ±2,..., ±n} and satisfying the condition that $|v(i) - v(-i)| \leq 1$ and $|e(i) - e(-i)| \leq 1$ for each i with $1 \leq i \leq n$, where v(i), e(i): $i \in \{\pm1, \pm2, \ldots, \pm n\}$ are the number of vertices and edges labeled with i respectively.

3 Main Results

3.1 *Edge Product Cordial Labeling of Quadratic Residue Digraph*

Here in Sect. 3.1 we explain the existence of Edge Product Cordial labeling for quadratic residue digraph of order q.

Algorithm 3.1.1
Input: Finite field elements of order q, with $q \equiv 3 \pmod 4$

Step 1: Using Definition 2.1 construct the quadratic residue digraph called Paley digraph P(q)
Step 2: Define f: E (G) $\rightarrow$ {0, 1} as follows

When $1 \leq i \leq p$

$$f(e_{ij}) = \begin{Bmatrix} 0 & \text{for} & j = 1, 3, \ldots, q-2 \\ 1 & \text{for} & j = 2, 4, \ldots, q-1 \end{Bmatrix} \tag{1}$$

When j = q

$$f(e_{ij}) = \begin{cases} 1 & \text{for} \quad i = 1, 2, \ldots, \frac{p+1}{2} \\ 0 & \text{for} \quad i = \frac{p+1}{2} + 1, \ldots, p \end{cases} \tag{2}$$

Step 3: Define the induced map

$$f^{*}: V \rightarrow \{0, 1\} \text{ as } \quad f^{*}(v) = f(e_1)f(e_2)\ldots f(e_n) \tag{3}$$

Output: Edge Product Cordial labeling of quadratic residue digraph P(q).

Theorem 3.1.2 The character difference digraph or quadratic residue digraph admits Edge Product Cordial labeling for all odd prime number congruent to 3 mod 4.

Proof Consider any prime number q which is congruent to 3 mod 4. Construct a quadratic residue digraph using Definition 2.1. To verify every quadratic residue

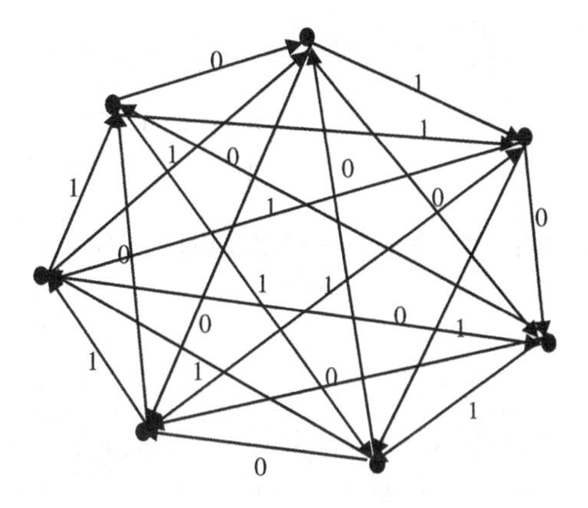

Fig. 1 Edge product cordial labeling of P(7)

digraph admits Edge Product Cordial labeling, we have to prove for every vertex v_i, the product of the labels of arcs incident at v_i is labeled as 0 or 1 and also satisfying the condition that $|v(0) - v(1)| \leq 1$ and $|e(0) - e(1)| \leq 1$. Let $v_i \in V$ be any vertex of the directed quadratic residue graph called Paley graph P(q). In Paley digraph, the incoming and outgoing arcs are same at each vertex. From the definition of f and f* of the algorithm, the total add up of edges labeled zero is $[((q - 1)/2) p + (p - 1)/2]$ and the total add up of edges labeled one is $[((q - 1)/2)p + (p + 1)/2]$ which differ by one. Correspondingly the total add up of vertices labeled zero is $(p + 1)$ and the total add up of vertices labeled one is p which also differ by one. Hence the directed quadratic residue graph admits Edge Product Cordial labeling.

Example 3.1.3 Consider the quadratic residue digraph over a finite field with odd prime number $7 \equiv 3 \pmod 4$. The Edge Product Cordial labeling is shown in Fig. 1.

3.2 *K Edge Graceful Labeling of Quadratic Residue Digraph*

In Sect. 3.2 we explain that for quadratic residue digraph of order q are K Edge Graceful graphs.

Algorithm 3.2.1
Input: Finite field elements of order q, with $q \equiv 3 \pmod 4$

Step 1: Using Definition 2.1 construct the quadratic residue digraph called Paley digraph P(q)

Step 2: Define f: E (G) $\rightarrow$ {k, k + 1, … , k + pq − 1} as follows

$$f_{r_k}(e_{ij}) = \begin{cases} kq - 2i + 2 & \text{for} \quad 1 \leq i \leq \frac{q+1}{2} \\ (k+1)q - 2i + 2 & \text{for} \quad \frac{q+1}{2} + 1 \leq i \leq q \end{cases} \tag{4}$$

Step 3: Define the induced map f*: V $\rightarrow$ {0, 1, 2, … , |v| − 1} as

$$f^*(v_i) = \sum_{j=1}^{p} f(e_{ij}) \tag{5}$$

where e_{ij} is an outgoing arc of *i*th vertex.
Output: K Edge Graceful labeling for quadratic residue digraph P(q).

Theorem 3.2.2 The character difference digraph or quadratic residue digraph admits K Edge Graceful labeling for all odd prime number congruent to 3 mod 4.

Proof Construct a quadratic residue digraph for odd prime number $q \equiv 3 \pmod 4$. To verify every quadratic residue digraph admits K Edge Graceful labeling, let us explain that there exists a function f: E(G) $\rightarrow$ {k, k + 1, k + 2, … k + pq − 1} such that for every vertex v_i, the addition of the labels of the leaving arcs are distinct modulo q. Consider a random vertex $v_i \in V$ of the directed Paley graph P(q). In Paley digraph, the incoming and outgoing arcs are same at each vertex. Use f and f* of the above algorithm.

We get for $1 \leq i \leq p + 1$

$$\begin{aligned} f^*(v_i) &= f(e_{1i}) + f(e_{2i}) + \cdots + f(e_{pi}) \\ &= (q + 2q + \cdots + pq) - p(2i - 2) \end{aligned}$$

For $p + 2 \leq i \leq q$

$$\begin{aligned} f^*(v_i) &= f\ (e_{1i}) + f\ (e_{2i}) + \cdots + f\ (e_{pi}) \\ &= (q + 2q + \cdots + pq) - p(2i - q - 2) \end{aligned}$$

Hence all v_i are distinct for all i = 1,2, …. , n and for k = 0, 1, 2, 3…
This shows every directed Paley graphs are K Edge Graceful.

Example 3.2.3 Consider the character difference digraph or quadratic residue digraph over a finite field with odd prime number $11 \equiv 3 \pmod 4$.The digraph P (11) and its K Edge Graceful labeling with k = 0 is shown in Fig. 2.

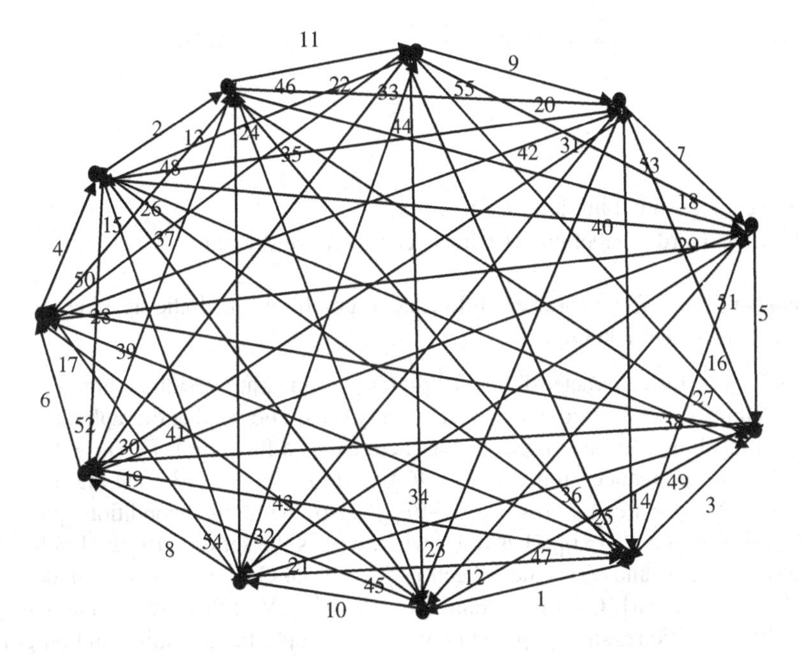

Fig. 2 K edge graceful labeling of P(11)

3.3 H_n Cordial Labeling of Character Difference Digraph

In Sect. 3.3 we illustrate that a Character difference digraph of order q is H_n Cordial graph.

Algorithm 3.3.1
Input: Finite field elements of order q, with $q \equiv 3 \pmod 4$

Step 1: Using Definition 2.1 construct the Character difference digraph called Paley digraph P(q)
Step 2: Define f: E (G) $\rightarrow$ {±1, ±2, …, ±n} as follows
When j is odd

$$f(e_{ij}) = \begin{cases} -i & \text{for} \quad i = 1, 3, \ldots, q \\ i & \text{for} \quad i = 2, 4, \ldots, q-1 \end{cases} \tag{6}$$

When j is even

$$f(e_{ij}) = \begin{cases} i & \text{for} \quad i = 1, 3, \ldots, q \\ -i & \text{for} \quad i = 2, 4, \ldots, q-1 \end{cases} \tag{7}$$

Step 3: Define the induced map $f^{*}: V \rightarrow \{\pm 1, \pm 2, \ldots, \pm |v|\}$ as

$$f^{*}(v_i) = \sum_{j=1}^{p} f(e_{ij}) \tag{8}$$

Where e_{ij} is an outgoing arc of *i*th vertex.
Output: H_n Cordial labeling for Character difference digraph.

Theorem 3.3.2 The character difference digraph or quadratic residue digraph admits H_n Cordial labeling.

Proof Construct a character difference digraph or quadratic residue digraph for odd prime number $q \equiv 3 \pmod 4$. To verify every quadratic residue digraph admits H_n Cordial labeling, we have to prove that there exists a function $f: E(G) \rightarrow \{\pm 1, \pm 2, \ldots, \pm n\}$ and the induced function $f^{*}: V(G) \rightarrow \{\pm 1, \pm 2, \ldots, \pm n\}$ such that for each vertex $v_{i,} f^{*}(v_i) = f(e_{1i}) + f(e_{2i}) + \cdots + f(e_{pi})$ also with the condition $|v(i) - v(-i)| \leq 1$ and $|e(i) - e(-i)| \leq 1$ is hold, where v(i) is the set of vertices of G having labeled i under f^{*} and e(i) is the collection of edges of G having labeled i under f for $i = \{\pm 1, \pm 2, \ldots, \pm n\}$. Consider a random vertex $v_i \in V$ of the directed Paley graph P(q). In quadratic residue graph namely Paley digraph, the incoming and outgoing arcs are same at each vertex. Use f and f* from above algorithm. We get for every vertex v_i

When i is odd

$$\begin{aligned} f^{*}(v_i) &= f\ (e_{1i}) + f\ (e_{2i}) + \cdots + f\ (e_{pi}) \\ &= ((p+1)/2)(i) + ((p-1)/2)(-i) \end{aligned}$$

When i is even

$$\begin{aligned} f^{*}(v_i) &= f\ (e_{1i}) + f\ (e_{2i}) + \ldots. + f\ (e_{pi}) \\ &= ((p+1)/2)(-\ i) + ((p-1)/2)(i) \end{aligned}$$

Hence for all v_i the above value belongs to $\{\pm 1, \pm 2, \ldots, \pm n\}$ for all i. The sum of edges labeled i is ((p + 1)/2) (i) and sum of edges labeled −i is ((p − 1)/2) (−i) when i is odd and the sum of edges labeled i is ((p − 1)/2) (i) and sum of edges labeled −i is ((p + 1)/2) (−i) when i is even. Therefore, $|e(i) - e(-i)| \leq 1$ is hold. Hence the vertices receives distinct labeling from $\{\pm 1, \pm 2, \ldots, \pm n\}$. Therefore $|v(i) - v(-i)| \leq 1$ is hold.

Hence every character difference directed graphs are H_n Cordial.

Example 3.3.3 Consider the Character difference digraph of order 7, where $7 \equiv 3$ (mod 4) The H_n—Cordial labeling of directed graph is shown in Fig. 3.

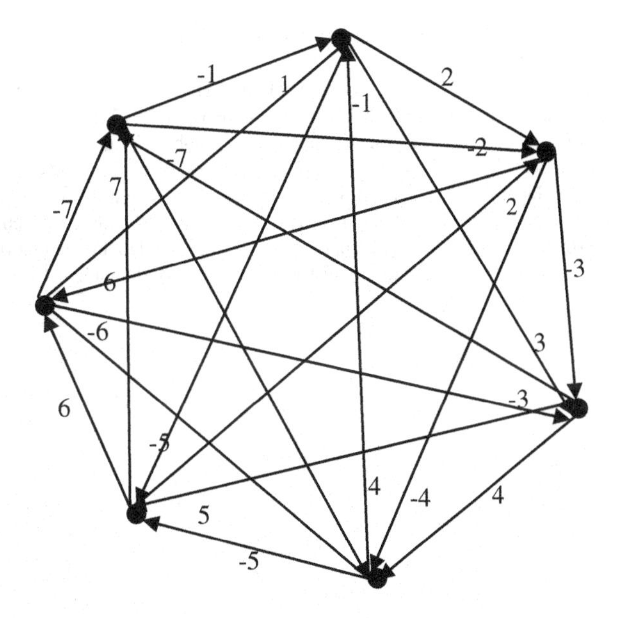

Fig. 3 H_n-cordial labeling of P(7)

4 Conclusion

In this paper we have discussed three labeling namely Edge product cordial, K edge graceful and H_n Cordial labeling for the quadratic residue graph called Paley digraph. In future for the same digraph we are going to discuss two more labeling called super vertex(a, d) antimagic and modulo magic labeling.

References

1. Erdos, P., Renyi, A.: Asymmetric graphs. Acta Math. Acad. Sci. Hun. **14**, 295–315 (1963)
2. Paley, R.E.A.C.: On orthogonal matrices. J. Math. Phys. **12**, 311–320 (1933)
3. Sachs, H.: Uber selbstkomplementare graphen. Publicationes Math. **9**, 270–288 (1962)
4. Ananchuen, W., Caccetta, L.: On the adjacency properties of Paleygraphs. Networks **23**(4), 227–236 (1993)
5. Graham, R.L., Spencer, J.H.: A constructive solution to a tournament problem. Canad. Math. Bull. **14**, 45–48 (1971)
6. Gallian, J.A: A dynamic survey of graph labeling. Electron. J. Combinatory **18**, #DS6 (2011)
7. Cahit, I.: Cordial graphs: a weaker version of graceful and harmonious graphs. Ars Combin. **23**, 201–208 (1987)
8. Cahit, I.: H-cordial graphs. Bull. Inst. Combin. Appl. **18**, 87–101 (1996)
9. Sundaram, M., Ponraj, R., Somsundaram, S.: Product cordial labeling of graphs. Bull. Pure Appl. Sci. (Mathematics and Statistics) **23E**, 155–163 (2004)
10. Vaidya, S.K., Barasara, C.M.: Edge product cordial labeling of graphs. J. Math. Comput. Sci. **2**(5), 1436–1450 (2012)

11. Ghebleh, M., Khoeilar, R: A note on H-cordial graphs. Bull. Inst. Combin. Appl. **31**, 60–68 (2001)
12. Freeda, S., Chellathurai, R.S.: H and H_2-cordial labeling of some graphs. Open J. Discrete Math. **2**, 149–155 (2012). doi:10.4236/ojdm.2012.24030
13. Lo, S.: On edge graceful labeling of graphs. Congr. Numer. **50**, 231–234 (1985)
14. Lee, S.M., Wang, L.: On K-edge graceful trees (preprint)
15. Gayathri, B., Devi, S.K.: K-edge-graceful labeling and k-global edge-graceful labeling of some graphs. Internat. J. Engin. Sci., Adv. Comput. Bio-Tech. **2**(1), 25–37 (2010)
16. Rajeswari, R., Thamizharasi, R.: H_n cordial labeling and Edge graceful labeling of Caley digraphs. In: International Conference on Mathematical Computer Engineering, ICMCE-13, VIT University

Comparative Performance Study of Various Content Based Image Retrieval Methods

Rushabh Shah, Jeetendra Vaghela, Khyati Surve, Rutvi Shah, Priyanka Sharma, Rasendu Mishra, Ajay Patel and Rajan Datt

Abstract With the increase in the data storage and data acquisition technologies there is an increase in huge image database. Therefore we need to develop proper and accurate systems to manage this database. Here in this paper we focus on the transformation technique to search, browse and retrieve images from large database. Here we have discussed briefly about the CBIR technique for image retrieval using Discrete Cosine Transform for generating feature vector. We have researched on the different retrieval algorithms. The proposed work is experimented over 9000 images from MIRFLIKR Database (Huskies ACM International Conference on Multimedia Information Retrieval (MIR'08), 2008 [1]). We have focused on

R. Shah (✉) · J. Vaghela · K. Surve · R. Mishra · A. Patel · R. Datt
Institute of Technology, Nirma University, Gota, Ahmedabad, India
e-mail: rushabh.shah@nirmauni.ac.in

J. Vaghela
e-mail: 14mca57@nirmauni.ac.in

K. Surve
e-mail: 14mca53@nirmauni.ac.in

R. Mishra
e-mail: rasendu.mishra@nirmauni.ac.in

A. Patel
e-mail: ajaypatel@nirmauni.ac.in

R. Datt
e-mail: rajandatt27@nirmauni.ac.in

R. Shah
Shri Chimanbhai Patel Institute of Computer Applications, Makarba, Ahmedabad, India
e-mail: rutvirshah@gmail.com

P. Sharma
Raksha Shakti University, Meghani Nagar, Ahmedabad, India
e-mail: pspriyanka@yahoo.com

S.C. Satapathy and S. Das (eds.), *Proceedings of First International Conference on Information and Communication Technology for Intelligent Systems: Volume 1*, Smart Innovation, Systems and Technologies 50, DOI 10.1007/978-3-319-30933-0_40

showing the difference between the precision and recall and also the time of different methods and its performance by querying an image from the database and a non-database image.

Keywords Content based image retrieval · Image retrieval methods · Image processing techniques · Digital image processing

1 Introduction

The old Chinese proverb "A picture speaks a thousand words" says that the images are far better in communication than text. As the world is moving towards digital India and also people are storing more and more images, retrieval of them becomes a challenge. Also we should bring the right image when queried. So here in this paper we will be focusing on Image retrieval methods or various metrics which will be used for finding out the best image from the database. Right now there is a heavy need of searching images from the database to match with different parameters Interest in digital image processing methods comes from two main applications: betterment of image information for human understanding and to process image for storage, transmission and to represent sovereign machine perception.

1.1 What Is Digital Image Processing?

Digital Image Processing means to process digital images with the help of a digital computer. This processing divided into three parts as a continuous series: *Bottom-level* processes, *Middle-level* processes and *Top-level* processes. Primary operations such as image preliminary process to decrease noise, intensifying contrast, and sharpen the image are called as *Bottom-Level* processes which inputs and outputs are images. Partition an image into segments, encompassing segments to make suitable for computer processing, and identification of individual segments are called *middle-level* processes which take image as input and produce image attributes fetched from that images. And, process which is sensible for grouping known segments is called *Top-level* processes [2–4].

2 Image Retrieval

In today's world there is a fast increase in the collection of image. Generally military and civilian sectors produce huge almost huge amount of data in gigabytes. The first and foremost system for image retrieval based on microcomputer was developed in 1990 by MIT.

There are two approaches for image retrieval:

1) Text Based Image Retrieval (TBIR), and
2) Content Based Image Retrieval (CBIR).

2.1 Text Based Image Retrieval (TBIR)

This approach started in the 1970s where the images are tagged manually by text descriptors which will be further used by DBMS to retrieve image. Technique is popular but needs very specific description, tedious and not always possible.

2.2 Content Based Image Retrieval (CBIR)

CBIR is retrieval of image based on color, shape and texture. These system try to retrieve images similar to a user-defined specification or pattern. Content based means the search is done based on the content of image rather than keywords and captions. This CBIR technique was introduced due to large size of image data and problem in indexing them and giving those numbers and description. In CBIR, images will be stored in the database by storing their main features and when the query image will be given, the images will be searched by matching the features of two images.

3 Various Methods Description for Image Retrieval

For Image Retrieval, we used an Image Retrieval application and examine several image retrieval methods. Various image retrieval methods used are:

1) Linear Distance Method (L1)
2) Linear-2 Distance Method (L2)
3) Standardized Linear-2 Distance Method (Standardized L2)
4) CityBlock Distance Method (CityBlock)
5) Minkowski Distance Method (Minkowski)
6) Chebyshev Distance Method (Chebyshev)
7) Cosine Similarity Distance Method (Cosine)
8) Pearson's Correlation Distance Method (Correlation)
9) Spearman's Rank Correlation Coefficient Distance Method (Spearman)
10) Normalized Linear-2 Distance Method (Normalized L2)
11) Relative Standard Deviation Distance Method (Relative Deviation).

3.1 Linear Distance Method (L1)

Manhattan distance [5] between 2 points measured with the axis at 90° is used by L1. It is the distance which is the sum of vertical and horizontal points whereas the distance diagonally may be computed by applying Pythagorean Theorem. The difference of similar components and then its addition is the Manhattan distance between 2 points. The formula that gives the distance between point $x = (x1, x2, \ldots, xn)$ and point $y = (y1, y2, \ldots, yn)$ is:

$$d = \sum_{i=1}^{n} |x_i - y_i| \tag{1}$$

where, n = number of variables. x_i and y_i are the values of ith variable at points x and y.

3.2 Linear-2 Distance Method (L2)

Distance between 2 points in metric space called Euclidean Distance is used by L2. Line segment which connect points p and q is the Euclidean Distance. Euclidean distance between point's p and q is the length of the line segment connecting them $(\overline{pq})$. If 2 points in Euclidean space p and q where p = (p1, p2, …, pn) and q = (q1, q2, … qn) the distance among them is recognized from Pythagoras theorem.

$$d(p,q) = d(q,p) = \sqrt{\sum_{i=1}^{n} (q_i - p_i)^2} \tag{2}$$

The Euclidean length measure the length of the vector.

$$|p| = \sqrt{p_1^2 + p_2^2 + \cdots + p_n^2} = \sqrt{p \times p} \tag{3}$$

A vector here can be described as the directed line segment from the vector tail to the vector tip. Simply Euclidean distance [6] between two points is the space which corresponds to the length of the straight line drawn between them.

3.3 Standardized Linear-2 Distance Method (Standardized L2)

The Euclidean distance [7] was standardized to make it independent of size and range of ratings. Standardized Euclidean Distance [8, 9] which is bigger than 1 are greater than expected, if lesser than 1 are smaller than expected. Standardized L2 have been used to compare the inter-element distances between grids.

$$d_{x,y} = \sqrt{\sum_{j=1}^{j} \left(\frac{x_j}{S_j} - \frac{y_j}{S_j} \right)^2}. \tag{4}$$

3.4 *CityBlock Distance Method (CityBlock)*

CityBlock Distance (CD) [10, 11] is distance matric which is the sum of absolute distance between two points X and Y, with their D dimensions is calculated using:

$$CD = \sum_{i=1}^{D} |X_i - Y_i| \tag{5}$$

Always, CityBlock distance ≥ 0. If distance = 0 then the points are identical and if the distance is high then the points may have some similarity.

3.5 *Minkowski Distance Method (Minkowski)*

By generalizing Manhattan and Euclidean distance the result we get is Minkowski distance which is a distance between two points $X = (x1, x2, x3, \ldots, xn)$ and $Y = (y1, y2, y3, \ldots, yn) \in \mathbb{R}^n$ calculated as:

$$M = \left(\sum_{i=1}^{n} |x_i - y_i|^p \right)^{1/p} \tag{6}$$

when $p \geq 1$ then the Minkowski inequality distance metric is formed. The distance between point (0, 0) and (1, 1) is $2^{1/p} > 2$ when p < 1, but distance is 1 for point is (0, 1) from both of these points which breach the triangle inequality [12], It is not a metric.

3.6 *Chebyshev Distance Method (Chebyshev)*

Chebyshev distance [13] is also known as the TChebyshev distance or maximum metric. Metric defined on a space of vector where the length between 2 vectors is greatest of difference with coordinate is Chebyshev. Chebyshev is also called as

Chessboard distance. Chebyshev distance between two points' p_i and q_i with standard coordinates and is:

$$D_{Chebyshev}(p, q) := max_i(|p_i - q_i|). \tag{7}$$

3.7 *Cosine Similarity Distance Method (Cosine)*

Cosine Similarity [14, 15] is used to determine similarity between 2 observations. Cosine similarity helps user to effectively find cosine of the angle between the two objects. If cosine similarity value is 0 the documents don't share any attributes or words because the angle between the objects is 90°. If cosine similarity value is 1 the observation share same attributes and if cosine similarity value is −1 then it means that the two vectors are diametrically opposed.

$$similarity(x, y) = \cos(\theta) = \frac{x \times y}{|x| \times |y|}. \tag{8}$$

3.8 *Pearson's Correlation Distance Method (Correlation)*

Correlation distance [16] is a distance R between X and Y with D dimensions is calculated as:

$$R = \frac{Cov(X, Y)}{std(X) \times std(Y)} \tag{9}$$

where,

$$cov(X, Y) = \frac{1}{D} \sum_{i=1}^{D} (X_i - \bar{X}) \times (Y_i - \bar{Y}) \tag{10}$$

$$std(X) = \sqrt{\frac{1}{D} \sum_{i=1}^{D} (X_i - \bar{X})} \tag{11}$$

$$\bar{X} = \frac{1}{D} \sum_{i=1}^{D} X_i \tag{12}$$

This is also known as Pearson's Correlation or Pearson's R which range from +1 to −1. When R is near or equal to +1 then it states highest correlation and in opposite when R is near or equal to −1 then it states lowest correlation.

3.9 Spearman's Rank Correlation Coefficient Distance Method (Spearman)

Spearman's rank correlation coefficient [17] denoted by Greek letter ρ (rho) or r_s is a statistical dependence measure between two ranked points. It examines relationship between two points. For N data value the N rows includes X_i, Y_i points are converted to ranks x_i, y_i and ρ is calculated from:

$$\rho = 1 - \frac{6\sum d_i^2}{n(n^2 - 1)} \tag{13}$$

where, $d_i = x_i - y_i$ is the difference between two ranks. Spearman's rank correlation coefficient lies between +1 and −1, is show perfect correlation. When rank is 0 or near to 0 it show minor or no correlation between two points.

3.10 Normalized Linear-2 Distance Method (Normalized L2)

If the covariance matrix is the diagonal then the resulted distance is the Normalized Linear-2 Distance.

$$d(\vec{x}, \vec{y}) = \sqrt{\sum_{i=1}^{N} \frac{(x_i - y_i)^2}{S_i}} \tag{14}$$

where, S_i = standard deviation and x_i, y_i = sample set.

3.11 Relative Standard Deviation Distance Method (Relative Deviation)

Relative Standard Deviation (RSD) [18] is also known as Coefficient of Variation (CV) [19] is a standardized measure of scattered ness of probability distribution. It is a defined as a ratio of the Standard deviation (σ) [20] and the mean (μ) in percentage. Relative Standard Deviation is formulated as:

$$C_v = \frac{\sigma}{\mu} \times 100\,\%, \textit{ where } \sigma = \sqrt{\frac{1}{N}\sum_{i=1}^{N}(x_i - \mu)^2}, \textit{ where } \mu = \frac{1}{N}\sum_{i=1}^{N} x_i. \tag{15}$$

4 Image Retrieval Methods Testing and Results

Image Retrieval Methods is evaluated in the three terms: First, *Image Retrieval Time* (in second) second, methods *Precision* and third, methods *Recall.*

Image Retrieval Time: is a time taken by method to retrieve query image from the image database.

$$\textbf{\textit{Precision}} = \frac{Number\ of\ Relevant\ Image\ Retrieved\ from\ Database}{Total\ Number\ of\ Image\ Retrieved\ from\ Database} \tag{16}$$

$$\textbf{\textit{Recall}} = \frac{Number\ of\ Relevant\ Image\ Retrieved\ from\ Database}{Total\ Number\ of\ Relevant\ Image\ in\ Database}. \tag{17}$$

4.1 Scenario-1: Calculate Average Image Retrieval Time, Precision and Recall of Every Methods with Same Image

Table 1 is the representation of precision, recall and average retrieval time for retrieving of image from a database of *9000 images*. Here we have taken same image as a query image for every method and *retrieved 20 images*. We have done this process *10 times for each method*. The best method whose average retrieval time is the lowest of this three methods and precision and recall is *highest is the Relative Deviation method* with average retrieval time *2.1976934, precision 15 % and recall 13.64 %.*

4.2 Scenario-2: Calculate Non-database Image Retrieval Time, Precision and Recall of Every Methods

Table 2 is the representation of precision, recall and average retrieval time for retrieving of image from a database of *9000 images*. We have taken same image not in database as a query image for every method and *retrieved 20 images*. The best method whose retrieval time is the lowest of this three methods and precision and recall is highest is the method with *retrieval time 1.758306, precision 10 % and recall 9.09 %.*

Figure 1 show the working application of Image Retrieval in which retrieval done using L1 method.

Table 1 Average image retrieval time, precision and recall of every methods with same images

Retrieval method	T1	T2	T3	T4	T5	T6	T7	T8	T9	T10	Average ret time	Precision (%)	Recall (%)
L1	2.38	2.27	2.32	2.15	2.18	2.19	2.27	2.05	2.13	2.25	2.22	15	13.64
L2	1.84	1.55	1.75	1.87	1.66	1.75	1.57	1.69	1.61	1.63	1.69	5	4.55
Standardized L2	1.92	1.72	1.58	1.75	1.67	1.59	1.63	1.78	1.57	1.58	1.68	5	4.55
Cityblock	3.78	2.13	2.04	2.21	2.18	1.97	2.00	1.94	1.96	2.12	2.23	15	13.64
Minkowski	1.68	1.47	1.46	1.41	1.43	1.48	1.53	1.58	1.52	1.50	1.51	5	4.55
ChebyChev	1.62	1.44	1.43	1.56	1.45	1.46	1.58	1.48	1.69	1.52	1.52	5	4.55
Cosine	1.65	1.56	1.63	1.51	1.56	1.51	1.73	1.64	1.56	1.73	1.61	5	4.55
Correlation	1.48	1.49	1.53	1.60	1.75	1.57	1.50	1.49	1.52	1.63	1.56	5	4.55
Spearman	3.78	3.56	3.93	4.00	3.35	3.83	3.69	3.63	3.37	3.33	3.65	5	4.55
Normalized L2	2.11	2.18	2.13	2.21	2.17	2.19	2.42	2.25	2.32	2.24	2.22	5	4.55
Relative deviation	2.03	1.96	2.44	2.49	2.24	2.21	2.36	2.03	2.01	2.16	2.19	15	13.64

Table 2 Non database image retrieval time, precision and recall

Method	Retrieval time	Precision (%)	Recall (%)
L1	2.523199	0	0
L2	2.113015	10	9.09
Standardized L2	2.072475	5	4.55
Cityblock	2.358247	5	4.55
Minkowski	1.758306	10	9.09
ChebyChev	1.989151	5	4.55
Cosine	2.234872	5	4.55
Correlation	1.869555	5	4.55
Spearman	4.024227	5	4.55
Normalized L2	2.436479	10	9.09
Relative deviation	1.868826	5	4.55

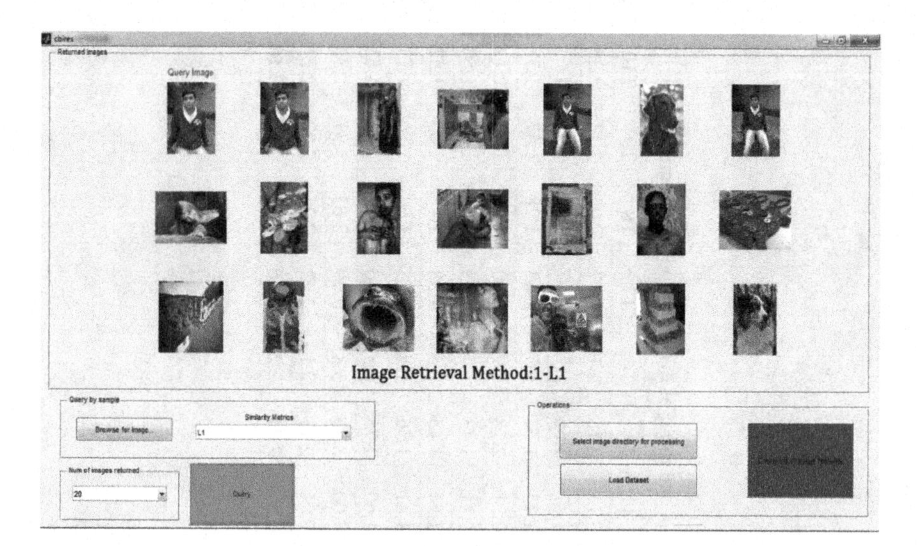

Fig. 1 Image retrieval application screenshot

5 Conclusion

For the image in the database whose feature vector was generated and then it was queried we found the result that the three methods naming L1, CityBlock and Relative Deviation have the highest precision and recall time but the average time is lower. The best method whose average retrieval time is the lowest of this three methods and precision and recall is highest is the Relative Deviation method with average retrieval time 2.1976934, precision 15 % and recall 13.64 %. For Non-database images, after performing the process we resulted that the three methods naming L2, Minkowski and Normalized L2 have the highest precision and recall time but the average time is lower. The best method whose retrieval time is the lowest of this three methods and precision and recall is highest is the method with retrieval time 1.758306, precision 10 % and recall 9.09 %.

References

1. M.J. Huskies, M.S. Lew: The MIR Flickr retrieval evaluation. In: ACM International Conference on Multimedia Information Retrieval (MIR'08), Vancouver, Canada (bib) (2008)
2. Gonzalez, R.C., Woods, R.E.: Digital Image Processing, 3rd edn. Pearson Education, India ISBN: 8131726959 (2009)
3. Gonzalez, R.C., Woods, R.E., Eddins, S.L.: *Digital Image Processing using MATLAB*, Pearson Education, India (2004). ISBN: 8177588982
4. Sridhar, S.: *Digital Image Processing*, Oxford University Press, Oxford (2011). ISBN: 0198070780, India
5. http://xlinux.nist.gov/dads/HTML/manhattanDistance.html
6. https://en.wikipedia.org/wiki/Euclidean_distance
7. http://www.pbarrett.net/techpapers/euclid.pdf
8. http://reference.wolfram.com/language/ref/NormalizedSquaredEuclideanDistance.html
9. http://stats.stackexchange.com/questions/136232/definition-of-normalized-euclidean-distance
10. https://docs.tibco.com/pub/spotfire/5.5.0-March-2013/UsersGuide/hc/hc_city_block_distance.htm
11. http://people.revoledu.com/kardi/tutorial/Similarity/CityBlockDistance.html
12. https://en.wikipedia.org/wiki/Triangle_inequality
13. https://en.wikipedia.org/wiki/Chebyshev_distance
14. http://reference.wolfram.com/language/ref/CosineDistance.html
15. https://en.wikipedia.org/wiki/Cosine_similarity
16. http://reference.wolfram.com/language/ref/CorrelationDistance.html
17. https://en.wikipedia.org/wiki/Spearman%27s_rank_correlation_coefficient
18. http://www.statisticshowto.com/relative-standard-deviation/
19. https://en.wikipedia.org/wiki/Coefficient_of_variation
20. Rui, Y., Huang, T.S., Chang, S.F.: Image retrieval: current techniques, promising directions, and open issues. J. Vis. Commun. Image Represent. **10**(1), 39–62 (1999)
21. https://en.wikipedia.org/wiki/Image_retrieval
22. https://en.wikipedia.org/wiki/Standard_deviation
23. http://in.mathworks.com/help/stats/pdist2.html#syntax
24. https://docs.tibco.com/pub/spotfire/5.5.0-March-2013/UsersGuide/hc/hc_correlation.htm
25. https://en.wikipedia.org/wiki/Minkowski_distance

GIS and Statistical Approach to Assess the Groundwater Quality of Nanded Tehsil, (M.S.) India

Wagh Vasant, Panaskar Dipak, Muley Aniket, Pawar Ranjitsinh, Mukate Shrikant, Darkunde Nitin, Aamalawar Manesh and Varade Abhay

Abstract The present research work has been done to assess the groundwater quality and its drinking suitability by developing a Water Quality Index (WQI) for Nanded Tehsil, Maharashtra. The representative groundwater samples (50) were collected from Dug/Bore wells during pre monsoon 2012. By clutching Hydrochemical analysis and GIS based IDW technique were used to represent spatial variation of WQI in study area. The physicochemical parameters viz. pH, EC, TDS, TH, Ca^{2+}, Mg^{2+}, Na^{+}, K^{+}, CO_3^{-}, HCO_3^{-}, Cl^{-}, NO_3^{-}, SO_4^{-}, PO_4^{-} were determined to assess the groundwater quality and compared with BIS standards 2003. Water quality index used to classify water into four categories. Water quality index shows that 14 % samples are excellent, 84 % samples are good and 2 % Poor for drinking purpose. The GIS tools and statistical techniques used for spatial distribution and representation of water quality.

Keywords Groundwater · Water quality index · Nanded · GIS

1 Introduction

Groundwater is an essential source of Irrigation, domestic and drinking all over the world. Now a day, the groundwater exploration has been increased for agricultural purpose, due to the failure of frequent monsoons in arid and semi arid environments.

W. Vasant (✉) · P. Dipak · P. Ranjitsinh · M. Shrikant · A. Manesh
School of Earth Sciences, Swami Ramanand Teerth Marathwada University, Nanded, Maharashtra, India
e-mail: wagh.vasant@gmail.com

M. Aniket · D. Nitin
School of Mathematical Sciences, Swami Ramanand Teerth Marathwada University, Nanded, Maharashtra, India

V. Abhay
Department of Geology, Rashtrasant Tukdoji Maharaj University, Nagpur, Maharashtra, India

S.C. Satapathy and S. Das (eds.), *Proceedings of First International Conference on Information and Communication Technology for Intelligent Systems: Volume 1*, Smart Innovation, Systems and Technologies 50, DOI 10.1007/978-3-319-30933-0_41

In the world about 301 million ha land is under agriculture of which 38 % uses groundwater for irrigation [1]. The 60 % groundwater used for irrigation and 85 % for drinking purpose in urban and rural population of India. It has resulted in the overexploitation of the groundwater resources. This ultimately leads to the depletion of the groundwater resources in the corresponding area. The interaction of the groundwater with the aquifer rock results in the dissolution of many minerals ultimately affecting the groundwater quality. Many reactions takes place within the aquifer may be understood by monitoring of different cation and anion contents of the groundwater. Some other sources which give valuable input for the contamination of groundwater include inadequately managed agriculture, expanding urban and Industrial activities etc. With the rapid development of economy, demand of water resources has been increased which leads to over-exploitation of groundwater. Drinking contaminated water leads to considerable effects on human health, soil environment, aquatic ecosystem and socioeconomic development [2].

WQI's can be very useful for the purpose of decision making and management of water resources. Water quality indices provide a single numerical value to the user. It is required for resolving analyzed physico-chemical parameter. The development of WQI of any area is an essential process for the development of that area. All over the world enormous study has been carried out to determine the WQI and GWQI for water suitability of their region [3–11].

GIS is a comprehensive computer based technique which plays significant role in water resource management and pollution study. It is also used to identify the hydrochemical characteristics and its suitability for different purposes. The objectives of the present study are application of GIS and statistical techniques for the delineation of groundwater quality Index. The physicochemical data of groundwater samples has been subjected to SPSS 22.0 [12] version for the statistical analysis and Inverse Distance Weightage (IDW) technique has been used for pictorial representation.

2 Study Area

The study area is located in Nanded District, South Eastern part of Maharashtra with latitudes 19° 03′ to 19° 17′N and longitudes 77° 10′ to 77° 25′E (Fig. 1). The Nanded Tehsil occupies an area about 1022.8 km^2 in which agricultural area is around 82,740 ha. The area receives an average rainfall 900 mm with 88 % of the annual rainfall received during the south-west monsoon. The temperature varies from 13 to 46 °C. The Godavari River flows in SW to SE direction having alluvial plain along its coast in the study area.

The geological formation are mainly composed of Deccan Basalt flows viz. Vesicular, Amygdoloidal, Weathered, Fractured, Jointed, Compact basalt etc. The rock dominant secondary minerals are Limestones, Calcite, Quartz and Zeololites

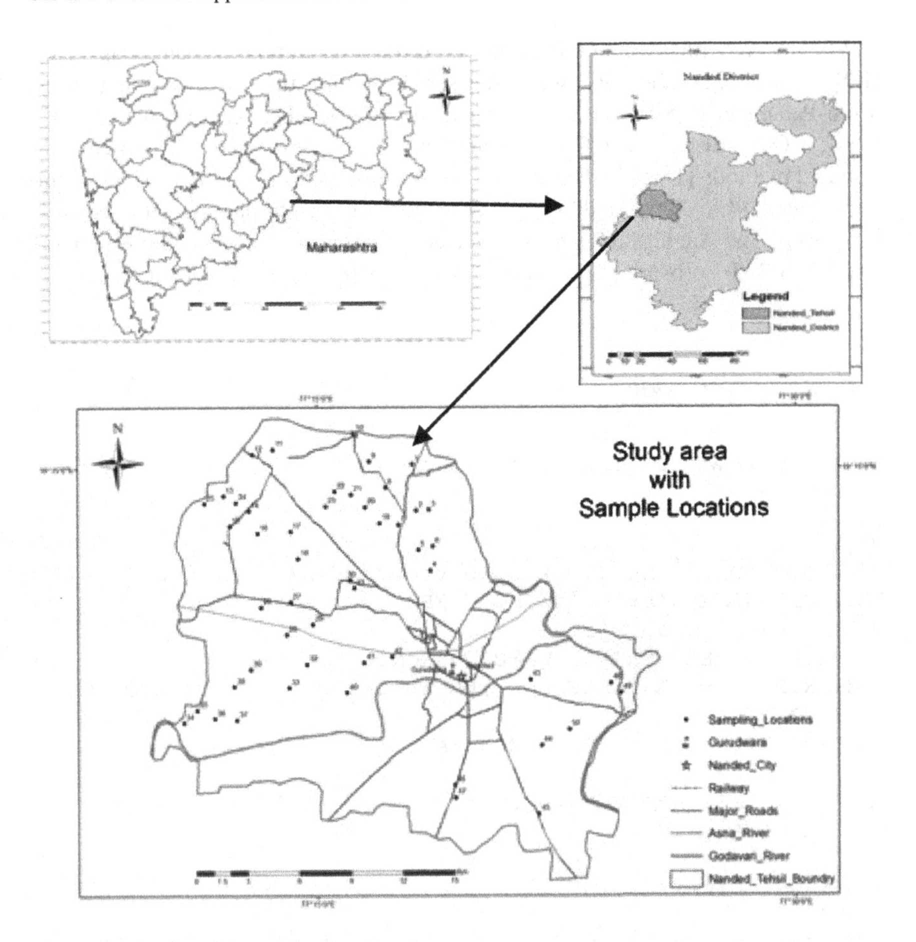

Fig. 1 Study area with samples location map

etc. [13]. In black cotton soil thickness ranges from 2 to 20 m, which is rich in calcium, magnesium, and carbonate but poor in nitrogen, potassium and phosphorous. The soil has a high moisture and humidity retention capacity.

3 Material and Methodology

In this study, 50 groundwater samples collected from different dug/bore wells in 2012 (Pre Monsoon season). The 1 liter plastic cans were pretreated with Acid and also earlier to the collection they were rinsed with water to avoid any contamination; afterwards these cans were sealed, labeled properly and brought to the laboratory for analysis.

We have recorded location coordinates for the preparation thematic maps using GPS device. pH and EC were measured at the time of sampling using Multi-Parameter PCS Tester 35. The remaining physicochemical parameters: Ca^{2+}, Mg^{2+}, CO_3^-, HCO_3^- and Cl^- measured by gravimetric analysis following the standard methods [14]. The Na^+ and K^+ ions determined using Elico CL 361 flame photometer. NO_3^-, SO_4^- and PO_4^- determined using spectrophotometer (Shimadzu UV-1800). The Arc GIS 9.3 was used for the preparation of color composite maps. The SPSS 22.0 software [12] was used for the statistical analysis of the data.

3.1 Program Code

SPSS 22.0 software program commands are,

```
GET
FILE='C:\Users\hp1\Desktop\env analysis\WQI.sav'.
DATASET NAME DataSet1 WINDOW=FRONT.
GETFILE='I:\wagh sir.sav'.
DATASET NAME DataSet2 WINDOW=FRONT.
DESCRIPTIVES VARIABLES=pH TDS TH Ca Mg Na K CO3 HCO3 Cl
NO3 SO4 PO4
/STATISTICS=MEAN STDDEV MIN MAX.

FREQUENCIES VARIABLES=WQI
/ORDER=ANALYSIS.

* Visual Binning.
*WQI.
RECODE  WQI (MISSING=COPY) (LO THRU 50=1) (LO THRU 150=2)
(LO THRU 250=3) (LO THRU 350=4) (LO THRU 450=5) (LO THRU
HI=6) (ELSE=SYSMIS) INTO wqi1.
VARIABLE LABELS  wqi1 'WQI (Binned)'.
FORMATS  wqi3 (F5.0).
VALUE LABELS  wqi1 1 '= 50.0000' 2 '50.0001 - 100.0000' 3
'100.0001 - 200.0000' 4 '200.0001 - 300.0000' 5 '300.0001
- 400.0000' 6 '400.0001+'.
VARIABLE LEVEL  wqi1 (ORDINAL).
EXECUTE.
FREQUENCIES VARIABLES=wqi1
/ORDER=ANALYSIS.2.5
```

3.2 WQI Calculation

WQI used to calculate the influence of different activities based on a number of parameters. The relative weights were assigned (i.e. 1–5) to the groundwater quality parameters according to their significance. The relative weight is calculated as follows (Eq. 1).

$$W_i = \frac{w}{\sum_{i=1}^{n} w_i} \tag{1}$$

where,
W_i Relative weight,
w_i Weight of each parameter
n Number of parameters

The quality rating scale for each parameter is calculated by following equation.

$$q_i = \left(\frac{c_i}{s_i}\right) * 1 \tag{2}$$

where,
q_i Quality rating,
c_i Concentration of each chemical parameter in each sample in mg/l,
s_i Bureau of Indian Standards [15]

Finally WQI is calculated as follows (Table 1):

$$\begin{aligned} \mathrm{WQI} &= \Sigma \mathrm{SI_i} \\ (\mathrm{SI_i} &= \mathrm{W_i} * \mathrm{q_i}) \end{aligned} \tag{3}$$

Table 1 W_i of physico-chemical parameters [10]

Parameters	Weight (w_i)	Relative weight (W_i)
pH	3	0.076923077
TDS	5	0.128205128
TH	3	0.076923077
Ca^{2+}	2	0.051282051
Mg^{2+}	2	0.051282051
Na^{+}	3	0.076923077
K^{+}	2	0.051282051
CO_3^{-}	2	0.051282051
HCO_3^{-}	2	0.051282051
Cl^{-}	5	0.128205128
NO_3^{-}	5	0.128205128
SO_4^{2-}	3	0.076923077
PO_4^{-}	2	0.051282051
Total	39	1

where,

SI_i Sub-index of ith parameter,

q_i Rating based on concentration of ith parameter

n Number of parameters

4 Results and Discussion

4.1 Physicochemical Properties of Groundwater

The descriptive statistics of groundwater quality parameters is compared with BIS standards [15] are as shown in Table 2. The nature of pH is alkaline and it varies from 7.1 to 7.8. The electrical conductivity (EC) represents the total dissolved salts, ranges from 843.0 to 64963.0 µS/cm, with an average 4889.58, sample number 12, 40 and 44 shows higher EC which signify more ion exchange and solublization of aquifer [16]. Total dissolved solids (TDS) indicate the solute concentration and salinity behavior of the groundwater which varies from 539.9 to 41576.32 mg/l. The average TDS concentration is 3129.331 mg/l. The Total Hardness (TH) as

Table 2 Descriptive statistics of physicochemical parameters

Parameters	BIS [15]	Minimum	Maximum	Mean	Std. dev.
pH	6.5–8.5	7.10	7.80	7.48	0.174
EC	–	843.00	64963.00	4889.58	13201.70
TDS	500–2000	539.52	41576.32	3129.33	8449.09
TH	300–600	89.00	1033.00	274.89	178.0
Ca^{2+}	75–200	13.17	106.20	46.37	22.31
Mg^{2+}	30–100	10.13	67.84	34.36	13.49
Na^{+}	–	29.10	349.90	115.82	80.26
K^{+}	–	1.60	3.90	2.76	0.61
CO_3^{-}	–	0.00	60.00	16.60	12.55
HCO_3^{-}	–	40.00	460.00	180.80	81.31
Cl^{-}	250–1000	59.70	495.10	163.13	83.49
NO_3^{-}	45	1.28	60.65	35.44	17.31
SO_4^{2-}	200–400	2.93	27.06	11.26	6.06
PO_4^{-}	–	0.00	0.30	0.08	0.054

All values in mg/l excluding pH and EC (µS/cm)

$CaCO_3$ varies from 89.0 to 1033.0 mg/l. The average TH content is 274.8 mg/l. The 28 % samples surpass the desirable limit but below the permissible limit and 4 % samples are above the permissible limit given by BIS [15]. The concentrations of Na^+ and K^+ ranges from 29.10 to 349.9 mg/l and from 1.60 to 3.90 mg/l respectively and the mean values are 115.81 and 2.76 mg/l respectively. The carbonate and bicarbonate content ranges from 0 to 60 mg/l and 40 to 460 mg/l. Bicarbonate affects on the alkalinity of groundwater. The CO_3^- and HCO_3^- are the products interaction of weathers silicate rock with groundwater, dissolution of carbonate precipitates and soil CO_2 gas [17, 18]. The Cl^- concentration ranges between 59.70 and 495.10 mg/l. The average of Cl^- is 163.13 mg/l. The 10 % samples exceed the desirable limit but below the maximum permissible limit given by, BIS [15]. Naturally, the chloride is originated from dissolution of halite and domestic sewage. Generally, excess chloride in water considered as index of groundwater pollution [19].

4.2 *Spatial Distribution of WQI*

In the study area, WQI varies from 39.88 to 116.5. The classification of WQI with their range and type of water is shown in Table 3. Based on the calculated groundwater quality index 14 % samples are excellent for drinking, 84 % groundwater samples are in good category and only 2 % samples falls in poor category. Figure 2 shows that, the sample no. 9, 12 and 23 represents the water quality is near to poor category (WQI values 97.7, 97.37 and 83.49 respectively). These samples are affected by the leachate accumulation due to the lineament in the NW to SE direction. The water quality index of sample no. 32 and 40 are mainly (86.304 and 91.173) deteriorated due to urban and Agricultural activities. The Sample number 44 (WQI 116.59) falls under poor type of water category; located near to Tuppa industrial area of Nanded city. It indicates that industrial activities are affecting the groundwater quality.

Table 3 Classification—range of WQI and its type with their percentage

WQI range	Type of water	Frequency	Percent	Valid percent	Cumulative percent
00–50	Excellent	07	14.0	14.0	14.0
50–100	Good	42	84.0	84.0	98.0
100–200	Poor	01	2.0	02.0	100.0
Total		50	100.0	100.0	

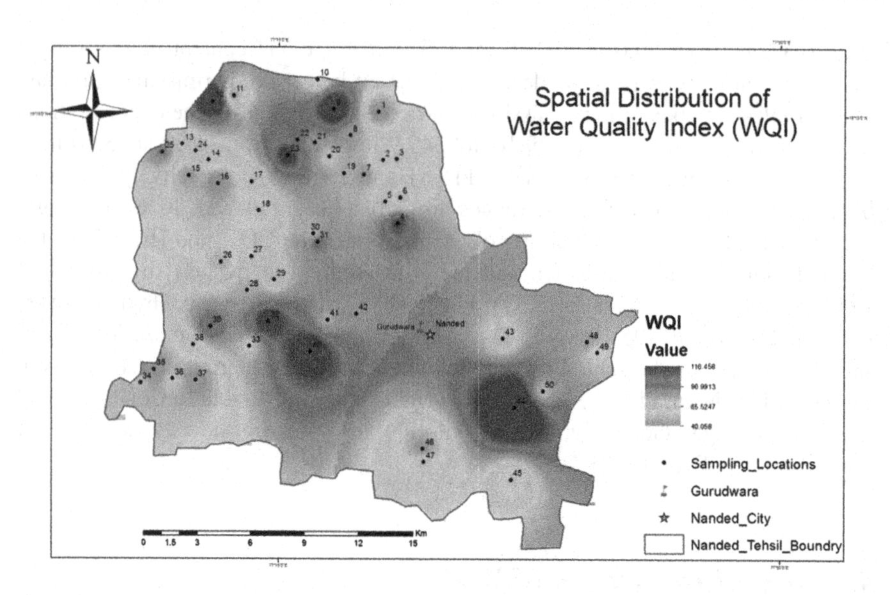

Fig. 2 Spatial distribution of WQI

Acknowledgements The Authors gratefully acknowledge the financial support by SRTM University, Nanded for sanctioned Minor Research Project and Director, school of Earth Sciences for providing necessary facilities during research work. Also, authors would like to thanks to reviewers for their valuable comments and suggestions.

References

1. Food and Agricultural organization of United States, http://www.fao.org/
2. Milovanovic, M.: Water quality assessment and determination of pollution sources along the Axios/Vardar River, Southeastern Europe. Desalination **213**, 159–173 (2007)
3. Brown, V. M., Shaw, T. L., and Shurben, D. G.: Aspects of water quality and the toxicity of copper to rainbow trout. Water Res. **8**(10), 797–803 (1974)
4. Reza, R., Singh, G.: Assessment of ground water quality status by using water quality index method in Orissa, India. World Appl. Sci. J. **9**(12), 1392–1397 (2010)
5. Sajil Kumar, P.J., Elango, L., James, E.J.: Assessment of hydrochemistry and groundwater quality in the coastal area of South Chennai, India. Arab. J. Geosci. **7**(7), 2641–2653 (2014)
6. Shweta, T., Bhavtosh, S., Prashant, S., Rajendra, D.: Water quality assessment in terms of water quality index. Am. J. Water Resour. **1**(3), 34–38 (2013)
7. Srinivas, J., Lokesh Kumar, P., Purushotham, A.V.: Evaluation of ambient air quality index status in industrial areas of Visakha-Patnam, Andhra Pradesh, India. J. Environ. Res. Dev. **7** (4A), 1501 (2013)
8. Arunprakash, M., Giridharan, L., Krishnamurthy, R.R, Jayaprakashl, M..: Impact of urbanization in groundwater of south Chennai City, Tamil Nadu, India. Environ. Earth Sci. **71**(2), 947–957 (2014)

9. Kumar, S.K., Bharani, R., Magesh, N.S., Godson, P.S., Chandrasekar, N.: Hydrogeochemistry and groundwater quality appraisal of part of south Chennai coastal aquifers, Tamil Nadu, India using WQI and fuzzy logic method. Appl. Water Sci. **4**(4), 341–350 (2014)
10. Pawar, R.S., Panaskar, D.B., Wagh, V.M.: Characterization of groundwater using groundwater quality index of Solapur industrial belt, Maharashtra, India. Int. J. Res. Eng. Technol. **2**(4), 31–36 (2014)
11. Mukate, S.V, Panaskar, D.B., Wagh, V.M., Pawar R.S.: Assessment of groundwater quality for drinking and irrigation purpose: A case study of Chincholikati MIDC area, Solapur (MS), India. SRTMUs J. Sci. **4**(1), 58–69 (2015)
12. SPSS 22.0. standard version, Statistical Package for Social Services, SPSS for windows, Release, SPSS Inc., Chicago, USA (1999)
13. Wadia, D.N.: Geology of India, 4th edn. Tata McGraw Hill, New Delhi (1973)
14. APHA, Awwa.: WEF, 1998. Standard methods for the examination of water and wastewater, 17th ed. Washington DC (1995)
15. BIS, 'Beauro of Indian Standard', Drinking Water Specifications, IS 10509:2003 (2003)
16. Sanchez-Pérez, J.M., Trémolières, M.: Change in groundwater chemistry as a consequence of suppression of floods: the case of the Rhine floodplain. J. Hydrol. **270**(1), 89–104 (2003)
17. Jeong, C.H.: Effect of land use and urbanization on hydrochemistry and contamination of groundwater from Taejon area, Korea. J. Hydrol. **253**(1), 194–210 (2001)
18. Magesh, N.S., et al. Groundwater quality assessment using WQI and GIS techniques, Dindigul district, Tamil Nadu, India. Arab. J. Geosci. **6**(11), 4179–4189 (2013)
19. Loizidou, M., Kapetanios, E.G.: Effect of leachate from landfills on underground water quality. Sci. Total Environ. **128**(1), 69–81 (1993)

Observation of AODV Routing Protocol's Performance at Variation in ART Value for Various Node's Mobility

Sachin Kumar Gupta and R.K. Saket

Abstract ART is a fixed parameter in AODV routing, which tells that how long route state information should be kept in the routing table. However, as per this simulation study, an optimal value of ART is the function of node's mobility. Therefore, the presented research work attempts to analyse the impact of variation in Active Route Timeout (ART) value on the Quality of Service (QoS) metrics for different values of node's mobility in AODV routing of the ad hoc network. This research work concludes that the lower values of ART perform better, especially at higher node's mobility. Moreover, it is also concluded that the network performs better in terms of throughput at ART = 1 s and in terms of delay as well as jitter at ART = 0.5 s, which is less than its default QualNet ART value (i.e. 3 s).

Keywords ART · Mobility · AODV · QoS metrics · Ad hoc network

1 Introduction

In an infrastructure-less network, the mobile wireless network is commonly known as an ad hoc network [1, 2]. It is a group of wireless mobile nodes creating a temporary network without the aid of any centralized administration [3]. Therefore, these kinds of networks are self-organizing, self-administering and self-configuring multi-hop wireless networks, where the topology of the network changes dynam-

S.K. Gupta (✉)
School of Electronics and Communication Engineering, Shri Mata
Vaishno Devi University, Kakryal, Katra, Jammu & Kashmir 182320, India
e-mail: sachin.rs.eee@iitbhu.ac.in

S.K. Gupta · R.K. Saket
Systems Engineering, Department of Electrical Engineering, Indian Institute
of Technology (Banaras Hindu University), Varanasi, Uttar Pradesh 221005, India
e-mail: rksaket.eee@iitbhu.ac.in

S.C. Satapathy and S. Das (eds.), *Proceedings of First International Conference on Information and Communication Technology for Intelligent Systems: Volume 1*,
Smart Innovation, Systems and Technologies 50, DOI 10.1007/978-3-319-30933-0_42

ically due to node's mobility [4]. In this network, due to the absence of infrastructure, the destination node might be out of the range of a source node for transmitting packets. Hence, an efficient routing procedure is always required to find a path between the source and the destination pairs for forwarding the packets. In addition, each node of the ad hoc network must be able to forward the data to other nodes. This creates additional problems along with the problem of dynamic topology, which are unpredictable connectivity changes.

However, in an ad hoc network environment, the route discovery and route maintenance process is the key concept that deals with the topology changes. Nevertheless, there also exist various factors, which influence the network performance more than the topology changes such as ART (i.e. holding time of route state information), network area, offer loads in the network, node's mobility, node's transmission range, etc. [5]. The main concern of this paper is to analyse how variation in ART value, influences the AODV routing protocol's performance under various node's mobility in an ad hoc network. In addition, this paper tries to find out an optimal relation between ART value and node's mobility by analysing the impact of variation of ART value on QoS metrics such as throughput, delay and jitter for different values of node's mobility. As per this article, an optimal value of ART is the dynamic function of node's mobility. Hence, by considering their proper value according to network situations, the network performance could be enhanced.

2 Overview of AODV Routing Protocol

The algorithm of AODV routing protocol offers the self-starting, dynamic and multi-hop routing among the participating mobile host that desires to deploy and maintain the ad hoc network [6]. AODV allows mobile nodes to adopt routes quickly for the new destination nodes. In AODV, mobile nodes do not maintain routes toward the destinations that are not in active communication and it allows mobile nodes to act fast in case of link breakages and changes in the network topology in a timely manner. In case of link breakages, AODV causes the affected set of nodes to be notified so that they are able to invalidate the routes using the lost link [7]. As, AODV routing protocol falls under a reactive type of routing, hence its main features are route discovery and route maintenance process. AODV forms its own concept by taking an idea of route discovery and route maintenance process from DSR, and the idea of sequence numbers and sending of periodic beacons from DSDV [8–10]. It is known for the prevention of loops by using a destination sequence number for proper route maintenance [11].

3 Active Route Timeout (ART)

It is a period of time during which the route is considered to be valid. When a route is not used for some period of time, the route state information is removed by nodes from the routing table. The time, until the node removes the route state information from the routing table, is called ART [12]. Mainly, it is a fixed parameter that tells how long route state information should be kept in the routing table after the last transmission of a packet from this route [13, 14]. The default value of this parameter is arbitrary set to 3 s in AODV routing. However, finding an optimal ART is not an easy task. It's a trade-off between opting a shorter ART resulting in a new route discovery process while a valid route is still there and a longer ART for sending the packets on an invalid route. The first case introduces the delay in the network that could be avoided. The consequence of the second case is the loss of one or more packets and initiation of the RERR process, instead of a new route discovery process. Consider Fig. 1, a connection between node A and node E has been established through intermediate nodes F and G after the route discovery process. At the time of route discovery process, other routes have also been searched along the destination rather than the main communication link and these are $A \rightarrow B \rightarrow C \rightarrow D \rightarrow E$ and $A \rightarrow I \rightarrow H \rightarrow C \rightarrow D \rightarrow E$. These other routes along the destination are known as active routes and they are valid up to 3 s by default. After 3 s, active routes will be invalid and this time period is called ART. Suppose, during communication one intermediate node G is moved to G′, then this session can be completed via other active routes without initializing the route discovery process.

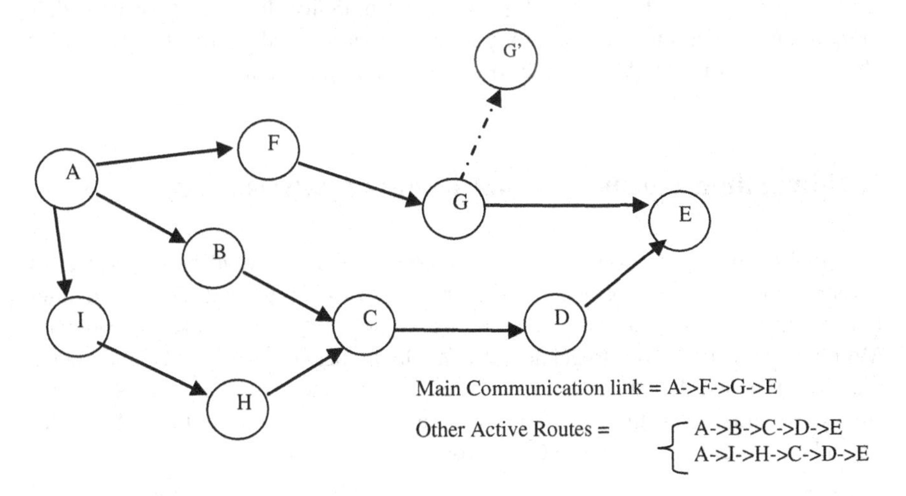

Fig. 1 Impact of ART on connectivity

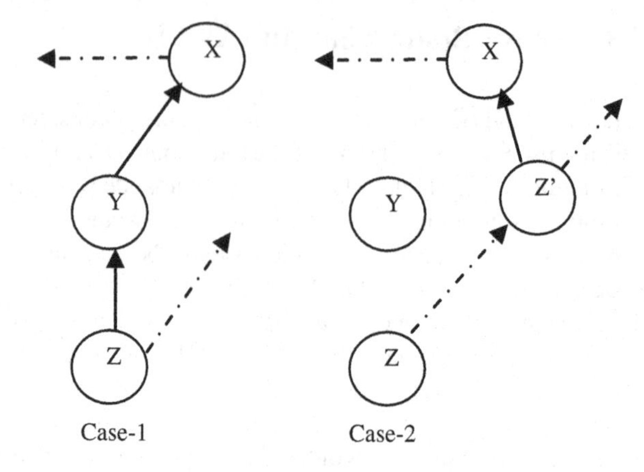

Fig. 2 Impact of node's mobility on connectivity (adopted from [5])

4 Mobility

Figure 2 depicts how the node's mobility affects the connectivity in the network. In case-1, Fig. 2 demonstrates the connection between nodes X and Z through node Y. In other words, node Y plays the role of an intermediate node between these two parties if the node's mobility is low. However, in case-2, direct connection between these two parties is possible, only if the node Z will move to Z' position when node's mobility is high. On the other side, a very high value of node's mobility could influence the connectivity of the node, negatively [5, 15]. In other words, it could be said that an immediate change in the network topology may reduce the network performance because of higher node's mobility. In case of lower node's mobility, node Z has to wait for a long time in order to get the direct link, or it may choose the intermediate/relay node to get the direct link than to wait.

5 Simulation Parameters and Results Analysis

The goal of this simulation study is to identify that what would be the impact of variation in ART value on the QoS metrics such as throughput, delay and jitter under various node's mobility conditions in an ad hoc network environment. Moreover, this article tries to get an optimal relation between ART value and node's mobility in which the ad hoc network gives its best outcome. The obtained results, from these simulation studies are significant to propose solutions to the QoS metrics degradation problem in the ad hoc network.

5.1 Simulation Parameters

In this paper, the QualNet 7.1 simulation tool [16] has been used for the analytical study of considered scenario and the following parameters have been used which are tabulated in Table 1.

Table 1 Parameters while creating scenario

Parameters set-up at wireless subnet and nodes/devices level	
Radio type	802.11b radio
MAC protocol	802.11
Channel bandwidth	11 Mbps
Transmission power at 11 Mbps channel bandwidth	15 dBm
Antenna model	Omni-directional
Network protocol	IPv4
Routing protocol	AODV
Network diameter	35
Node traversal time	40 ms
Maximum_route_request_retries	2
Maximum_number_of_buffer_packets	100
Route deletion constant	5
Mobility model	Random waypoint
Active route timeout (ART)	0.5, 1, 3, 5, 7, 9 and 11 s
Node's mobility	0.5, 5, 10 and 15 mps
Node's transmission range	200 m
Network load density (NLD)	80
Offer loads in the network	8 (10 % of 80)
Pause time	5 s
Applications	CBR
Packet size	512 bytes/packet
Packet inter-arrival time	0.25 s or 4 packets/second
Data rate	2 Kbps
Packets sent during the whole simulation	1200 packets or 600,000 bytes
General parameters set-up	
Simulation time	300 s
Terrain size (area)	1500 m × 1500 m
Number of channels	1
Channel frequency	2.4 GHz in ad hoc mode
Pathloss model	Two ray
Node placement strategy	Randomly

5.2 Analysis of Simulation Results

- ***Scenario***: *QoS* metrics *(i.e. Throughput, Delay and Jitter) Vs ART for various Node's Mobility at constant network*

In this scenario, 80 nodes are spread randomly over a constant area size of 1500 m × 1500 m and moving according to the Random Waypoint Mobility model. In addition, offer loads in the network are 10 % of total nodes. Moreover, the throughput, delay and jitter are taken into account to compare the QoS metrics for proposed scenario. This article uses a graphical method to analyse the impact of variation in ART value from 0.5 to 11 s on the QoS metrics for various node's mobility.

Figure 3 shows the throughput as a function of ART for different values of node node's mobility like 0.5, 5.0, 10.0 and 15.0 mps. From Fig. 3, one can observe that, as expected, the throughput decreases as the average node's mobility increases. When there is zero mobility or very less mobility, such as less than 0.5 mps, routes never expire. Hence, for lower node's mobility, throughput is higher than the other node's mobility and nearly unchanged for higher values of ART. This result is expected since the node's mobility is stationary and the changes in ART value do not influence the throughput. At higher values of node's mobility, such as 5.0, 10.0, and 20 mps, the values of throughput decreases with an increase in ART. The main reason of this consequence is the continuous change in node positions, which make

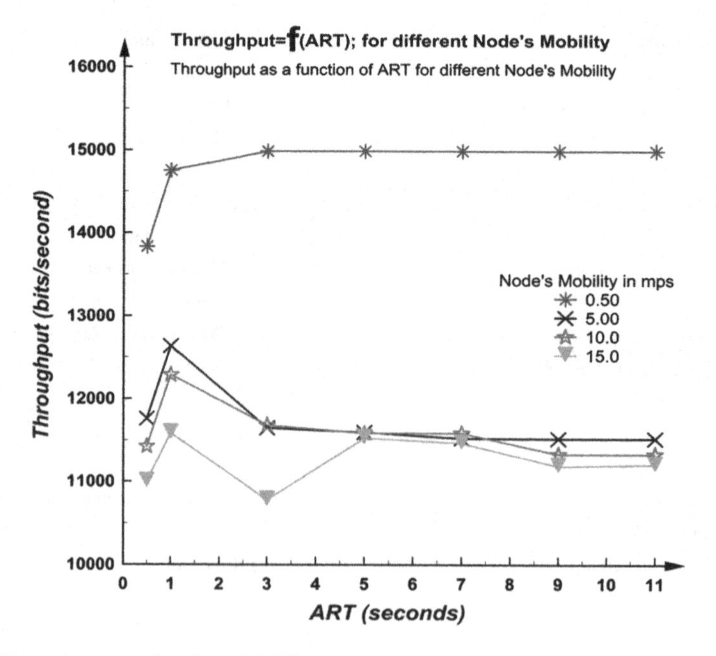

Fig. 3 Throughput as a function of ART

it difficult to establish a connection between them. From Fig. 3, one can also see that at a very low values of ART (i.e. at less than 1 s), the throughput is not good. This is because, the node holds the route state information for a very less time after it has been used, which causes a node to repeat the route discovery process most of the time, after each use of the route. From the above discussion, it is concluded that the higher throughput value is achieved at a lower value of ART than the default QualNet value (i.e. 3 s). It is also noticed that the ART value at 1 s gives the highest throughput.

Figures 4 and 5 reflects the delay and jitter as a function of ART for different values of node's mobility respectively. In both cases, one can easily understand that the value of delay as well as jitter increases as node's mobility increases in the network. It happens because, with the increase in node's mobility, the changes in network topology become more common. Therefore, the route discovery process becomes very difficult. Also at high node's mobility, the route breaking issues are more general among established connection. At the very high node's mobility, such as 10 or 15 mps, drastic changes can be seen in both cases (delay as well as jitter), as the ART value increases. In case of high node's mobility, due to the fast changes in topology, the route becomes invalid very quickly. Hence, holding the route state information for a longer time is not good. From Figs. 4 and 5, it is clear that the

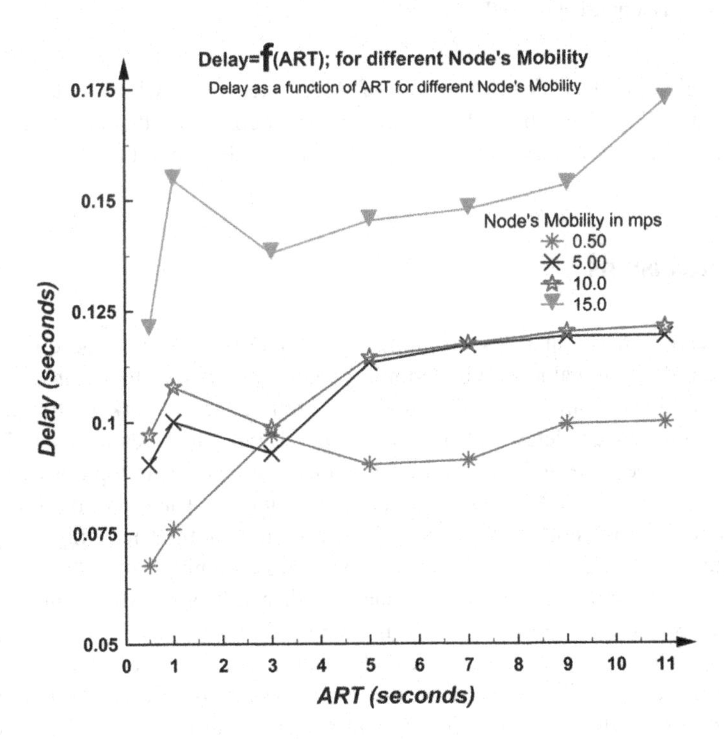

Fig. 4 Delay as a function of ART

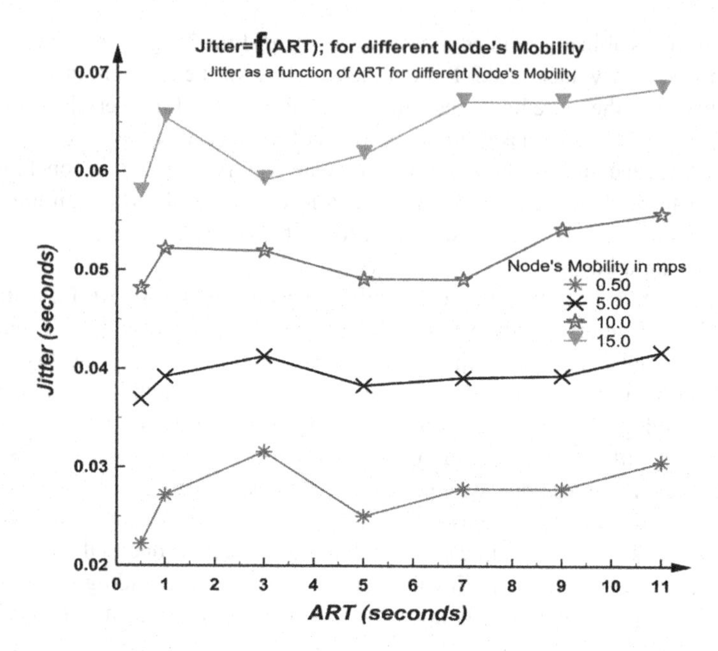

Fig. 5 Jitter as a function of ART

network gives its best at a low value of ART, (i.e. 0.5 s) which is less than its default value. From the above discussion, the paper concludes that an optimal value of ART is the dynamic parameter function of the node's mobility.

6 Conclusions

This research article analyses the impact of variation in ART value on the QoS metrics for different values of node mobility in the ad hoc network environment. As per the results obtained, at higher node's mobility, the network performance in terms of throughput, delay and jitter is not good with an increase in ART value. This happens because, at higher node's mobility, the changes in topology are more frequent and hence, routes become invalid very quickly. Therefore, the route state information should not be held for a longer period of time in a highly mobile environment. If there is zero mobility or very less mobility, the network performance is good and it is nearly constant for other values of ART. This result is obvious because the node's mobility is almost stationary and hence, changes of ART value do not affect the network performance. This section concludes that the best QoS metrics are achieved at the lower value of ART than the default QualNet ART value, which is 3 s. In other words, ART value at 1 s gives the highest throughput and at ART = 0.5 s, delay and jitter can also be seen at the lowest. In future, the above simulation studies could be extended for other reactive routing protocols like DSR, DYMO, etc.

References

1. Johansson, M.P., Larsson, P.: Wireless ad hoc networking: the art of networking without a network. Ericsson Rev. **4**, 248–263 (2000)
2. IETF Working Group: Mobile Ad hoc Networks (MANET).: http://www.ietf.org/html.charters/manet-charter.html
3. Johnson, D.B., Maltz, D.A.: Dynamic source routing in ad hoc wireless networks, In: Imielinski, T., Korth, H. (eds.) Mobile Computing, Chapter 5, pp. 153–181. Kluwer Academic Publishers, Berlin (1996)
4. Hong, X., Kaixin, X., Gerla, M.: Scalable routing protocols for mobile ad hoc networks. IEEE Netw. Mag. Global Internetw. **16**(4), 11–21 (2002)
5. Chen, Y., Wang, W.: The measurement and auto-configuration of ad-hoc. In: 14th IEEE 2003 International Symposium on Personal, Indoor and Mobile Radio Communication, vol. 2, pp. 1649–1653, Sept 2003
6. Perkins, C.E., Das, S.: Ad hoc on demand distance vector routing. In: Internet Engineering Task Force, Request for Comments (Proposed Standard) 3561 (2003)
7. Zahedi, K., Ismail, A.S.: Route maintenance approaches for link breakage prediction in mobile ad-hoc networks. Int. J. Adv. Comput. Sci. Appl. **2**(10), 23–30 (2011)
8. Reina, D.G., Toral, S.L., Johnson, P., Barrero, F.: A survey on probabilistic broadcast schemes for wireless ad hoc networks. Ad Hoc Netw. **25**(part A), 263–292 (2015)
9. Jain, R., Shrivastava, L.: Study and performance comparison of AODV & DSR on the basis of path loss propagation models. Int. J. Adv. Sci. Technol. **32**, 45–52 (2011)
10. Perkins, C.E., Royer, E.M.: Ad-hoc on-demand distance vector routing. In: Proceeding of 2nd IEEE Workshop, Mobile Computing System Applications, New Orleans, LA, pp. 90–100, Feb 1999
11. Gupta, S.K., Saket, R.K.: Performance metric comparison of AODV and DSDV routing protocols in MANETs using NS-2. Int. J. Res. Rev. Appl. Sci. **7**(3), 339–350 (2011)
12. Gupta, S.K., Sharma, R., Saket, R.K.: Effect of variation in active route timeout and delete period constant on the performance of AODV protocol. Int. J. Mobile Commun. (IJMC) **12**(2), 177–191 (2014)
13. Al-Mandhari, W., Gyoda, K., Nakajima, N.: Performance evaluation of active route time-out parameter in ad-hoc on demand distance vector (AODV). In: 6th WSEAS International Conference on Applied Electromagnetic, Wireless and Optical Communications (Electro science'08), Trondheim, Norway, pp. 47–51 (2008)
14. Tseng, Y.-C., Li, Y.-F., Chang, Y.-C.: On route lifetime in multi-hop mobile ad hoc networks. IEEE Trans. Mob. Comput. **2**(4), 366–377 (2003)
15. Amjad, K., Stocker, A.J.: Impact of node density and mobility on performance of AODV and DSR in MANET. In: 7th International Symposium on Communication Systems Networks and Digital Signal Processing (CSNDSP), pp. 61–65 (2010)
16. Scalable Network Technologies, Inc. QualNet simulator Version 7.1, http://www.scalable-networks.com QualNet 7.1User's Guide, Scalable Network Technologies (2013)

PCA Based Optimal ANN Classifiers for Human Activity Recognition Using Mobile Sensors Data

Kishor H. Walse, Rajiv V. Dharaskar and Vilas M. Thakare

Abstract Mobile Phone used not to be matter of luxury only, it has become a significant need for rapidly evolving fast track world. This paper proposes a spatial context recognition system in which certain types of human physical activities using accelerometer and gyroscope data generated by a mobile device focuses on reducing processing time. The benchmark Human Activity Recognition dataset is considered for this work is acquired from UCI Machine Learning Repository, which is available in public domain. Our experiment shows that Principal Component Analysis used for dimensionality reduction brings 70 principal components from 561 features of raw data while maintaining the most discriminative information. Multi Layer Perceptron Classifier was tested on principal components. We found that the Multi Layer Perceptron reaches an overall accuracy of 96.17 % with 70 principal components compared to 98.11 % with 561 features reducing time taken to build a model from 658.53 s to 128.00 s.

Keywords Principal component analysis (PCA) · Human activity recognition (HAR) · Multi-layer perceptron (MLP) · Smartphone · Sensor · Accelerometer · Gyroscope

---

K.H. Walse (✉)
Department of CSE, Anuradha Engineering College, Chikhli 443201, India
e-mail: walsekh@acm.org

R.V. Dharaskar
DMAT-Disha Technical Campus, Raipur 492001, India
e-mail: rajiv.dharaskar@gmail.com

V.M. Thakare
Department of CS, S.G.B.A. University, Amravati 444601, India
e-mail: vilthakare@gmail.com

S.C. Satapathy and S. Das (eds.), *Proceedings of First International Conference on Information and Communication Technology for Intelligent Systems: Volume 1*, Smart Innovation, Systems and Technologies 50, DOI 10.1007/978-3-319-30933-0_43

1 Introduction

Days have gone when Mobile Phone used to be matter of luxury, now it has become a significant need for rapidly evolving fast track world [1]. Intelligent aspects of the computing device enhance the importance of the interface development since; efficient collaboration always relies on the good communication between man and machine [2]. To enhance the interaction between the user and the device, the information gathered about device and user's activity, the environment, other devices in proximity, location and time can be utilized in different situations [3]. Various applications in various areas in healthcare, senior care, personal fitness, security, entertainment and daily activities will be possible because of recognition of human activity [4].

2 Related Work

Human Activity Recognition (HAR) is an active research area since last two decades. Image and video based human activity recognition has been studied since a long time [5]. Increasingly powerful mobile phones, with built-in sensors Smartphone becomes a sensing platform. While studying the related literature within past few decades, it is found that various approaches for human activity recognition such as from video sequences, from wearable sensors or from sensors on a mobile device has been used [6].

Human Activity Recognition through environmental sensors is one of the promising approach. In this approach, it uses motion sensors, door sensors, RFID tags and video cameras etc. to recognize human activities. This method provides high recognition rate of human activity and can be more effective in indoor environments (i.e., smart home, hospital and office), but requires costly infrastructure [7, 8]. In another approach, multiple sensors wear on human body which uses human activity recognition. This approach has also proved to be effective sensors for human activity recognition [5]. But this type of sensing having drawbacks that user has to wear lot of sensors on his body.

In the literature extensive study also found on the approach in which available smart mobile devices used to collect data for activity recognition and adapt the interfaces to provide better usability experience to mobile users. We found few similar studies to the one proposed in the paper [9, 10]. Mobile device becomes an attractive platform for activity recognition because they offer a number of advantages including not requiring any additional equipment for data collection or computing. These devices saturate modern culture; they continue to grow in functionality but increases security and privacy issues [11].

3 Data Collection

3.1 Benchmark Data Set

The benchmark HAR dataset is considered for this work is acquired from the UCI machine learning repository which is available in public domain [12]. This dataset is recorded with the help of accelerometer and gyroscope sensors. In this dataset, the experiments were performed with 30 subjects. Each subject wearing a smartphone on the waist and performed six activities. Six activities are: Walking, Sitting, Standing, Stair-Up, Stair-Down and Laying. With the sampling rate of 50 Hz, the raw data from built-in 3-axial accelerometer and gyroscope captured. In each record of the dataset following attributes are included:

- Total acceleration from the 3-axis accelerometer and the estimated body acceleration.
- Angular velocity from the 3-axis gyroscope.
- Total 561 features generated from raw data in time and frequency domain.
- Label for corresponding activity

  1. Features were normalized and bounded within [−1, 1].
  2. Each row is representing feature vector.
  3. The unit for acceleration is m/s^2.
  4. The gyroscope unit is rad/s.

3.2 Feature Extraction

The features selected for this dataset come from raw data captured from the accelerometer and gyroscope raw signals. These 3-axial raw signals are time domain signals prefix 't' to denote time. Before selecting features, raw sensor signals were pre-processed by using a median filter and a 3rd order low pass Butterworth filter to remove noise. Another low pass Butterworth filter with a corner frequency of 0.3 Hz had been used to separate body and gravity acceleration signals from the acceleration signal. Finally, frequency domain signals were obtained by applying a Fast Fourier Transform (FFT) [10]. To get feature vector for each pattern, these time domain and frequency domain signals were used. The set of variables that were estimated from these signals are as shown in Table 1.

Table 1 Summary of features extraction methods used for accelerometer and gyroscope signals

Group	Methods
Time domain	Mean, SD, MAD, Max, Min, SMA, energy, entropy, IQR, auto regression coefficient, correlation, linear acceleration, angular velocity, kurtosis, skewness
Frequency domain	FFT, mean frequency, index of frequency component with largest magnitude

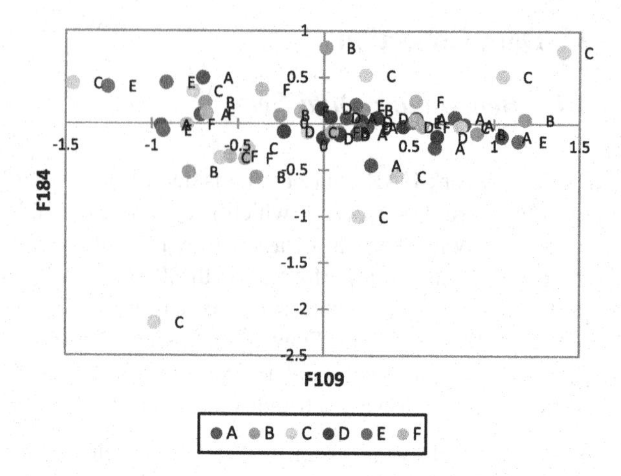

Fig. 1 A typical scatter plot of activities showing various categories

Most finer features which provide dominant activity related information should be selected before a feature set is fed into a classifier to improve the performance of a classifier. So, irrelevant or redundant features can be discarded to improve the classifier performance. This help to reduce dimensionality of features set. According to activities Walking, Sitting, Standing, Stair-Up, Stair-Down and Laying denoted by A, B, C, D, E and F respectively, a typical scatter plot for a few samples is shown in Fig. 1 from this plot, it is seen that the decision boundaries discriminating between different classes are quite complex, nonlinear and overlapping.

4 Method

In the proposed research work we have molded a context recognition problem as a six class classification problem. After surveying the extensive literature, the methodology adopted for the system. The system consists of acquisition, preprocessing, feature extraction, feature selection, classification.

In the feature extraction, we have used statistical and transformed based features. The Classifier is designed with the help of Neural Network like Multi Layer Perceptron (MLP).

In this work, Principal Component Analysis (PCA) technique is used to reduce the dimension of original features by extracting optimal features using Pearson rule [13].

4.1 Performance Measures

To assess the neural network performance well known confusion matrix, overall accuracy, precision and recall performance measures are used.

5 Experiment

For designing a classifier for activity recognition, a Simple Multilayer Perceptron (MLP) Artificial Neural Network is proposed. While designing a MLPNN classifier, in the input and output layers, the number of processing elements are equal to the number of inputs and outputs respectively. Six Processing Elements are used in the output layer for the six activity namely Walking, Sitting, Standing, Stair-Up, Stair-Down and Laying. In standard UCI dataset, number of input features are 561. For dimensionality reduction, we used Principle Component Analysis (PCA). Finally, as principal components given, on each input set, an average of minimum Mean Square Error (MSE) and average of the classification accuracy are computed and checked to decide number of inputs. As shown in the results of Fig. 2, we selected 70 principal components. Hence the number of PEs in input layer is seventy. Usually, the MLPs with one or two hidden layers are commonly used for human activity recognition applications. So, in this experiment, one hidden layer is used.

During training, cross validation and testing, the performance of various Transfer functions and learning rules are obtained. The performance of transfer functions, learning and step size are compared on training and CV data set. It is found that Tanh transfer function and Momentum learning rule give the optimum results as shown in Figs. 3 and 4. With above experimentation finally, the MLP NN classifier is designed with following specifications.

Number of Inputs : 70
Number of Output : 70
Number of Hidden Layers : 01
Number of PEs in Hidden Layer : 26
Hidden Layer Transfer Function : tanh
Learning Rule : Momentum
Step size: 0.1 Momentum : 0.7
Output Layer transfer Function : tanh

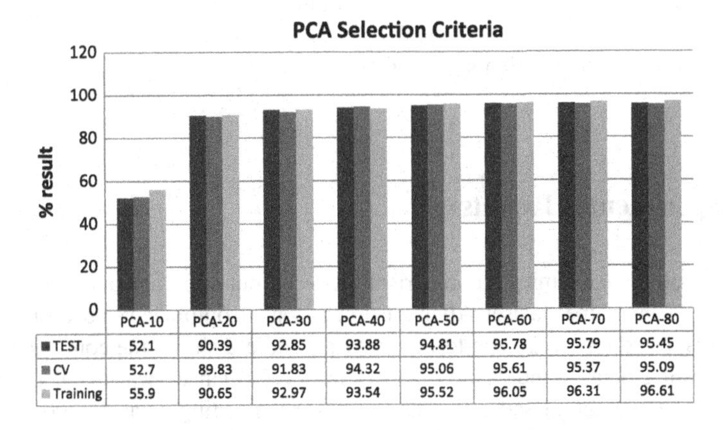

	PCA-10	PCA-20	PCA-30	PCA-40	PCA-50	PCA-60	PCA-70	PCA-80
TEST	52.1	90.39	92.85	93.88	94.81	95.78	95.79	95.45
CV	52.7	89.83	91.83	94.32	95.06	95.61	95.37	95.09
Training	55.9	90.65	92.97	93.54	95.52	96.05	96.31	96.61

Fig. 2 Variation of average classification accuracy with number of PCs as input

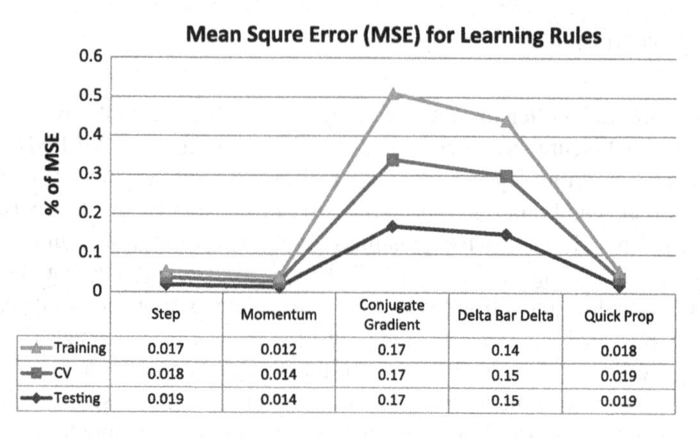

Fig. 3 % of root mean square (MSE) for learning rules

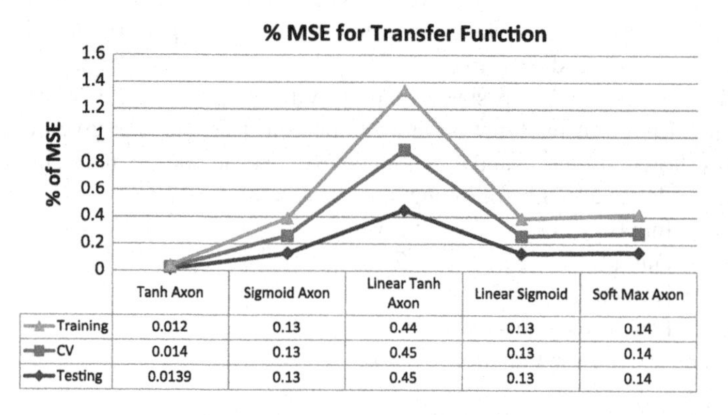

Fig. 4 % of accuracy of transfer function

Learning Rule	: Momentum
Step size: 0.1 Momentum	: 0.7
Number of connection weights	: 2008

6 Experimental Results

In this section, we examine a comparison between our approach and other existing methods on the same dataset After the formation of optimized and cost effective MLP, it is tested on the unseen testing dataset. Table 2 shows the confusion matrix for the testing dataset and Table 3 shows the performance measure of the testing data set. The average classification accuracy for the testing dataset is 96.16 % and the average mean square error is 0.014.

Table 2 Confusion matrix for Test dataset

Output /desired	Stair-up	Laying	Walking	Stair-down	Standing	Sitting
Stair-up	164	0	0	15	0	0
Laying	1	180	0	0	0	0
Walking	0	0	164	0	0	1
Stair-down	16	0	0	174	0	0
Standing	0	0	0	0	156	3
Sitting	0	0	2	0	3	151

Table 3 Performance measure on testing data set

Performance	Stair-up	Laying	Walking	Stair-down	Standing	Sitting
MS error	0.0280	0.0023	0.0077	0.0253	0.0076	0.0125
NMS error	0.1930	0.0161	0.0569	0.1686	0.0580	0.0974
MA error	0.0923	0.0337	0.0588	0.0899	0.0547	0.0670
Min Abs error	0.0000	0.0000	0.0008	0.0001	0.0002	0.0001
Max Abs error	1.0449	0.6158	0.7618	1.0376	0.9080	1.0268
r	0.9008	0.9932	0.9773	0.9164	0.9750	0.9572
Percent correct	90.60	100.00	98.80	92.06	98.11	97.41

We have compared the proposed PCA feature selection method with the results of other techniques such as RF (95/5) PCA (186 features), correlation (232 features, 93 % threshold), and stepwise LDA (60 features) using CHMM used in previous experiments by other authors. MLP with all 561 features we received. 98.11 % of overall accuracy while MLP with PCA we have received 96.16 % of overall accuracy. While studying the previous work by Anguita [12], Devenport [14] and Ronao [15], we found Anguita received overall accuracy of 89.3 % with MC-SVM with all 561 features, Devenport received 90.28 % overall accuracy using Random Forest with 95/5 reduction and Ronao got overall accuracy 91.76 %.

7 Conclusions

In this work, we have investigated human activity recognition by using built-in accelerometer and gyroscope data on standard dataset. For this purpose, we have designed optimized MLP based classifier. For reducing the dimension, we have used PCA. Principal Component Analysis used for dimensionality reduction brings 70 principal components from 561 features of raw data. Multi Layer Perceptron Classifier was tested on principal components and with various time domain and frequency domain features. We found that Multi Layer Perceptron reaches to an overall accuracy of 96.17 % with 70 principal components compared to 98.11 % with 561 features by using MLP, reducing time taken to build a model from 658.53

s to 128 s. We found Anguita received the overall accuracy of 89.3 % with MC-SVM with all 561 features, Devenport received 90.28 % overall accuracy using Random Forest with 95/5 reduction and Ronao got overall accuracy 91.76 % [14, 15]. Though the classification accuracy for the MLP with PCA is slight less than MLP with all 561 features but time to build the MLP with PCA model is 6 times less than the MLP with all 561 features. While comparing our results with existing, we conclude that overall it is tradeoff to select the classifier with PCA.

References

1. Central Investigation Agency.: World Fact Book. https://www.cia.gov/library/publications/the-world-factbook/rankorder/2151rank.html (2014)
2. Acay, L.D.: Adaptive user interfaces in complex supervisory tasks. M.S. thesis, Oklahoma State University (2004)
3. Korpipaa.: Blackboard-based software framework and tool for mobile. Ph.D. thesis. University of Oulu, Finland (2005)
4. Zeng, M.: Convolutional neural networks for human activity recognition using mobile sensors. In: 6th International Conference on Mobile Computing, Applications and Services (MobiCASE) doi:10.4108/icst.mobicase.2014.257786 (2014)
5. Rao, F., Song, Y., Zhao, W.: Human activity recognition with smartphones
6. Yang, J.: Toward physical activity diary: motion recognition using simple acceleration features with mobile phones. In: ACM IMCE'09, Beijing, China, 23 Oct 2009
7. Jalal, A., Kamal, S., Kim, D.: Depth map-based human activity tracking and recognition using body joints features and self-organized map. In: 5th ICCCNT—2014, 2 Hefei, China, 11–13 Jul 2014
8. Dernbach, S., Das, B., Krishnan, N.C., Thomas, B.L., Cook, D.J.: Simple and complex activity recognition through smart phones. In: Intelligent Environments (IE) 8th International Conference, pp. 214–221 (2012). doi:10.1109/IE.2012.39
9. Walse, K.H., Dharaskar, R.V., Thakare, V.M.: Frame work for adaptive mobile interface: an overview. In: IJCA Proceedings on National Conference on Innovative Paradigms in Engineering and Technology (NCIPET 2012) vol. 14, pp. 27–30 (2012)
10. Walse, K.H., Dharaskar, R.V., Thakare, V.M.: Study of framework for mobile interface. In: IJCA Proceedings on National Conference on Recent Trends in Computing NCRTC9, pp. 14–16 (2012)
11. Rizwan, A., Dharaskar, R.V.: Study of mobile botnets: an analysis from the perspective of efficient generalized forensics framework for mobile devices. In: IJCA Proceedings on National Conference on Innovative Paradigms in Engineering and Technology (NCIPET 2012) ncipet 15, pp. 5–8 (2012)
12. Anguita, D.: A public domain dataset for human activity recognition using smartphones. In: 21th European Symposium on Artificial Neural Networks, Computational Intelligence and Machine Learning, ESANN-2013. Bruges, Belgium (2013)
13. Kharat, P.A., Dudul, S.V.: Daubechies wavelet neural network classifier for the diagnosis of epilepsy. Wseas Trans. Biol. Biomed. **9**(4), 103–113 (2012)
14. Devenport, K.: Samsung phone data analysis project, 19 Mar 2013. Blog http://kldavenport.com/samsung-phone-data-analysis-project/
15. Ronao, C.A., Cho, S.-B.: Human activity recognition using smartphone sensors with two-stage continuous hidden Markov models. In: 10th International Conference on Natural Computation (ICNC), pp. 681–686, 19–21 Aug 2014. doi:10.1109/ICNC.2014.6975918

Higher Order Oriented Feature Descriptor for Supporting CAD System in Retrieving Similar Medical Images Using Region Based Features

B. Jyothi, Y. MadhaveeLatha, P.G. Krishna Mohan and V. Shiva Kumar Reddy

Abstract The technological advancements in the medical field generate a massive number of medical images and are saved in the database in order to make easy in the future. This paper presents a new approach for retrieving the similar medical images from a huge database for supporting computer aided diagnosing system to improve the quality of treatment. This approach is a two step process based on the concept of multi-scale orientation structure, firstly the region of the object is detected with the facilitation of segmentation method and secondly the texture patterns are extracted with the help of higher order steerable texture description. The related medical images are retrieved by computing similarities matching among the given queries of medical image feature vector and the consequent database image feature vector by means of Euclidian distance. The efficiency of the projected scheme is tested and exhibited with a variety of medical images. With the investigational outcome, it is understandable that the higher order oriented steerable features yields better results than the classical retrieval system.

Keywords Content based image retrieval · Segmentation · Feature extraction

B. Jyothi (✉) · V. Shiva Kumar Reddy
Department of ECE, MallaReddy College of Engineering and Technology, Maisammaguda, India
e-mail: bjyothi815@gmail.com

V. Shiva Kumar Reddy
e-mail: Mrcet2004@gmail.com

Y. MadhaveeLatha
Department of ECE, MallaReddy Engineering College for Women, Maisammaguda, India
e-mail: madhaveelatha2009@gmail.com

P.G. Krishna Mohan
Department of ECE, Institute of Aeronautical Engineering, Dundigal, India
e-mail: pgkmohan@yahoo.com

S.C. Satapathy and S. Das (eds.), *Proceedings of First International Conference on Information and Communication Technology for Intelligent Systems: Volume 1*, Smart Innovation, Systems and Technologies 50, DOI 10.1007/978-3-319-30933-0_44

1 Introduction

Medical images are important diagnosing evidence to make available important information regarding each and every peculiar and complex disease. There have been new revolutionary improvements in medical knowledge, thus a great number of medical images are generated in data base. Computer-aided diagnosis (CAD) is the division of computer vision in medical field imaging concerned with the assistance of the computer in the diagnostic process. This field grows quickly due to a growing array of methodologies which performs better than the individual diagnosis in specificity and in the volume of medical data in diagnostic area and necessitates support. These medical images also help us in imparting medical education and Training and assisting CAD applications [1]. To satisfy the above needs, it is absolutely imperative to retrieve similar medical images from a large data base. Basically, there are two methods for searching required images. They are Text based searching method and Content based searching method. The former is confronted with many challenges [2] because of its complex nature. It is cumbersome to save database images which involve large manual annotations, consumption of too much time and huge expenses. The more the number of database images are the more complex problems will be. Since there are innumerable images in a given database, even a medical expert has to struggle to retrieve required images which facilitate the diagnosing process in time.

To address these confines, content-based medical image retrieval (CBMIR) methods have been explored in the recent decade [3]. Content based methodology answers the problems involved in the text based methodology. We can retrieve the images required by the medical expert by providing the query image whose content is similar to that of retrieved images. In CBMIR System the characteristics of the image is outlined with the help of visible features such as Color, Texture and Shape. H. Greenspan and A.T. Pinhas have discussed various CBMIR approaches in [4] and noticed that these approaches are effective for radiographic image retrieval task which is still a problem. The characteristics of the medical image can be examined more successfully by adapting multiple features as compared to the single feature [5].

Shape features provide useful information for identifying entities which carries semantic information [6]. Basically, the two shape extraction methods are boundary-based and region-based techniques. The first one exploits the boundary information and second one extracts the entire region content.

Texture is an instinctive surface chattels of an object in which its relationship to the neighboring environment is calculated. Gabor representations have been particularly successful in many computer vision and applications of image processing [7].

This paper concentrates on developing a novel and efficient procedure for describing the content about region. We expand the structure of image segmentation by using edge following method which detects the boundaries of the image [8]. A boundary limits the edges and separates it from the surroundings. Medical images have imperfect boundaries due to some realistic problems while obtaining the medical images such as poor illumination, low contrast and noise.

The content of the region is extracted by using higher order texture followed by describing the content using the statistical components. Finally, analogous medical images are retrieved by corresponding region features and various similarity measurement techniques which are discussed in the literature [9].

Further paper is structured in this way. Section 2 deals with CBMIR system. Section 3 deals with the Segmentation Algorithm Sect. 4 focuses on Features Extraction. Section 5 discusses the investigational results and Sect. 6 concludes the paper.

2 CBMIR System

The prominence of image processing in medical is increasing day by day; various developments are enhancing in medical images, tumor detection, de-noising the medical images, contrast enhancing and medical image fusion.

A computer system is used to retrieve, browse and search images from large data bases of digital images through online and offline modes.

A classic conceptual CBMIR system is shown in Fig. 1. In online mode, the query image is submitted as an input to the system for searching relevant images

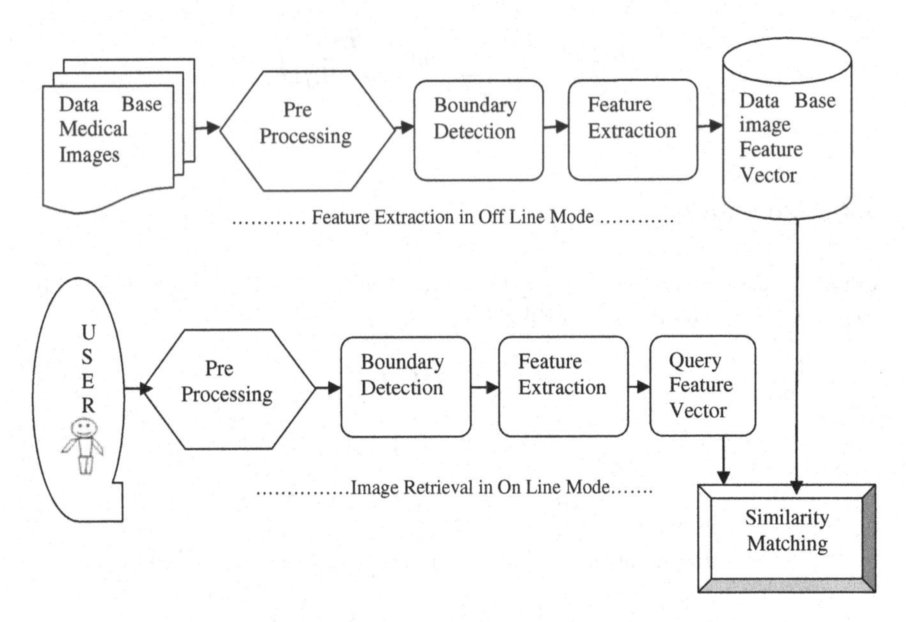

Fig. 1 Content based image retrieval system

and the system returns the relevant images by quantifying the correspondence between the input image and data base images.

3 Boundary Detection

Medical images are incorporated with lots of objects which can be recognized by means of appropriate boundaries. The boundary of the medical image is detected by adapting the edge following method integrated with intensity gradient and edge map features which retains principal features of the input image.

3.1 Intensity Gradient Vector Model

The intensity gradient for the given input image F (i j), calculated using the following mathematical expressions.

$$H(i,j) = \frac{1}{H_r} \sum_{(i,j) \in N} \sqrt{H_x^2(i,j) + H_y^2(i,j)} \tag{1}$$

$$A(i,j) = \frac{1}{H_r} \sum_{(i,j) \in N} \tan^{-1}\left(\frac{H_y(i,j)}{H_x(i,j)}\right) \tag{2}$$

3.2 Edge Map

Edges are exploited by convolving the texture mask and query image followed by canny edge detection [10] defined as follows.

$$TI(x,y) = TM \cdot TM^T \tag{3}$$

$$TM = (1,4,6,4,1)^T \tag{4}$$

$$R(i,j) = F(i,j) * TM(x,y) \tag{5}$$

The edge is tracked at the position (i, j) of an image as shown in Fig. 2.

$$L_{ij}(i,j) = H_{ij}(i,j) + A_{ij}(i,j) + R_{ij}(i,j) \tag{6}$$

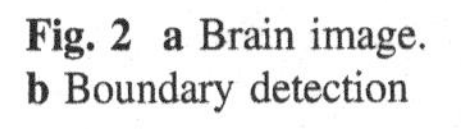

Fig. 2 **a** Brain image. **b** Boundary detection

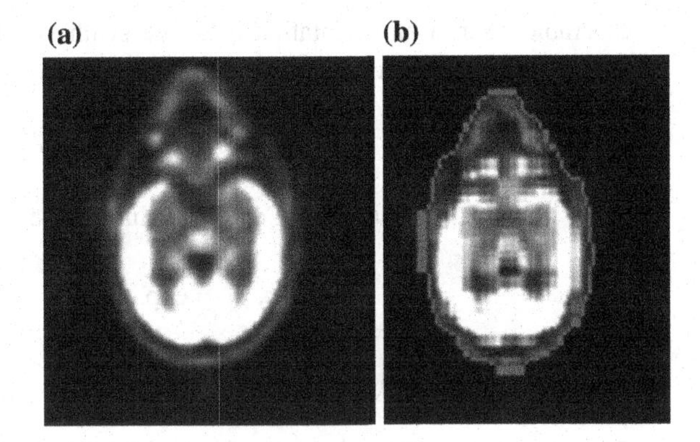

4 Features Extraction

Feature detectors plays significant role in computer visualization. This has to enable a computer to identify objects in medical images with the help of feature description. A feature is confined to definite visual characteristics of an image.

The corresponding region content of the image F(x, y), is described using higher order steerable filter at various orientations (Fig. 3). The propose scheme uses first derivative and second derivative orthogonal steerable filter. First derivative operators detect odd-symmetric gray level changes alone and second derivatives detect even-symmetric gray level variations. The second derivative operators detect only a range of symmetric shapes. This leads to confusing false shapes in the medical image. In our approach we are applying together first and second derivative operators so that both information can be extracted.

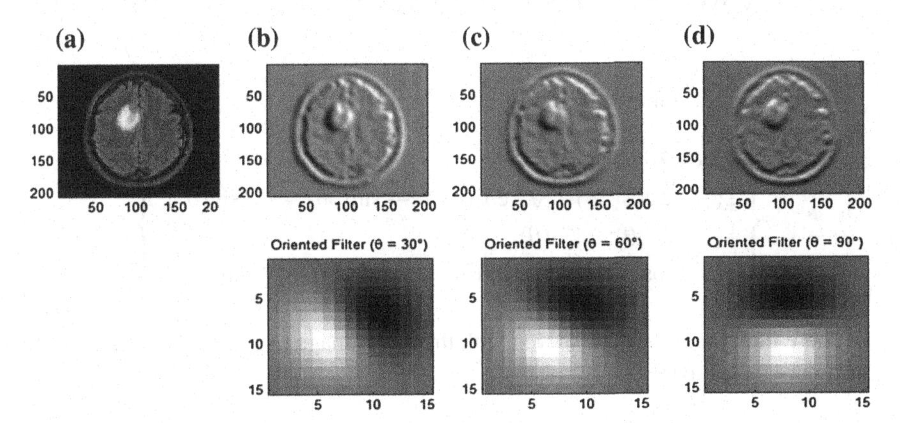

Fig. 3 The Steerable filter image at different orientations. (**a**) The input image (**b**) Image at 30° orientation (**c**) Image at 60° orientation (**d**) Image at 90° orientation

Steerable filter at an orientation θ can be synthesized by linear combination of $G_1^{0^\circ}$ &$G_1^{90^\circ}$ and interpolation functions given as follows.

$$G_1^{\theta} = \cos(\theta)G_1^{0^\circ} + sin(\theta)G_1^{90^\circ} \quad (7)$$

$G_1^{0^\circ}$ &$G_1^{90^\circ}$ are set of basis filtersCos (θ) & Sin (θ) are the interpolation functions

$$\mathrm{G(x, y) = exp^{-\left(x^2+y^2\right)}} \quad (8)$$

$$\mathrm{G_1^{0^\circ} = -2x\left(exp^{-\left(x^2+y^2\right)}\right) \text{ and } G_1^{90^\circ} = -2y\left(exp^{-\left(x^2+y^2\right)}\right)}$$

The second derivative operators are given by the following equations

$$G_{xx} = \frac{\partial^2}{\partial_x^2}\exp\left(-x^2 - y^2\right) \quad (9)$$

$$G_{yy} = \frac{\partial^2}{\partial_y^2}\exp\left(-x^2 - y^2\right) \quad (10)$$

$$G_{xy} = \frac{\partial^2}{\partial_x\partial_y}\exp\left(-x^2 - y^2\right) \quad (11)$$

Second order steerable functions

$$\left\{\begin{array}{l} \frac{1}{3}(1+\cos(\theta), \\ \frac{1}{3}(1-\cos(2\theta)+\sqrt{3}\sin(2\theta), \\ \frac{1}{9}((-1)^{\frac{1}{6}}\sqrt{3}-(-1)^{\frac{5}{6}}\sqrt{3}-(-1)^{\frac{1}{6}}\sqrt{3}\cos(2\theta)+ \\ (-`1)^{\frac{5}{6}}\sqrt{3}\cos(2\theta)-2\sqrt{3}\sin(2\theta)-(-1)^{\frac{1}{3}}\sqrt{3}\sin(2\theta)+(-1)^{\frac{2}{3}}\sqrt{3}\sin(2\theta) \end{array}\right\} \quad (12)$$

Third order steerable functions

$$\left\{\begin{array}{l} \frac{1}{2}(\cos(3\theta)+\cos(\theta), \\ \frac{1}{4}(-\sqrt{2}\cos(3\theta)+\sqrt{2}\cos(\theta)+\frac{1}{\sqrt{2}}\sin(3\theta)+\sqrt{2}\sin(\theta)); \\ \frac{1}{2}(-\sin(3\theta)+\sin(\theta), \\ \frac{1}{4}(\sqrt{2}\cos(3\theta)-\sqrt{2}\cos(\theta)+\frac{1}{\sqrt{2}}\sin(3\theta)+\sqrt{2}\sin(\theta)) \end{array}\right\} \quad (13)$$

These filter masks are convolved with the input image to generate the second and third derivative images.

The second order orientation is a occupation of L_{xx}, L_{yy} and L_{xy}.

$$\cos^2(\theta) \cdot L_{XX} + \sin^2(\theta) \cdot L_{YY} + \sin^2(2\theta) \cdot L_{XY} \tag{14}$$

Similarly third order orientation is a occupation of L_{xxx}, L_{yyy} L_{xxy} and L_{xyy}.

$$\sin^3(\theta) \cdot L_{XXX} + 3\sin(\theta) \cdot \cos^2(\theta) \cdot L_{XYY} + 3\sin^2(\theta) \cdot \cos(\theta) \cdot L_{XYY} + \cos^3(\theta) \cdot L_{YYY} \tag{15}$$

Texture information extracted by applying steerable filter at first and second order are from 10 oriented sub-bands. The statistical components are adapted for describing the content of the region called feature vector.

$$\text{Energy (E)} \quad E = \sum_{x=0}^{N-1} \sum_{y=0}^{M-1} \{F(i,j)\}^2 \tag{16}$$

$$\text{Contrast (C)} \quad C = \sum_{i=0}^{N-1} \sum_{j=0}^{M-1} |i-j| F(i,j) \tag{17}$$

$$\text{Inverse Difference Moment (IDM)} \quad IDM = \sum_{i=0}^{N-1} \sum_{j=0}^{M-1} \frac{1}{1+(i-j)^2} F(i,j) \tag{18}$$

$$\text{Entropy (E)} \quad E = \sum_{i=0}^{N-1} \sum_{j=0}^{M-1} F(i,j) X \log(F(i,j)) \tag{19}$$

$$\text{Correlation (Cr)} \quad Cr = \sum_{i=0}^{N-1} \sum_{j=0}^{M-1} \frac{\{iXj\} X \left(F(i,j) - \{\mu_i X \mu_j\}\right)}{\sigma_i X \sigma_j} \tag{20}$$

$$\text{Variance (V)} \quad V = \sum_{i=0}^{N-1} \sum_{j=0}^{M-1} (i-\mu)^2 F(i,j) \tag{21}$$

$$\mathrm{Fv} = [\mathrm{E}, \mathrm{C}, \mathrm{IDM}, \mathrm{E}, \mathrm{Cr}, \mathrm{V}] \tag{22}$$

5 Experimental Results

We have tested the effectiveness of the projected scheme with huge data base that consists of 5000 medical images collected from Frederick national laboratory and evaluated the effectiveness of the retrieval system with the help of Recall Rate as a performance measure and related medical images are retrieved by performing the

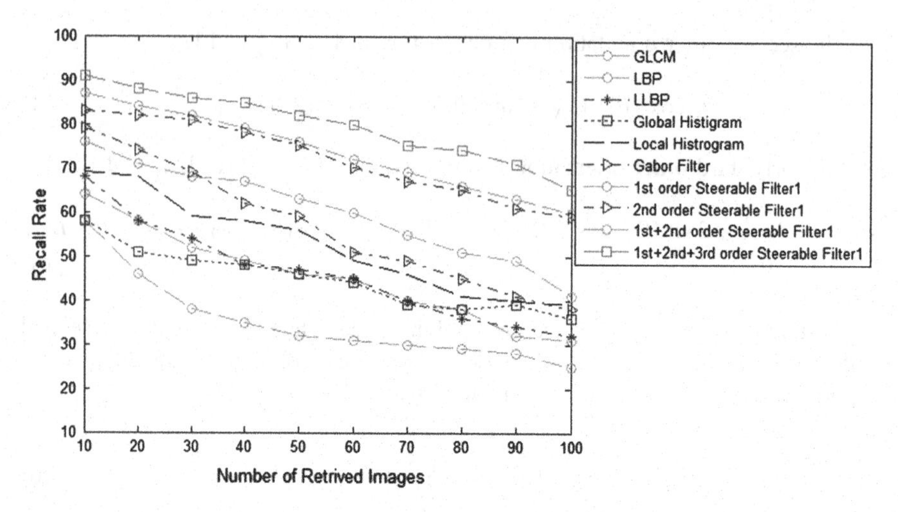

Fig. 4 Comparison of the retrieval results using various methods

Euclidian distance between the input image feature vector and consequent database image feature.

$$Euclidian - Distance - D(Fv_1, Fv_2) = \sqrt{\sum_{\forall i} (Fv_1(i) - Fv_2(i))^2} \quad (23)$$

From the investigational results it is very clear that the projected approach is outstanding for the CBMIR system in comparing with the existing system like GLCM, Histogram etc. shown in Fig. 4.

$$\text{Recall Rate} = \frac{\text{Number of Relavant Images Retrieved}}{\text{Total Number of Relavant Images in Database}} \quad (24)$$

6 Discussion and Conclusion

The proposed approach extracted the features from the interior of the object and it shows that it is more effective in which boundary of an image is detected using Edge Following algorithm followed by adapting higher order oriented region method for representing the content of the image by means of steerable filter. The results show that the presented method increases the retrieval performance when compared to the existing retrieval techniques and it can be promoted by incorporating relevance feedback.

References

1. Khoo, L.A., Taylor, P., Given-Wilson, R.M.: Computer-aided detection in the United Kingdom national breast screening programme: prospective study. Radiology **237**, 444–449 (2005)
2. Müller, H., Michoux, N., Bandon, D., Geissbuhler, A.: A review of content-based image retrieval systems in medical applications-clinical benefits and future directions. Med. Inform. **1**, 73 (2004)
3. Greenspan, H., Pinhas, A.T.: Medical image categorization and retrieval for PAC Suing the GMM-KL framework. IEEE Trans. Inf. Techol. Biomed. **11**(2), 190–202 (2007)
4. Chun, Y.D., Kim, N.C., Jang, I.H.: Content-based image retrieval using multiresolution color and texture features. IEEE Trans. Multimedia **10**(6), 1073–1084 (2008)
5. Jyothi, B., MadhaveeLatha, Y., Krishna Mohan, P.G.: Multidimentional feature vector space for an effective content based medical image retrieval. In: 5th IEEE International Advance Conputing Conference (IACC-2015), BMS College of engineering Bangalore (2015)
6. Samanta, S., Ahmed, S.S., Salem, M.A.M.M., Nath, S.S., Dey, N., Chowdhury, S.S.: Haralick features based automated glaucoma classification using back propagation neural network. Springer International Publishing Switzerland, vol. 1, p. 327. Advances in Intelligent System and Computing (2015). doi:10.1007/978-3-319-11933-5-38
7. Jyothi, B., MadhaveeLatha, Y., Krishna Mohan, P.G.: Steerable texture descriptor for effective content based medical image retrieval system using PCA. In: 2nd International Conference on Computer & Communication Technologies (IC3T-2015) published by proceedings of IC3T-2015, Springer-Advanced in Intelligent System and Computing Series 11156, vol. 379, 380 & 381 (2015)
8. Somkantha, K., Theera-Umpon, N.: Boundary detection in medical images using edge following algorithm based on intensity gradient and texture gradient features. IEEE Trans. Biomed. Eng. **58**(3), 567–573 (2011)
9. El-Naga, I., Yang, Y., Galatsanos, N.P., Nishikawa, R.M., Wernick, M.N.: A similarity learning approach to content-based image retrieval: application to digital mammography. IEEE Trans. Med. Imaging **23**(10), 1233–1244 (2004)
10. Navaz, A.S.S., Sri, T.D., Mazumder, P.: Face recognition using principal component analysis and neural networks. Int. J. Comput. Netw. Wireless Mobile Commun. (IJCNWMC) **3**(1):245–256, ISSN 2250-1568 (2013)

Enhancing Fuzzy Based Text Summarization Technique Using Genetic Approach

R. Pallavi Reddy and Kalyani Nara

Abstract Automatic text summarization provides a solution to the problem of information overload. Summarization technique preserves important information while reducing the original document. This paper focuses on enhancing Fuzzy text summarization technique by Genetic approach. The Genetic approach is used for optimizing the feature set given to fuzzy system. The optimization is achieved on feature set by natural evolution. The optimized input features are entrusted to fuzzy system which concentrates on the membership function and fuzzy rules. The analysis is performed on documents related to Earth, Nature, Forest and Metadata. The comparative study shows that using genetic approach would improve recall and F-measure.

Keywords Text summarization · Genetic approach · Fuzzy system · Information overload

1 Introduction

In recent years much focus is given on developing system that facilitates the use of information available on internet. In the new era of globalization, the production of information on internet has increased because of applications such as machine translation on the web, spelling and grammar correction in word processors, automatic question answering, fraud mail detection, extracting information from email, and detecting people's opinions about products or services. All the applications lead to the growth of information due to which the problem of information overload raised. Information overload refers to the problem a person can have

R. Pallavi Reddy (✉) · K. Nara
G. Narayanamma Institute of Technology and Science—For Women, Hyderabad, India
e-mail: reddygaripallavi@gmail.com

K. Nara
e-mail: kalyani_nara@rediffmail.com

S.C. Satapathy and S. Das (eds.), *Proceedings of First International Conference on Information and Communication Technology for Intelligent Systems: Volume 1*,
Smart Innovation, Systems and Technologies 50, DOI 10.1007/978-3-319-30933-0_45

understanding an issue and making decisions that can be caused by the effect of too much information, so has interest in automatic summarization.

One of the tasks of Natural Language Processing (NLP) is Automatic summarization which is based upon machine learning. Natural Language Processing is a discipline of computer Science that mainly addresses the issues relating to interactions between human languages and computers using the paradigms developed in statistics, artificial intelligence and language technologies. NLP is used to design and build programs that will examine, interpret and develop languages which humans naturally use. A computer program can also be written to retain important points as summary of the original document. This process is known as automatic summarization.

Text Summarization mainly follows two approaches: Abstraction and Extraction. In Extractive methods selection of relevant sentences or paragraphs form the original document is the primary focus. For identifying suitable sentence the knowledge of statistical and linguistic features play important role. Abstractive methods attempt to develop a view of the main concepts in a document and express them in clear natural language.

The first summarization system [1] was created by Luhn in 1958. He proposed to derive statistical information from word frequency and a relative measure of significance is computed by the machine based on distribution. The significance sentences are calculated for individual words and then it is done for sentences. Sentences with highest score become the auto-abstract. Word frequencies and distribution of words is used by Rath et al [2] in 1961 to generate the summary.

Edmundson [3] in 1969 uses cue words and phrases in generating the summary. During 1990s, with the arrival of machine learning techniques in NLP, a set of influential publications come into sight that uses statistical techniques [4–7] to produce an extract of a document. Initially most systems assumed feature independence and depended on naive-Bayes methods [8]. Other notable approaches include Hidden Markov Models [9] and log-linear models [10] to achieve extractive summarization.

A Fuzzy-Neural Network model is also proposed for pattern recognition [11]. The proposed model consists of three layers. The input layer is the first layer. The input of first layer is mapped to the corresponding fuzzy membership values in the second layer, and the implementation of the inference engine is done in the third layer. As fuzzy logic is used for pattern recognition, a further study on its application to text summarization is needed. The ideas of Fuzzy Sets proposed by Zadeh [12] in 1965 are elaborated in a 1973 paper equating a variable to a fuzzy set. Fuzzy set is a mathematical tool for handling ambiguity, imprecision, vagueness and ambiguity.

Bergler studies [12] proposed a fuzzy-theory based approach and its application to text summarization. Suanmali et al. [13] in 2009 proposed a fuzzy approach on the features extracted. The most attractive method for improving the text summarization systems quality is the Genetic Algorithm. It is a process of combining fuzzy logic, genetic algorithm (GA) and genetic programming (GP) [14] was proposed by

Kiani and M.R. Akbarzadeh, Genetic Semantic Role labeling is proposed by Suanmali et al. in 2011 [15].

This paper focuses on improving the summary of Fuzzy system with the help of Genetic algorithm. The primary steps for these techniques are Preprocessing and Feature Extraction given in Sect. 2.

The Components of Fuzzy system and their arrangement are elaborated in Sect. 3. The Fuzzy rules and the calculation of membership function are described in influencing the summary generated.

The optimization technique and the Genetic-Fuzzy System architecture is elaborated in Sect. 4. The Experiments conducted and the outcomes are specified in Sect. 5.

The conclusions in Sect. 6 are drawn based upon the experimental evaluations.

2 Preprocessing and Feature Extraction

The document that is taken as input is of text format. The activities that are performed in this stage are: sentence segmentation, tokenization, stop word removal, and word stemming. Sentence segmentation is done by detecting the boundary and separating input text into sentences. Tokenization is achieved by dividing the source document into individual words. Words which rarely contribute to useful information in terms of document relevance and appear frequently, but present less meaning in recognizing the important content of the document are removed. These words include articles, prepositions, conjunctions and other high-frequency words such as '*a*', '*an*', '*the*', '*in*', '*and*', '*I*', *etc…* The final step is word stemming which reduces the inflected or derived words to their root form.

After Preprocessing, eight features are obtained for each and every sentence. The extracted features contain the relevant information from the source text, so that the desired task can be done by using this reduced representation instead of the complete initial data. Each Feature value ranges between 0 and 1. The eight features are described as follows:

1. **Title Feature**: In a sentence, the title words are important as they relate more to the topic being summarized. These sentences have a greater chance of getting incorporated into the summary. The title feature (TF) could be calculated as below:

$$\mathrm{TF} = \frac{Number\ of\ title\ Words\ in\ sentence}{Number\ of\ words\ in\ title}$$

2. **Sentence Length**: Sentence Length (SL) is important in generating summary. Short sentences such as names, date lines etc., are to be eliminated from the summary.

$$\text{SL} = \frac{\textit{Number of Words occuring in sentence}}{\textit{Number of words occuring in longest sentence}}$$

3. **Term Weight**: In a document, the frequency of occurrences of a term has been used for calculating the weight of each sentence. The sentence score can be calculated as the sum of the score of individual words in the sentence. The weight of term can be given by

$$W_t = tf_i * isf_i = tf_i * \log \frac{N}{n_i}$$

tf_i Term Frequency of word
N Number of sentences in a document
n_i Number of sentences in which the Word i occur

The Total Term Weight (TW) of a sentence could be given by the following formula

$$\text{TW} = \frac{\sum_{i=1}^{k} W_i(S)}{MAX\left(\sum_{i=1}^{k} W_i(S_i^N)\right)}$$

K Number of Words in Sentences

4. **Sentence Position**: The sentences that are at the beginning of the document are more important than the others. If there are 5 lines in document the sentence positions (SP) are given by

$$\text{SP} = 5/5 \text{ for1st}, 4/5 \text{ for 2nd}, 3/5 \text{ for 3rd}, 2/5 \text{ for 4th}, 1/5 \text{ for 5th.}$$

5. **Sentence to Sentence Similarity** (**STSS**): The similarity between the sentences helps in removing similar sentences in the summary generated.

$$\text{STSS} = \frac{\sum Sim(S_i, S_j)}{MAX(\sum Sim(S_i, S_j))}$$

S_i: Sentence i
S_j: sentence j
Sim (s_i, s_j): is the similarity of 1 to n terms in sentence s_i and s_j

6. **Proper Noun**: The sentences which contain more proper nouns (names) are included in the summary. The Proper noun feature (PN) is calculated as below:

$$\mathrm{PN} = \frac{\textit{Number of Proper Nouns in Sentence}}{\textit{Sentence Length}}$$

7. **Thematic word**: The terms that occur more frequently have more relativity to the topic. The top most 10 frequent words are taken as thematic words. Thematic words (TW) are calculated as below:

$$\mathrm{TW} = \frac{\textit{Number of Thematic Word in Sentence}}{\textit{MAX}(\textit{number of Thematic word})}$$

8. **Numerical Data**: This Feature (ND) is used to identify the statistical data in every sentence. This is based on the ratio of number of numeric's present in the sentence to the length of sentence and is calculated as follows:

$$\mathrm{ND} = \frac{\textit{Number of Numerical data in Sentence}}{\textit{Sentence Length}}$$

3 Fuzzy System

The Fuzzy system comprises of: Fuzzifier, Inference Engine, Rule Base, and Defuzzifier respectively. The Fuzzy system involves in selecting salient fuzzy rules and membership functions. The fuzzy rules selection and membership function governs the execution of the system.

The feature values which are obtained as output of feature extraction are given as input to the Fuzzy System. The first component, Fuzzifier takes the input and converts into linguistic values using a triangular membership function. The input membership function is divided into five fuzzy sets based on MFT.

Mean Frequency Term (MFT) = (max − min)/5.0

The five fuzzy sets are classified as very low (VL), low (L), medium (M), high (H), very high (VH). The range for feature Term weight can be graphically shown in Fig. 1:

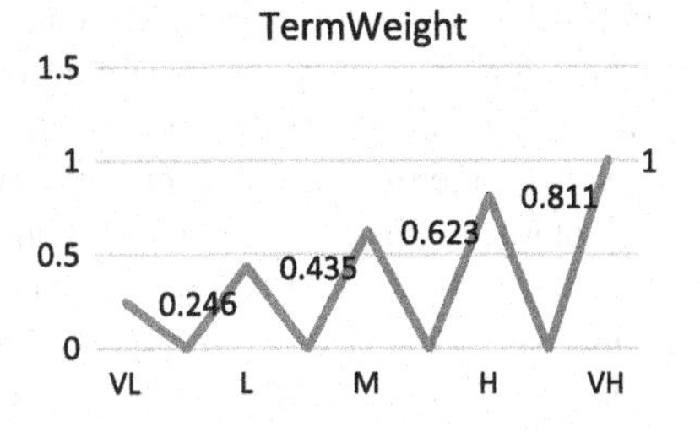

Fig. 1 Fuzzy sets for feature Term Weight

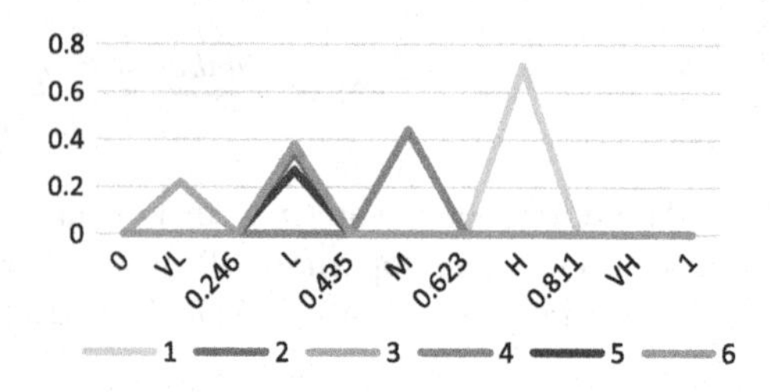

Fig. 2 Fuzzy set range for first six sentences of feature term weight

The Input feature can be categorized depending upon the range it falls in fuzzy set. Sample set of rules that compute weights to sentence are elaborated. The sentence with high weight are extracted which are found to be important according to the feature set (Fig. 2).

The main objective of the inference engine is to refer to the rules of fuzzy rule base. The fuzzy rule base involves the rules of IF-THEN. The extraction of important sentences is done by considering these rules according to the features criteria. Sample set of these rules are listed which assigns weight to each set.

IF (TW is VH) and (SL is H) and (TF is VH) and (SP is H) and (STSS is VH) and (PN is H) and (TW is VH) and (ND is H) THEN (Sentence is important)

The rules can be framed using logical operators like AND, OR and NOT. The minimum weight of all the antecedents is achieved by AND, while OR uses the maximum weight. Multiple number of rules are written to extract important sentences according to the features criteria. All the rules are executed sequentially and the defuzzifier generates the summary with dominant sentences.

The Computational cost of Fuzzy system is more as more number of computations are involved in fuzzification, fuzzy operators, fuzzy rules, selection of membership function i.e., triangular versus trapezoidal etc. The selection of membership function can affect the other parameters in a fuzzy rule as it is a multi-parameter problem. In order to reduce the overall computational cost and the number of parameters, optimization is required.

4 Genetic Fuzzy Technique

The process of extracting summary is made effective by including genetic approach in fuzzy system. The overall architecture of the proposed system is shown in Fig. 3. One of the specializations of genetic algorithms (GA) is Genetic Programming where a computer program is considered for each individual. These genetic algorithms provide powerful tools to identify the fuzzy membership function. This is based on identifying the input and output variables of fuzzy system and by using pre-defined rules. It also has limitations relating to identification of required variables. The basic idea of genetic approach is to maintain a population of candidate solutions to the concrete problem being solved. The genetic approach mutates and alters the candidate solutions to provide a better solution. Each candidate solution is represented by a chromosome.

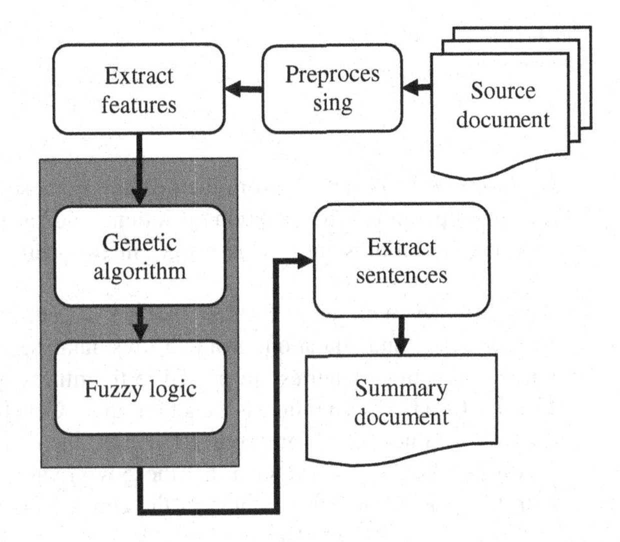

Fig. 3 Genetic Fuzzy technique

Chromosome:

A chromosome is basically represented in two ways: floating point representation or bit representation. The representation of each chromosome is done by a vector of dimension F (F being the total number of features). For most of genetic operations the floating representation is converted into bit representation (<0.5 means 0 and ≥0.5 means 1). A bit chromosome is represented as below:

The Pseudo code for the genetic approach is given as follows:

Procedure for Genetic Approach

1. Initialize Chromosomes
2. for Each Evolution

   {

   Perform one point crossover
   Perform mutation
   Add the chromosomes to result set

}

3. Find the fitness of result set

1. **Initial Population**: Each chromosome is represented by 0 and 1 binary bits. Let N be the population of chromosome. We generate 10 chromosomes of 8 bit length.
   The evolution process begins with the initially generated chromosomes. The population generated in each evolution is called as generation. In this paper we use 100 evolutions for generating the intermediate chromosomes. For each iteration the following steps are performed until a termination condition is met.
2. **Crossover and Mutation**: On the initial population that is generated, a one-point crossover is performed. The one point crossover starts with selecting

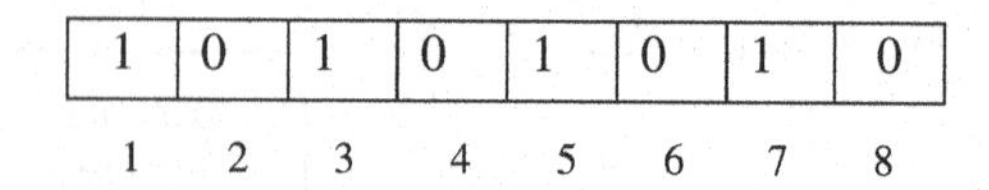

Fig. 4 Chromosome representation

two parent chromosomes. From the parent chromosomes a point is selected for mutation operation. The mutation operation includes exchanging of bits between two parent chromosomes. This results in two child chromosomes. These are added to the result set.

3. **Selection**: There are eight features that are present in each chromosome. After the crossover and mutation, from the resultant set the chromosomes with a minimum of three features are passed to the fitness function.
4. **Fitness function**: The fitness function in a Genetic Algorithm is problem dependent. Since the chromosomes represent the eight features with 8 bits, the fitness function is devised such that those with value zero are not fit and those with value 1 are fit. For example in the chromosome representation of Fig. 4: features 1, 3, 5, 7 are selected.
   Fuzzy system is used to formulate the ambiguous, imprecise values of text features to be admissible. The introduction of genetic approach to optimize the input features given to the fuzzy system reduces the number of input features to formulate i.e., the number of input fuzzy sets are minimized, increasing the probability of sentence to be included in summary.

5 Analysis of Results

Experiments are conducted on data sets relating to four different domains that include Earth, Nature, Forest, and Metadata. For every document a manually generated relevant summary is extracted to obtain the precision and recall values. Summary is generated using Fuzzy and Genetic-Fuzzy approaches. For each document the Precision, Recall and F-measure are calculated. The output of precision and recall are shown in the Table 1.

Table 1 Precision, recall and F-Measure for Fuzzy and Genetic-fuzzy system

	Fuzzy			Genetic-Fuzzy		
Document	Precision	Recall	F-Measure	Precision	Recall	F-Measure
Forest	56.25	10.46	17.46	60.33	84.4	70.53
Earth	71.42	16.12	26.31	68.88	40	50.06
Nature	80	5.714	10.66	63.33	13.57	22.532
Metadata	88.3	20.99	33.92	77.06	92.81	84.02
Average	73.99	13.321	22.0875	67.4	57.695	56.785

The analysis of recall suggests that Genetic Fuzzy gives better results in extracting the summary compared to fuzzy technique. Figures 5, 6, 7 and 8 shows the significance of improvement relating to four different domains.

The F-Measure in these graphs strengthens the observation of improvement in the Genetic Fuzzy technique over fuzzy technique.

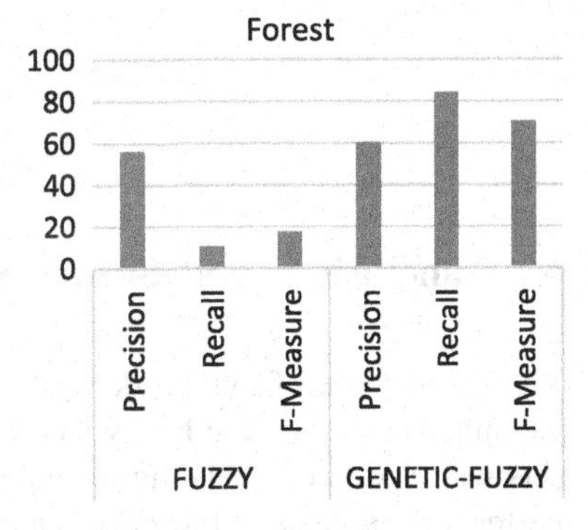

Fig. 5 Precision, recall and F-Measure for the domain Forest

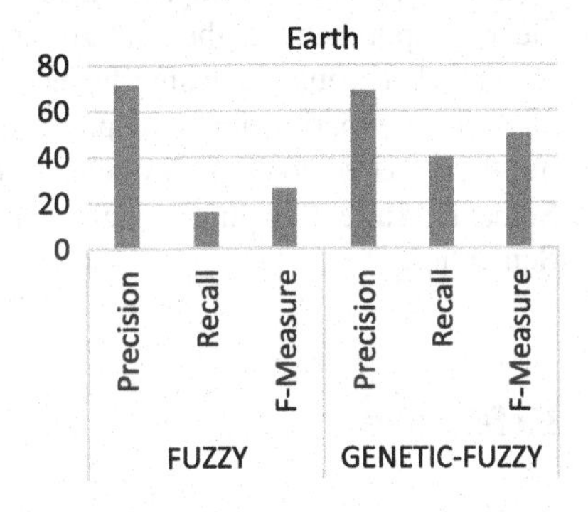

Fig. 6 Precision, recall and F-Measure for the domain Earth

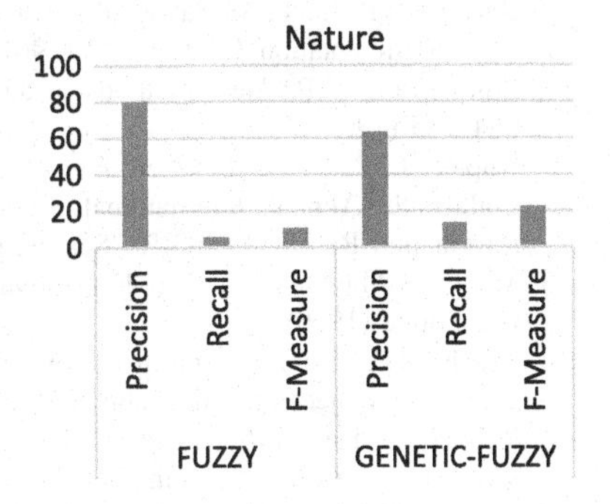

Fig. 7 Precision, recall and F-Measure for the domain Nature

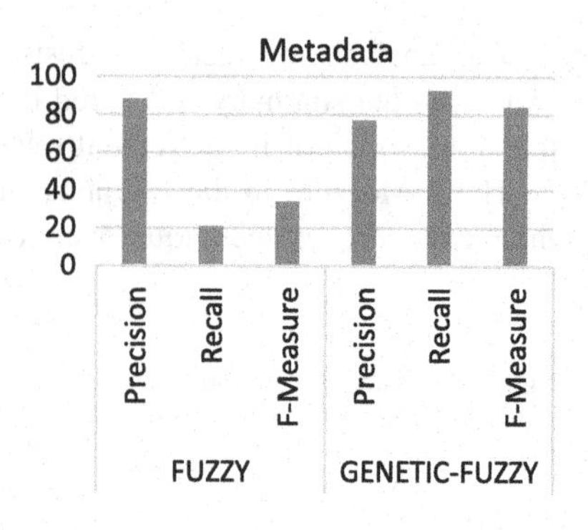

Fig. 8 Precision, recall and F-Measure for the domain Metadata

6 Conclusions and Future Scope

The main scope of this work was to enhance the performance of summarization technique using Genetic Fuzzy technique. The experiments were conducted on data sets with documents relating to four different domains and a comparative study is performed on Genetic Fuzzy over Fuzzy technique and it was observed that using Genetic approach will improve the extracted summary. It is also observed that using Genetic approach on the feature set would result in optimizing the number of features that would have significance role in the rules framed by Fuzzy system. The analysis of experiment results proves that recall and F-Measure of Genetic Fuzzy is more compared to fuzzy system. The system can be further enhanced by using Semantic Role Labeling which helps in eliminating the sentences with similar Semantics.

References

1. Luhn, H.P.: The automatic creation of literature abstracts. IBM J. Res. Dev. **2**, 159–165 (1958)
2. Rath, G.J., Resnick, A., Savage, T.R.: The formation of abstracts by the selection of sentences. Am. Documentation **12**, 139–143 (1961)
3. Edmundson, H.P.: New methods in automatic extracting. J. Assoc. Comput. Mach. **16**(2), 264–285 (1969)
4. Kupiec, J., Pedersen, J., Chen, F.: A trainable document summarizer. In: Proceedings of the Eighteenth Annual International ACM Conference on Research and Development in Information Retrieval (SIGIR), Seattle, pp. 68–73 (1995)
5. Mani, I., Maybury, M.T. (eds.): Advances in automatic text summarization. MIT Press, Cambridge (1999)
6. Fattah, M.A., Ren, F.: Automatic text summarization. In: Proceedings of World Academy of Science, Engineering and Technology, vol. 27, pp. 192–195 (2008)
7. Yeh, J.Y., Ke, H.R., Yang, W.P., Meng, I.H.: Text summarization using a trainable summarizer and latent semantic analysis. Special Issue Inf. Process. Manage. An Asian Digit. Libraries Perspect. **41**(1), 75–95 (2005)

8. Aone, C., Okurowski, M.E., Gorlinsky, J., Larsen, B.: A trainable summarizer with knowledge acquired from robust nlp techniques. In: Mani, I. Maybury, M.T. (eds.) Advances in Automatic Text Summarization, pp. 71–80. MIT Press, Cambridge (1999)
9. Conroy, J.M., O'leary, D.P.: Text summarization via hidden markovsmodels. In: Proceedings of SIGIR'01, pp. 406–407. New York (2001)
10. Osborne, M.: Using maximum entropy for sentence extraction. In: Proceedings of the ACL'02 Workshop on Automatic Summarization, pp. 1–8, Morristown (2002)
11. Kulkarni, A.D., Cavanaugh, D.C.: Fuzzy neural network models for classification. Appl. Intell. **12**, 207–215 (2000)
12. Zadeh, L.: Fuzzy sets. Information Control, vol. 8, pp. 338–353 (1965)
13. Witte, R., Bergler, S.: Fuzzy coreference resolution for summarization. In: Proceedings of 2003 International Symposium on Reference Resolution and Its Applications to Question Answering and Summarization (ARQAS), pp. 43–50. Università Ca' Foscari, Venice (2003)
14. Kiani, A., Akbarzadeh, M.R.: Automatic text summarization using: hybrid Fuzzy GA-GP. In: Proceedings of 2006 IEEE International Conference on Fuzzy Systems, pp. 977–983. Sheraton Vancouver Wall Center Hotel, Vancouver (2006)
15. Suanmali, L., Salim, N., Binwahlan, M.S.: Fuzzy genetic semantic based text summarization. In: Ninth IEEE International Conference on Dependable, Autonomic and Secure Computing, pp. 1185–1191 (2011)

Application of Rule Based and Expert Systems in Various Manufacturing Processes—A Review

M.R. Bhatt and S. Buch

Abstract The era of modern developments enhanced the technological advancements in all domains like medical, transportation, manufacturing, space and aviation, accounting, agriculture etc. The rule based and expert systems have proven their capabilities in all the well established and emerging domains. Manufacturing is one of the emerging fields, having major impact on global market. Rules and expert systems are mainly governed by knowledge, facts, empirical theorems etc. These systems are very useful to user for better prediction/approximation of the outputs. It also servers certain advantages like easy to operate, reduced human error, low and semi skilled person can handle etc. Here, a review has been done to converge and highlight the major applications of such systems for different manufacturing processes.

Keywords Application · Rule based system · Expert system · Manufacturing process

1 Introduction

Expert system is an AI technique, used to solving a problem. Normally expert systems used to transform human knowledge and expertise into computer programs using different software's. Such way, system can be developed to enhance the working environment for human being. Many systems have been developed and used in various areas viz. medical, transportation, manufacturing, space and aviation etc. These expert systems are developed based on industrial need and real life

M.R. Bhatt (✉)
RK University, Rajkot, Gujarat, India
e-mail: mrs.mrbhatt1412@yahoo.in

S. Buch
Reliance Industries, Ahmedabad, Gujarat, India
e-mail: drshbuch@gmail.com

S.C. Satapathy and S. Das (eds.), *Proceedings of First International Conference on Information and Communication Technology for Intelligent Systems: Volume 1*, Smart Innovation, Systems and Technologies 50, DOI 10.1007/978-3-319-30933-0_46

problems. Expert System comprises many types of systems based on rules, frames and fuzzy sets.

In this review, authors have exposed to the most popular system, expert system that is based on rules, knowledge, facts etc. Basically expert system comprises of analytical skill and knowledge (by learning or by experience) of single or multiple persons for the desired sector. The rules and knowledge in the system are framed in such a fashion that it is very easy to be used by semi skilled or unskilled user. In present review, some significant work are been highlighted which were done in the manufacturing sector. Figures 1 and 2 shows the block diagram of expert system (ES) and development of life cycle of expert system respectively. It depicts the flow chart to design the system and execution of the rules and knowledge for the particular system.

Manufacturing or production can be defined as to produce or create a worthy or useful product/component. Such practice requires some external work to be done on raw material to convert it into finished product. Normally products are designed and fabricated for fitness of its use. It includes dimensional as well as manufacturing accuracy. Generally, there are several manufacturing processes. Some of them are conventional and some of them are advanced. Both of them include material subtraction or material addition type of phenomenon to reach to the final product.

There are mainly four traditional manufacturing processes used for producing various products.

- Casting
- Welding
- Machining
- Forming

These major processes having their sub modules and each of them are differently termed. But the fundamental principles will remain same. All of them are having different process physics and various governing parameters. Many researchers have developed expert systems for these crude processes considering different process improving conditions.

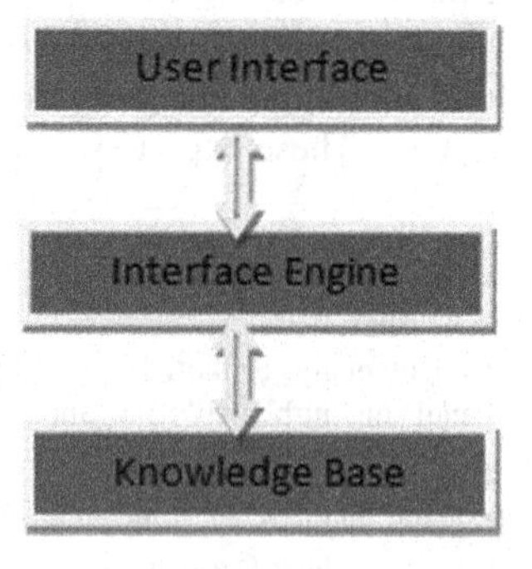

Fig. 1 Basic block diagram of expert system

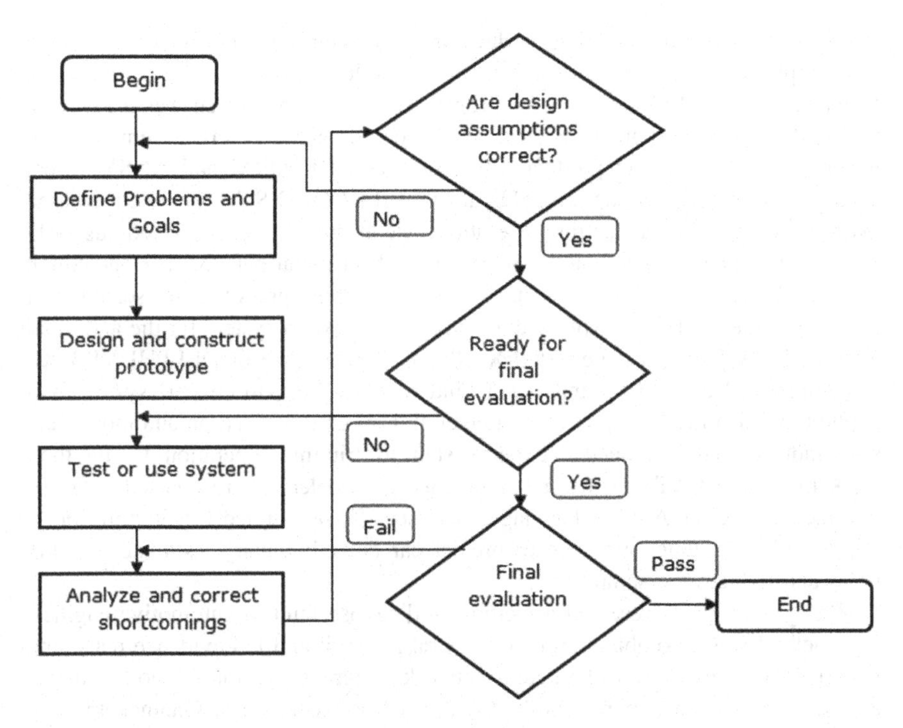

Fig. 2 Development of life cycle of expert system

2 Literature Review

Xie et al. [1] developed knowledge based CAPP using CAD environment and global and concurrent data integration platform. This real time knowledge based system integrates manufacturing and design. The pro/INTRALINK and STEP platform were used for this compound cutting and punching method. Ramana and Rao [2] developed a system to evaluate manufacturability automatically. The system was designed for shearing and bending operation to evaluate design, process planning, data and knowledge modeling etc. Giannakakis et al. [3] developed expert system which includes tool selection, die and press selection, process planning along with initial calculation modules. Ghatrehnaby and Arezoo [4] concentrated on progressive die for automated nestiong and piloting using CAD system. Kim and Park [5] developed a rule based system for hot steel forging of axi-symmetric components. The system was able to design process automatically. FORTRAN and AutoLISP of AUTOCAD were used to develop the system. Kim et al. [6] used the Autolisp to generate automated process planning and die design system. It was developed for quasi-axi-symmetric cold forging products. Ravi et al. [7] examined and developed Expert Systems for Hot-Forging Process which emphasized material's workability. Ohashi et al. [8] examined a computer aided die design system for axis-symmetric cold forging products by feature elimination. System generated process plans and die profile design from the product's shape to its raw materials. Kumar and Singh [9] described for progressive die design which was a low cost knowledge based system. Kim et al. [10] worked on an expert system for cold forging of axi-symmetric product. FEM software, DEFORM and ANSYS were used in this system. Cavdar et al. [11] developed a CORO Solve knowledge-based expert system for metal cutting operation. Metal cutting is a very tedious task. System predicted the maximum utilization of sheet metal part. Kumar and Singh [12] developed an intelligent system for metal stamping press tool for selection of die set. Cavdar et al. [13] explained an approach of expert system for die and mold making. They have presented DieEX system with the use of DELPHI VISUAL programming language for making a kind of software. This expert system was applied to finish machining of steel. Lee et al. [14] concentrated on automotive car body industry and developed an expert system for trimming operation. During their work they used CATIA software as geometric modeler and rule based data. To interface with CATIA, C++ language was used. It was limited for timming operation. The main functions were to predict car body boundary, estimate piercing holes and reduces the scrap.

Zhang et al. [15] presented a method to diagnose fault in automotive engines. The method was a combination of BP neural network and DS evidence reasoning model. It was nothing but the neural network integration fusion method to detect and solve low precision rate problems in automotive engines. Giannakakis and Vosniakos [16] gathered knowledge from theory and empirical formulas, handbooks and industry persons to develop an expert system. This expert system was developed for cutting and piercing operation to estimate process planning and die design. Zhang et al. [17] developed an Active Knowledge Support System for Design of automobile ball. During their work they used J2SDK 1.5, Dreamweaver MX 2004 and Microsoft SQL Server 2000 DB software to create such kind of system. Jhonston et al. [18] used six sigma and neural network techniques in the case study of hard disc drive manufacturing. The developed system was able to control the mechanisms, read write head recording and overall construction along with function of hard drive. A system was developed by Dale et al. [19] which was based on rules and knowledge to predict forging volumes. This volume of forging helped to estimate cost of component. It was developed for axi symmetric part which is working on geometry of product. Veera Babu et al. [20] produced an expert system using PAM-STAMP FE code for tailor welded blanks forming operation. Naranje and Kumar [21] discussed about the AI applications in sheet metal stamping die. Bhatt and Buch [22] developed an intelligent model for formability prediction for sheet metal component. During their work they used expert system as an AI technique. They used stainless steel, steel and aluminum as a sheet metal component. Based on experiment work they show that steel sheet component work much more without any failure. Khalajzadeh et al. [23] reviewed on an applicability of expert system in designing and control of autonomous cars. Even, an expert system was designed and developed for automobile dampers by Samuel and Bhagat [24]. Mobile vehicle expert system (MVES) was proposed by

Yaw and Kusi [25] for online vehicle trouble shooting using driver's mobile. An expert system was developed by Mumtaz et al. [26] to integrate material selection with manufacturing. Moreover, a system was developed for turning and milling operations in metal cutting by Luis et al. [27]. The system was capable to predict machine setting parameters. Agarwal and Goel [28] discussed Expert system and its requirement engineering process.

Ang et al. [29] discussed Requirement engineering techniques in developing expert systems. An automated expert system was developed by Tudor and Moise [30] to control robot trajectory in joint space using fuzzy control. Lin et al. [31] constructed automated structural design system (for progressive die) for bending and punching operations using CATIA V5 software. Ahmed and Taani [32] established an expert system for diagnosis of car failure. A neural network based expert system was proposed by Jiang et al. [33]. The system was developed for fault diagnosis of automobile engine. Horikoshi et al. [34] concentrates on expert system of deep drawing process using high velocity water jet. The fluid dynamics concept was incorporated along with Reynold's equation. However some local issues of the micro, small and medium scale industries are still not addressed.

Hence in the present review, applications of rule based and expert system in manufacturing sector is highlighted. That will encourage new researchers to bridge the gap and enhance the capabilities of manufacturing process using the potential of different AI techniques.

3 Summary

In a nutshell, it have been noticed that the expert systems with different other tools of AI and modeling are found very useful to resolve many problems in manufacturing sector. Also many new systems had developed which are effectively working. This reduces indirect labor cost and improves productivity. Further micro, small and medium scale enterprise got benefited with such systems. Moreover, there will be increased accessibility to these systems by people who are not computer-oriented through the use of natural language and explanation facilities. Such way many problems of shop floor can be reduced up to great extent.

References

1. Xie, S.Q., Liu, J.Q.: Integrated concurrent approach for compound sheet metal cutting and punching. Int. J. Prod. Res. **39**(6), 1095–1112 (2001)
2. Ramana, K.V., Rao, P.V.M.: Automated manufacturability evaluation system for sheet metal components in mass production. Int. J. Prod. Res. **43**(18), 3889–3913 (2005)
3. Giannakakis, T., Vosniakos, G.C.: Sheet metal cutting and piercing operations planning and tools configuration by an expert system. Int. J. Adv. Manuf. Technol. **36**(7–8), 658–670 (2008)

4. Ghatrehnaby, M., Arezoo, B.: A fully automated nesting and piloting system for progressive dies. J. Mater. Process. Technol. **209**(1), 525–535 (2009)
5. Kim, D.Y., Park, J.J.: Development of an expert system for the process design of axisymmetric hot steel forging. J. Mater. Process. Technol. **101**(1–3), 223–230 (2000)
6. Kim, M.S., Choi, J.C., Kim, Y.H., Huh, G.J., Kim, C.: An automated process planning and die design system for Quasi-Axisymmetric cold forging products. Int. J. Adv. Manuf. Technol. **20**(3), 201–213 (2002)
7. Ravi, R., Prasad, Y.V.R.K., Sarma, V.V.S.: Development of expert systems for the design of a hot-forging process based on material workability. J. Mater. Eng. Perfor. **12**, 646–652 (2003)
8. Ohashi, T., Imamura, S., Shimizu, T., Motomur, M.: Computer-aided die design for axis-symmetric cold forging products by feature elimination. J. Mater. Process. Technol. **137**(1–3), 138–144 (2003)
9. Kumar, S., Singh, R.: A low cost knowledge base system framework for progressive die design. J. Mater. Process. Technol. **153–154**, 958–964 (2004)
10. Kim, C., Park, C.W.: Development of an expert system for cold forging of axisymmetric product. Int. J. Adv. Manuf. Technol. **29**(5), 459–474 (2006)
11. Cemalcakir, M., Cavdar, K.: Development of a knowledge-based expert system for solving metal cutting problems. Mater. Des. **27**, 1027–1034 (2006)
12. Kumar, S., Singh, R.: An intelligent system for selection of die-set of metal stamping press tool. J. Mater. Process. Technol. **164–165**, 1395–1401 (2005)
13. Cemalcakir, M., Ozgur, I., Cavdar, K.: An expert system approach for die and mold making operations, Robot. Comput. Integr. Manuf. **21**:175–183 (2005)
14. Lee, S.J., Kim, T.S., Lee, S.S., Park, K.S.: Development of an expert system for the trim die design in automotive industry. In: IEEE Proceedings of the 10th International Conference on Computer Supported Cooperative Work in Design (CSCWD) (2006)
15. Zhang, X., Lu, M., Su, P., Xu, G., Zhao, H.: Research on neural network integration fusion method and application on the fault diagnosis of automotive engine. In: Second IEEE Conference on Industrial Electronics and Applications (2007). doi:10.1109/ICIEA.2007.4318455
16. Giannakakis, T., Vosniakos, G.C.: Sheet metal cutting and piercing operations planning and tools configuration by an expert system. Int. J. Adv. Manuf. Technol. **36**(7–8), 658–670 (2008)
17. Zhang, X., Han, C.Y., Cui, Y., Lu, Y.P., Liu1, E.F., Cui, H.B.: An active knowledge support system for design of automobile ball. In: IEEE International Workshop on Intelligent Systems and Applications ISA 2009, pp. 1–4 (2009)
18. Johnston, A.B., Maguire, L.P., McGinnity, T.M.: Downstream performance prediction for a manufacturing system using neural networks and six-sigma improvement techniques. Robot. Comput. Integr. Manuf. **25**(3), 513–521 (2009)
19. Dale, T.M., Young, W.A., Judd, R.P.: A rule-based approach to predict forging volume for cost estimation during product design. Int. J. Adv. Manuf. Technol. **46**(1–4), 31–41 (2010)
20. VeeraBabu, K., Ganesh Narayanan, R., Saravana Kumar, G.: An expert system for predicting the deep drawing behavior of tailor welded blanks. Expert Syst. Appl. **37**, 7802–7812 (2010)
21. Naranje, V., Kumar, S.: AI applications to metal stamping die design—a review. World Acad. Sci. Eng. Technol. **4**, 08–22 (2010)
22. Bhatt, M.R., Buch, S.: Prediction of formability for sheet metal component using artificial intelligent technique. In: 2nd International Conference on Signal Processing and Integrated Networks (SPIN), pp. 388–393 (2015)
23. Khalajzadeh, H., Dadkhah, C., Mansouri, M.: A review on applicability of expert system in designing and control of autonomous cars. In: IEEE Fourth International Workshop on Advanced Computational Intelligence (IWACI), pp. 280–285 (2011)
24. Samuel, G.L., Bhagat, A.: Development of an expert system for designing of automobile dampers. In: Proceedings of the 2011 Fourth International Conference on Emerging Trends in Engineering and Technology (2011). doi:10.1109/ICETET.2011.20

25. Yaw, N.A., Simonov, K.S.: MVES—a mobile vehicle expert system for vehicle troubleshooting through a driver's mobile device. Int. J. Eng. Res. Appl. **2**(6), 1108–1123 (2012)
26. Mumtaz, I., Selvi, I.H., Findik, F., Torku, O., Cedimoglu, I.H.: An expert system based material selection approach to manufacturing. Mater. Des. **47**, 331–340 (2013)
27. Luis, M., Trevino, T., Indira, G., Salazar, E., Ortiz, B.G., Alejo, R.P.: An expert system for setting parameters in machining processes. Expert Syst. Appl. **40**(17), 6877–6884 (2013)
28. Agarwal, M., Goel, S.: Expert system and it's requirement engineering process. In: IEEE International Conference on Recent Advances and Innovations in Engineering (ICRAIE), pp. 1–4 (2014)
29. Ang, J., Leong, S.B., Lee, C.F., Yusof, U.K.: Requirement engineering techniques in developing expert systems. In: Computers & Informatics (ISCI), pp. 640–645 (2011)
30. Tudor, L., Moise, A.: Automatic expert system for fuzzy control of robot trajectory in joint space. In: IEEE International Conference on Mechatronics and Automation (ICMA), pp. 1057–1062 (2013)
31. Lin, B.T., Huang, K.M., Su, K.Y., Hsu, C.Y.: Development of an automated structural design system for progressive dies. Int. J. Adv. Manuf. Technol. **68**, 1887–1899 (2013)
32. Ahmad, T., Taani, A.: An expert system for car failure diagnosis. World Acad. Sci. Eng. Technol. **1**, 12–20 (2007)
33. Jiang, L.L., Yong, N., Tang, L.H., Yong, H.: Fault diagnosis expert system of automobile engine based on neural networks. Key Eng. Mater. **460–461**, 605–610 (2011)
34. Horikoshi, Y., Kuboki, T., Murata, M., Matsui, K., Tsubokura, M.: Die design for deep drawing with high-pressured water jet utilizing computer fluid dynamics based on Reynolds equation. J. Mater. Process. Technol. **218**, 99–106 (2015)

Reinforcement Learning with Neural Networks: A Survey

Bhumika Modi and H.B. Jethva

Abstract Reinforcement learning (RL) comes from the self-learning theory. RL can autonomously get optional results with the knowledge obtained from various conditions by interacting with dynamic environment. It allows machines and software agents to automatically determine the ideal behavior within a specific context, in order to maximize its performance. Neural network reinforcement learning is most popular algorithm. Advantage of using neural network is that it regulates RL more efficient in real life applications. In this paper, we firstly survey reinforcement learning theory and model. Then we present various main RL algorithms. Then we discuss different neural network RL algorithms. Finally we introduce some application of RL and outline some future research of RL with NN.

Keywords Reinforcement learning · Q-learning · Artificial neural networks · Recurrent networks

1 Introduction

Reinforcement learning is learning what to do how to map situations to actions so as to maximize a numerical reward signal. The learner is not told which actions to take, as in most forms of machine learning, but instead must discover which actions yield the most reward by trying them. In the most interesting and challenging cases, actions may affect not only the immediate reward but also the next situation and, through that, all subsequent rewards. These two characteristics—trial-and-error search and delayed reward—are the two most important distinguishing features of

B. Modi (✉) · H.B. Jethva
Computer Science and Engineering Department, LD College of Engineering, Ahmedabad, India
e-mail: modibhumika31@gmail.com

H.B. Jethva
e-mail: hbjethva@gmail.com

S.C. Satapathy and S. Das (eds.), *Proceedings of First International Conference on Information and Communication Technology for Intelligent Systems: Volume 1*, Smart Innovation, Systems and Technologies 50, DOI 10.1007/978-3-319-30933-0_47

reinforcement learning. Reinforcement learning is defined not by characterizing learning methods, but by characterizing a learning problem [1].

There are supervised and unsupervised learning methods. But reinforcement learning differs from both. The signal called reinforcement signal provided by environment is given to intelligent agent. Intelligent agents have to take correct and good action. It does not tell which action to be taken. So intelligent agent depends on previous experience actions and results. To obtain a lot of reward, a reinforcement learning agent must prefer actions that it has tried in the past and found to be effective in producing reward. But to discover such actions, it has to try actions that it has not selected before. The agent has to exploit what it already knows in order to obtain reward, but it also has to explore in order to make better action selections in the future [2].

Neural Networks are very useful and suitable in pattern recognition, clustering, classification. So main advantage of neural network is capability of learning. Noisy data, untrained pattern are also classified using Neural Network. So Neural Networks are very popular for part of RL [3].

In this paper, we will survey RL algorithms with neural network function approximation for complex and real application. So we will study some algorithm and model of Neural Network (NN) and Reinforcement Learning (RL). We will study concepts and model of RL in Sect. 2. Then we present major RL algorithms including TD, Q-learning and in Sect. 3. Various neural network reinforcement learning are introduces in Sect. 4. In Sect. 5 we present some application. In the end, we conclude the paper and point out some future research in Sect. 6.

2 Reinforcement Learning Framework and Concept

Basic model of reinforcement learning is described in Fig. 1.

In Reinforcement Learning agent sense the environment and then choose appropriate action to get good reward. Agents will continuously interacting with environment and get bigger reward. So agent predicts action and future reward in order to increase reward with value function.

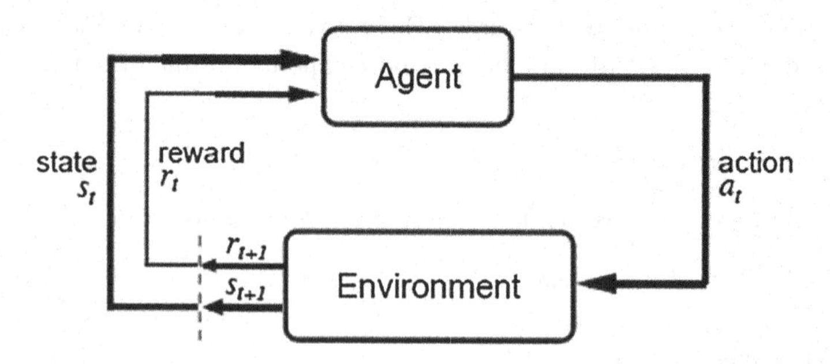

Fig. 1 RL learning model

So, here agent accept environment state **s** as an input and gives appropriate action a as an output. Each time RL takes decision to get big reward and change new state **s′**. Then again agent accept input state **s′** and get reward or penalty signal **r** for system. Reward function describes reward value. It describe sum of reward all over time also called Total reward. If Reward function is increased then action leads to goal otherwise its value decreased. So main goal of RL system is to get largest reward value of environment system. It can be defined as following formula (1).

$$\sum_{i=0}^{\infty} \gamma^i r_{t=i} \quad \text{Where } (0 < \gamma \leq 1) \tag{1}$$

Here γ is called discount factor of RL system. So reward function is based on assumed discount factor.

If in RL system future state of environment is only depend on current state then that is called markov property and interaction between agent and environment is called markov decision process (MDP). Reinforcement learning can solve markov decision problem. So in MDP state transition function $\mathbf{P_a}$**(s, s′)** is defined. Which shows probability that state **s** is changed to new state **s′** by choosing action **a**. So next state is dependent on current state **s** and chosen action **a** not on previous states and actions.

Learning a Markov Decision Process has a 5-tuple **(S, A, P, R, γ)**, Where

- **S** is used to describe finite number of states,
- **A** denote finite number of action. So $\mathbf{A_s}$ describe action which is chosen from state **s**,
- **Ra(s, s′)** describe reward which is derived through change of state **s′** from **s**,
- γ It is called discount factor. It is in [0, 1].

MDP can be solved and algorithms calculate maximized reward. Following formula (2, 3) are used and that gives maximized reward. Those calculations are recursively called for all states until no further changes are needed or take place.

$$\pi(s) = \arg\max_a \left\{ \sum\nolimits_{s'} P_a(s, s')(R_a(s, s' + \gamma V(s')) \right\} \tag{2}$$

$$V(s) = \sum_{s'} P_{\pi(s)}(s, s')(R_{\pi(s)}(s, s' + \gamma V(s')) \tag{3}$$

Here this algorithm uses two arrays called **V(s)**, which contains real values and $\boldsymbol{\pi}$, which is used to describe policy. And $\boldsymbol{\pi}$ contains chosen appropriate actions for state change and to get reward. So by formula (2) $\boldsymbol{\pi}$**(s)** will contain which actions are chosen and by formula (3). **V(s)** will contain total output sum of all the reward which is derived by state transfer and optimal action. So we can say that MDP is a delayed reinforcement problem.

3 Major Reinforcement Learning Algorithms

Generally various algorithms are used in various type of reinforcement learning method. If it is search-based learning then evolution directly takes place on policy. So for that type genetic algorithms are used. If it is model-based learning then model for environment is build and then we can solve algorithm by dynamic programming. SARSA uses model-based method. If method is model-free then directly policy is learned without model. TD, Q-Learning uses model-free method.

3.1 Temporal Difference Algorithm

Temporal Difference algorithm is proposed by Sutton [4]. It is mixing of Monte Carlo and dynamic programming ideas. So that it uses idea that future prediction state is based on current estimated state. Policy iteration of TD is called Actor-Critic learning. TD method is used to estimate value function V(s). So value of policy V (s) is learned using Sutton's TD(0) algorithm as follow Eq. (4) [4],

$$V(s) = V(s) + \alpha(r + \gamma V(s') - V(s)) \tag{4}$$

Here $\boldsymbol{\alpha}$ is learning rate. $\boldsymbol{r + \gamma V(s')}$ is a sample of V(s) and it is updated by γ and taking reward when transferring state $\mathbf{s'}$. So $\mathbf{r}$ is current reward and $\mathbf{V(s')}$ is estimated value of next state. TD(0) convergence is slower because value of $\boldsymbol{\alpha}$ is adjusted as it decreased slowly. The TD(0) rule is instance of algorithm called TD (λ). By putting $\lambda = 0$ into TD(λ) we get TD(0). So general rule of TD(λ) is same as above. It is written as follows equation.

$$V(u) = V(u) + \alpha(r + \gamma V(s') - V(s))e(u) \tag{5}$$

Here **e(s)** defines as eligibility of state. So from above Eq. (5), rule is applied according to eligibility **e(u)** not only to the previous state. So eligibility state can defined to be,

$$e(s) = \sum_{k=0}^{t} (\lambda\gamma)^{t-k} \delta_{s,s_k}$$

where $\delta_{s,s_k} = 1 \; if \;\; s = ss_k,$ otherwise $\delta_{s,s_k} = 0$

Eligibility of state is degree of convergence which is recently visited states. So if $\lambda = 1$ then it is same as updating states according to number of time they visited. Eligibility can update as follows Eq. (6) [5].

$$elligibility\ of\ state = e(s) \tag{6}$$

If **s** = current state then $e(s) = \gamma\lambda e(s) + 1$ otherwise $e(s) = \gamma\lambda e(s)$

So in TD method value is updated using partly existing approximation not on final reward, so this is called bootstrapping.

3.2 Q-Learning Algorithm

The Q–learning algorithm [6] is a well-know and widely used RL technique. It is model-free method. Value iteration of TD is called Q-learning. Agent change state current to next by taking proper action. So main goal of agent is to maximize its total reward. So that Q-learning uses strategy to learn optimal policy **π*** via taking values of learning actions. Now System can be modeled as MDP, so by choosing action every state-action pairs are visited continuously [7].

For that **Q*(s, a)** is defined. It describe expected discount for taking action an in state **s**. Now **V*(s)** is also value function. It is defined as follows Eq. (7).

$$V^*(s) = max_a Q^*(s,a) \tag{7}$$

Initially **V*(s)** assumed that it chooses best action. Now **Q*(s, a)** can be defines as follows Eq. (8).

$$Q^*(s,a) = R(s,a) + \gamma \sum_{s' \in s} P(s,a,s) \max_{a'} Q^*(s',a') \tag{8}$$

By Eq. (7) we have Eq. (9) act as an optimal policy.

$$\prod\nolimits^*(s) = max_a Q^*(s,a) \tag{9}$$

So that initially Q value is given and then function is updated according following Q-learning rule (10).

$$Q(s,a) = Q(s,a) + \alpha(r + \gamma\ max_{a'} Q^*(s',a') - Q(s,a)) \tag{10}$$

So main goal is to obtain maximized reward. So action is executed infinitely number of times to decay value of **α**. So Q-value converge with probability **1** to **Q*** [6].

Under below conditions Q-learning will converge to the correct Q-function [8].

- States and actions are finite
- learning rate decays with visits to state-action pairs
- environment model doesn't change
- Exploration method would guarantee infinite visits to every state-action pair over an infinite training period

4 Neural Network Reinforcement Learning

In real applications, Agent does not have proper and complete information about environment. So when Environment is unknown or dynamic then learning becomes difficult. So to predict state of unknown environment Neural Networks are used in Reinforcement Learning [9]. Most of RL algorithms uses value function as describe above section. So according to state-action pair value function is updated. So for large number of states this task becomes more complex. So to reduce complexity Neural Networks can be used with RL [10]. In RL problems reward are obtained by environment and it is given one by one (one state to next state). So interference (disruption) can arise in RL problem. To overcome or solve that problem incremental learning Neural Network can applied to RL problem [11].

Another thing is that neural networks can classify various pattern which they have not be trained, they are fault tolerance to noisy data [12]. So it has scaling and learning abilities. So neural network can be applied to RL system so that NN can generate new idea for agent during learning phase [13]. Recently many neural networks are coupled with various RL learning methods because it reduces complexity. Many studies have been done in this area. So various NN like Feed-Forward NN, Recurrent NN, Boltzman NN are coupled with Q-Learning etc.

4.1 *Feedforward Neural Network Reinforcement Learning*

Feedforward NN (FFNN) is one of most NN which is used with RL model. FFNN consist of an input and an output layer. Multi-layer feedforward neural networks consist of an input, a number of hidden, and an output layer. So Weighted sum is forwarded from input layer to over the hidden to the output layer. There is no backward loop or connections or self-loop between or within layers. So output of current layer cannot give to previous layer or same layer [14]. So FFNN is learned by comparing different output and can achieve minimize error. So FFNN is able to deal with complex value function approximation. And it can achieve value function more accurate. But limitation of FFNN is that it has not more repeatable property. So it cannot handle inter-dependency.

4.2 *Recurrent Neural Network Reinforcement Learning*

Recurrent Neural Network (RNN) is a one type of neural network. In this directed connections are used between nodes. And this connection can made direction cycle. Each node has activation value which is varied according to time. And each connection has a weight value which is also modified in training rule. So thre are input, hidden and output nodes So that RNN has a benefit to use internal memory to train nd use sequence of inputs. RNN is trained various similar type of training patterns to identify dynamic problem. In RNN reward error can be back propagated. So that network tends to learn control policy that can give good or maximum total reward. So for dynamic and optimal policy identification the training of RNN reinforcement learning continues until a good result is achieved [10]. In RL no label or teacher provided for RNN. So reward or value function is used for decision taking and for provide to check performance.

Following are various NN which used recurrent Neural Network model [10, 15].

- Hopfield network
- Elman networks and Jordan networks
- Echo state network
- Long short term memory network
- Bi-directional RNN
- Continuous-time RNN
- Hierarchical RNN
- Recurrent multilayer perceptron
- Second Order Recurrent Neural Network
- Multiple Timescales Recurrent Neural Network (MTRNN) Model

Drawback of RNN is training capability. RNN cannot be easily trained for large number of nodes and large number of input.

5 Applications

Reinforcement Learning and Neural Networks are used in many fields. RL is used mainly in intelligence control and robotics. RL can be used in game of Othello. In that three algorithm Q-Learning, Sarsa and TD-Learning [16]. Multi-agent reinforcement learning is used in many applications like traffic control system. In this multi-agent Q-learning with either ε-greedy method and agent predict state to reduce traffic signal delay to road users [17]. Neural Network are used in Pattern Classifications Control, time series modeling, estimation Optimization etc. So Neural Networks with RL are used in many fields to decrease complexity, error and delay of system. NN with increment ability is applied to RL problem to solve disruption. So this method can applied to Extended problems like Random-Walk Task and Extended Mountain-Car Task [11]. Dynamic Neural Network is used to help agent and agent should learn during real time operation. It eliminate the problem of look-up-table (Q table) [13]. In an adaptive light seeking robot ANN is as function approximation and ANN is combined with Q-Learning to decrease complexity [18]. Feedforward neural networks can be applied to high-dimension problems like motor control. So when continuous model-based reinforcement learning with feedforward neural network is used in motor control problem then it reduces complexity than linear function approximation [19]. Reinforcement Learning, Neural Networks and Pi control can be applied to heating coil. NN is trained to reduce error between coil output and set point. And RL is used to reduce error of all over time [20]. In unknown environment multi-agent path planning is done using RL and NN. NN is used to predict unknown environment and RL is applied to get maximum reward. This can solve path planning problem in multi-agent system [9].

6 Conclusions

In this paper, we have reviewed basic RL algorithms and basic neural network RL models. Various applications that use both NN and RL are stated. So the combination of RL with NN becomes key research area in many fields like Robotics, Dynamic path planning, and control theory. In real work there are still further future work needed to solve problems. Best choice of NN to combine RL model is needed skill for selection. In multi-agent RL, slow training speed is main issue. There for how to combine NN with RL in multi agent system to speed up learning is need to explore. Selection of best parameter for the policy is main task. This gives performance of system. So it is main research focus in this area. For dynamic environment it is difficult to learn agent so using NN and RL models speed up and optimal policy can be achieved. That can be applied to robotics system when decisions are taken dynamically. New neural network reinforcement learning model are also developed to increase convergence capability, So research in this area for high dimensional and complex problems needs to explore. So for multi-agent system neural network with reinforcement learning is important area for future work.

References

1. Sutton, R.S., Barto, A.G.: Reinforcement Learning: An Introduction. MIT Press, Cambridge (1998)
2. Qiang, W., Zhongli, Z.: Reinforcement learning model, algorithms and its application. In: 2011 International Conference on Mechatronic Science, Electric Engineering and Computer (2011)
3. Asadi, R., Mustapha, N., Sulaiman, N.: A framework for intelligent multi agent system based neural network classification model. (IJCSIS) Int. J. Comput. Sci. Inf. Secur. (2009)
4. Sutton, R.S.: Learning to predict by the method of temporal differences. Mach. Learn. **3**(1), 9–44 (1988)
5. Littman, M.L., Moor, A.W.: Reinforcement Learning: A Survey. Leslie Pack Kaelbling (1996)
6. Watkins, C.J.C.H., Dayan, P.: Q-learning. Mach. Learn. **8**(3), 279–292 (1992)
7. Celiberto, L.A., Jr., Matsuura, J.P., de Màntaras, R.L.: Using transfer learning to speed-up reinforcement learning: a cased-based approach. In: 2010 Latin American Robotics Symposium and Intelligent Robotics Meeting (2010)
8. Torrey, L.: Reinforcement Learning. University of Wisconsin, Madison HAMLET (2009)
9. Luviano Cruz, D., Yu, W.: Multi-agent path planning in unknown environment with reinforcement learning and neural network. In: 2014 IEEE International Conference on Systems, Man, and Cybernetics, San Diego, CA, USA, Oct 5–8 2014

10. Ghanbari, A., Vaghei, Y., Reza, S.M., Noorani, S.: Reinforcement learning in neural networks: a survey. Int. J. Adv. Biol. Biomed. Res. **2**(5), 1398–1416 (2014)
11. Shiraga, N., Ozawa, S., Abe, S.: A reinforcement learning algorithm for neural networks with incremental learning ability
12. Han, J., Kamber, M.: Data Mining: Concepts and Techniques. Simon Fraser University, Academic Press. Information Quality Issues. International Institute for Advanced Studies in Systems Research and Cybernetic, IIAS (2001)
13. Yadav, A.K., Sachan, A. K.: Research and application of dynamic neural network based on reinforcement learning. In: Satapathy, S.C., et al. (eds.) Proceedings of the InConINDIA 2012, AISC, vol. 132, pp. 931–942
14. Haykin, S.: Neural Networks: A Comprehensive Foundation. Macmillan, New York (1994)
15. Rojas, R.: Neural Networks: a Systematic Introduction, pp. 336. Springer, Berlin (2013). ISBN 978–3-540-60505-8
16. van der Ree, M., Wiering, M.: Reinforcement learning in the game of Othello: learning against a fixed opponent and learning from self-play
17. Prabuchandran, K.J., Hemanth Kumar, A.N., Bhatnagar, S.: Multi-agent reinforcement learning for traffic signal control. In: 2014 IEEE 17th International Conference on Intelligent Transportation Systems (ITSC), Qingdao, China, 8–11 Oct 2014
18. Dini, S., Serrano, M.: Combining Q-learning with artificial neural networks in an adaptive light seeking robot, 6 May 2012
19. Coulom, R.: Feedforward neural networks in reinforcement learning applied to high-dimensional motor control. In: Cesa-Bianchi, N., et al. (eds.) ALT 2002, LNAI 2533, pp. 403–413 (2002)
20. Anderson, C.W., Hittle, D.C., Katz, A.D., Matt Kretchmar, R.: Reinforcement learning, neural networks and PI control applied to a heating coil

Part V
ICT for Information Sciences

Automatic Metadata Harvesting from Digital Content Using NLP

Rushabh D. Doshi, Chintan B. Sidpara and Kunal U. Khimani

Abstract Metadata Harvestings is one of the prime research fields in information retrieval. Metadata is used to references information resources. Metadata play an significant role in describing and searching document. In early stages of metadata harvesting was manually. Later on automatic metadata harvesting techniques were invented; still they are human intensive since they require expert decision to identify relevant metadata also this is time consuming. Also automatic metadata harvesting techniques are developed but mostly works with structured format. We proposed a new approach to harvesting metadata from document using NLP. As NLP stands for Natural Language Processing work on natural language that human used in day today life.

Keywords Metadata · Extraction · NLP · English grammars

1 Introduction

Metadata is data that describes another data Metadata provides help to access information resources. A group of metadata can be helpful for one or many information resources. For example, a library catalogue record is a collection of metadata, which are linked to the book, journals, news papers, magazines, CD etc. in the library collection. In web page HTML tag "META" stored elements which are metadata for that page.

The main idea of metadata is to ease and get better the repossession of information. Many places (library, college etc.) metadata is chosen from the different features

R.D. Doshi (✉) · C.B. Sidpara · K.U. Khimani
V.V.P Engineering College, Rajkot, Gujarat, India
e-mail: rushabh.doshi@vvpedulink.ac.in

C.B. Sidpara
e-mail: chintan.sidpara@gmail.com

K.U. Khimani
e-mail: khimanikunal1988@gmail.com

S.C. Satapathy and S. Das (eds.), *Proceedings of First International Conference on Information and Communication Technology for Intelligent Systems: Volume 1*,
Smart Innovation, Systems and Technologies 50, DOI 10.1007/978-3-319-30933-0_48

of the information: author, subject, title, publisher and so on. Various metadata harvesting techniques are developed to extract the data from digital libraries.

NLP is a area of computer science, artificial intelligence and linguistics concerned with the interactions between computers and human (natural) languages. Recent research has progressively more focused on rule based or semi-supervised learning algorithms. Such algorithms are able to learn from data that has been non-annotated or combination of annotated and non-annotated data. The objective of NLP is to compute one or more merits of an algorithm or a system, to find out whether its meets the desires of its users.

2 Method

In this paper we proposed automatic metadata harvesting algorithm using natural language (i.e. humans used in day today works). Our system is rule based. So it does not require any training dataset for it.

We harvest metadata based on English Grammar Terms. We identify the possible set of metadata then compute their frequency then applying weight term based on their position or format that apply to it.

The rest of the paper is organized as follows. The next section review some related work concerning to metadata harvesting from digital content. Section gives the detailed description of proposed idea presented here. At last paper is concluded with summary.

3 Related Work

Existing Metadata harvesting techniques are either machine learning method or ruled based methods. In machine learning method set of predefined model that contains dataset are given to machine for training. Then trained machine is used to harvest metadata from document based on that dataset. While in rule based method most of techniques set ruled that are used to harvest metadata from documents.

In machine learning approach harvested keywords are given to the machine from training documents to learn specific models then that model are applied to new documents to harvest keyword from them. Many techniques used machine learning approach like automatic document metadata extraction make use of support vector machine.

In rule based techniques some predefined set of rules are given to machine, based on that machine harvest metadata from documents. “Positions of word in document”, “identified keyword are used as types of record” & etc. These are the references, that set in various metadata harvest techniques. In few case metadata classification is based on document types (e.g. purchase order, sales report etc.) and data context (e.g. customer name, order date etc.) [1].

Other statistical methods include word frequency [2], TF*IDF [3], word co-occurrences [4]. Later on some techniques are used to harvest key phrase based on TF*PDF [5]. Other techniques use TDT (Topic Detection and Tracking) with aging theory to harvest metadata from news website [6]. Some techniques used DDC/RDF editor to characterize and extraction of metadata from document and authenticate by thirds parties [7]. Several techniques are developed to harvest metadata from corpus. Currently most of the methods are using models, that depends on corpus.

4 Proposed Theory

We focused on rule based techniques. Our method focused on harvesting a metadata from document based on English grammar. English grammar has many categories which classified the word in statement. Grammar categories such as NOUN, VERB, ADJECTIVES, ADVERB, NOUN PHRASE, VERB PHRASE etc. each and every grammar category has a priority in statement. So our approaches to extract out the Metadata extraction based on its priority in grammar. Priority in grammar component is as follows:

1. Noun
2. Verb
3. Adjective
4. Adverb
5. Noun phrase

5 Proposed Idea

In Fig. 1 we give our system architecture.

5.1 Article Pre-Processing

This step is used to remove irrelevant contents (i.e. tags, header-footer details etc.) from documents.

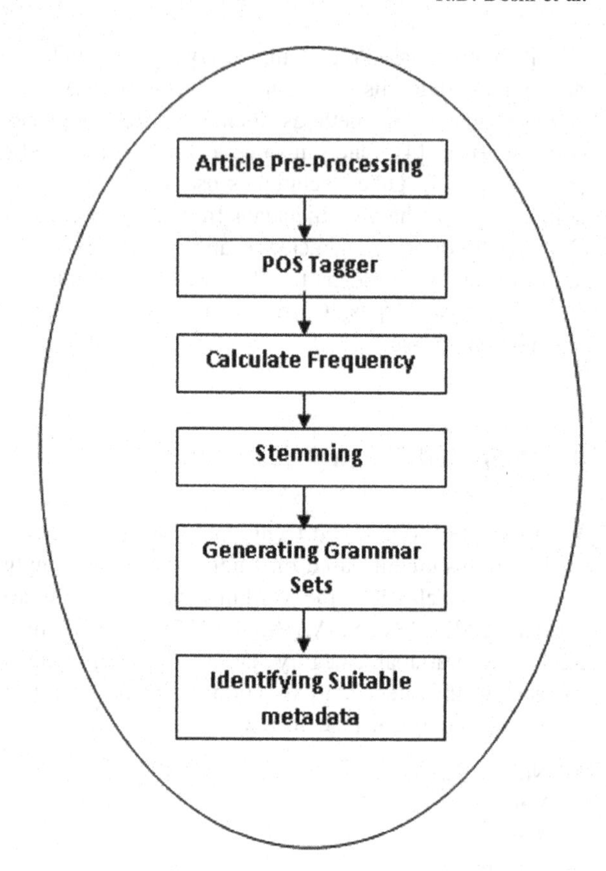

Fig. 1 System architecture

5.2 *POS Taggers*

A Part-Of-Speech Tagger is a piece of software that reads assigns parts of speech to each word such as noun, verb, adjective, etc.

5.3 *Stemming*

In most cases, morphological variants of words have similar semantic interpretations can be considered as equivalent for the point of IR applications. For our system we used port stemmer.

5.4 Generating Grammar Set

Here each text in documents classified with its Grammar Type. Here Different Grammar category set is prepared such as NOUN set contains all the noun from the given documents.

5.5 Calculate Frequency

Here each word frequency is calculated i.e. how many occurrence of each word in document.

5.6 Identify Suitable Metadata

Now metadata is extracted from word set based on their frequency, grammar and their positions.

6 Experiments and Results

In this study we take a corpus with 100 documents. Documents contain the news article about various categories. Here we first extract the metadata manually from each & every documents. Then apply our design to corpus. We measure our result from following parameter.

Precision = No of terms identified correctly by the system/Top N terms out of total terms generated by the system.

Recall = Number of key-terms identified correctly by the system/Number of key-terms identified by the authors. F-measure = F = 2* ((precision* recall)/(precision + recall)).

We tested our results with different number of terms. We tested our idea with term number 10, 20, 30, our precisions are 0.43, 0.42, 0.32, Recalls are 0.36, 0.63, 0.72 and F-measures are 0.40, 0.51, 0.49 respectively.

As per Evaluation result our F—measure is low. Now we will See results in other point of view. We will explain our method with one article which is given below. Here we set 30 different key-term from the document (Table 1).

> *To preserve integrity of data, the database system must ensure:*
>
> *Atomicity. Either all operations of the transaction are properly reflected in the database or none are.*
>
> *Consistency. Execution of a transaction in isolation preserves the consistency of the database.*

Table 1 Evaluation result

Actual metadata	Our system metadata
Transaction, preserve, integrity of data, atomicity, reflected, isolation, durability, consistency, concurrent transaction, system failure	Transaction, database, transactions, system, isolation, execution, finished, started, preserves, execute, Tj, intermediate, atomicity, reflected, executing, hidden, consistency, operations, data, other, multiple, concurrently, properly, database system, database isolation, multiple transactions, transaction results, system failures

Isolation. Although multiple transactions may execute concurrently, each transaction must be unaware of other concurrently executing transactions. Intermediate transaction results must be hidden from other concurrently executed transactions.

That is, for every pair of transactions Ti and Tj, it appears to Ti that either Tj, finished execution before Ti started, or Tj started execution after Ti finished.

Durability. After a transaction completes successfully, the changes it has made to the database persist, even if there are system failures.

As you see in Evaluation Result-2, our system derived metadata more than actual metadata. Our system metadata are arranged in their index position inside the documents. By reading our system metadata we can draw approximate inference regarding to the documents. For example by reading our system metadata we can assume that this document contains some information regarding to transaction execution in database.

7 Conclusion and Future Works

This method based on grammar component Our Aim to use this algorithm to identifying metadata in bigram, trigram tetragram. After that we used this algorithm to generate summary of documents.

References

1. Manning, C.D., Raghavan, P., Schtze H.: An introduction to information retrieval book (2008)
2. Luhn, H.P.: A statistical approach to mechanized encoding and searching of literary information. IBM J. Res. Dev. **1**(4), 309–317 (1957)
3. Salton, G., Yang, C.S., Yu, C.T.: A theory of term importance in automatic text analysis. J. of the Am. Soc. for Inf. Sci. **26**(1), 33–44 (1975)
4. Matsuo, Y., Ishizuka, M.: Keyword extraction from a single document using word co-ocuurrence statistical information. Int. J. on Artif. Intell. Tools. 13(1), 157–169 (2004)
5. Gao, Y., Liu, J.: Peixun ma the hot keyphrase extraction based on TF*PDF. In: IEEE conference (2011)

6. Wang, C., Zhang, M., Ru, L., Ma S.: An automatic online news topic keyphrase extraction system. In: IEEE conference (2006)
7. Yahaya, N. A., Buang R.: Automated metadata extraction from web sources. In: IEEE conference (2006)

Gradient Descent with Momentum Based Backpropagation Neural Network for Selection of Industrial Robot

Sasmita Nayak, B.B. Choudhury and Saroj Kumar Lenka

Abstract Fast development of industrial robots and its utilization by the manufacturing industries for many different applications is a critical task for the selection of robots. As a consequence, the selection process of the robot becomes very much complicated for the potential users because they have an extensive set of parameters of the available robots. In this paper, gradient descent momentum optimization algorithm is used with backpropagation neural network prediction technique for the selection of industrial robots. Through this proposed technique maximum, ten parameters are directly considered as an input for the selection process of robot where as up to seven robot parameter data be used in the existing methods. The rank of the preferred industrial robot evaluates from the perfectly the best probable robot that specifies the most genuine benchmark of robot selection for the particular application using the proposed algorithm. Moreover, the performance of the algorithms for the robot selection is analyzed using Mean Square Error (MSE), R-squared error (RSE), and Root Mean Square Error (RMSE).

Keywords Industrial robot selection · Gradient descent algorithm · Backpropagation algorithm · Neural network

S. Nayak (✉)
Department of Mechanical Engineering, Government College of Engineering, Kalahandi, Odisha, India
e-mail: sasmitacet@rediffmail.com

B.B. Choudhury
Department of Mechanical Engineering, IGIT, Sarang, Odisha, India
e-mail: bbcnit@gmail.com

S.K. Lenka
Department of Information Technology, Mody University, Lakshmangarh, India
e-mail: lenka.sarojkumar@gmail.com

S.C. Satapathy and S. Das (eds.), *Proceedings of First International Conference on Information and Communication Technology for Intelligent Systems: Volume 1*, Smart Innovation, Systems and Technologies 50, DOI 10.1007/978-3-319-30933-0_49

1 Introduction

In engineering and technology robots are the commonly used device and tool in the advanced manufacturing facilities. Day by day the robot manufacturers are increasing and the grades of manufacturing are also increasing. The robot selection suitable for a specific implication and also the manufacturing environment from the very vast number of robots excerption in the market becomes a difficult task. Diverse thoughtfulnesses such as accessibility, production, and economic need to be studied. Furthermore, many of the properties are conflicting in nature as well as have a different unit [1]. Moreover, none of the above solutions may not take care all the necessitates and constraints of particular applications. Paul and Nof [2] weigh against robots with humans. Vukobratovic [3] states that the spherical configuration was better-quality to other arrangements. Khouja [4] discussed the applications of Data Envelopment Analysis (DEA) of the 1st phase and a multi-attribute decision making method in the second step. Furthermore, DEA needs more computation resources and the number of factors that the decision maker willing to consider is large as well as the number of alternative robots are smaller than DEA which may be a poor discriminator. Here the Author's quote example of twenty-seven alternative robots with four attributes robot selection. Again, the DEA may has disfavor in terms of its rationale, suppose the decision making is unfair using linear programming technique. Liang and Wang [5] discussed the robot selection algorithm, which was used to find out policy makers' fuzzy assessments about robot selection factor weightings. The Chu and Lin [6] showed the limitations of the method proposed by Liang and Wang [5] and proposed a new technique fuzzy-TOPSIS method for robot selection process. However, Liang et al. had modified the objectives for the robot selection factors into fuzzy values which actually violates the fundamental rule of fuzzy logic. Further, a 5 point scale was utilized for the rating of robots under the subjective factors. Furthermore, the fuzzy logic method is really complicated and needs huge processing power. In the similar context, Agrawal et al. [7] have proposed a multiple-attribute-decision-making (MADM) approach with 'TOPSIS' for the selection of a industrial robot i.e. by following four attributes as well as five alternative robots. Similarly, Rao et al. [8] have proposed a digraph with matrix method for industrial robot selection process. Mainly, four attributes have been justified by Agrawal et al. [7] for a given industrial application as well as five robots have been shortlisted. In this paper, the required attributes used are same as of the method proposed by Agrawal et al. [7]. Moreover, these parameters are load capacity, repeatability error, vertical reach, degree of the freedom and higher quantitative values. However, for qualitative attributes smaller values are desirable. This was obtained from the robot selection digraph, which was based on various selection attributes as well as their relative importance. This method will be uncomfortable if the decision maker is unfamiliar with the use of graph theory and matrix methods. Parkan and Wu [9] made particular emphasis on a performance analysis technique called as Operational Competitiveness RAting (OCRA). The ultimate selection was made by averaging the results of TOPSIS, OCRA, and utility based model. Suprakash Mondal and S. Chakraborty, presented [10], four models of data envelopment analysis (DEA), specified additive, and cone-ratio models with respect to cost and process optimization. Also, multi-attribute decision-making concept has been employed in arriving at the best robot selection. The main objective of the industrial robot selection method is to identify the robot selection factors and to obtain the most appropriate combination. Efforts need to be extended using a suitable logical technique to eliminate unsuitable type of robots and to choose the most appropriate robot. In this article, we have proposed the robot selection methodology using the gradient descent with momentum based back-propagation neural network technique.

2 Proposed Methodology

In this section, we have discussed two things. In the first part, we have discussed the proposed method gradient descent with momentum based back-propagation neural network algorithm used for selection of industrial robots. In the second part proposed workflow for the optimized way of selecting the rank of the robot using the proposed method gradient descent with momentum based back-propagation neural network algorithm. Proposed gradient descent with momentum based back-propagation neural network algorithm for robot selection.

2.1 Proposed Gradient Descent with Momentum Based Back-Propagation Algorithm for Robot Selection

Industrial robot selection models are complex nonlinear systems that can be solved using robust estimation methodologies like neural network algorithms. In this work, neural network pattern classification technique is proposed for the prediction of manipulator attributes, i.e. quantitative attributes as well as qualitative attributes. A feedforward neural network is trained using different training functions that update weight and bias values [11]. In this work training algorithms such as gradient descent with momentum are studied for the implementation of robots selection prediction techniques.

Stochastic Gradient Descent (SGD) addresses the negative gradient of the target after considering only few training sets. This makes use of SGD optimization technique in the neural network by the costliest of implementing back propagation on the complete training sets.

The standard SGD updates the parameter θ of the required objective J(θ) as:

$$\theta = \theta - \alpha \nabla \theta E\{J(\theta)\} \tag{1}$$

Where, the probability of the above mentioned equation approximates by estimating the cost as well as gradient over the complete training sets. SGD merely does away with the anticipation in the bring up to date and compute the gradient of the parameters utilizing only a single or an only some training cases. The latest update is given by, $\theta = \theta - \alpha \nabla \theta J(\theta; x(i), y(i))$ with a pair of (x(i), y(i)) from the training set.

Momentum is another technique for obtaining the purpose more rapidly with the shallow abyss.

Momentum update is specified by the following equation,

$$v = \gamma v + \alpha \nabla \theta J(\theta; x(i), y(i)) \tag{2}$$

$$\theta = \theta - v \tag{3}$$

Here, v is the current velocity vector which is same dimension as parameter vector θ.

The learning rate α is expressed above, even though when using momentum α might need to be smaller because the magnitude of the gradient shall be larger. Lastly $\gamma \in (0,1)$ resolves the how many iterations the preceding gradients integrated into the current update. In general γ should be chosen 0.5 until the initial learning becomes stable and then it should be increased to 0.9 or higher.

2.2 Proposed Workflow Design for Selection of Robot Rank

In general, a realistic robot must have minimum specifications which are equal to or better than equals to the minimal requirements for the desired application. The ranges of specifications of robot listed as shown in Table 1. It has been noticed that a robot with specifications all equal to or better than the minimal prerequisites of the application that may nevertheless fail to present the required performance during the complete process. This failure occurs as discussed in the beginning due to improper handling of the manufacturer's specifications. Table 1 summarizes the principal parameter requirements with its values for the selection of an industrial robot.

The comprehensive activities for selecting the rank of the robot are presented with a workflow diagram in Fig. 1. We have developed a robot ranking chart by considering the standard specifications for each robot rank. As per the industrial

Table 1 Principal parameters required for a robot

Sl. no.	Parameter	Values
1	Working envelope	Minimum 500 mm
2	Payload	MAXIMUM 100 kg
3	Repeatability	±0.1 mm
4	Work lot size (production rate per hour)	$\geq$ 25 tasks
5	Maximum tip speed	Minimum 255 mm/s
6	Degrees of freedom	$\leq$ 7
7	Controller type	$\leq$ 4
8	Actuator type	$\leq$ 3
9	Arm geometry	$\leq$ 10
10	Robot type-programming	$\leq$ 5

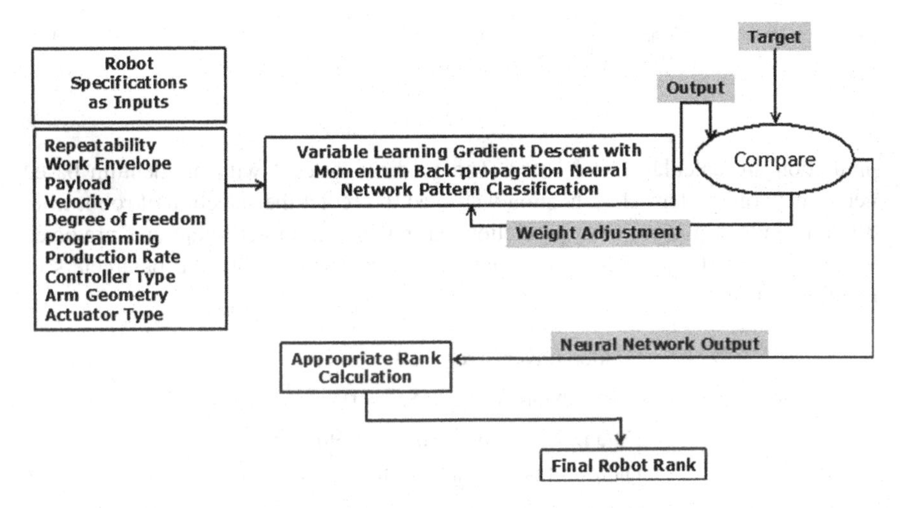

Fig. 1 Proposed architecture of the selection of robot rank methodology

requirement the ten numbers of principal parameters taken as inputs with different values. The result provides the output as robot rank. Following is the proposed robot classification only for eight categories of robot ranking listed in Table 2.

3 Results and Discussions

The overall performance of the optimization techniques for the prediction of selection of industrial robots is examined by considering ten manipulator attributes. The name of the inputs and output parameters used are listed in Table 2. Mean Square Error (MSE), Root Mean Square Error (RMSE), and R-squared error of the

Table 2 Proposed industrial robot ranking

Sl. no.	Name of robot parameter	Unit	Rank 1	Rank 2	Rank 3	Rank 4	Rank 5	Rank 6	Rank 7	Rank 8
1	Repeatability	±mm	5.5	5	4.5	4	3.5	3	2.5	2
2	Work envelop	mm	500	1000	1500	2000	2100	2200	2300	2400
3	Payload	kg	10	20	30	40	50	60	70	80
4	Velocity	mm/s	500	1000	1500	2000	2500	3000	3500	4000
5	Degrees of freedom	Nos	1	2	3	4	5	6	7	7
6	Production rate	Task/hour	100	200	250	300	350	400	450	500
7	Arm geometry	8 major types	1	2	3	4	5	6	7	8
8	Controller type	3 major types	1	1	1	2	2	3	3	3
9	Actuator types	3 major types	1	1	2	2	2	2	3	3
10	Programming	4 major types	1	1	2	2	3	3	4	4

prediction are calculated and listed. The gradient descent with momentum based neural network pattern classification was used to predict the selection of robot from different industrial data. The prediction error with the target value and predicted value are plotted here. The neural network is configured with the parameters as mentioned below.

$$\begin{aligned} &\text{Hidden Layer Size} = 10 \\ &\text{Training Objective MSE} = 0 \\ &\text{Data Division for Tarining} = 90/100 \\ &\text{Data Validation for Tarining} = 1/100 \\ &\text{Data Testing for Tarining} = 1/100 \end{aligned}$$

We observed from the 2nd plot of Fig. 2 that the actual result is matched correctly with the predicted values. The only rank 1, and 2 are having a minimum error, but others have all freest from error. The ranking prediction response and error curve is in Fig. 2.

The gradient descent with momentum based neural network (NN) pattern classification training performance is shown in Fig. 3. The best training performance is at 700 epochs.

Training state after neural network training based on the gradient descent with momentum technique in below Fig. 4. The best training performance obtained at 700 epochs.

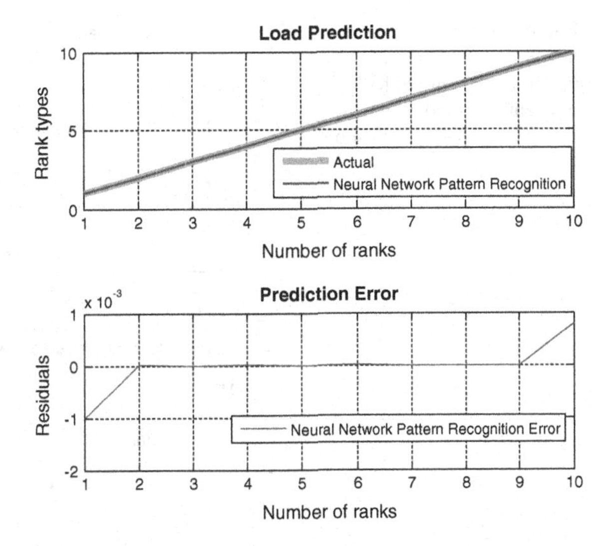

Fig. 2 Actual ranktype and predicted rank type v/s number of ranks. The error graph plotted between residuals and number of rank (Gradient descent with momentum based neural network pattern classification)

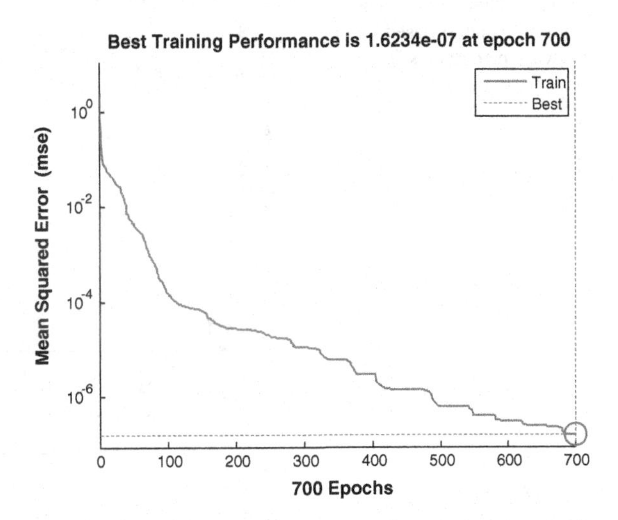

Fig. 3 Gradient descent with momentum based neural network (NN) pattern classification training performance

Error histogram after training performance by gradient descent with momentum based neural network pattern classification is in below Fig. 5. Instances vs. Error Histogram are 20 Bins.

The confusion matrix of gradient descent with momentum based neural network pattern classification in Fig. 6. The errors obtained from the simulations are RMSE = 4.0291e-04, MSE = 1.6234e-07, R-squared error = 1.0000. The test inputs and outputs of the proposed method are given in Table 3.

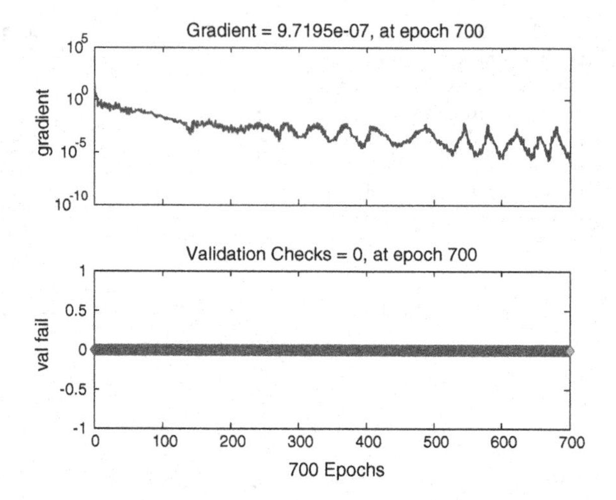

Fig. 4 Training state after neural network training (Gradient descent with momentum based neural network pattern classification)

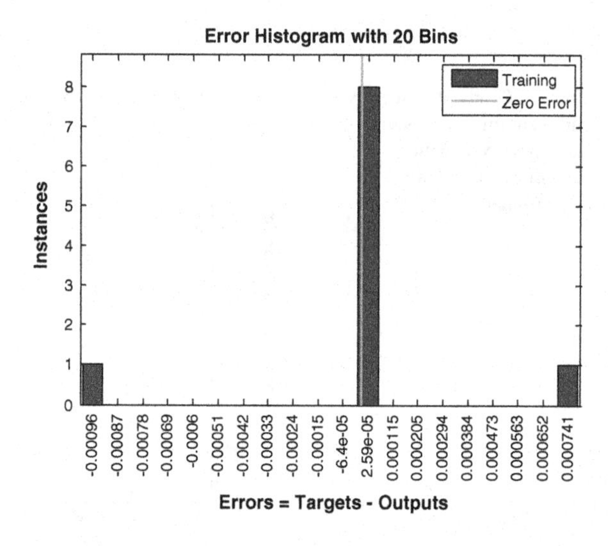

Fig. 5 Error histogram after training (Gradient descent with momentum based neural network pattern classification)

We have observed that our proposed method is very much consistent and produces qualitative results in comparison with the published methods. In the published papers, the methodology uses the least number of parameters as compared with the proposed method. In addition to the above facts, the proposed method also offers a more feasibility, easy to use, as well as simple robot selection approach with maximum numbers of the major parameters of the robot.

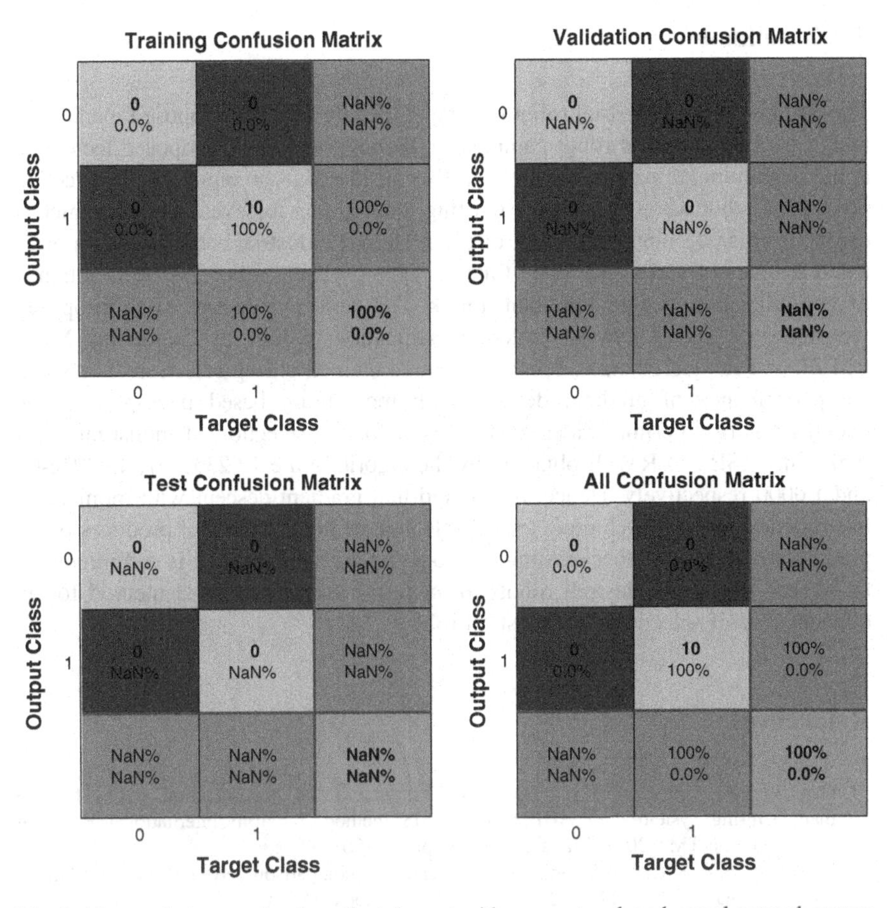

Fig. 6 The confusion matrix of gradient descent with momentum based neural network pattern classification training

Table 3 Test inputs and outputs considered for the proposed methodology

Test inputs	Desired rank (R)	Neural network pattern classification output
Repeatability (±mm) = 4	4	Neural network training type: gradient descent with momentum
Work envelope (mm) = 2000		
Payload (kg) = 40		Neural network classified rank: 4
Velocity (mm/s) = 2000		Final robot rank: 4
Freedom = 4		
Production rate (Task/hour) = 300		
Controller = 2		
Actuator = 2		
Arm geometry = cylindrical light		
Programming = task-oriented program		

4 Conclusion

The ranking of the industrial robot is done efficiently with the proposed method by taking various industrial robot parameters. Through the above-proposed technique at the maximum ten parameters are directly considered as an input for the selection process of robot whereas with the existing methods up to seven robot parameters can be used as an input. The rank of the preferred industrial robot has been evaluated seamlessly and at the same time the best probable robot has been obtained that specifies the most genuine benchmark. The performance analysis of proposed technique for neural network pattern classification is done by calculating MSE, RMSE, and R-squared error. From the errors obtained during selection shows that the performance of gradient descent with momentum based back-propagation neural network algorithm gives abetter result for the selection of industrial robot rank. The MSE and RMSE obtained by the algorithm are 1.6234e-07, 4.0291e-04 and 1.0000 respectively. Hence, it is found that gradient descent with momentum based neural network technique for the selection of industrial robot produces better prediction result than other existing methods. At the same time, it is recommended to all the clients of industrial robots to make use of the proposed method for an efficient way of selecting the industrial robots.

References

1. Özgürler, Ş., Güneri, A.F., Gülsün, B., Yılmaz, O.: Robot selection for a flexible manufacturing system with AHP and TOPSIS method. In: 15th International Research Conference on TMT-2011, Prague, Czech-Republic (2011)
2. Paul, R.P., Nof, S.Y.: Work methods measurement: a comparison between robot and human task performance. IFPR **17**(3), 277–303 (1979). Taylor&Francis
3. Vukobratovic, M.: Scientific Fundamentals of Industrial Robots1: Dynamics of Manipulator Robots Theory and Applications. Springer-Verlag Publication, New York (1982)
4. Khouja, M.: The use of data envelopment analysis for technology selection. Comput. Ind. Eng. **28**, 123–132 (1995)
5. Liang, G.H., Wang, M.J.: A fuzzy multicriteria decision-making approach for robot selection. Robot. Comput.-Aided Manuf. **10**, 267–274 (1993)
6. Chu, T.C., Lin, Y.C.: A fuzzy-TOPSIS method for robot selection. Int. J. Adv. Manuf. Technol. **21**, 284–290 (2003)
7. Agrawal, V.P., Kohli, V., Gupta, S.: Computer-aided robot selection: the multiple-attribute decision making an approach. Int. J. Prod. Res. **29**(8), 1629–1644 (1991)
8. Rao, R.V., Padmanabhan, K.K.: Selection, identification and comparison of industrial robots using digraph and matrix methods. Elsevier Robot. Comput. Integr. Manuf. **22**(4), 373–383 (2006)
9. Parkan, C., Wu, M.L.: Decision-making and performance measurement models with applications to robot selection. Comput. Ind. Eng. **36**(3), 503–523 (1999)
10. Mondal, S., Chakraborty, S.: A solution to robot selection problems using data envelopment analysis. Int. J. Ind. Eng. Comput. **4**(3), 355–372 (2013)
11. Rehman, M.Z., Nawi, N.M.: The Effect of Adaptive Momentum in Improving Accuracy of Gradient Descent Backpropagation Algorithm on Classification Problems, vol 179, pp. 380–390. Springer, Berlin (2011)

Network Security Analyzer: Detection and Prevention of Web Attacks

Nilakshi Jain, Shwetambari Pawar and Dhananjay Kalbande

Abstract In today's technology world one may have been attacked or witnessed cyber-attacks on their applications. Currently there are many systems that help you detect as well as prevent various kinds of attacks that your application may be vulnerable to. There is dire necessity to protect your Projects from these kinds of attacks. Using NSA tool, security can be implemented, one can detect and analyze if there is any attack taking place or there has been an attack. NSA helps in detecting all sorts of attacks ranging from databases to network. Cross-Site Scripting, SQL injection, URL rewriting, Buffer Overflow and Cross-Site Request Forgery are amongst the few that are found by NSA. Studies have also showed rapid rise in these attacks, it has become necessary to provide solution to protect the web applications against them. Use of firewalls along with NSA is one of the solutions to mitigate these attacks along with others.

Keywords URL rewriting · Cross-site scripting · SQL injection · Buffer overflow · Cross-site request forgery · Attacks · Security · IDS · IPS

1 Introduction

SQL injection is a technique which exploits the database layer of the web application through some loop hole that is the security vulnerability. When the parameters of the request are adjusted but maintaining the URL's syntax and slightly changing its semantic meaning you can break into any system. The websites that are CGI driven has be attacked the most by SQL Injection. Cookie poisoning is a similar attack that involves web browsers cookies [1]. As we know the database is the most important part of the application, as it contains data of immense use.

N. Jain (✉) · S. Pawar · D. Kalbande
Mumbai University, Mumbai, India
e-mail: nilakshijain1986@gmail.com

S. Pawar
e-mail: shwetambari.pawar.2014@gmail.com

S.C. Satapathy and S. Das (eds.), *Proceedings of First International Conference on Information and Communication Technology for Intelligent Systems: Volume 1*, Smart Innovation, Systems and Technologies 50, DOI 10.1007/978-3-319-30933-0_50

Cross-Site Scripting is a type of vulnerability that will allow the hacker to run his own script on the target's application. In a typical Cross-Site Scripting vulnerability the attacker injects the legitimate website with his malicious JavaScript. This bug called as Cross-Site Scripting (XSS) has affected many websites [2]. XSS can also steal personal data, and can also perform actions that may represent the actions of some third person.

URL rewriting is a feature in many web servers that modifies the URL or redirect to the requested URL to new URL [3]. The attackers use this technique to get data which they are not authenticated to access.

Buffer Overflow is another kind of attack that was discovered by the hackers and their community. A completely harmless application that is not implemented properly and which has rooted or administrative privileges are more prone to attacks like this.

Cross-Site Request Forgery is one of the most un-noticed attacks that prevail amongst top 5 attacks of OWASP. It is very easy for the attacker to exploit the victim in this attack. Cross-Site Request Forgery attacks are also known as Cross-Site Reference Forgery, XSRF, and Session Riding and Confused Deputy attacks [4].

2 SQL Injection

SQL Injection Interactive web applications that have employed database services are common for SQL injection attack. A request to the database would be made by the attacker by providing a SQL query that is malicious and get access to the application. Information that is confidential or sensitive can thus reach the hands of attacker causing him to modify the same [5]. When the developer does not have much of training and experience in development, the developer may tend to do mistakes in the use of SQL query which might lead to attack by the attacker [6].

The parameters that access the database should be secured properly else if they are accessed by the attacker exploitation using SQL injection can take place. The malicious query will be sent to the database by the attacker by appending the malicious SQL commands into the parameters [7]. By observing the behaviour of the application and by trying different patterns, the attacker can reconstruct the data [8].

A typical SQL query looks like this:

Select id, email, name from members

This statement will retrieve the 'id', 'email', and 'name' from the table called 'members'.

Attacker with a very good SQL knowledge can insert a SQL into the main query by inputting a value like:

Name: 'an'dy'

The query string will look like this:

Select id, email, name from members where name = ‘an’dy’

This query in the database server will return an error as such given below and the fact that this will happen is due to the fact that the single quote character will break the single quote delimited data in the main query.

Line 1: Incorrect syntax near ‘dy’.

By looking at this the attacker will understand that this application is SQL injection vulnerable and he/she will launch the main attack by inputting the following:

Name: ‘an’; drop table members—

With this input the single quote after ‘an’ will close the quote that was open in the original query and “drop table” will remove the table members from the database. The two dashes that are there after the query will comment the remaining query and it will not be executed [9].

3 Cross-Site Scripting

If Cross-Site Scripting is a computer security vulnerability that is very well known which is associated with web applications. This vulnerability allows the hackers to inject client-side scripts onto the web applications which are malicious in nature [10]. Computer Emergency Response Team (CERT) was the first to discuss about Cross-Site Scripting in the year 2003 [11]. According to the severity of the threats, Cross-Site Scripting was at the second number as per the findings of the OWASP (Open Web Application Security Project). Cross-Site Scripting scenario is depicted in the Fig. 1 [12].

Cross-Site Scripting (XSS) has two types of attacks:-

(1) Persistent XSS Attack.
(2) Non-Persistent XSS Attack.

(1) Persistent XSS Attack: Code that will run permanently on the web pages because the malicious code that was submitted by the attacker was saved in the database. This type of attack is termed into Persistent XSS Attack [3]. For instance the attacker had code the script to steal cookie. With this the attacker can gain control of one’s account by the use of cookie. Persistent XSS Attack is also called as Stored XSS Attack [13].
(2) Non-Persistent XSS Attack: This type of attack makes use of the HTTP Request to send the injected code to the server. The file that is the HTTP Response is returned to the browser by the server with the input with the HTML file. If the server has not properly validated the input then the result will be the injected script. This type of attack is also called as Reflected XSS Attack [13].

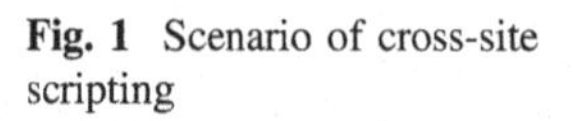

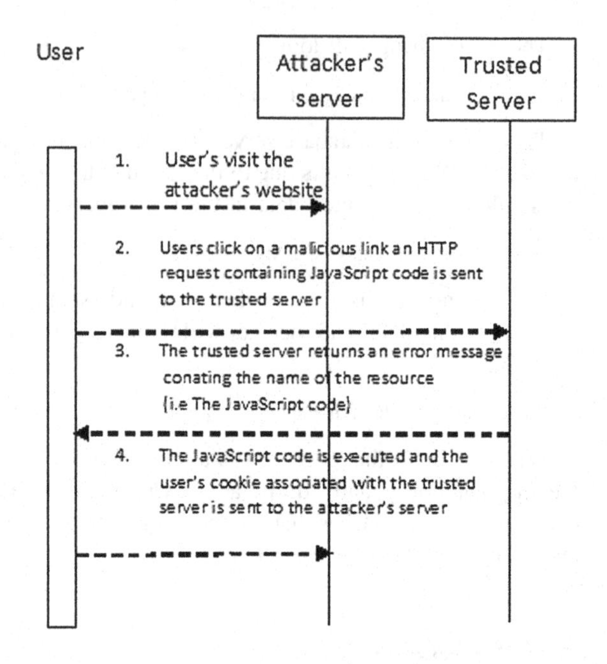

Fig. 1 Scenario of cross-site scripting

DOM (Document Object Model)—based Attack: Document Object Model based Attack is also called as "Type-0 XSS". Here the already present JavaScript file while loading sends the URL parameter value. When the website that is on the user's browser suffers DOM modification it results in XSS.

4 URL Rewriting

A technique that can change unsightly URLs into the URL those give vital or required data. Here by slightly modifying the URL the attacker can obtain the required information of his interest. Suppose we open a particular application over the network, we get the URL as follows:

http://www.xyz.com/page3.php?id=14

Now this URL will take us to the page whose name is page3.php and will fetch the details whose id is 14 and which is permissible to be viewed by that particular user. The attacker will take advantage of this condition and modify the URL to

http://www.xyz.com/page3.php?id=1

Now in this URL the attacker is not authenticated to get the details of id = 1. The attacker is authorized to only get the details of id = 14, but because of URL rewriting the attacker gets the information of id = 1 which might be the admin data or data of great importance. Advantage of this situation will be taken by the attackers.

Different types of URL that can be compromised are as follows:

1. Action URL
   a. URLs for client facing actions.
   b. Portlet environment makes use of this term more frequently.
2. Render URL
   a. URLs for client facing renders.
   b. Again Portlet environment makes use of this term frequently.
3. Service URL
   a. URLs that are generally used to reach services for server-facing.
4. Resource URL

Resources like images, JavaScript and CSS are used as client-facing resources.

5 Buffer Overflow

Buffer overflow attack is one of the most dangerous category of attacks because of its ability of inject or execute the attack code [14]. Buffer Overflow attacks are very easy to perform and there are many tools and techniques that demonstrate Buffer Overflow attacks. To understand Buffer Overflow attack it is necessary to understand basics about stack.

Stack Overflow and Heap Overflow attacks are part of buffer overflow attacks. In Stack overflow attack, the attacker will try to place the address of the top stack in place of the return address. The level of privilege that these lines of code will execute will be the same as that of the program. This is an advantage for the attacker as they have to just transmit small script of program [14].

Along with the use of stack some programs make use of dynamically allocated memory. Once this behavior of the program is noticed by the attacker, they will try various types of inputs to corrupt the stack.

6 Cross-Site Request Forgery

Cross-Site request Forgery is an attack that forces a user's browser to send request that the victim did not intend to make. But here that attacker must have some prior knowledge of which applications are vulnerable.

Cross-Site request Forgery attack can take by invisible image tags and by forms.

1. Invisible Image Tags:

```
<img
scr=http://xyzbank.com/transfer?accountfrom=ABC&accountto=ATTACKER&amount=100
00 width="1" height="1">
```

2. Forms:

```
<form name=form1 method="POST" action="http://xyzbank.com/transfer">
<input type="hidden" name="accountfrom" value="ABC">
<input type="hidden" name="accountto" value="ATTACKER">
<input type="hidden" name="amount" value="10000">
</form>
<script>document.form1.submit()</script>
```

Then the user visits a site hosting Gmail Cross-Site request Forgery attack code.

7 The Proposed Approach

The proposed approach is with the use of NSA software. These signatures are already so widely used that they can be easily coded into NSA. One can code new identified signatures also in XML.

The system architecture shows a central manager receiving events from the agents and system logs from remote devices. Active responses can be executed, when something is detected, then the admin is notified. Architecture is as shown below in Fig. 2.

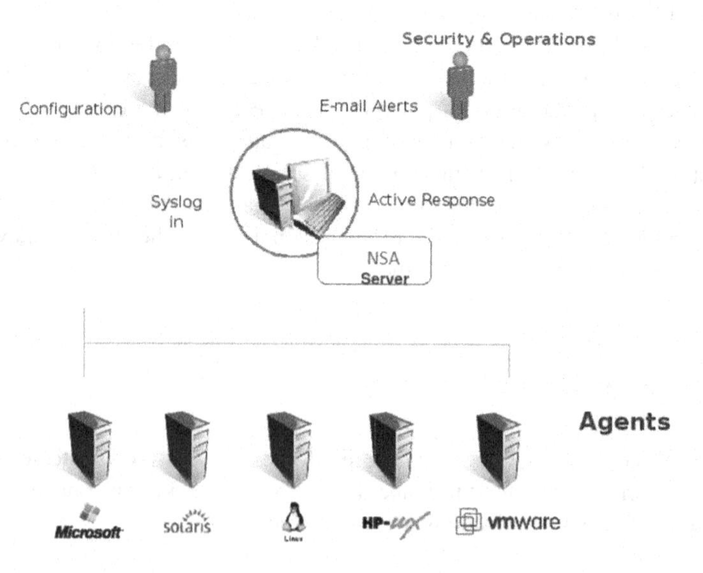

Fig. 2 Architecture of the system

The above system is composed of multiple pieces. Anything and everything is monitored by the central manager and information is received from the databases, syslog, agents and also from the agentless devices.

NSA also comprises of Manager, Agents, Agent Security, Agentless systems and systems with Virtualization/Vmware.

1. Manager: The manager is the central piece of the NSA deployment. Manager stores the events and system auditing entries, file integrity checking databases, the logs.
2. Agents: The systems you desire to monitor are monitored by a small program which is installed on the systems is called agents. Agent will collect information on real time and this information is forwarded for analysis and correction to the manager. Agent has CPU footprint and a very small memory by default, not affecting with the system's usage.
3. Agent Security: Agent Security runs with a low privilege user (created during the installation) and inside a chroot jail which is isolated from the system. If in case these local options are changed, this information will be received by the manager and will generate an alert.
4. Agentless: For systems that you can't install an agent, File integrity monitoring is allowed by NSA even without the agent installed. These systems can be very useful to monitor routers, firewalls and also Unix systems which do not allow to install the agent.
5. Virtualization/Vmware: Agent on guest operating system or even inside the host (Vmware ESX) is allowed to install by NSA. NSA performs the CIS checks for Vmware, and if there is any issue or any insecure configuration option is enabled or anything else like this, then alerts are sent.

The basic Web Attack Detection Algorithm:

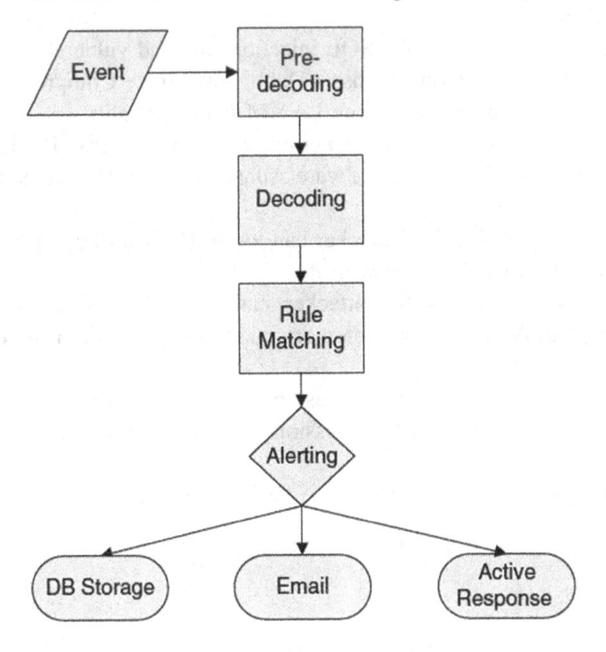

As soon as an event is received, the Security Analyzer System tries to decode and extract any relevant information from it. Decoding or normalization of the event is done in two parts, called pre-decoding and decoding.

Basic algorithm for attack detection using NSA is as follow:

1. Auto start the detector.
2. Run the detector in background using '&' operator.
3. Monitor the URL.
4. Monitors the GET and POST requests.
5. Use URL encoding to avoid simple respective attacks requests.
6. Match every request with XML file using pattern matching.
7. Look for patters of respective attacks.
8. If success for a particular attack, go to 9 or go to 3.
9. Block potential harmful request.
10. Take IP address from HTTP request.
11. Make entry into the log for that particular attack.
12. Notify on admin homepage for attack alert.
13. Go to 3.

8 Conclusion

From the above discussion we can see that various attacks ranging from web to databases can be tackled with the help of NSA and signature matching and can be easily notified to the admin. NSA is a system which identifies many attacks together with the help of signatures.

If these attacks are not taken care of then SQL injection can lead vulnerabilities like Bypassing Authentication, Compromise Data Integrity, Compromise Availability of Data, Remote Control Execution, Loss of Confidentiality.

If Cross-Site scripting attack takes place attacker can Steal the Identity, Bypass restrictions in websites, Session Hijacking, Malware Attack, Website Defacement, Denial of Service Attacks with it.

If URL rewriting attack takes place the attacker can loose the data of company, Redirect without authenticity, Steal Identity with it.

If Buffer Overflow attack happens the attacker can have Access Control, Availability of vital data, and Access to main memory and –program code, Loss of Confidentiality.

If Cross-Site Request Forgery attacks takes place attacker can Bypass Authentication Mechanism, Identity theft, Session Hijacking and Loss of Confidentiality.

NSA is one of the simplest yet effective ways to handle the attacks.

References

1. Sun, Y., He, D.: Model checking for the defense against cross-site scripting attacks. In: International Conference on Computer Science and Service (CSSS), 2161–2164 August 2012
2. Malviya, V.K., Saurav, S.: On security issues in web applications through cross site scripting (XSS). In: 20th Asia-Pacific Conference on Software Engineering (2013)
3. Singh, A., Ranjan, R.: A dynamic web caching tecquine for using URL Rewriting. Vol. 14, 181–189 December 2010
4. Zeller, W., Felten, E.W.: Cross-site request forgeries: exploitation and prevention (2010)
5. Wei, K., Muthuprasanna, M., Kothari, S.: Preventing SQL Injection attacks in stored procedures. In: Conference on Software Engineering, Australian, April 2006
6. Shar, L.K., Tan, H.B.K.: Defeating SQL Injection, 69–77 March 2013
7. Sadeghian, A., Zamani, M., Ibrahim, S.: SQL Injection is still alive:a study on SQL Injection signature evasion techniques. In: International Conference on Informatics and Creative Multimedia (ICICM), 265–268 September 2013
8. Focardi, R., Luccio, F.L., Squarciana, M.: Fast SQL blind injections in high latency networks. In: European Conference on Satellite Telecommunictaions (ESTEL), 1–6 October 2012
9. Dabbour, M., Alsmadi, I., Alsukhni, E.: Efficient assessment and evaluation for websites vulnerabilities using SNORT. Int. J. of Secur. and Its Appl. **7**, 7–16 (2013)
10. CERT advisory CA-2000-02.Malicious HTML Tags Embedded in Client Web Requests, February 2000
11. Yusof, I., Pathan, A.S.K.: Preventing presistent cross-site scripting (xss) attack by applying pattern filtering approach. Information and Communication Technology for the Muslim World (ICT4M), 1–6, November 2014
12. OWASP Top 10: The ten most critical web application security risks (2013)
13. Sonewar, P.A., Mhetre, N.A.: A novel approach for detection of SQL injection and cross site scripting attacks, In: International Conference on Pervasive Computing (ICPC), 1–4 January 2015
14. Homoliak, I., Ovsonka, D., Koranda, K., Hanacek, P.: Characteristics of buffer overflow attacks tunneled in HTTP Traffic, In: Internatinational Carnahan Conference on Security Technology (ICCST), 1–6, 2014

Diagnosis of Glaucoma Using Cup to Disc Ratio in Stratus OCT Retinal Images

Kinjan Chauhan and Ravi Gulati

Abstract Glaucoma is one of the second most important cause of blindness after cataracts. Detection of glaucoma is essential to prevent visual damage. In India, Glaucoma is the third leading cause of blindness. This paper discusses an algorithm developed for detection and diagnosis of glaucoma. The algorithm calculates the Cup to Disc Ratio (CDR) which is obtained through image processing of the retinal images obtained from stratus OCT from Sudhalkar Eye Hospital, Vadodara. Image processing has been carried out on 120 retinal images obtained from Stratus OCT out which 79 are glaucomatous eyes and 41 are normal eyes. OTSU histogram and watershed algorithm have been used for image segmentation. Image analysis has been carried out through image segmentation, which is the process of dividing an image into regions or object. Morphological operations have been performed for the enhancement of the images to extract optic cup and optic disc region from the eye. The accuracy of the algorithm is 94 %. MATLAB software has been used for image processing on the retinal images.

Keywords Glaucoma · Cup to disc ratio (CDR) · Image processing · OTSU histogram · Segmentation · Morphological operation

K. Chauhan (✉)
Shree Ramkrishna Institute of Computer Education and Applied Sciences, Surat, Gujarat, India
e-mail: kinjanchauhan99@gmail.com

R. Gulati
Department of Computer Science, Veer Narmad South Gujarat University, Surat, Gujarat, India
e-mail: rmgulati@vnsgu.ac.in

S.C. Satapathy and S. Das (eds.), *Proceedings of First International Conference on Information and Communication Technology for Intelligent Systems: Volume 1*, Smart Innovation, Systems and Technologies 50, DOI 10.1007/978-3-319-30933-0_51

1 Introduction

Glaucoma is a condition that causes damage to your eye's optic nerve and gets worse over time. Glaucoma tends to be inherited and may not show up until later in life. The increased pressure, called intraocular pressure, can damage the optic nerve, which transmits images to the brain. If damage to the optic nerve from high eye pressure continues, glaucoma will cause permanent loss of vision. Without treatment, glaucoma can cause total permanent blindness within a few years.

Diagnosis of glaucoma is dependent on various findings such as elevated Intra Ocular Pressure (IOP > 23 mm Hg is considered as a suspicious case for glaucoma), optical nerve cupping and visual field loss. Detection and diagnosis of Glaucoma is performed through various tests such as Tonometry, Ophthalmoscopy, Perimetry, OCT, Gionoscopy and Pachymetry [1].

2 Research Background

In diagnosing glaucoma, IOP (Intra Ocular Pressure and CDR-Cup to Disc Ratio) are considered. IOP is not considered to be an accurate factor for glaucoma diagnosis. Glaucoma is currently diagnosed by considering CDR. CDR is calculated as shown below:

$$\text{CDR} = \text{vertical area of optic cup}/\text{vertical area of the optic disc}$$

As shown in Fig. 1b, the optic cup progresses as the cup becomes larger in comparison to the optic disc as shown in Fig. 1a. Generally if CDR is greater than 0.3 then it is considered to be a positive case of glaucoma [2].

Figure 2 shows retinal colour images of normal and glaucoma affected eyes.

Hatanaka et al. [3] in their paper have discussed an efficient method for detection of optic disc in colour retinal image. They have measured varying optic disc to cup diameter ratio for detection of the glaucoma in fundus images [4, 5].

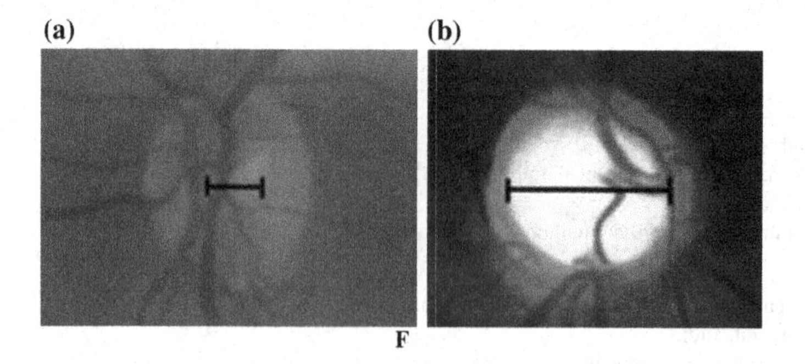

Fig. 1 **a** Normal optic nerve head. **b** Glaucomatous optic nerve head

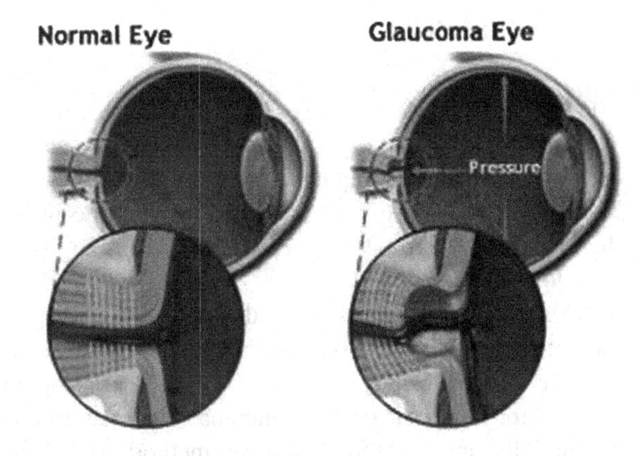

Fig. 2 A retinal image of normal and Glaucoma affected eye (*Source* http://yourbrainonbliss.com/Blog/?p=1467)

Optic Disc is a bright region where the optic nerve and blood vessels enter the retina while the cup is a depressed area inside the Optic Disc The optic disc position is used as a reference length for measuring the distances. Glaucoma progression has been evaluated by understanding the shape of deformation within Optic Disc [2, 6, 7].

Chauhan et al. [1] have proposed a framework for diagnosis and detection of glaucoma. The proposed framework will help the ophthalmologist in early detection of glaucoma thus helping the people for a better treatment before glaucoma progresses substantially.

Caroline Viola Stella Mary et al. [8] have segmented optic disc using active contour. The size of tested data was 169 and they achieved 94 % accuracy rate.

Dehghani et al. [9] have used DRIVE data base. Testing size for the system has been 40 images. Accuracy achieved is 91.36 %. Running time for the algorithm is 27.6 s which is quite high compared to the small dataset.

Burana-Anusorn et al. [10] have calculated CDR using variation level set method. They had tested on 44 eyes and their proposed method gave an accuracy rate of 89 %. They have used edge detection for extracting the optic disc. A colour component analysis method and threshold level-set method have been used to segment optic cup [11].

Tariq et al. [12] have used STARE and MESSIDOR database. The tests were performed on 1281 patients. They have used SVM for classification of glaucomatous and normal eyes with an accuracy rate of 94.37 %.

Arturo Aquino et al. [13] detected optic disc using circular Hough transformation. They have used two independent methodologies to detect optic disc in retina images. The optic disc has been located using circular approximation obtained through morphological operation, edge detection and circular hough transformation. Accuracy rate has been 86 %.

There are many other algorithms proposed in the literature that localizes the optic disc using vessel segmentation methods, though transformation for detecting geometric shapes using thresholding and morphological operations [2, 4, 5, 7, 14–17].

3 Methodology

The diagnosis of Glaucoma is carried out with the help of Perimetry and OCT. These test are more preferred as they have high sensitivity and specificity. OCT is an non-invasive technique which uses light to capture micrometer resolution. Whereas, Standard Automated white-on-white Perimetry (SAP) is used to examine the visual field. Perimetric tests aide ophthalmologist to diagnose significant visual field loss [1].

This research paper discusses the algorithm to diagnose and detect glaucomatous patient which can help ophthalmologists take necessary steps to prevent further damage to eyes. Cup to Disc ratio (CDR) is considered to be one of the important features for diagnosis of the glaucoma. Our method proposes the detection of optic cup and disc using OTSU histogram method [18], feature extraction has been done through morphological operations [19–21].

3.1 *Proposed Method Flow Chart*

The following flow chart shows the flow of the proposed method to distinguish between a normal eye and an eye affected by glaucoma.

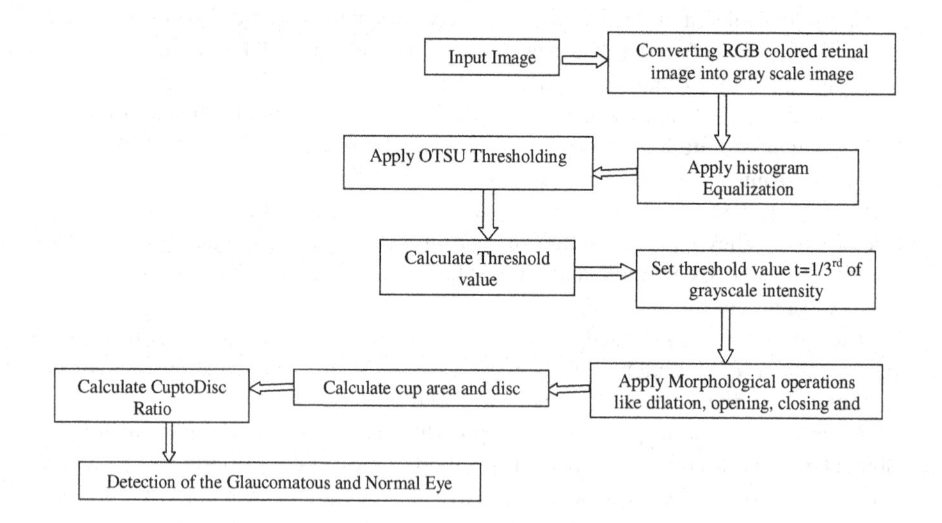

3.2 Image Pre-processing

To obtain uniform intensity in an image, retinal images are converted into grayscale image. Histogram equalization on each image has been applied to increase the contrast in image. To obtain a binary image from a grayscale image thresholding techniques are applied. Several popular methods are used such as: Maximum entropy, OTSU, K-means clustering [22].

In our image processing we have applied OTSU thresholding because Otsu's method automatically performs clustering-based image thresholding. It finds the threshold that minimizes the weight within-class variance and operates on the gray level histogram directly [e.g. 256 numbers, P(i)], so it's fast [9].

Thresholding is the simplest method for converting a color image or grayscale image into a binary image. In thresholding, pixels are assigned two levels which are above or below the threshold value specified [23].

The proposed algorithm calculates the optimum threshold value by separating the two pixel classes. Threshold is given as the total variance, which is the sum of the within-class variances (weighted) and the *between class variance*, which is the sum of weighted squared distances between the class means and the grand mean [9, 12].

The total variance is given as:

$$\sigma 2 = \sigma w2(t) + q1(t)[1 - q1(t)][\mu 1(t) - \mu 2(t)]2. \tag{1}$$

Threshold value t has been set to 1/3rd of the grayscale intensity.

3.3 Feature Extraction of Cup and Disc from the Images

To segment cup and disc from the retinal image, extracting features from the image has to be carried out. To obtain this features grayscale image is converted to binary image which holds two pixel value 0 or 1. After obtaining the binary image, various feature extraction technique such as blob analysis, connectivity analysis, morphological image processing.

We have applied morphological operations such as closing, erosion and opening to obtain a segment cup and disc area from the retinal image.

Morphological closing is performed on ROI to calculate the magnitude gradient of edge detection and fill the vessels according to [1].

$$\mathrm{f} \cdot \mathrm{B} = (\mathrm{f} \oplus \mathrm{B}) \Theta \mathrm{B}. \tag{2}$$

For removing any peaks, morphological erosion is applied according to (2).

$$f \cdot B = (f \Theta B) \oplus B \quad (3)$$

where f is the grayscale image, B is binary structuring element; $\oplus$ is dilation and Θ is erosion operator.

To remove noise and preserve the edges in an image median filtering has been applied to image. After filtering, thresholding has been carried out manually by selecting a threshold value, as the prior information about the region to be segmented is known. The optimum threshold value used is the top 1/3rd of the normalized grayscale intensity. Morphological opening is repeated in order to remove the unwanted pixels around the segmented optic disc [1, 3, 18, 19, 24].

3.4 *Cup to Disc Ratio (CDR)*

CDR is defined as the ratio of vertical cup diameter to vertical disc diameter. There are many indicators for diagnosis of glaucoma but CuptoDisc ratio is considered to be one the most important indicator. As glaucoma progresses the cup enlarges until it occupies most of the disc area. Thus increase in CDR indicates the progression in glaucoma condition. For normal eye it is found to be 0.3–0.5. As the neuro-retinal degeneration occur the ratio increases and at the CDR value of 0.8 the vision will be lost completely. CDR is obtained by taking the ratio of cup diameter and disc diameter in the vertical direction [1, 3].

$$\mathrm{CDR} = \text{area of cup}/\text{area of disc}$$

If CDR is greater than 0.3 then retinal image of the patient is diagnosed as Glaucoma else if CDR is less than 0.3 than retinal image of the patient is normal.

Psuedo code to diagnose Glaucoma using calculate CDR ratio:

```
START
        Step 1: I= Input the retinal image
        Step 2: Convert I into gray scaled image.
        Step 3: Add salt and pepper for noise removal.
        Step 4: Calculate threshold value t using grayscale intensity value.
                Set t= 1/3rd of the gray intensity value.
        Step 5: Disc area = segmented disc area.
        Step 6: Apply dilation and erosion morphological operation on the image.
        Step 7: Repeat step 6 until segmented cup is obtained
        Step 8: CDR = area of cup/ area of disc
        Step9: if CDR>0.3 and CDR <0.8 then
                GLAUCOMA IS DETECTED
         IF CDR <0.3
                NORMAL EYE DETECTED
```

4 Experimental Result

Total 120 retinal images obtained from stratus OCT were tested. Out which 79 were glaucomatous patient tested clinically by the ophthalmologist and 41 were normal patient. We have been able to achieve 94 % accuracy. Accuracy has been measured as per the following formulae.

$$\text{Accuracy} = (\text{TP} + \text{TN})/\text{Sample size}$$

where,

TP True Positive (A true positive is an observation, identified by your algorithm, which is a real instance of a feature).

TN True Negative (A true negative is the case where the observation is not a real instance of the feature and your algorithm identifies it as such).

The graphs are plotted for the CDRcalc from the proposed algorithm and CDR value obtained from Stratus OCT machine. Graphs indicate the sensitivity of the CDR value calculated from the algorithm. Thus diagnosis of glaucoma from the CDRcalc detects the glaucoma positively even for the lower value of CDR (Figs. 3 and 4; Table 1).

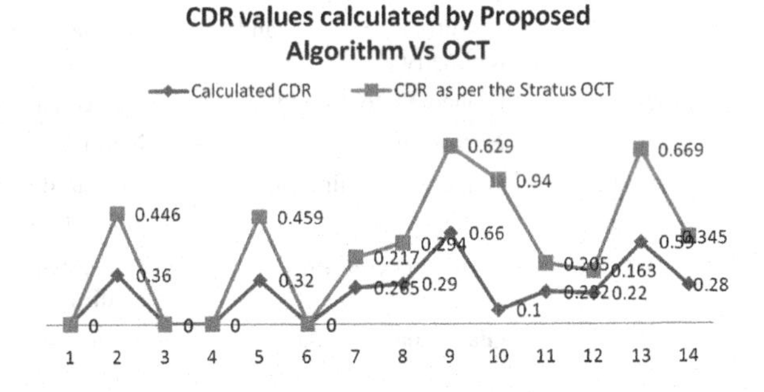

Fig. 3 Graph indicating sensitivity of CDRcalc from algorithm and Stratus OCT CDR value

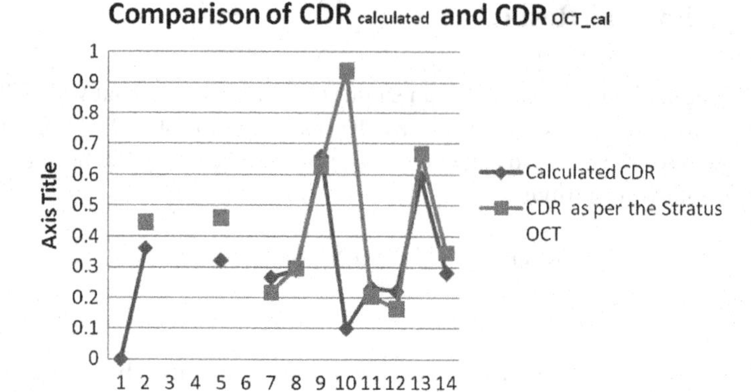

Fig. 4 Graph of CDRcalc versus CDR stratus OCT

Table 1 Comparative result of the algorithm and result obtained from OCT and perimetry test of the patient

Patient sr. no.	Calculated CDR as per proposed algorithm	CDR as per the stratus OCT	Clinically diagnosis of ophthalmologist based on OCT and perimetry results	Results of algorithm
1	0.36	0.446	Glaucoma confirmed	Glaucoma detected
2	0.32	0.459	Glaucoma confirmed	Glaucoma detected
3	0.265	0.217	Normal eye	Normal detected
4	0.290	0.294	Result vary in OCT and perimetry	Normal detected
5	0.66	0.629	Glaucoma confirmed	Glaucoma detected
6	0.1	0.94	Normal eye	Normal detected
7	0.232	0.205	Glaucoma confirmed	Normal detected
8	0.22	0.163	Normal eye	Normal detected
9	0.59	0.669	Glaucoma confirmed	Glaucoma confirmed
10	0.28	0.345	Glaucoma confirmed	Normal detected

5 Conclusion

An efficient algorithm (technique) for glaucoma diagnosis is proposed in this work. The image based analysis increases the potential for classifying the retinal images inputted. The pre-processing of image and feature extraction of optic cup and disc has been carried out. After obtaining the result, Cup to Disc ratio has been calculated. The efficiency of the algorithm has been calculated to 94 %. The future work is to work on the improvisation of the proposed algorithm. Classification of the Glaucomatous eyes based on degree of the damage due to glaucoma can be done so that preventive measures can be taken well in advance.

References

1. Chauhan K., Chauhan P., Gulati R.: Diagnosis system using data mining approach for Glaucoma (a social threat). In: 2012 IEEE Conference on Technology and Society in Asia (T&SA), pp. 1, 4, 27–29 Oct 2012. doi: 10.1109/TSAsia.2012.6397985. url: http://ieeexplore.ieee.org/stamp/stamp.jsp?tp=&arnumber=6397985&isnumber=6397962
2. Pruthi, J., Mukherjee, S.: Computer based early diagnosis of glaucoma in biomedical data using image processing and automated early nerve fiber layer defects detection using feature extraction in retinal colored stereo fundus images. Int. J. Sci. Eng. Res. **4**(4), Apr 2013. ISSN 2229-5518
3. Hatanakaet, Y., et al.: Vertical cup-to-disc ratio measurement for diagnosis of glaucoma on fundus images. In: Karssemeijer, N., Summers, R.M. (eds.) Medical Imaging 2010: Computer-Aided Diagnosis, Proceedings of SPIE, vol. 7624, 76243C © 2010 SPIE CCC code: 1605-7422/10/$18. doi:10.1117/12.843775
4. Vasanthi, S.: Segmentation of optic disc in fundus images. Indian J. Comput. Sci. Eng. (IJCSE), **3**(2), 230–234 (2012). ISSN: 0976-5166
5. Jacob, E., Venkatesh, R.: 2A method of segmentation for glaucoma screening using superpixel classification. Int. J. Innov. Res. Comput. Commun. Eng. **2**(1) (2014). ISSN (Online): 2320-9801 ISSN (Print): 2320-9798
6. Lu, S., Lim, J.H.: Automatic optic disc detection from retinal images by a line operator. IEEE Trans. Biomed. Eng. **58**(1), 88–94, (2011). doi: 10.1109/TBME.2010.2086455. Epub 2010 Oct 14.
7. Welfer, D., Scharcanski, J., Kitamura, C.M., Dal Pizzol, M.M., Ludwig, L.W.B., Marinho, D. R.: Segmentation of the optic disk in color eye fundus images using an adaptive morphological approach. Comput. Biol. Med. **40**, 124–137 (2010)
8. Mary, C.V.S.M., Marri, J.S.B.: Automatic optic nerve head segmentation for glaucomatous detection using hough transform and pyramidal decomposition. In: IJCA Proceedings on International Conference in Recent trends in Computational Methods, Communication and Controls (ICON3C 2012), ICON3C(1), 33–37, Apr 2012.
9. Dehghani, A., et al.: Optic disc localization in retinal images using histogram matching, EURASIP J. Image Video Process., **2012**(19), (2012). doi: 10.1186/1687-5281-2012-19, http://jivp.eurasipjournals.com/content/2012/1/19
10. Burana-Anusorn, C., et al.: Image processing techniques for glaucoma detection using the cup-to-disc ratio. TIJSAT
11. Tariq, A., Akram, M.U., Shaukat, A., Khan, S.A.: Automated detection and grading of diabetic maculopathy in digital retinal images. Soc. Imaging Inf. Med. J Digit Imaging **26**, 803–812 (2013). doi:10.1007/s10278-012-9549-4
12. Niemeijer, M., Abramoff, M.D., Ginneken, B.V.: Segmentation of the optic disc, macula and vascular arch in fundus photographs. IEEE Trans. Med. Imaging **26**(1), 116–127 (2007)
13. Youssif, A.A.H.A.R., Ghalwash, A.Z., Ghoneim, A.S.A.R.: Optic disc detection from normalized digital fundus images by means of a vessels' direction matched filter. IEEE Trans. Med. Imaging **27**(1), 11–18 (2008)
14. Jayanthi, G., et al.: Glaucoma detection in retinal image using medial axis detection and level set method. Int. J. Comput. Appl. (0975–8887) **93**(3) (2014)
15. Darsana, S., Nair, R.M.: Mask image generation for segmenting retinal fundus image features into ISNT quadrants using array centroid method. IJRET Int. J. Res. Eng. Technol. **03**(04) (2014) ISSN: 2319-1163|pISSN: 2321-7308, Available @ http://www.ijret.org
16. Meier, J., Bock, R., Michelson, G., Nyull, L.G., Hornegger, J.: Effects of preprocessing eye fundus images on appearance based glaucoma classification
17. Burana, C., et al.: Image processing techniques for glaucoma detection using the cup-to-disc ratio. Thammasat Int. J. Sci. Technol. **18**(1) (2013)
18. Chauhan, K., Gulati, R.: Pre-processing of retinal image and image segmentation using OTSU histogram. Int. J. Adv. Inf. Sci. Technol. (IJAIST), **29**(29) (2014). ISSN: 2319:2682

19. Li, H., Chutatape, O.: Automated feature extraction in color retinal images by a model based approach. IEEE Trans. Biomed. Eng. **51**(2), 246–254 (2004)
20. Chauhan, K., Gulati, R.: A survey on various image processing techniques for glaucoma diagnosis. Int. J. Manage. Inf. Technol., **5**(1), 393–396, (2013). ISSN 2278–5612
21. Chauhan, K., Gulati, R.: A proposed framework for diagnosis of Glaucoma—a data mining approach. Int. J. Eng. Res. Dev **3**(5), 06–09 (2012). e-ISSN: 2278-067X, p-ISSN: 2278-800X
22. Otsu, N.: A threshold selection method from gray level histograms. IEEE Trans. Syst. Man Cybern. **9**(1), 62–66 (1979)
23. Aquino, A., Gegundez-Arias, M.E., Marin, D.: Detecting the optic disc boundary in digital fundus images using morphological, edge detection, and feature extraction techniques. IEEE Trans. Med. Imaging **29**(11), 1860–1869 (2010)
24. Yu, H., Barriga, E.S., Agurto, C., Echegaray, S., Pattichi, M.S., Bauman, W., Soliz, P.: Fast localization and segmentation of optic disk in retinal images using directional matched filtering and level sets. IEEE Trans. Inf Technol. Biomed. **16**(4), 644–657 (2012)

Improvement Power System Stability Using Different Controller in SMIB System

Ruchi Sharma, Mahendra Kumar and Kota Solomon Raju

Abstract The paper presents various controllers for damping low frequency oscillations in a single-generator infinite-bus (SMIB) electrical power system. The intent of the Fuzzy Logic based UPFC controller systems are to dampout low frequency oscillations. UPFC controller based ahead amplitude modulation index of exciter m_E has been intended. System response with Damped-UPFC controller and PI-fuzzy logic based UPFC controllers are compared at various loading environment. Relevant models have been designed and simulated in Matlab/Simulink version R2013a. The hybrid PI-Fuzzy Logic based UPFC controller is developed by selecting suitable controller parameters based on the experience of the power system performance. This paper also presents the TLBO (Teacher Learner based Optimization) based PSS design. The simulation results of these models show that the TLBO based PSS modeled has an excellent capability in damping low frequency oscillations on power systems.

Keywords Dampout controller · Hybrid controller · Hybrid PI-fuzzy logic controller · Fuzzy logic · Low frequency oscillations · PI controller · TLBO (Teacher learner based optimization), UPFC

R. Sharma (✉)
Electronics Department, Vivekanand Global University, Jaipur, India
e-mail: ruchisharma2k6@gmail.com

M. Kumar
Electronics Department, Rajasthan Technical University, Kota, India
e-mail: miresearchlab@gmail.com

K.S. Raju
Electronics Department, CSIR-CEERI, Pilani, India
e-mail: kota_solomonraju@yahoo.co.uk

S.C. Satapathy and S. Das (eds.), *Proceedings of First International Conference on Information and Communication Technology for Intelligent Systems: Volume 1*, Smart Innovation, Systems and Technologies 50, DOI 10.1007/978-3-319-30933-0_52

1 Introduction

Now recent years, in the power system design high efficiency operation and reliability of the power systems have been considered more than previous. Due to the growth in consuming electrical energy, the maximum capacity of the transmission lines should be increased. Therefore in a normal condition also the stability as well as the security is the major part of discussion. Several years the power system stabilizer act as a common control approach to damp the system oscillations [1, 2].

However, in some operating conditions, the PSS may fail to stabilize the power system, especially in low frequency oscillations [3]. As a result; other alternatives have been suggested to stabilize the system accurately. It is proved that the FACTS devices are very much effective in power flow control as well as damping out the swing of the system during fault. Recent years lots of control devices are implemented under the FACTS technology [1, 2].

By implementing the FACTS devices gives the flexibility for voltage stability and regulation also the stability of the system by getting proper control signal [4]. The FACTS devices are not a single but also collection of controllers which are efficiently not only work under the rated power, voltage, impedance, phase angle frequency but also under below the rated frequency. Among all FACTS devices the UPFC most popular controller due to its wide area control over power both active and reactive, it also gives the system to be used for its maximum thermal limit.

It's primarily duty to control both the powers independently. It has been shown that all three parameters that can affect the real power and reactive power in the power system can be simultaneously and independently controlled just by changing the control schemes from one type to other in UPFC. Moreover, the UPFC is executed for voltage provision and transient stability improvement by suppressing the sub-synchronous resonance (SSR) or LFO [5]. For example, in it has been shown that the UPFC is capable of inter-area oscillation damping by means of straight controlling the UPFC's sending and receiving bus voltages.

Therefore, the main aim of the UPFC is to control the active and reactive power flow through the transmission line with emulated reactance. It is widely accepted that the UPFC is not capable of damping the oscillations with its normal controller. A well-modeled UPFC controller can not only improve the transmission capability but also increase the power system stability and evaluate the performance of UPFC controllers with and without fuzzy logic controller.

Rest of this paper is organized as follows. System description contains SMIB, UPFC fundamentals are explained in Sect. 2. Section 3 gives detailed explanation about the Fuzzy logic controller. In Sect. 4, the experimental setup, results and corresponding performance evaluations are given. Finally, Sect. 5 concludes the proposed system and results are acceptable.

2 System Description

2.1 *Single Machine Infinite Bus System*

This system consist of a synchronous generator which is linked via two transformer to n infinite bus system through a transmission line. It is seen that the SM connected to the infinite bus always concerned with the frequent load change and it may leads to be serious stability problem and should be discussed.

2.2 *Unified Power Flow Controller*

Unified power flow controller (UPFC) consist of two voltage source converters (VSC), one shunt connected and the other series connected. VSC1 along with its transformer operates as STATCOM and VSC2 along with its transformer operates as SSSC. UPFC enables control of both real and reactive power flow in the line. UPFC can be connected either at sending end or receiving end of the line [6] (Figs. 1 and 2).

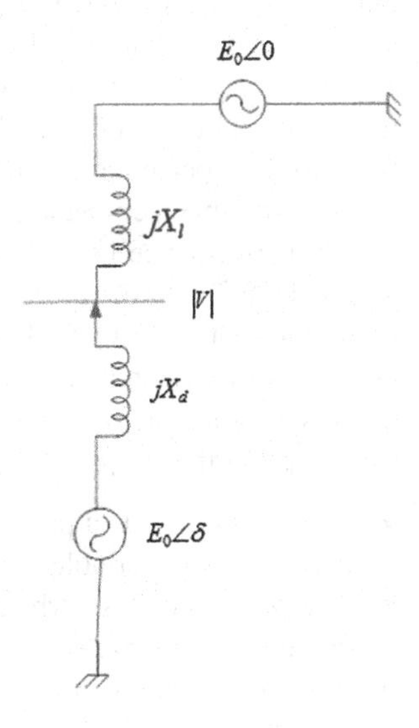

Fig. 1 SMIB power system

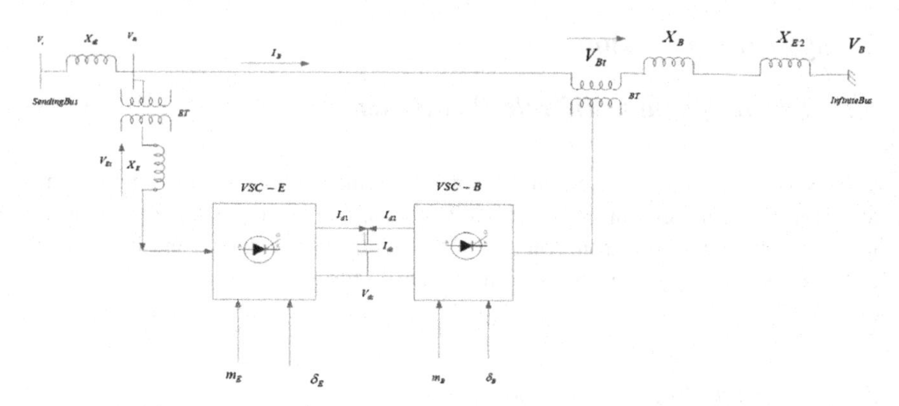

Fig. 2 UPFC installed in SMIB system

3 Fuzzy Logic Controller

In order to providing stabilizer signal, the output of obtained model reference of power system is compared with output of real power system and the error signal is fed to a fuzzy controller [7]. The Fuzzy controller provides stabilizer signal in order to damping system oscillations. In order to providing stabilizer signal, the output of obtained model reference of power system is compared with output of real power system and the error signal is fed to a fuzzy controller [7].

The Fuzzy controller provides stabilizer signal in order to damping system oscillations. FLC work on the principle of simple understanding of the system behavior of a person and simple rule based "If x and y then z", this rule base again defined by some membership function of FLC with proper argument to enhance the system performance [8–10].

The UPFC with Fuzzy controller is shown in the figure. Inaccurate, noisy, or lost input information FLC work on the principle of simple understanding of the system behavior of a person and simple rule based "If x and y then z", this rule base again defined by some membership function of FLC with proper argument to enhance the system performance [8–10] (Fig. 3).

- A Fuzzyfication is a process or platform in which we can convert the input data into linguistic variable.
- A Knowledge Base which contains the data base with the required linguistic definitions and control rule set.
- A Defuzzyfication interface which yields a non-fuzzy control action after an incidental fuzzy control action.

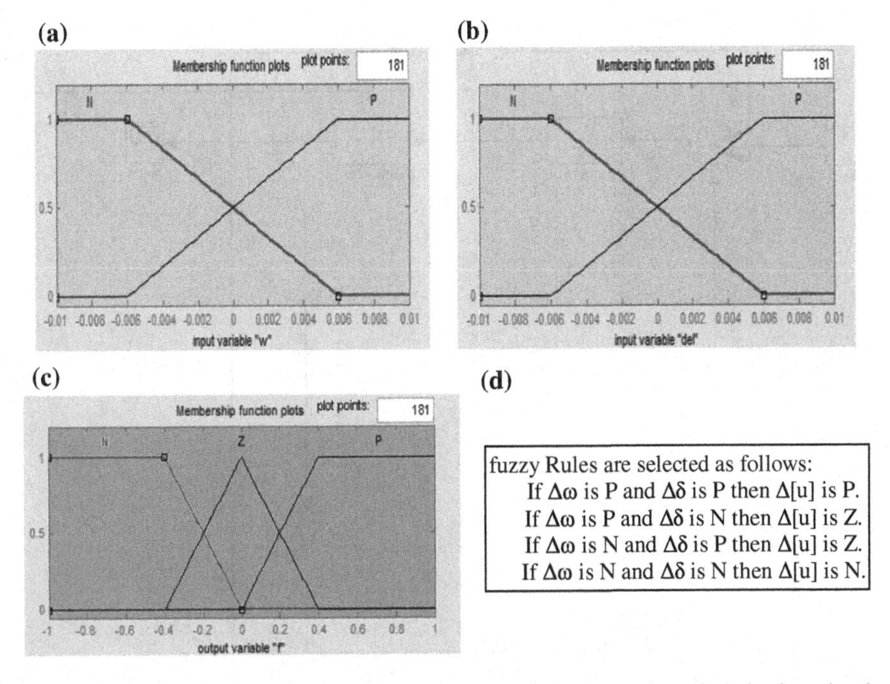

Fig. 3 **a** Membership function of error signal; **b** member function of derivative signal; **c** membership function of output; **d** fuzzy rule

4 Simulation Results

4.1 Modeling of SMIB Without UPFC

See Figs. 4, 5 and 6.

4.2 Modeling of SMIB with UPFC

See Figs. 7, 8, 9 and 10.

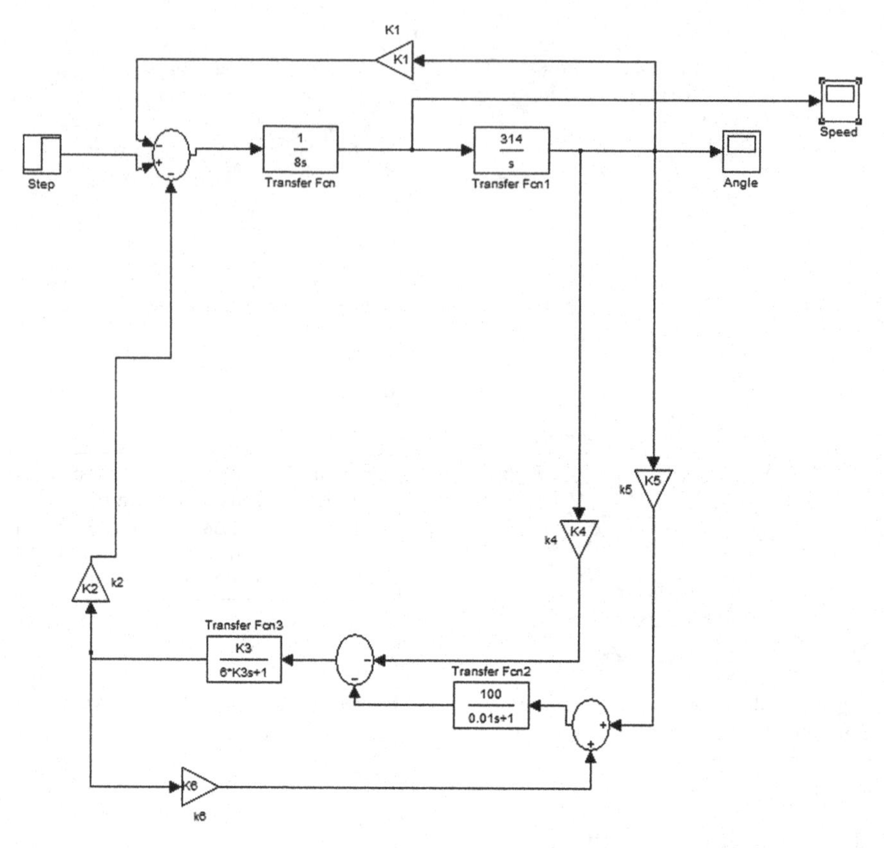

Fig. 4 Simulink model for SMIB power system without UPFC

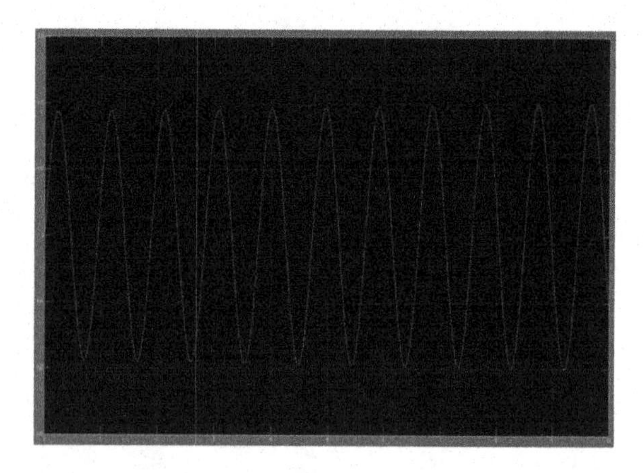

Fig. 5 Speed deviation for SMIB power system without UPFC

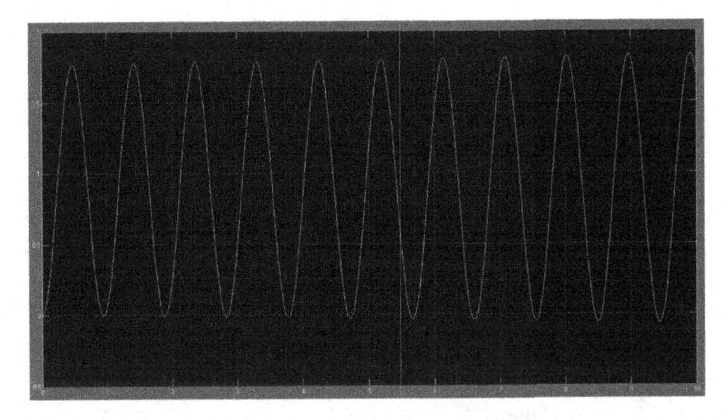

Fig. 6 Angle deviation for SMIB power system without UPFC

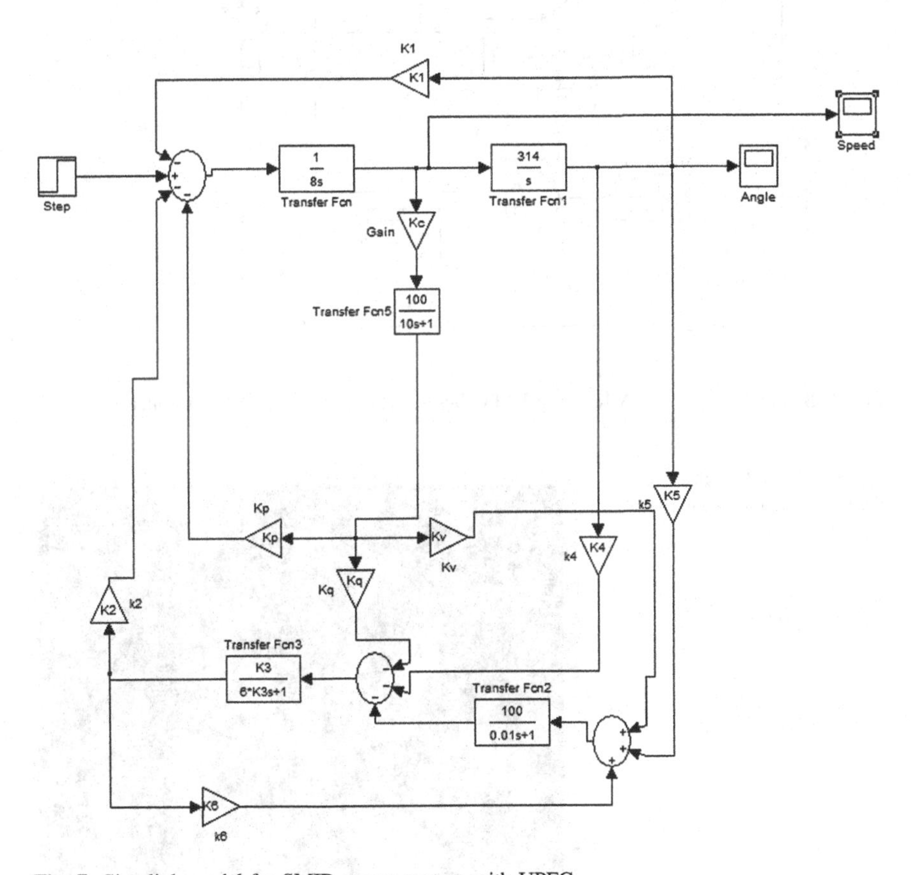

Fig. 7 Simulink model for SMIB power system with UPFC

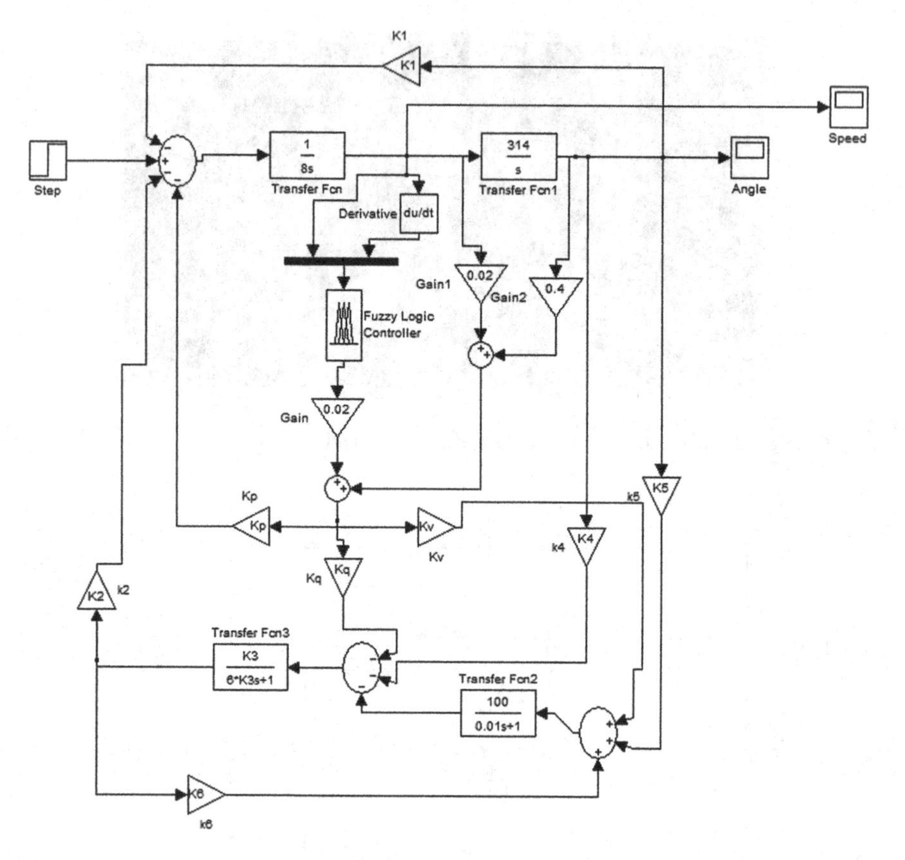

Fig. 8 Simulink model of SMIB using UPFC based hybrid fuzzy logic-PI controller

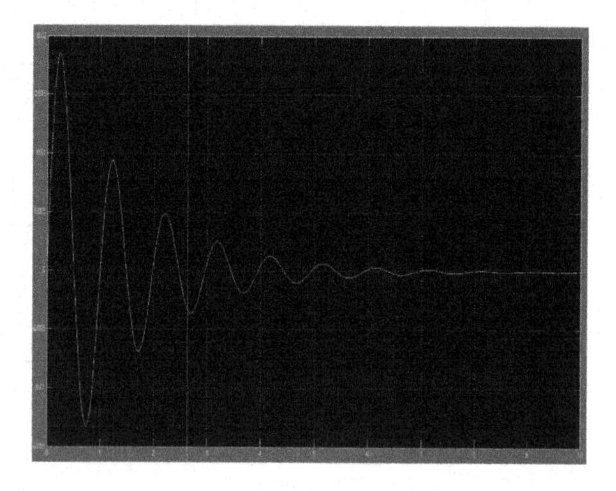

Fig. 9 Speed deviation for SMIB power system with UPFC

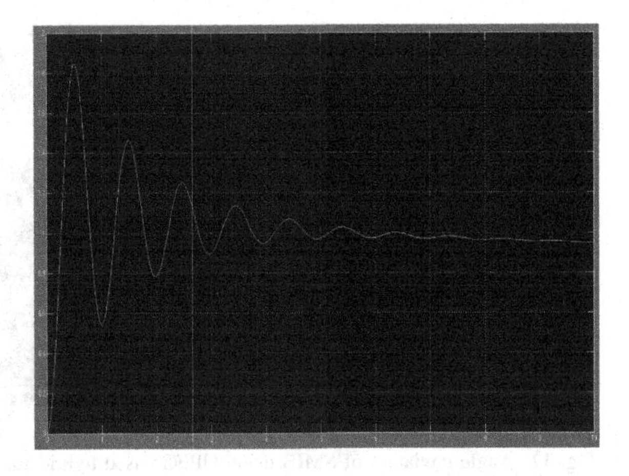

Fig. 10 Angle deviation for SMIB power system with UPFC

4.3 Modeling of SMIB Using UPFC Based Hybrid Fuzzy Logic-PI Controller

See Figs. 11 and 12.

4.4 TLBO Algorithm Based PSS Design

It is population based evolutionary method [11]. Teachers and students are the two essential components of this algorithm and discusses two basic modes of the learning, through teacher phase and cooperating with the student phase. In this

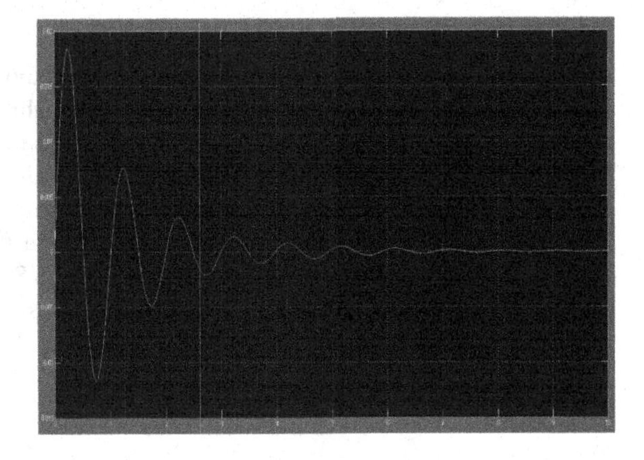

Fig. 11 Speed deviation of SMIB using UPFC based hybrid fuzzy logic-PI controller

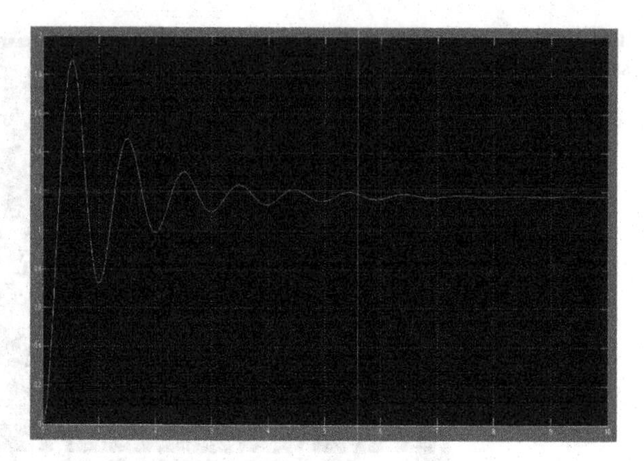

Fig. 12 Angle deviation of SMIB using UPFC based hybrid fuzzy logic-PI controller

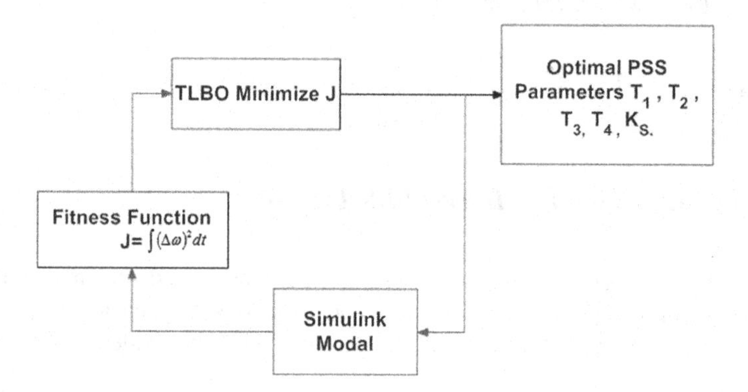

Fig. 13 PSS design with TLBO

optimization algorithm a group of students are well thought out as population and various design variables are measured as various subjects offered to the students and students result are analogous to the 'fitness' value of the optimization algorithm problem. Figure 14 shows the flowchart of TLBO algorithm and working fundamental of algorithm [12]. In this section the PSS parameters tuning based on the Teacher learning based optimization is presented. In this study the performance index J. In fact, the performance index is the Integral Square Error (ISE). It should be noted that TLBO algorithm is execute many times and then optimal set of PSS parameters are obtained (Fig. 13; Table 1).

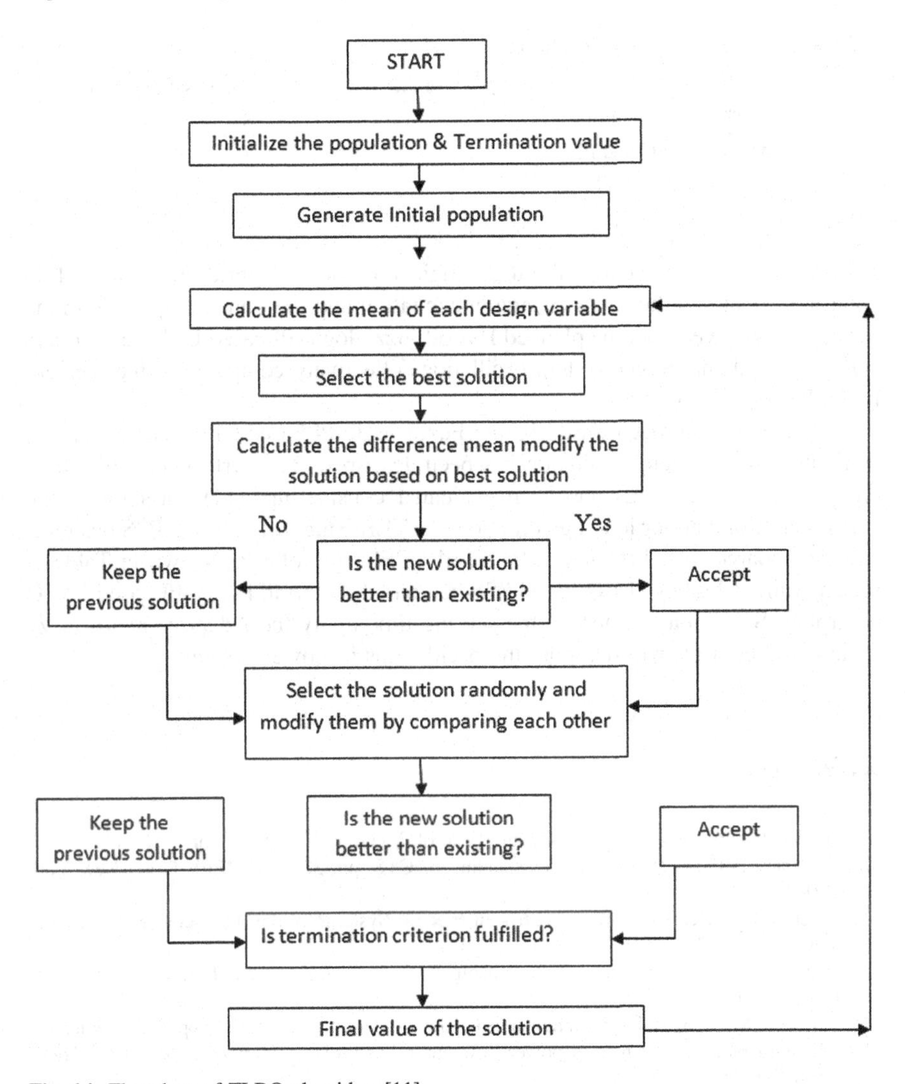

Fig. 14 Flowchart of TLBO algorithm [11]

Table 1 Optimize TLBO-PSS parameter

Kc	*T1*	*T2*	*T3*	*T4*
24.5218	0.0111	0.1968	1.7590	1.1546

5 Conclusion and Future Aspects

In the paper, a brief discussion is made about the LFO and small signal stability of a system. A linearize Haffron-Philips model is considered and a dynamic behavior of the system was examined by using different controllers like UPFC with power

Table 2 Comparatively study of results

S. no.	Techniques	Speed settling time (s)	Angle settling time (s)
1	UPFC damp controller	8	8
2	Hybrid UPFC and PI	7	7
3	PSS based on TLBO	5	5

system stabilizer, conventional Fuzzy logic controller, Hybrid fuzzy logic-PID controller … etc. to the small change in excitation and mechanical input. From the figure it is observed, that the planned Hybrid fuzzy logic-PI based UPFC controllers significantly damp power system oscillations effectively compared to the conventional Fuzzy logic UPFC.

A systematic approach for designing Fuzzy logic-PI based UPFC controllers for damping power system oscillation has been described. The performance of UPFC damping controller (mE) has been evaluated considering wide variation in the various environment or loading condition. As TLBO algorithm based PSS design is better compared to hybrid Fuzzy-PI based UPFC controller as shown in Table 2, settling time is less in TLBO based PSS compared to hybrid Fuzzy-PI based UPFC controller. So in future improve the settling time apply the Adaptive neuro-fuzzy logic based controller to dampout the oscillations in power system.

References

1. Kundur, P.: Power System Stability and Control. McGraw-Hill, New York (1999)
2. Hingorani, N.G., Gyugyi, L.: Understanding FACTS, pp. 323–387. IEEE Press, New York (2000)
3. Miller, T.J.E.: Reactive Power Control in Electric Systems. Wiley Interscience Publication, New York (1982)
4. Haque, M.H.: Damping improvement using FACTS devices. Electr. Power Syst. Res. **76**, 9–10 (2006)
5. Zarghami, M., Crow, M.L., Sarangapani, J., Liu, Y., Atcitty, S.: A novel approach to interarea oscillation damping by unified power flow controller utilizing ultracapacitors. IEEE Trans. Power Syst. **25**(1), 404–412 (2010)
6. Hingorani, N.G., Gyugyi, L.: Understanding Facts. IEEE PRESS, New York (2000)
7. Eldamaty, A.A., Faried, S.O., Aboreshaid, S.: Damping power system oscillations using a fuzzy logic based unified power flow controller. In: IEEE Conference, CCECE/CCGEI, Saskatoon, 0-7803-8886-0/05/$20.00@2005, IEEE, May 2005
8. Adware, R.H., Jagtap, P.P., Helonde, J.B.: Power System Oscillations Damping using UPFC Damping Controller. In: IEEE Conference, Third International Conference on Emerging Trends in Engineering and Technology (2010)
9. Bigdeli, N., Ghanbaryan, E., Afshar, K.: Low frequency oscillations suppression via CPSO based damping controller. J. Oper. Autom. Power Eng. **1**(1) (2013)
10. Manrai, R., Khanna, R., Singh, B., Manrai, P.: Power system stability using fuzzy logic based unified power flow controller in SMIB power system. IEEE Conference, 978-1-4673-0449-8/12/$31.00©2012 IEEE (2012)

11. Rao, R.V., Savsani, V.J., Vakharia, D.P.: Teaching learning based optimization: a novel method for constrained mechanical design optimization problems. Comput. Aided Des. **43**(3), 303–315 (2011)
12. Rao, R.V., Savsani, V.J., Vakharia, D.P.: Teaching-learningbased optimization: a novel optimization method for continuous non-linear large scale problems. Inform. Sci. **183**, 1–15 (2012)

Authors Biography

Ruchi Sharma is currently Pursuing Ph.D. in Electronics Engg. From Vivekananda Global University, Jaipur. Her research interests are electronics design, softcomputing, UPFC etc.

Mahendra Kumar has obtained M. Tech (Control and Instrumentation Engg.) and B.Tech (Electronics and Communication Engg.) from Rajasthan Technical University, Kota in 2012 and 2010 respectively. He is working as Director at MI Tech Society, Kota and Guesty Faculty in Electronics Engineering Department at RTU, Kota. His research interests are power system stabilizer, Multi-rate output feedback (fast output sampling feedback) techniques and model order reduction method, Evolutionary optimization techniques, Optimal control system, time interval (uncertain) system analysis, signal processing, image processing, Non-linear control system.

Kota Solomon Raju is currently working as Principal Scientist at CSIR-CEERI, Pilani. His research interests are electronics design, softcomputing, UPFC etc.

Secured Data Storage and Computation Technique for Effective Utilization of Servers in Cloud Computing

Manoj Tyagi and Manish Manoria

Abstract Cloud computing delivers digital services over the internet by using various applications which were carried out at distributed datacenters by computer systems. It provides protocol based high performance computing which permits shared storage and computation over long distances. This proposed work bridges the efficient computation and secure storage in cloud environment. Secure cloud storing includes receiving the data by cloud server for storage after applying security steps like authentication, encryption of data and allocation of storage space. In cloud computing, a secured computing infrastructure is provided to cloud user through computing request and commitment generation. Dynamic server stipulation by cuckoo algorithm is utilized after the completion of successful user access. The uncheatable computation, secure access and storage of proposed work achieve confidentiality. It improves the efficiency and manages concurrent users' requests.

Keywords Authentication · Secure storage · Efficient computation · Cuckoo algorithm

1 Introduction

Cloud computing provides resources like processors, storage, operating system, application software, network etc. through internet based service provider that permits accessing the services on demand as per the client request [1]. Client request can be promptly provisioned and released with less service provider interaction or management effort. The cloud model enhances availability and is consists of five characteristics like resource pooling, wide network access, scalability, measured

M. Tyagi (✉)
Mahatma Gandhi Chitrakoot Gramodaya Vishwavidyalaya, Chitrakoot, India
e-mail: manojtyagi80.bhopal@gmail.com

M. Manoria
Truba Institute of Engineering and Information Technology, Bhopal, India
e-mail: manishmanoria@gmail.com

S.C. Satapathy and S. Das (eds.), *Proceedings of First International Conference on Information and Communication Technology for Intelligent Systems: Volume 1*,
Smart Innovation, Systems and Technologies 50, DOI 10.1007/978-3-319-30933-0_53

service and on-demand self-service. Three conventional services like Platform as a service, Software as a Service and Infrastructure as a Service is given by cloud computing [2]. User as a client to the cloud computing can access the above services in effective and efficient way [3]. Enterprise users, data-bounded applications and academic users are moving towards cloud computing for their computing demand where they accessing applications and resources as a utility services [4, 5]. In present scenario, user access resources across the Internet independently without developing its own hosting infrastructure. User share the cloud Infrastructure and service provider supervise and manage this infrastructure [6]. Cloud service providers earn benefits by charging users for accessing these services [7]. Users can access their data and also put their tasks into cloud for computational processing or store their data in cloud. To handle data requirements in business environment and other domains, cloud introduces an attractive data storage and interactive model with clear advantage [8]. User can store their data remotely in cloud storage and access it whenever required without any overhead of local hardware and software management. Along with these beneficial services, there are also many security threats regarding privacy of data because cloud service provider has physical possession of users' data [9]. As we know there are several advantages to adopting Cloud Computing, but there are also some important issues to adoption [10]. Security is one of the most significant issues to adoption towards fulfillment, confidentiality and permissible matters [11]. The growth of cloud computing is always challenged by security issue like privacy and protection of data [12].

2 Related Work

To meet privacy requirement Cao et al. [13] proposed a MRSE scheme having secure inner product computation, and significantly enhanced to accomplish the privacy requirements for two levels of threat models. A proposed scheme is given the detail analysis of both privacy and efficiency assurance. Yang et al. [14] have proposed a framework for efficient auditing and privacy for cloud storage systems. The experimental results showed that above auditing protocols are secure as well as efficient, and also cost effective for auditing.

According to Wei et al. [15], SecCloud is a secure auditing protocol along with privacy-cheating discouragement in the cloud. They said that it jointly consider data storage security and computation auditing in the cloud. Moghaddam et al. [16] have proposed an Algorithm [HE-RSA] which includes strengths of RSA small-e as well as efficient-RSA for enhancing the security in cloud environment. According Yuwen et al. [17], it is possible to improve the efficiencies of AES by reconfiguring its operation.

Kemal et al. [18] provide the solution for Man-In-The-Middle (MITM) using secure authentication. Xingquam et al. [20] proposed an efficient scheduling of resources for cloud using SLPSO algorithm. Zineddine [21] proposed cuckoo optimization algorithm to moderate an known set of vulnerabilities by optimal set of solutions. Singh et al. [22] performs a scheduling of task using modified version of Genetic approach in cloud environment.

3 Problem Formulation

Increasing volume of data and its management has become a challenge for all IT Industry today. In turn, it is also a big issue for cloud service providers to maintain multi-tenancy of data and its efficient retrieval. It is becoming complex when provide cloud services in a heterogeneous computing environment. Privacy protection and data security are the main factors in the cloud technology with respect to users' perspective. Users store and access their data in cloud storage remotely with the help of cloud services. It is essential to prevent unauthorized access of users' outsourced data by other users. To assure data availability and confidentiality, a scalable and efficient user authentication algorithm along with secured data storage has been a mandatory issue here. Due to virtualization of resources, global migration and replication, remotely situated data and machine, it is not easy for cloud user to manage and maintain trust on cloud server for the stored data and the computation results. Previous works on the cloud security have more concerns about the storage security, not for the computation. Therefore in proposed work, storage security as well as computation taken into account for provides security and efficiency. This paper bridges efficient computation and secured storage in cloud by authentication, server selection and stored data using efficient encryption algorithm. Dynamic server stipulation is the phase carried out after the completion of successful user access. For obtain an optimized result in server selection, cuckoo algorithm is utilized. After the selection of eminent servers, the proposed system should be ready to serve user request for data access by utilizing a cloud service provider.

4 System Model

This model includes three entities: User (U_y), a cloud server (S), and an Interface Layer.

- **User**: A person or group of persons, which uses cloud for online data storage and computing.
- **Cloud Server**: Cloud service provider managed the server for data storage and computing services. The cloud server is an entity with unlimited storage and computational resources.
- **Interface Layer**: This layer works as an intermediary between cloud server and client where it takes the requests from the clients and forwards it to the server.

5 Data Storage Security

Here security layer is used as an interface for private cloud architecture. First client connect to this security layer and then connect to server through this layer. Client is able to store their data after verification by predefined authentication scheme. Then it stored their data in secured way by using security approach at the server end. Additional users along with all the clients which are the part of that network are connected to the same security architecture. For downloading and viewing the data requires having access through same framework to maintain the security and privacy of the data. The technique secures the data using server selection, authentication and effective encryption-decryption techniques. This technique prevents from brute force attack, timing attack, Mathematical attack, man in the middle attack and also provides efficient computation (Fig. 1).

5.1 *Authentication Process*

User requires install an extension of provider's website on browser with unique access code for accessing the server. Users choose an optional key with the help of AES-128 and encrypt/decrypt to unique access code at client side and remove the dependency on service provider. User authentication process verifies the identity of user before accessing to cloud server. Here user has to register on website of service provider and set user-id and password for accessing the cloud. Each time when user login to the website, he enters their user id and password. This information

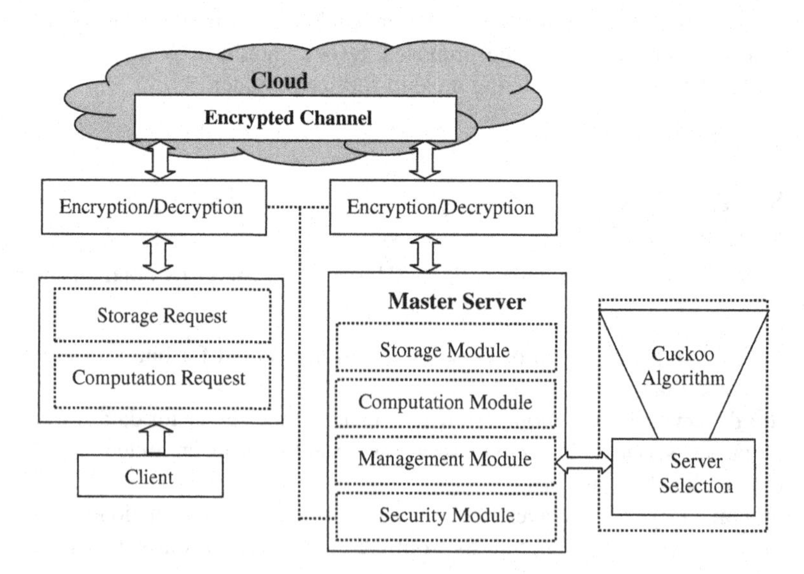

Fig. 1 Proposed technique

forwarded to server in encrypted form using secured channel. After verification of password, user receives an OTP (one time password) in his registered Mobile phone, which is used as second level verification. For transmission of information, we use secured SSL layer. In this way the cloud server authenticates the user. It is suggested to use the separate servers for keys and data.

5.2 *Secured Data Storage Process*

After authentication client is ready to store their data on cloud. Cloud server stores all data in encrypted form. There are three ways to store their data on cloud in secure way.

- If user wants to store their private data then he uses AES-128 approach.
- When he stores sharable data then he should use HE-RSA approach.
- Combined HE-RSA and AES-128 for enhance the security strength. First encrypt the data using secret key of AES-128 for storing purpose and then this secret key is encrypted by HE-RSA for sharing purpose.

Key generation algorithm. The key generation algorithm of HE-RSA [16] is given:

1. Randomly choose two large prime numbers: m, n and compute $p = mn$
2. Compute $\varphi(p) = (m-1)(n-1)$
3. Compute
$$\gamma(p,h) = (m^h - m^0)(m^h - m^1)\ldots(m^h - m^{h-1}) + (n^h - n^0)(n^h - n^1)\ldots(n^h - n^{h-1})$$
4. Select random integer: r such as $1 < r < p$ and $\gcd(r,\varphi) = 1$ and $\gcd(r,\gamma) = 1$ (r is small integer)
5. Compute e such as $re \equiv 1 \bmod \varphi(p)$ and $1 < e < \varphi(p)$
6. Compute d such as $de \equiv 1 \bmod \gamma(p)$ and $1 < d < \gamma(p)$
7. Public Key: (e, p)
8. Private Key: (r, d, p)

Encryption process. User X encrypts the message S using the public key of Y before sending to Y.

$$C = ((S^e \bmod p)^e \bmod p)$$

Decryption process. User Y will decrypt the received message using own private key.

$$S = \left((C^r \bmod p)^d \bmod p\right)$$

6 Efficient Server Selection Using Cuckoo Algorithm

The cuckoo is a process which checks all the Servers to evaluate its Quality and Fitness. The fitness of a server is estimated based on the fitness of the virtual machines running under that server. The fitness of the virtual machines is illustrated as VM_{fit}. From the efficient VMs are selected but here we need to cull a new set of VMs from the VMs selected by VM_m in need the power of a VM change after the VM has been deployed and is in use by resource allocation process. To determine Fitness for a VM from the generated population, cuckoo algorithm is utilized. In cuckoo optimization algorithm there are many Virtual Machines present to complete user request these VM_m are populated initially as '*m*' the value of *m* ranges from 0 to 1000. The objective function corresponds to VM_m, the population of VMs are derived as VM_m. The population generation of VMs is factored from L'evy Flight. The L'evy Flight process executes the cuckoo until the VM population reaches maximum generation. After populating the maximum VMs then the cuckoo checks all the VMs to evaluate its Quality and Fitness VM_{fit}. The selected VMs are capable for Resource Allocation the Remaining VMs are discarded.

6.1 *Cuckoo Optimization Algorithm*

Quality and Fitness VM_{fit} for a VM should be measured by using various parameters such as Total Number of Jobs, Number of successes Jobs, Number of Failure Jobs, Remaining Memory, power/speed of Execution. The user specifies the job *J*, deadline *D* of the job to be completed, memory requirement *Mem* of the job for the job *J*. First the task *J* has to be sorted according to the deadline requirement *D* of the job. The information about the available resource in the grid has to be collected from GIS. For each (VM) resource *R*, we have to check whether memory requirement *Mem* of the task *J* is satisfies by resource *R*. If it satisfies, the resource *R* has to be added to the list *L*. If the list is not empty, we have to calculate the original capability *OC*, utilized capability *UC* and completion time *CT* of the job *J*. The original capability of the VM, *OC* has calculated using MIPS.

The pseudo-code of Cuckoo Search according to Yang and Deb [19]

```
begin
   Objective function f(y), y = ( y 1 , y2 ,……,yd)^T
   Total number of Virtual machine VMj,
   Create initial population of n Virtual machine VMm (m = 1, 2, ..., n)
        while(t <Max Generation) or (stop condition)
        select  a cuckoo randomly by L´evy flights
        calculate its quality/fitness in eq. (7).
        Select  a nest among n (say, r) randomly
        if(Fh>Fr ),
         Substitute r by the new results;
         end
Fractions (qf ) of of poorer quality nests are discarded and built the new one;
                Hold the best Results (or nests with quality);
                Grade the Results and pick current best
        end while
Pass the best results to next generation.
end
```

6.2 Evaluation Parameters of Fitness

Total Number of Jobs (*J*) is collected from Grid Information system (GIS)

$$VM_m = \mathrm{VM_{GIS}} \tag{1}$$

The utilized capability of the VM_m is calculated using Eq. (2)

$$VM_{UC} = \frac{OC}{ET} \tag{2}$$

For all the jobs in the queue WT from total job T. The difference of estimated execution time (EET) and the completed execution time (CET) is computed using Eq. (3)

$$VM_{WT} = EET - CET \tag{3}$$

The execution time (ET) of the job determines the speed of executing a job J. It can be calculated by finding difference between the current time (*CuT*) and entry time of job (*ETJ*) into VM using Eq. (4)

$$VM_{ET} = CuT - VM_{ETJ} \tag{4}$$

CuT Current Time
VM_{ETJ} Entry time of Job into VM

The completion time of the job is sum of waiting time (WT) and execution time (*ET*)

$$VM_{CT} = WT + ET \tag{5}$$

Remaining memory *(RM) is* calculated using difference between original capability *(OC)* and utilized capability (*UC*)

$$VM_{RM} = OC - UC \tag{6}$$

A candidate habitat matrix of size $N_{pop} \times N_{var}$ is generated in the beginning of optimization algorithm. Then it assume some random number of eggs for every initial cuckoo habitat. In general, cuckoo lays from 5 to 20 eggs and this count works as upper and lower bounds of egg for each cuckoo at various iterations. In reality cuckoos lay their eggs in between the maximum distance with respect to their habitat, termed as (ELR) egg laying radius. Here upper and lower bound is denoted by the variables var_{hi} and var_{low} respectively. The cuckoo was denoted as scheduler, cuckoo eggs as Jobs and cuckoo nest as VM. Each cuckoo has an egg laying radius that is proportional to the total eggs, eggs of current cuckoo and the variables var_{hi} and var_{low}. The Fitness of each VM is evaluated by VM_{fit} it is defined by:

$$VM_{fit} = \alpha \times \frac{nsj}{tj} \times \text{var}_{hi} - \text{var}_{low} \tag{7}$$

where, tj is Total number of Jobs and α is an integer, supposed to hold the maximum value of VM_{fit}. Number of successes job $nsj \propto VM_{Rmem}$, VM_{UC} and VM_{ET}. From the above equations Quality and Fitness VM_{fit} for a VM could be measured.

7 Experimental Results and Discussion

Proposed Cuckoo Algorithm gives better utilization of resources compare than Genetic Algorithm (GA) and SLPSO. It is proved by experimental result using MATLAB 2013a. So we prefer the cuckoo search for server selection. Initial Population = 20, No. of CPU = 20, memory size = 40 GB, Maximum No. of Iteration = 60 are used as parameter.

Utilization Rate:

At the ith time, VM's utilization rate or memory's utilization rate for the private cloud is calculated as,

Table 1 VM and memory utilization rates

Utilization rate	Proposed	SLPSO	GA
Average VM utilization rate	0.911	0.892	0.709
Average memory utilization rate	0.822	0.774	0.682

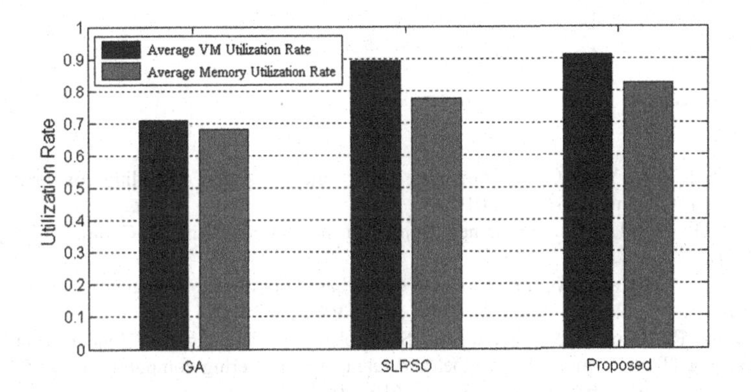

Fig. 2 Average VM and memory utilization rate

$$c(i) = \sum_{n=1}^{S} \frac{RT_n a_{ni}}{TR}, \quad i \in \{1, 2, \ldots, I\} \tag{8}$$

where,

RT Number of VM or the size of memory requested by task t_n
TR Total VM or memory in the private cloud

If task t_n runs in the ith time slot, then $a_{ni} = 1$; otherwise, $a_{ni} = 0$.

The average utilization rate of VM or memory is calculated by (Table 1; Fig. 2),

$$A_v = \sum_{i=1}^{I} \frac{c(i)}{I} \tag{9}$$

8 Conclusion

This paper identified and rectified the problem of data security and improves the utilization of resources in cloud storage system. This work achieves the integration of an effective user verification, secure data storage and efficient utilization of resources in cloud-based environments. Here, a user authentication process is defined to authenticate the user at client-side. Cuckoo search is used for server selection. Comparison of Cuckoo Algorithm with GA and SLPSO prove that

Cuckoo gives better utilization of resources that improve the efficiency of cloud. Overall analysis of the proposed work shows that, designing the user authentication, secured storage and server stipulation approach will improve the reliability, efficiency and belief in cloud computing environments as an upcoming and efficient technology in different industries.

References

1. Kumar, K., Lu, Y.H.: Cloud computing for mobile users: can offloading computation save energy. J. Comput. **4**, 51–56 (2010)
2. Pallis, G.: Cloud computing the new frontier of internet computing. J. Comput. **14**(5), 70–73 (2010)
3. Koehler, M.: An adaptive framework for utility-based optimization of scientific applications in the cloud. J. Cloud Comput. Adv. Syst. Appl. **3**(4), 1–12 (2014)
4. Buyyaa, R., Yeoa, C.S, Venugopala, S., Broberg, J., Brandic, I.: Cloud computing and emerging IT platforms: vision, hype, and reality for delivering computing as the 5th utility. J. Future Gener. Comput. Syst. **25**, 599–616 (2009)
5. Sultan, N.: Cloud computing for education: a new dawn. J. Inf. Manag. **30**, 109–116 (2010)
6. Marc, M., Stephan G., Alexander, S.: User-controlled resource management in federated clouds. J. Cloud Comput. Adv. Syst. Appl. **3**(10), 1–18 (2014)
7. Voss, A., Barker, A., Targhi, M.A., Ballegooijen, A.V., Sommerville, I.: An elastic virtual infrastructure for research applications (ELVIRA). J. Cloud Comput. Adv. Syst. Appl. **2**(20), 1–11 (2013)
8. Lombardi, F., Pietro, R.D.: Secure virtualization for cloud computing. J. Netw. Comput. Appl. **34**(4), 1113–1122 (2011)
9. Ali, M., Khan, S.U., Vasilakos, V.A.: Security in cloud computing: opportunities and challenges. J. Inf. Sci. **305**, 357–383 (2015)
10. Sheikh, M.H., Sascha, H., Sebastian, R., Max, M.: Trust as a facilitator in cloud computing: a survey. J. Cloud Comput. Adv. Syst. Appl. **1**, 1–18 (2012)
11. Hashizume, K., Rosado, D.G., Medina, E.F., Fernandez, E.B.: An analysis of security issues for cloud computing. J. Internet Serv. Appl. **4**, 1–13 (2013)
12. Gonzalez, N., Milers, C., Redigolo, F., Simplicio, M., Carvalho, T., Naslund, M., Pourzandi, M.: A quantitative analysis of current security concerns and solutions for cloud computing. J. Cloud Comput. Adv. Syst. Appl. **1**, 1–18 (2012)
13. Cao, N., Wang, C., Li, M., Ren, K., Lou, W.: Privacy-preserving multi-keyword ranked search over encrypted cloud data. J. Trans. Parallel Distrib. Syst. **25**(1), 222–233 (2014)
14. Yang, K., Jia, X.: An efficient and secure dynamic auditing protocol for data storage in cloud computing. J. Parallel Distrib. Syst. **24**, 1717–1726 (2013)
15. Wei, L., Zhu, H., Cao, Z., Dong, X., Jia, W., Chen, Y., Vasilakos, A.V,: Security and privacy for storage and computation in cloud computing. J. Inf. Sci. **258**, 371–386 (2014)
16. Moghaddam, F., Alrashdan, M., Karimi, O.: A hybrid encription algorithm based on RSA small-e and efficient-RSA for cloud computing environments. J. Adv. Comput. Netw. **1**(3) (2013)
17. Yuwen, Z., Hongqi, Z., Yibao, B.: Study of the AES realization method on the reconfigurable hardware. In: International Conference on Computer Sciences and Applications, pp. 72–76. IEEE, Wuhan (2013)
18. Kemal. B., Devrim, U., Nadir, A., Oktay, A.: Mobile authentication secure against man-in-the-middle attacks. In: 2nd International Conference on Mobile Cloud Computing, Services and Engineering, pp. 273–276. IEEE, Oxford (2014)

19. Yang, S.X., Deb, S.: Cuckoo search via Levy flights. In: World Congress on Nature and Biologically Inspired Computing, pp. 210–214. IEEE, Coimbatore (2009)
20. Xingquan, Z., Guoxiang, Z., Wei, T.: Self-adaptive learning PSO-based deadline constrained task scheduling for hybrid IaaS Cloud. J. Autom. Sci. Eng. **11**(2), 564–573 (2014)
21. Zineddine, M.: Vulnerabilities and mitigation techniques toning in the cloud: a cost and vulnerabilities coverage optimization approach using cuckoo search algorithm with Levy flights. J. Comput. Secur. **48**, 1–18 (2015)
22. Singh, S., Kalra, M.: Scheduling of independent tasks in cloud computing using modified genetic algorithm. In: International Conference on Computational Intelligence and Communication Networks (CICN), pp. 565–569, IEEE, Bhopal (2014)

Review of Security and Privacy Techniques in Cloud Computing Environment

Rutuja Mote, Ambika Pawar and Ajay Dani

Abstract Cloud is an ironic solution of the ultimate globalization which is a bunch of dedicated servers, networking elements such as software and hardware networked with the internet. Cloud computing enables the user to utilize the applications, storage solutions, and resources, likewise it also authorizes the data access, data management, and connectivity. All the activities of the cloud infrastructure seem to be transparent to ensure it without any notion of the locale to its user. The rising technology aids to deliver on-demand web access speedily to the computing resources. Cloud user can access and release the computing resources with minimum management effort and least interaction with cloud service provider. A consolidated cloud service i.e., hybrid cloud renders a federation, bridge, secure encrypted connection, information placement decision between the public and private cloud and utilizes the services of the two clouds those are security, availability, and cost effectiveness. Though cloud computing services have bags of inherent benefits, there are likelihood risks in privacy and security heeding's that should be thought over before collecting, processing, sharing or storing enterprise or individual's data in the cloud. This survey paper explores the contemporary techniques and methods to furnish a trustworthy and foolproof cloud computing environment.

Keywords Cloud computing · Privacy · Security · Encryption · Hybrid cloud · Map—reduce

R. Mote (✉) · A. Pawar
Computer Science Symbiosis Institute of Technology (SIT),
Symbiosis International University (SIU), Lavale, Pune 412115, India
e-mail: rutuja.mote@sitpune.edu.in

A. Pawar
e-mail: ambikap@sitpune.edu.in

A. Dani
G. H. Raisoni Institute of Engineering and Technology, Wagholi, Pune,
Maharashtra 412207, India
e-mail: ardani_123@rediffmail.com

S.C. Satapathy and S. Das (eds.), *Proceedings of First International Conference on Information and Communication Technology for Intelligent Systems: Volume 1*, Smart Innovation, Systems and Technologies 50, DOI 10.1007/978-3-319-30933-0_54

1 Introduction

Cloud computing is at its revision phase, but even so it fascinates the individuals and enterprise level users to wield it. Substantially, its vital purpose is to integrate economy based utility model and metamorphic development in many of those computing approaches and technologies. It again refers to the infrastructural details, applications, services, etc. When it comes to availability and accessibility, cloud environment provides an abrupt provision with the readily available pursuit of key elements namely infrastructure, applications, and features [1]. Cloud computing routine the resources and services over the networked connection anywhere, at any time, to strengthen the functionality and enhance the ease with minimal cost.

Though the widespread recourse of cloud computing entangles cloud user, the salient trait is to prefer the apt service considering the benefits and possible risks [1]. The pitfalls lie at the recognizing the security and privacy risks and necessary assessment by developing powerful and coherent solutions. Cloud computing is an important model, which helps to notably reduce costs through optimization and increased operating. It can be named as truly global computing model that could remarkably strengthen agility, collaboration and scale over the internet infrastructure [2].

In the cloud environment, we cling on third-party authorities to form the decision regarding our data selection and platforms. It seems there is exigency to unify technical moreover non-technical means to avert the third parties from embezzling user data in all cases.

The rest of the paper is categorized as follows: Sect. 2 depicts the review of security and privacy technologies for cloud environment with few sub-entries. Section 3 exploits the comparison of reviewed research papers. Section 4 enumerates future work and Sect. 5 outline the conclusion.

2 Review of Security and Privacy Technologies

This section confers the core idea about the security and privacy problems concerning with cloud computing. The foremost aspect of research is accomplishing the security to preserve data inside the cloud. The security issues encompass many technologies comprising data, resource mapping, transaction logs, memory management, etc. [3].

Cloud environment differs in some of the features viz. cloud architectures, the sensitivity level of information, and security measures which prolong security risks related to every cloud delivery model. The several surveys manifests that, many researchers are focusing on novel algorithms and encryption techniques to enhance the security and privacy of the data in the cloud [4, 5].

Privacy preservation of cloud data includes securing identity information, policy agreements while establishing the integrated environment and transaction records.

Many private organizations disagree to keep their private data and applications which inhabit outside to their own premise data storages. Cloud come up with workload relocation solution in shared infrastructure, but the sensitive information may get accessed by an unauthorized user and leads to increased risk exposure. High level of transparency is provided to the cloud user without any perception of computation and privacy assurance. To seek the security and privacy of data in the cloud environment, the following taxonomy is provided with notable techniques and the type of data handled in a cloud environment in Fig. 1.

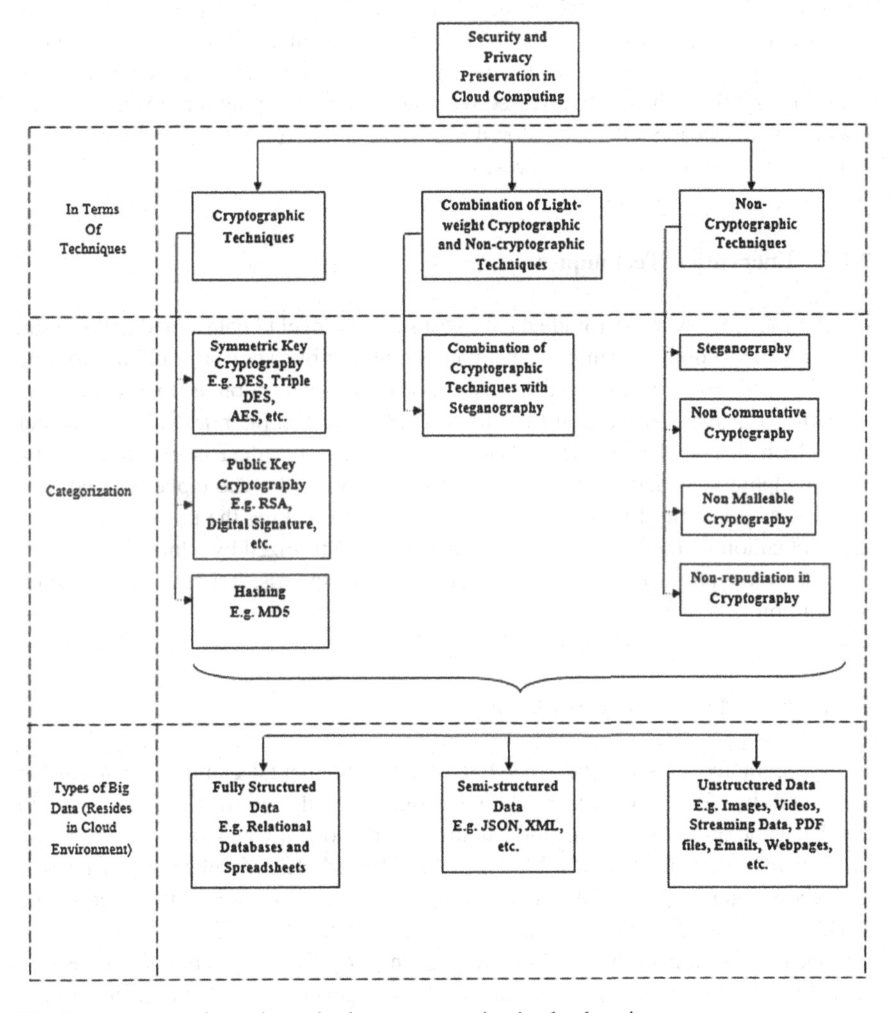

Fig. 1 Taxonomy of security and privacy preservation in cloud environment

2.1 *Cryptography*

The basic idea of cryptography is to provide the relationship between input and output constraint. To secure the transactional and stored data in the cloud, cryptography brings the encryption techniques which merely helps a user to securely access the data within the cloud environment. Cloud provider barely hosts the encrypted data. Cryptography preserves the sensitive data stored in the cloud and also the data in a transaction without any information exchange [6].

Cryptography in the cloud environment secures sensitive data beyond the user's corporate IT environment, where the data is no longer under the user's control. Cryptography expert Ralph Spencer Poore [7] stated that "Cryptographic security measures pre-eminently protects the data in motion and the data at rest. Additionally, the user is not bestowed with the luxury of having the control over the operations on data. So the data stored at a site can be cryptographically protected with the help of the cryptographic key."

2.1.1 Encryption Techniques

Encryption is the way to provide the highest security level to data stored at the cloud storages. Encryption assembles the data unintelligible so it is difficult for the unauthorized user to attack. The basic motive of encryption is to secure users valuable information by converting the readable data i.e., plain text into the cipher text which is not in readable format. Consequently, the data is encrypted using cryptographic keys and decrypted using the keys only [8]. The cipher can be considered as a blueprint for encryption offering keys to making the users data unique. Authentication method decides whether the user is authorized by a long sequence of characters or numbers, such as secret keys, passwords, tokens, or biometric recognition.

2.1.2 Key Management Mechanism

In cryptography keys can be considered as a fundamental element, so it can be generated by following ways: (1) Cryptographic algorithm (2) Auto random number generation. Key generation is the groundwork for generating the keys on random request, whereas Key Management Method (KMM) plays a crucial part with the greater responsibility of managing the keys. KMM keeps the track of the cloud users, cloud service providers and flow of the keys. The KMM helps to establish and maintain the keying relationship between the authorized people. KMM has greater importance than that of the key generation since more attacks are likely to happen at key management level [9]. Intruders keep their eye to make unauthorized access and tamper the data. And so the KMM must be seriously taken into account.

2.1.3 Why Use AES Technique

AES encryption is a sort of symmetric encryption which is the most secure and preferred all the time to secure the valuable information. Vincent Rijmen and Joan Daemen device AES over the old symmetric key encryption standard i.e., Data Encryption Standard (DES). DES adopts identical key to achieving an encryption as well as decryption method. To all probable cloud environment, for both hardware along with software platform, AES is more superior. It is the case of parallelism to improve the performance of the software. It assures quick key setup, least memory is required for implementation, and good key agility.

There is no such restriction on the key and block size, it varies accordingly in multiple of 32-bit as 128, 160, 192, 224, 256 bits. But AES uses the persistent block size with 128 bits and allows to choose among the available three keys- 128, 192, 256 bits [10]. AES processing performs substitutions and permutations and the parameters totally depend upon the key length. AES protects the data from differential and linear cryptanalysis attacks.

2.2 Privacy Within Hybrid Cloud Environment for Performance Improvement

The hybrid cloud model is an enhanced way to capture the values from cloud computing as it combines benefits extended by both public and private cloud. Hybrid cloud is an integrated cloud which utilizes the services of both private and public clouds. Though they are on single premises, they perform distinct functions and remain as unique entities. Hybrid cloud provides greater flexibility and it permits the movement of workload between private and public cloud, simply it will again depend upon cost changes and computing needs [11]. The hybrid environment helps to guard sensitive data in a private cloud as it is more secure and non-sensitive data in the public cloud which provides scalable and cost effective options.

Corporate IT environment find themselves in sometimes the lead of the security curve as they use all the security and operation aspects provided by cloud service provider. Besides, it makes the infrastructure plus computation transparent in by appealing various hybrid cloud offerings. Cloud Service Provider is proficient to possess more security expertise and to gain more current technology to adequately protect the data and applications. Cloud providers help to last the data out by current intruders and can also offer firewalls as part of their service [12]. Notably, all the necessary actions and computations can be untaken with one solution i.e., hybrid cloud, which provides both the solutions for storage and privacy.

2.3 *Hadoop Map—Reduce*

Hadoop is an open source framework which contains different systems heading towards file storage, analysis and processing of large amounts of data ranging from only a few gigabytes to thousands of petabytes. Hadoop's software is written in Java language for distributed storage including processing for vast data sets. The three major subdivisions of Hadoop are given as follows:

2.3.1 Hadoop Distributed File System (HDFS)

The Hadoop Distributed File System (HDFS) is a specialized file system to store the huge amount of data across a distributed system of computers with very high throughput and multiple replications on a cluster [13]. It provides reliability between the different physical machines to support a base for very fast computations on a large dataset.

2.3.2 Map—Reduce

This programming model is used for analyzing, it also processes huge datasets in a fast, scalable and distributed way. It is originally conceived by Google as a way of handling the enormous amount of data produced by their search bots, it has been adapted in a way that it can run on different clusters of normal machines.

The Map-Reduce framework is split up into two major phases: (1). Map phase and (2). Reduce phase. The entire framework is built around Key-Value pairs and the only thing that is communicated between the different parts of the framework is Key-Value pairs. The keys and values can be user-implemented, but they are required to be serialized since they are communicated across the network. Keys and values can range from simple primitive types to large data types. When implementing a Map-Reduce problem, the input is split into n parts, where n is the least number of Hadoop nodes in the cluster [13].

2.3.3 Common Factor

The Hadoop provides interfaces and components built in Java to support distributed file systems and I/O. This is more of a library that has all the features that HDFS and Map-Reduce use to handle the distributed computation. It has the code for persistent data structures and Java RPC that HDFS needs to store clustered data [14].

3 Comparison of Reviewed Research Paper

(See Table 1).

Table 1 Summary of the reviewed research paper

Sr. no.	Idea of the paper	Advantages	Future scope
1.	Security transparency	Stimulate the research to fill the gap between security transparency and mutual audit ability in the cloud [15]	Researcher should focus on more existing gaps in achieving security transparency [15]
2.	Symmetric encryption for privacy preserving keyword search	User is practiced to securely accumulate an encrypted data with the provision. Outsourcing and search are feasible over the encrypted data without unveiling any proof of information or query [16]	Need to exercise faster encryption process [16]
3.	Trust mechanisms for cloud computing	In this paper author identifies the mechanisms which are based on evidence, attribute certification, and validation. It helps to overcome the drawbacks using the above mechanism [17]	In future, the mathematical frameworks need to be developed to gain the trust [17]
4.	Meta-Map Reduce	Reduces the computational complexity significantly when the number of computing nodes increases [18]	Parallelize the original dataset working [18]
5.	Security as a service	The paperwork helps to address the prediction related learnings. This can be done with the help of Partial Least Squares (PLS), presented in the paper [19]	Many assumptions are comparatively considered regarding the security services hosted by cloud. The paper gives the survey on security applications and explores the relevant application fields [19]
6.	Web application partitioning	It mainly focuses on two improvements particularly to integer programming which is related to the application partitioning: (1) An asymmetric cost-model for optimizing data transfer in environments where ingress and egress data-transfer have differing costs, such as in many infrastructures as a service platforms. (2) A new encoding of database query plans as integer programs, to enable simultaneous optimization of code and data placement in a hybrid cloud environment [20]	Segregation of web applications to deploy the hybrid model [20]

4 Future Scope

Cloud computing is an inception and so forth numerous reinforcement in previous findings is obligated as the security and privacy landscape changes, with the aid of supplementary and advanced techniques are required. It assures the worthiness and utility of cloud computing environment are fully perceived as its widespread advocacy accelerates. There is some recent work on hybrid cloud architectures for security conscious Map—Reduce applications that use public clouds for storing and processing non sensitive data while using a secure enterprise site for storing and processing private data. So, we have investigated that there is a need of integrating Map—Reduce with hybrid cloud environment to achieve a higher throughput and efficiency with privacy-preserving of cloud data.

5 Conclusion

The cloud pledges as a global phenomenon, the virtualized landscape probably assembles computing as well as infrastructural resources which are located far apart however it is possible to access the services from anywhere and anytime. In this paper, the review of security and privacy pitfalls in the cloud environment feasibly fine-tuned, also the issues proposed by researchers potentially renders effective security management and threat evaluation services. The various technologies are identified shows that crucial reevaluation of prevailing security and privacy disciplines required to be done to improve the performance concerning their worthiness for clouds.

References

1. Wang, Q., Wang, C., Ren, K., Lou, W., Li, J.: Enabling public auditability and data dynamics for storage security in cloud computing. Trans. IEEE Parallel Distrib. Syst. **22**(5), 847–859 (2011)
2. Wang, C., Chow, S.S., Wang, Q., Ren, K., Lou, W.: Privacy-preserving public auditing for secure cloud storage. IEEE Trans. Comput. **62**(2), 362–375 (2013)
3. Ghogare, S., Pawar, A., Dani, A.: Revamping optimal cloud storage system. In: Proceedings of 3rd International Conference on Advanced Computing, Networking and Informatics, pp. 457–463. Springer India (2016)
4. Pawar, A.V., Dani, A.A.: Design of privacy model for storing files on cloud storage. J. Theor. Appl. Inf. Technol. **67**(2), (2014)
5. Yao, X.: Anonymous credential-based access control scheme for clouds. Cloud Comput. IEEE J. Mag. **2**, pp. 34–43, (2015)
6. Wang, C., Ren, K., Lou, W., Li, J.: Toward publicly auditable secure cloud data storage services. Netw. IEEE **24**(4), 19–24 (2010)

7. Lord N.: Cryptography in the Cloud: Securing Cloud Data with Encryption. Retrieved from https://digitalguardian.com/blog/cryptography-cloud-securing-cloud-data-encryption 28 Sept 2015
8. Seo, S.H., Nabeel, M., Ding, X., Bertino, E.: An efficient certificateless encryption for secure data sharing in public clouds. IEEE Trans. Knowl. Data Eng. **26**(9), 2107–2119 (2014)
9. Ningning, S., Zhiguo, S., Yan, S., Xianwei, Z.: Multi-matrix combined public key based on big data system key management scheme (2014)
10. Rahulamathavan, Y., Phan, R.C.W., Veluru, S., Cumanan, K., Rajarajan, M.: Privacy-preserving multi-class support vector machine for outsourcing the data classification in cloud. IEEE Trans. Dependable Secure Comput. **11**(5), 467–479 (2014)
11. Zhang, H., Jiang, G., Yoshihira, K., Chen, H.: Proactive workload management in hybrid cloud computing. IEEE Trans. Netw. Serv. Manage. **11**(1), 90–100 (2014)
12. Lu, P., Sun, Q., Wu, K., Zhu, Z.: Distributed online hybrid cloud management for profit-driven multimedia cloud computing. IEEE Trans. Multimedia, **17**, 1297–1308 (2015)
13. Jiang, D., Tung, A.K., Chen, G.: Map-join-reduce: toward scalable and efficient data analysis on large clusters. IEEE Transactions Knowl. Data Eng. **23**(9), 1299–1311 (2011)
14. Dahiphale, D., Karve, R., Vasilakos, A.V., Liu, H., Yu, Z., Chhajer, A., Wang, C.: An advanced mapreduce: cloud mapreduce, enhancements and applications. IEEE Trans. Netw. Serv. Manage. **1**, 101–115 (2014)
15. Ouedraogo, M., Mignon, S., Cholez, H., Furnell, S., Dubois, E.: Security transparency: the next frontier for security research in the cloud. J. Cloud Comput. **4**(1), 1–14 (2015)
16. Salam, M.I., Yau, W.C., Chin, J.J., Heng, S.H., Ling, H.C., Phan, R.C., Yap, W.S.: Implementation of searchable symmetric encryption for privacy-preserving keyword search on cloud storage. Hum.-Centric Comput. Inf. Sci. **5**(1), 1–16 (2015)
17. Huang, J., Nicol, D.M.: Trust mechanisms for cloud computing. J. Cloud Comput. **2**(1), 1–14 (2013)
18. Liu, X., Wang, X., Matwin, S., Japkowicz, N.: Meta-MapReduce for scalable data mining. J. Big Data **2**(1), 1–21 (2015)
19. Senk, C.: Adoption of security as a service. J. Internet Serv. Appl. **4**(1), 1–16 (2013)
20. Kaviani, N., Wohlstadter, E., Lea, R.: Partitioning of web applications for hybrid cloud deployment. J. Internet Serv. Appl. **5**(1), 1–17 (2014)

Detection of Bundle Branch Block Using Bat Algorithm and Levenberg Marquardt Neural Network

Padmavathi Kora and K. Sri Rama Krishna

Abstract Abnormal Cardiac beat identification is a key process in the detection of heart ailments. This work proposes a technique for the detection of Bundle Branch Block (BBB) using Bat Algorithm (BA) technique in combination with Levenberg Marquardt Neural Network (LMNN) classifier. BBB is developed when there is a block along the electrical impulses travel to make heart to beat. The Bat algorithm can be effectively used to find changes in the ECG by identifying best features (optimized features). For the detection of normal and Bundle block beats, these Bat feature values are given as the input for the LMNN classifier.

Keywords Bundle block · Bat algorithm · LMNN classifier · MIT-BIH Arrhythmia database

1 Introduction

Electro-Cardiogram is used to access the electrical activity of a human heart. The diagnosis of the heart ailments by the doctors is done by following a standard changes. In this project our aim is to automate the above procedure so that it leads to correct diagnosis. Early diagnosis and treatment is of great importance because immediate treatment can save the life of the patient. BBB is a type of heart block in which disruption to the flow of impulses through the right or left bundle of His, delays activations of the appropriate ventricle that widens QRS complex and makes changes in QRS morphology.

BBB is developed when there is a block along the path that electrical pulses travel to make the heart to beat. Since the electrical impulses cannot travel through the preferred pathway across the BBB, it may travels through muscles that slows

P. Kora (✉)
GRIET, Hyderabad, India
e-mail: padma386@gmail.com

K. Sri Rama Krishna
VRSEC, Vijayawada, India

S.C. Satapathy and S. Das (eds.), *Proceedings of First International Conference on Information and Communication Technology for Intelligent Systems: Volume 1*, Smart Innovation, Systems and Technologies 50, DOI 10.1007/978-3-319-30933-0_55

down the electrical movement and changes the directional path of the pulses. A condition in which there is a delay in the in the hearts lower chambers may be observed through changes in the ECG. ECG is a cost effective tool for analyzing cardiac abnormalities. The diagnosis of heart ailments by doctors is done by standard changes. In this project, our aim is to automate the above procedure so that it leads to correct diagnosis. Early diagnosis and treatment is of great importance because immediate treatment can save the life of the patient. Good performance depends on the accurate detection of ECG features.

Detection of BBB using ECG involves three main steps: preprocessing, feature optimization and classification as shown in Fig. 1. The first step in preprocessing mainly concentrates in removing noise from the signal using filters. Next step in the preprocessing is the 'R' peak detection and then segmentation of ECG file into beats. The samples that are extracted from each beat contains non uniform samples. The non uniform samples of each beat are converted into uniform samples of size 200 by using a technique called re-sampling.

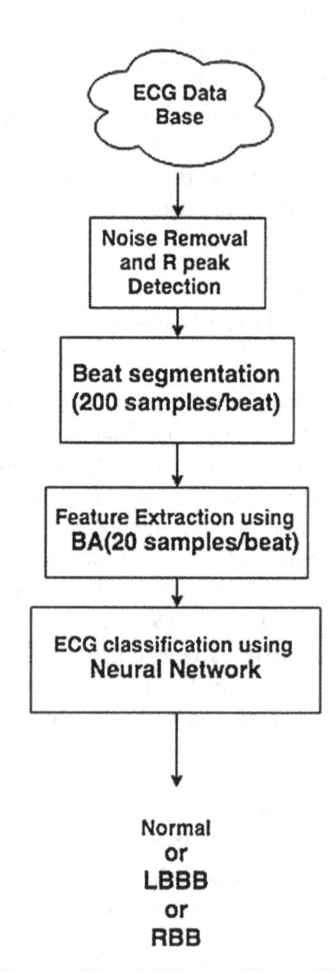

Fig. 1 ECG classification using BA

Feature extraction in the present paper is based on the extracting key features using nature inspired algorithm [1, 2] called BA. In recent years many models are developed based on the behaviors of living beings and have been applied for solving the sensible real world issues. Among them, BA may be a population based optimization technique. BA is used extensively as a model to solve many engineering applications. In Recent years, BA has been applied with success to some engineering concepts like harmonic estimation, optimum management, reduction machine learning and transmission loss and so on. The proposed BA [3] scheme has been compared to Particle Swarm Optimization (PSO) [4, 5] which is a popular algorithm for optimization of ECG features with respect to the following performance measures like convergence speed, and the accuracy in the final output.

2 Pre Processing

2.1 Data Collection and Noise Removal

To prove the performance of BA, the usual MIT BIH arrhythmia database [6] is considered. The data used in this algorithm confines to 11 recordings that consists of 5 normal, 3 LBBB and 3 RBBB for a duration of 60 min at 360 Hz sampling rate. Denoising of ECG data is a preprocessing step that removes noise and makes ECG file useful for subsequent steps in the algorithm. The Sgolay FIR smoothing filter is used for filtering.

3 Feature Extraction

In the feature extraction procedure, a fraction of signal around the R peak is extracted as the time-domain features since the R peaks of ECG signal are an important index for cardiac diseases. To ensure the important characteristic points of ECG like P, Q, R, S and T are included, a total of 200 sampling points before and after the R peak are collected as one ECG beat sample. The samples that are extracted from each beat contains non uniform samples. The non uniform samples in each beat are converted into uniform samples of size 200 by using a technique called resampling. The resampled ECG beat samples/features is shown in Fig. 2.

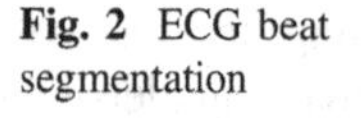
Fig. 2 ECG beat segmentation

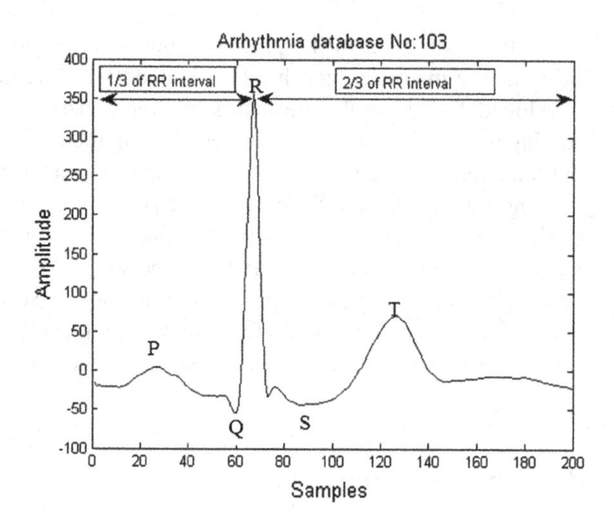

4 Feature Optimization

The customary feature extraction methods generally yields a large number of features, and many of these might be insignificant. Therefore, the effective technique in this study is to extract the key features useful in the classification of ECG beats.

In this sub-section, 20 key features from 200 samples are extracted using BA algorithm.

4.1 Bat Algorithm(BA)

BA [7] was developed by Yang. Bats use eco-location technique to find their food/prey.

Echolocation capability of Bats: Generally bats have the ability to identify many bugs, recognize distance of the quarry and move gently and can circumvent the obstacle in the whole duskiness and for communication, they use a special type of sonar known as echo-location. All creatures including bats issue out some pulses that have the capability of using echo-location. These pulses ranges from frequency of high pitch (>200 kHz) to low pitch (10 kHz). When these pulses strike the obstacles, they reach the bats again in the form of echoes. The echo-location features are conceptualized by following rules.

Most of the bats use a sophisticated sense of hearing. They release sounds that bounce (echoes) back from the insects or objects in their path. From these echoes, the bats can identify how far the insects or objects are from their current position and the also estimate the size of insects or objects within a fraction of second.

Though they can vary the loudness in many forms, we assume that the loudness varies from a large (positive) A0 to a minimum constant value Amin.

Bat algorithm structure

(a) Initialization: The fitness of the initial bat population is evaluated using the Eq. (1) and the values of pulse rate r_i, loudness A_i and frequency f_{min}, f_{max} are initialized.

$$f(x) = \sum_{i=1}^{j-1} \left[100(x_{i+1} - x_i^2)^2 + (x_i - 1)^2 \right] \quad (1)$$

where j is the dimension and x(i) is the ith bat.

(b) Movement of Virtual Bats:
The new bat population is generated by adjusting the position x_i and the velocity v_i for each bat in the population as given in Eqs. (2) and (3) respectively.

$$f_t = fmin + (fmax - fmin) * rand$$

$$v_i(t+1) = v_i(t) + (x_i(t) - x_{Gbest})f_t \quad (2)$$

$$x_i(t+1) = x_i(t) + v_i(t) \quad (3)$$

The velocity and position updates of the ith bat are calculated using Eqs. (2) and (3). The wavelength λ and loudness A are varied according to the location and size of the food. The v_i and x_i are initialized with some initial random values and a fitness function f is calculated, using the particles positional coordinates as input values.

(c) Local search capability of the algorithm:

$$x_{new} = x_{Gbest} + \epsilon A^{-t} \quad (4)$$

In order to enhance the local search capability of the algorithm, Yang [8] has created the best solution using the Eq. (4) where x_{Gbest} is a high quality solution chosen by some mechanism. A^{-t} is the average loudness value of all the bats at tth time step and the ϵ is generated by some random mechanism ranging between −1 and 1.

(d) Loudness(A):
Loudness A_i has been updated using Eq. (5) as the iterations proceed. As bats approach their food, the loudness usually decreases. Generally, the loudness value will decrease when the bat starts approaching the best solution as shows in the following equation:

$$A_i(t+1) = \alpha A_i(t) \quad (5)$$

The amount of decrease is determined by α where $0 < \alpha < 1$.

(e) Pulse emission rate(r):
The pulse emission rate r_i has been updated using the Eq. (6) as the bats approach their food. r increases. Pulse rate r yields a better solution near X_{Gbest}, higher pulse rate doesn't yield best solution in local search space.

$$r_i(t+1) = r_i^0[1 - e^{-\gamma t}] \tag{6}$$

where $\gamma > 0$
The r and A are updated only when the new solution is made better than the previous solution where γ is constant.

The original BA has been demonstrated in the following algorithm. In this algorithm, the bat behavior has been analyzed based on the its fitness function. It consists of the following points:

Algorithm 1:Bat Algorithm

1.The fitness(objective) function needs to be defined here.
2. The bat population is generated randomly.
3.fi, r, A must be initialized.
4. The iterations need to be continued up to t < 30.
5. New solution has to be generated using eq. (2) and eq. (3).
6. if(rand > ri)
7. Local solution should be Selected around the best solution using Eq. (4)
8.end if
9.if (rand< Ai and f(xi) < f(x))
10. The new solution needs to be stored.
11. r rand A are updated using eq. (5) and eq.(6)
12. end if
13. The bats must ranked and the current best solution is obtained.
14. The algorithm is ended.
15. The optimum solution should be displayed.

5 Classification of BBB with Optimized BA Features

The extracted features from BA algorithm (20 features) are classified using different types of classification techniques such as KNN, SVM, Neural Network classifiers.

5.1 Levenberg-Marquardt Neural Network (LM NN)

In this work for the detection of BBB, back propagation Levenberg-Marquardt Neural Network (LMNN) was used. This NN provides rapid execution of the network to be trained, which is the main advantage in the neural signal processing applications [9].

The NN was designed to work well if it was built with 20 input neurons, 10 neurons in the hidden layer and 3 neurons in the output layer.

The performance of this algorithm is compared with Scalar Conjugate Gradient (SCG) NN. The LMNN algorithm is a robust and a very simple method for approximating a function. SCG NN method provides conjugate directions of search instead of performing a linear search. The network is trained with 1800 ECG beats, and tested with 1006 ECG beats. The total number of iterations are set to 1000 and mean square error less than 0.001. The main advantage of this algorithm is that the time required to train the network is less.

6 Results

ECG features before optimization = [1 2 3………200];

The optimized ECG features (20 features) after BA algorithm are given below

Optimized features (column numbers) using BA = [41, 14, 198, 17, 189, 139, 22, 81, 177, 1, 171, 82, 134, 40, 49, 38, 80, 86, 129, 138];

These reduced feature are given as input for the Neural Network so that its convergence speed and final accuracy can be increased.

The ECG beats after segmentation are re-sampled to 200 samples/beat. Instead of using morphological feature extraction techniques, in this paper BA is used as the feature optimization technique using BA ECG beat features are optimized to 20 features. The BA gives optimized features (best features) for the classification. The performance of BA is compared with classical PSO technique. The BA, PSO features are classified using SVM, KNN, SCG NN, LM NN as in the Table 1.

- Count of Normal beats used for classification-9193.
- Count of RBBB beats user for classification-3778.
- Count of LBBB beats user for classification-6068.

Table 1 Classification with LM NN classifier

Classifier	Sensi (%)	Speci (%)	Accuracy (%)
PSO + SCG NN	86.1	85.3	86.0
BA + SCG NN	97.42	92.28	97.13
PSO + LM NN	91.2	89.2	80.9
BA + LM NN	99.97	98.7	98.9

- Total number of beats used for classification-19,039.
- Count of correctly classified beats-18,800.
- Total misclassified beats-239.

For measuring accuracy two parameters sensitivity and specificity are calculated using the following equations.

$$Specificity = \frac{True\_Negative}{True\_Negative + False\_Positive} \times 100 \quad (7)$$

$$Sensitivity = \frac{True\_Positive}{True\_Positive + False\_Negative} \times 100 \quad (8)$$

$$Accuracy = \frac{TP + TN}{TP + TN + FP + FN} \times 100 \quad (9)$$

- TP(True_Positive) = Count of all the correctly classified Normal beats.
- TN(True_Negative) = Count of all beats the correctly classified Abnormal beats.
- FP(False_Positive) = Count of Normal beats which are classified as Abnormal.
- FN(False_Negative) = Count of Abnormal beats which are classified as Normal.

In the training mode we applied multilayer NN and checked the network performance and decided if any changes were required to the training process or the data set or the network architecture. First, check the training record, `trainlm' Matlab function.

7 Conclusion

The results showed that the proposed BA method extracts more relevant features for ECG analysis. These extracted features are fed into a three layer LM NN to classify the beats into normal or LBBB or RBBB. The LM NN clearly distinguishes the Left and Right bundle blocks by taking features from the BA. The BA is compared to PSO algorithm. If this procedure helps us automate a certain section or part of the diagnosis then it will help the doctors and the medical community to focus on other crucial sections. This has also increased the accuracy of diagnosis.

References

1. Yang, X.S.: Bat algorithm for multi-objective optimisation. Int. J. Bio-Inspired Comput. **3**(5), 267–274 (2011)
2. Yang, X.S., Gonzalez, J.: A new metaheuristic bat-inspired algorithm. In: Nature Inspired Cooperative Strategies for Optimization (NICSO 2010), pp. 65–74. Springer, Berlin (2010)

3. Baziar, A., Rostami, M.A., Akbari-Zadeh, M.R.: An intelligent approach based on bat algorithm for solving economic dispatch with practical constraints. J. Intell. Fuzzy Syst. **27**(3) (2014)
4. Melgani, F., Bazi, Y.: Classification of electrocardiogram signals with support vector machines and particle swarm optimization. IEEE Trans. Inf. Technol. Biomed., **12**(5), (2008)
5. Kora, P., Rama Krishna, K.S.: Hybrid bacterial foraging and particle swarm optimization for detecting bundle branch block. Springerplus, **4**(9), (2015)
6. Mark, R., Moody, G.: MIT-BIH Arrhythmia Database, May. 1997. Available http://ecg.mit.edu/dbinfo.html
7. Yang, X.S.: A new metaheuristic bat-inspired algorithm. In: Nature inspired cooperative strategies for optimization (NICSO 2010), pp. 65–74. Springer, Berlin (2010)
8. Yang, X.S., He, X.: Bat algorithm: literature review and applications. Int. J. Bio-Inspired Comput. **5**(3), 141–149 (2013)
9. Sapna, S., Tamilarasi, A., Kumar, M.P.: Backpropagation learning algorithm based on Levenberg Marquardt Algorithm. Comput. Sci. Inf. Technol. (CS and IT) **2**, 393–398 (2012)

Bot Detection and Botnet Tracking in Honeynet Context

Saurabh Chamotra, Rakesh Kumar Sehgal and Sanjeev Ror

Abstract Here in this paper we have proposed a framework for Bot detection and Botnet tracking. The proposed system uses a distributed network of Honeynets for capturing malware samples. The captured samples are processed by Machine learned model for their classification as bots or not-bots. We have used the Native API call sequences generated during the malware execution as feature set for the machine learned model. The samples identified as Bot are clustered based upon their network and system level features, each such cluster thus obtained represents a Botnet family. The Bot samples belonging to such clusters are executed regularly in the sandbox environment for the tracking of botnets.

Keywords Honeypots · Honeynets · Bot detection · Botnet tracking

1 Introduction

The term malware abstractly refers to a wide range of malicious software classes. Among such malicious software classes, Bot is a class that has recently gained a lot of attention among hackers as well as the research's community. Reason behind this popularity is the effectiveness and the coverage of attacks launched using Botnets. A single Botnet having millions of Bots could even take whole network infrastructure of a country on its knees.

Honeynet based Botnet mitigation techniques are among some of the most popular approaches used for the detection and response against botnets [1, 2, 3].

S. Chamotra (✉) · R.K. Sehgal · S. Ror
Syber Security Technology Group CDAC, Mohali, India
e-mail: saurabhc@cdac.in

R.K. Sehgal
e-mail: rks@cdac.in

S. Ror
e-mail: sanjeev@cdac.in

S.C. Satapathy and S. Das (eds.), *Proceedings of First International Conference on Information and Communication Technology for Intelligent Systems: Volume 1*,
Smart Innovation, Systems and Technologies 50, DOI 10.1007/978-3-319-30933-0_56

In all such approaches Honeynets are used as a tool for capturing malwares and their corresponding logs. The effectiveness of any Honeynet deployment greatly depends upon the system configuration and the deployment strategy followed. For efficiently configuring Honeynets, one has to understand the propagation strategy employed by the targeted malware class and accordingly configure the Honeynet. Sehgal et al. [4] have categorized malwares in two broad classes based upon their propagation strategy the (1) Autonomously spreading malware and (2) Malware which requires medium to propagate (email, USB drives, websites etc.). In the work presented in this paper we have focused on autonomously spreading Bot malwares, which spreads by infecting vulnerabilities in operating system's network services. For the large scale capturing of these autonomously spreading malwares we have configured Passive Honeypots in Honeynets and deployed Honeynets in a geographically distributed fashion. Further Honeynet nodes are deployed within broadband subnets as the home user segment using broadband networks is the least aware user segment and hence greatly targeted by the Botnet attacks.

The malware samples captured at Honeynet are logically fused with system, network logs and the contextual information of the environment (i.e. Operating system, service pack installed, opened ports, software running etc.) in which they were captured. This data is later processed by the Bot detection engine which segregates them in Bot and non Bot classes. The feature sets extracted from malware samples labeled as Bot and the corresponding logs are clustered by Botnet detection and tracking engine. Each such cluster thus obtained represents a Botnet family and the corresponding attributes (i.e. Egg download IP, DNS, C&C IP, Commands, usernames and passwords) collectively characterize that Botnet family. For tracking of these bonnets and to counter the Botnet detection evasion techniques (i.e. fast flux and the server migration) [5]. Bots belonging to identified botnet families are executed in sandbox on regular basis.

2 Related Work

There are two basic approaches employed by the researchers for malware analysis (1) Static analysis [6, 7, 8] and (2) Dynamic analysis [9] a third approach named as hybrid approach is proposed by Madou et al. [10] which is a combination of the above two approaches. In the work presented in this paper we have employed dynamic analysis approach for the detection and tracking of Botnets.

There are challenges such as self-modifying code, malware detecting analysis environment [11] and conditional obfuscation [12] in the Dynamic analysis approach. To counter these anti-analysis techniques we have used fixes available in the literature [13] which implements some light weight fake artifacts in the virtual environment to give it the appearance of actual environment. Also we use contextual information of the malware to create an ideal execution environment in the sandbox, which ensures optimum behavior coverage during the malware execution in sandbox.

The Problem of detecting generic malware have been addressed by many researchers [14, 15, 16, 17]. Some of them were specifically focused on the detection of Bot malware [1, 18, 19]. Most of these approaches for Bot detection were based upon active or passive network monitoring. [3, 9, 20]. Such detection approaches were quite effective in most of the cases but they failed in cases where Bot malware uses encrypted channels or either employs a complex communication protocols (i.e. P2P) [21] for communication with C&C servers. To address this problem researchers explored approaches involving the host level data capturing. Elizabeth Stinson [22] introduced a taint analysis based technique to detect the remote control behavior of the Bot malwares. Al hammid [23] was the first to introduce the concept of API call hooking and API call sequence mining for Bot detection. Different researchers have used different host level features such as API call sequences, key logging and file system activities for the detection of Bot malwares [18].

In the work presented in this paper we have used windows native API call sequences as feature set for the development of malware classifier. Our technique for the detection of bot malware is similar to the technique used by Wang et al. [24]. Wang et al. have used Native API call sequence mining with SVM for the intrusion detection in windows environment where as we have used this technique for the detection of bot malwares in the Honeynet context. The Honeynet context is slightly different from the context in which the conventional malware detection system operates. As in the Honeynet context one has to detect Bot malware from a set of other malicious software instead of detecting malware from a set of benign files.

3 System Overview and Logical Design

Figure 1 shows an abstract overview for Honeynet based Bot detection and Botnet tracking framework. The whole framework could be divided in to three major sections (1) Distributed Honeynet system (2) Bot Detection engine (Malware classification engine) and (3) Botnet tracking Engine

Distributed Honeynet system: Is a geographically distributed network of Honeynet systems having a combination of low interaction and high interaction Honeypots and is used as for malware capturing. The captured malware samples with information about the environment in which they were captured are pushed from the Honeynet nodes to the central collection server on a regular basis using a secured channel. At the central collection server theses malware samples are logically fused with corresponding metadata (i.e. Source IP, MD5 hash of malware sample, operating system etc.) and converted in to relational data base format.

Bot Detection Engine: The Malware database at central server acts as an input source for the Malware classification\Bot Detection engine. The data from malware

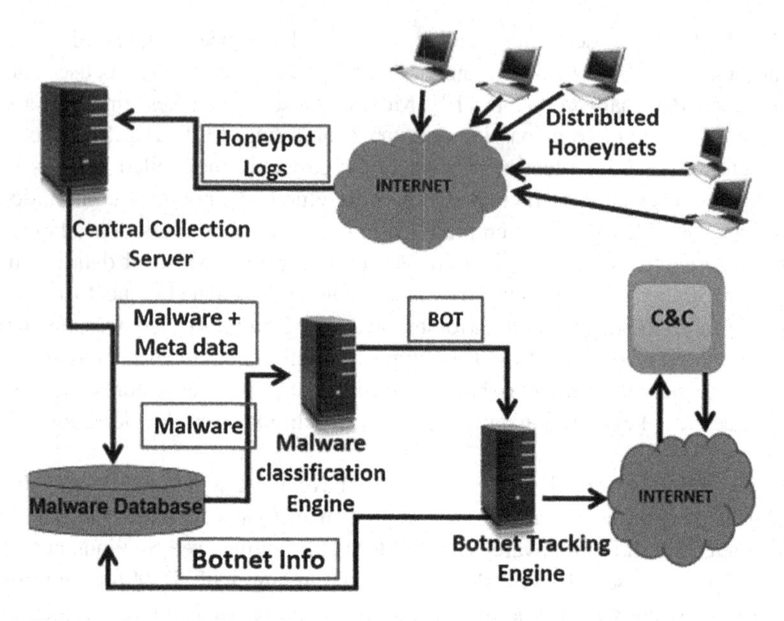

Fig. 1 Bot detection and Botnet tracking framework

database containing malware samples and the corresponding contextual information is used to recreate environment and sandbox execution. During the sandbox execution traces of Native API call sequences are logged. These API call traces are processed and 6 length API sequences are generated. These six length Native API call sequences are taken as input by the Bot detection engine. The Bot detection engine uses an SVM based classifier to label the malware samples as Bot or non Bot.

Botnet tracking engine: The botnet tracking engine detects and track the botnet families, it employs the clustering technique for the detection of botnet families. The network features along with the Binary syntactic feature are used as an input for botnet clustering engine. Each botnet clusters thus obtained represents a different bot family. For the tracking of these Botnets the malware samples belonging to corresponding botnet families are executed on regular basis in the sandbox environment and the botnet attributes are logged.

3.1 Detailed System Design

Figure 2 is a modular representation of the Bot detection and Botnet tracking framework. Each module below shown performs following roles in the system:

Data fetcher: fetches the malware sample and corresponding environment information from databases and submits this data to the sandbox manager.

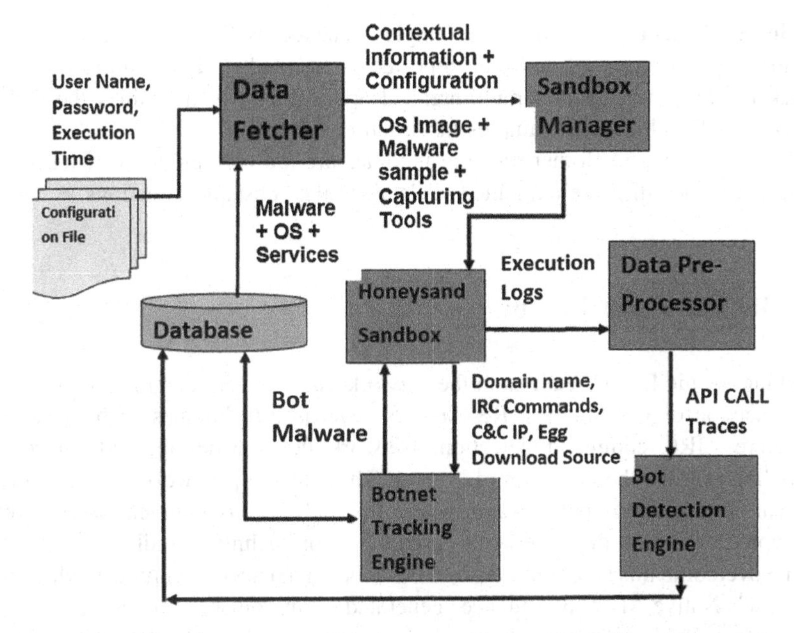

Fig. 2 System design

Sandbox Manager: It takes Honeypot system configuration template as an input with malware sample. The template consist of details such as (1) Operating system version, (2) Service pack, (3) System software details, (4) Running Network services. The template submitted to the sandbox manager having above mentioned information is an xml file. The sandbox manager parses this xml file use it as a blueprint for creation of the sandbox environment and then executes the malware sample.

Sandbox Environment: controlled environment where the malwares are executed. The malware contextual information is used to create the execution environment for the malware sample. The malware sample is executed and network, host level execution traces are captured. We have used Honeysand sandbox [25], which is a specially designed open source sandbox for Bot malware execution. It provides features such as limited internet connectivity, fake DNS query response with provision for fake input feeds for accelerating the malware execution.

Data Pre-Processor: The data captured during sandbox execution acts as an input for data Pre-processor module. This module provides input to Bot detection engine for performing classification and later to Botnet tracking engine for performing the clustering. The input provided to Bot detection engine is the 6 length sequences of the native API calls and the input provided to Botnet tracking engine is network, system execution traces and malware syntactic attributes.

Bot Detection Engine (Bot Detector): is an SVM based binary classifier which uses the Native API call traces collected during the Malware sample execution as a feature set for the classification of malware sample as Bot or not Bot.

Botnet Tracker: The malware samples detected as Bot by the Bot detection engine are first clustered and then repeatedly executed in the sandbox by Botnet Tracker. Botnet tracker is a multithreaded application which enables the parallel execution of multiple Bot samples in virtual environment.

Bot detection and Botnet tracking modules are the core modules of the framework and hence they are explained in detail in the subsequent sections.

4 Bot Detection Engine

Looking for the Botnet features in the network traces is very common approach for the detection of Bot binaries. Researchers have used techniques such as network signatures, IRC command detection, Network flow clustering, and Snort alert modeling [13] for the detection of botnets. These techniques works successfully in normal circumstances but in cases where the malware communication channel is encrypted or covert, these network based detection techniques fails.

To overcome this shortcoming we have adopted a detection approach which uses window's Native API call sequences generated during binary execution as a feature set for the BOT detection. Figure 3 shows the process flow for Bot detection engine. The details of each sub process is given below.

Behavior Mining: For the Bot behavior mining following steps are performed (1) Quantification of the feature sets, (2) Feature Scaling, (3) Extracting most discriminating feature set. For the development of Bot detection engine we have used Native API call sequence's frequency as a feature set. Native API calls sits between the windows user space APIs and the windows system calls. It is not a very well documented API set and is internally used by the Windows NT family of operating systems. As no official documentation is available on the windows Native APIs hence it's difficult for the hackers to manipulate these APIs and implement malware evasion and bypass techniques. We have used NTtrace [26] tool for the

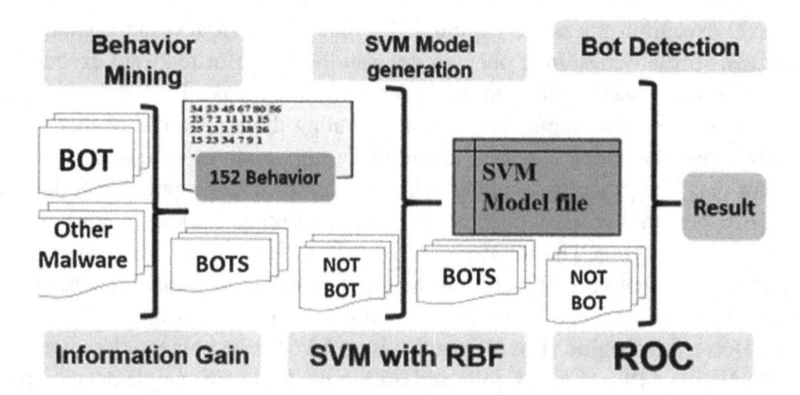

Fig. 3 Process flow

capturing of Native API call traces. NtTrace is a windows based tool which uses the windows debug interface to place breakpoints in NtDll around native Windows-calls into the kernel. Each time a breakpoint is hit the NtTrace reads the arguments passed to/values returned by the associated call. There are around 949 Native APIs present in Windows XP and out of these 949 APIs we have shortlisted 80 security related Native APIs calls (excluding APIs related to i.e. graphics calls, generated by Win32k.sys etc.). During the sandbox execution of binary sample the Nttrace captures Native API call traces generated. These captured Native API traces are filtered for extracting the selected 80 security related API call traces. The captured traces are converted in to numeric equivalent, as each API function out of the 80 APIs is arbitrarily assigned a numeric value from 1 to 80. The resulting long numerical sequence of APIs is broken in to k length sequences using sliding window technique. Based upon the literature survey (shown by wanke lee in his work [27]) we have chosen 6 as the value of K.

Selection of Most discriminating features: The feature selection process reduces the number of features either by eliminating the parameters or by transforming the parameters. There are various feature selection methods available in the literature. The most frequently used among them are Information gain, CHI square technique and Mutual Information, document frequency. Information gain is one of the most extensively used feature selection algorithm. We have used the Information gain algorithm for finding out most discriminative features. The information gain for a given attribute X with respect to the class attribute Y is the reduction in uncertainty about the value of Y when we know the value of X, I(Y; X). In our experiment we had taken 50 Bot samples and 50 non Bot samples. The non Bot samples consist of malware of category worms, viruses, Trojans, backdoors. During the execution of these 100 samples we got 1900 six length Native API call sequences. We applied the Information gain formula on above features with a cutoff of 0.13266 for IG value, which reduced this set in to a subset of 152 API call sequences. These 152 Native API call sequences are the one which are the most discriminating API call sequences. We have used the frequency of their occurrence as a feature set.

Model generation: For the selection of appropriate Machine learning algorithm we tested the performance of various popular classifiers. The results of the experiment is listed in Table 1. The SVM based classifier has shown low percentage

Table 1 Results for varrious classifiers

S. no	Classifier	Error percentage	Standard deviation after performing 10 fold cross validation
1	Linear discriminant classifier (LCD)	18.79	0.000236
2	Quadratic discriminant classifier (QDC)	17.89	0.000238
3	KNN classifier	14.17	0.001603
4	SVM classifier	10.145	0.000263

Fig. 4 ROC curve

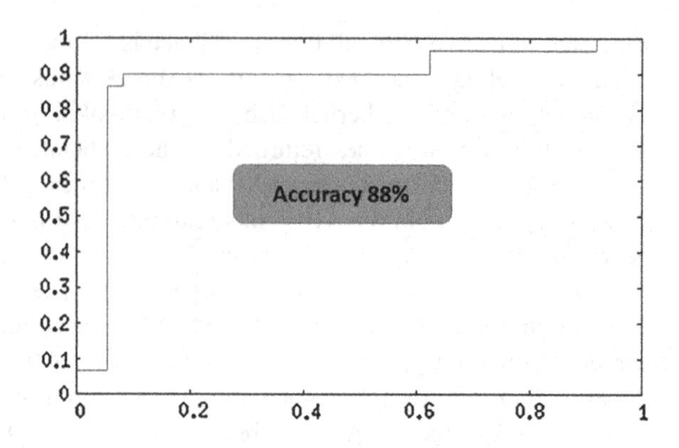

error and better performance, hence we have selected SVM based classifier for the Bot detection engine. Also SVM has generalization capability due to which the classifiers can generate more accurate results even with the smaller training data sets. We have used a radial basis function as a kernel function for the classifier. For the selection of optimum values for the penalty parameters C and Gama we have used grid search based tenfold cross validation algorithm. Based upon the results of the algorithm the value of C has been fixed to 64 and the value of Gama is set to 0.015625.

4.1 Bot Detection Engine's Results

We have evaluated the Bot detection engine with a test data set consisting 45 Bot samples (35 IRC Bots and 10 HTTP Bots) and 68 non-Bot samples (comprising of other malware such as worms, backdoors, virus and Trojans) collected form Honeynet setup. This test data is executed in the sandbox environment and the six length Native API sequences were obtained.

The frequency of occurrence of 152 prominent API sequences selected is noted for all binary samples. The hence created features are scaled and processed through the SVM model which predicts the target class of the malware samples. Figure 4 shows the ROC curve obtained which shows an overall accuracy of 88 %.

5 Botnet Tracking

The Botnet tracking module takes input from the honeypot sensors, sandbox and the Bot detection engine. The Bot detection engine forwards the malware samples labeled as Bot. The Botnet tracking engine exports network and system behavioral attributes corresponding Bot sample from the databses. Freiling et al. [3] had proposed a three step approach for botnet tracking which involves (1) infiltrating the remote control network (2) Analyzing the network in detail (3) shutting down the remote control server. In the work presented in this paper we have adopted somewhat similar approach with some improvements.

Currently we have identified total 15 attributes extracted from network logs, system logs and Malware structure that are used to characterize the Botnet family. We have implemented K-means cluttering algorithm on the feature.

In the experiment performed by us we processed 150 Bot samples. While discarding the clusters having less than 2 Bot samples we obtained 3 Bot clusters. Table 2 shows the Botnet\clusters obtained the sixth column shows the number of Bot malware that got clustered under the corresponding Botnet class. Rest of the columns are for attributes such as attacking IP (IP used to infect the remote machine), C&C IPs, Domain names and domain name count. The Bot samples labeled under these three Botnet families are further executed in sandbox on regular basis for tracking of the botnets. As every Bot class have multiple Bot sample hence the duration up to which these samples should be executed in sandbox is critical parameter effecting the system performance.

To determine the optimal value of malware execution time window we have performed an experiment with 100 Bot samples. In the experiment 100 Bot samples were executed in sandbox environment and the time taken to generate first communication with their C&C server was observed. Figure 5 shows the results of experiment which predicts that most of the Bot samples communicated with their C&C server within the 7 min. Based upon this observation we have fixed the execution time for each malware to be 7 s.

Table 2 Bot families obtained after clustering

Botnet	Count	Attacker IP	Download IP	C&C	Malware	DNS
B1	8	122.160.33.x 122.160.19.x 122.160.51.x 122.160.115.x	146.185.xx 60.10.179.xx	146.185.246.x 60.10.179.x 89.248.168.x	20	Bitcity.org, Pr0.net, Uhmm. svaravinoplaks.com Check. strongsearch.com loader.strongsearch. net Love.swastikano.net
B2	20	122.160.115.x 203.81.224.x 203.99.173.x 122.160.127.x	92.241.184.x 119.153.30.x 203.99.173.x	188.132.163.× 122.226.202.x 58.221.60.x 203.81.224.x 122.160.52.x	10	Tv.yaerwal.com API.WIPMANIA. COM shaimenal.com hubbing. blaztikwas.com .raniastyle.com j.symtec.us j.idolmovies.com d. theimmagebook.com
B3	10	82.165.149.x	203.81.224.x 203.100.70.x 122.160.127.x	124.232.137.x 111.123.180.x	8	ppppnipponp.r1m.us zaber.zaberhmar.com

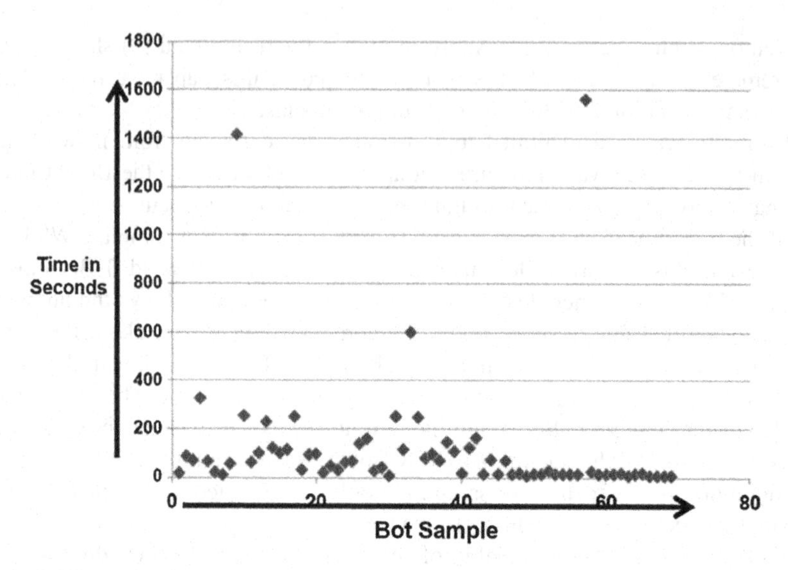

Fig. 5 BOT sample

6 Conclusion and Future Work

In the work presented in this paper we have proposed a system for Bot detection and tracking of botnets. We have used a distributed network of Honeynets for capturing the malwares followed by Native API based Bot detection approach which segregates the Bot malwares form rest of the collected malware sample. Bot samples identified are clustered based upon their system and network level features. The Bot clusters thus obtained represents a Botnet family. The Bot sample of thus identified Botnet families are executed on a regular basis in the sandbox environment for the tracking of these botnets.

More work needs to be done on the tracking of the botnets. More refined feature sets could be look in to for better Bot crusting results. In future we will also look for the possibility of implementing some time series analysis to monitor the trends in the evolution of the botnets.

References

1. Zeidanloo, H.R., Shooshtari, M.J.Z., Amoli, P.V., Safari, M., Zamani, M.: A taxonomy of Botnet detection techniques, IEEE (2010) 987–1-4244-5540-9
2. Vrable, M., Ma, J., Chen, J., Moore, D., Vandekieft, E., Snoeren, A.C., Voelker, G.M., Savage, S.: Scalability, fidelity and containment in the potemkin virtual honey farm. In: Proceedings of ACM SIGOPS Operating System Review, vol. 39(5), pp. 148–162 (2005)

3. Freiling, F., Holz, T., Wicherski, G.: Botnet tracking: exploring a root-cause methodology to prevent distributed denial-of-service attacks. In: Proceedings of 10th ESORICS. Lecture Notes in Computer Science, vol. 3676, pp. 319–335, Sept 2005
4. Sehgal, R.K., Bhilare, D.B., Chamotra, S.: An integrated framework for malware collection and analysis for Botnet Tracking. Int. J. Comput. Appl. (0975–8887) Commun. Secur. **10** (2012)
5. Gu, G., Yegneswaran, V., Porras, P., Stoll, J., Lee, W.: Active Botnet probing to identify obscure command and control channels. In: Proceeding of Annual Computer Security Application Conferences (ASAC), pp. 241–253 2009 (Botprobe)
6. Christodorescu, M., Jha, S: Static analysis of executable to detect malicious patterns. In: Proceedings of the 12th USENIX Security Symposium (2003)
7. Kinder, J., Katzenbeisser, S., Schallhart, C., Veith, H.: Detecting malicious code by model checking. In: Conference on Detection of Intrusions and Malware and Vulnerability Assessment (DIMVA) (2005)
8. Kruegel, C., Robertson, W., Vigna, G.: Detecting kernel-level rootkits through binary analysis. In: Annual Computer Security Application Conference (ACSAC) (2004)
9. Cohen, F.: Computer virus: theory and experiments. Comput. Secur. **6**, 2235 (1987)
10. Madou, M., Anckaert, B., De Sutter, B., De Bosschere, K.: Hybrid static-dynamic attacks against software protection mechanisms. In: ACM Workshop on Digital Rights Management. Alexandria, VA, Nov 2005
11. Chen, X., Andersen, J., Mao, Z., Bailey, M., Nazario, J.: Towards an understanding of anti-virtualization and anti-debugging behavior in modern malware. In: IEEE International Conference on Dependable Systems and Networks with FTCS and DCC. DSN 2008, pp. 177–186 (2008)
12. Sharif, M., Lanzi, A., Giffin, J., Lee, W.: Impeding malware analysis using conditional code obfuscation. In: 15th Annual Network and Distributed System Security Symposium (NDSS08) (2008)
13. Brumley, D., Hartwig, C., Liang, Z., Newsome, J., Poosankam, P., Song, D., Yin, H.: Automatically identifying trigger-based behavior in malware. In: Botnet Analysis and Defense (2007)
14. Christodorescu, M., Jha, S., Seshia, S.A., Song, D.: Semantics-aware malware detection. In: Proceedings of the 2005 IEEE Symposium on Security and Privacy 1081–6011/05
15. Zhang, B., Yin, J., Hao, J., Zhang, D: Using support vector machine to detect unknown computer viruses. Int. J. Comput. Intell. Res. ISSN 0973–1873, pp. 100–104
16. Wang, C., Pang, J., Zhao, R., Fu, W., Liu, X: Malware detection based on suspicious behavior identification. In: First International Conference on Education Technology and Computer Science, IEEE Computer Society, 987-0-7695-3557-9/09
17. Dai, J., Guha, R., Lee, J.: Efficient virus detection using dynamic instruction sequences. J. Comput. 4(5) (2009)
18. Liu, L., Chen, S., Yan, G.: BotTracer: execution based Bot-like malware detection. In: Volume 5222 of the Series Lecture Notes in Computer Science, pp. 97–113
19. Gu, G., Porras, P., Yegneswaran, V., Fong, M., Lee, W.: BotHunter: detecting malware infection through IDS-driven dialog correlation. In: 16th USENIX Security Symposium (2008)
20. Goebel, J., Holz, T.: Rishi: identify Bot contaminated hosts by IRC nickname evaluation. In: USENIX Workshop on Hot Topics in Understanding Botnets (HotBots'07) (2007)
21. Stover, S., Dittrich, D., Hernandez, J., Dietrich, S.: Analysis of the storm and Nugache Trojans: P2P is here. USENIX; login **32**, 18–27 (2007)
22. Stinson, E., Mitchell, J.C.: Characterizing Bots' remote control behavior. In: International Conference on Detection of Intrusion and Malware and Vulnerability Assessment (2007)
23. Al-Hammadi, Y., Aickelin, U.: DCA for Bot detection. In: Proceedings of the IEEE World Congress on Computational Intelligence (WCCI) Hong kong, pp. 1807–1816 (2008)

24. Wang, M., Zhang, C., Yu, J.: Native: API based windows anomaly intrusion detection method using SVM. In: Proceedings of the IEEE International on Sensor Networks, Ubiquitous, and Trustworthy Computing (SUTC'06)
25. Chamotra, S., Sehgal, R.K., Kamal, R.: HoneySand: an open source tools based sandbox environment for Bot analysis and Botnet tracking. Int. J. Comput. Appl. Commun. Secur. (Special issue) **7** (2012)
26. www.howzatt.demon.co.uk/NtTrace/
27. Lee, W., Dong, X.: Information-theoretic measures for anomaly detection. In: Needham, R., Abadi, M. (eds.) Proceedings of the 2001 IEEE Symposium on Security and Privacy (2001)

An Approximation of Forces and Tooling Configuration During Metal Forming Process Using Artificial Intelligence Technique

M.R. Bhatt and S. Buch

Abstract In the present work, a module is developed to approximate the forces during deep drawing operation. It also predicts the tooling configurations i.e. blank size, punch and die diameters, profiles, number of draws etc. during each stage of operation. Deep drawing is metal forming process, normally used to produce several components like automotive body panel, household utensils etc. Generally, during the deep drawing process, it is required to measure forces for tooling design. Traditionally it is done using hit and miss method. It consumes time and increase indirect costs. Hence in the present study, a module has been made using rules and knowledge based system (if then rules) to estimate forces during deep drawing operation. The codes are famed using VB and interfaced with the AUTOCAD to generate the 2D drawing of the tooling for production. It will help small and medium scale industries to predict the forces before actual operation. Thus it is easier for them to build tools (punch, die, blank holder etc.) for different geometrical and material conditions.

Keywords Approximation · Force · Tooling configuration · Deep drawing · VB · AUTOCAD

1 Introduction

Deep drawing is one of the metals forming process which is widely used in production of automobile and aerospace components, food containers, computer and washing machine body parts and other household utensils. In the deep drawing process, a blank (workpiece) is placed over the die and gripped with blank holder

M.R. Bhatt (✉)
R K University, Rajkot, Gujarat, India
e-mail: mrs.mrbhatt1412@yahoo.in

S. Buch
Reliance Industries, Ahmedabad, Gujarat, India
e-mail: drshbuch@gmail.com

S.C. Satapathy and S. Das (eds.), *Proceedings of First International Conference on Information and Communication Technology for Intelligent Systems: Volume 1*, Smart Innovation, Systems and Technologies 50, DOI 10.1007/978-3-319-30933-0_57

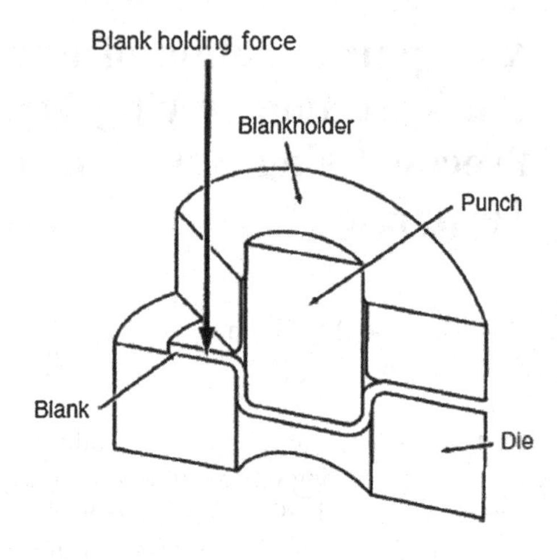

Fig. 1 Mechanics of deep drawing of cylindrical cup [1]

then the punch travels towards the blank and deform it under contact zone as shown in Fig. 1. Such a way, the finished components produced. The process requires certain tools i.e. punch, die, blank holder etc. The design of tooling changes based on the blank material and geometrical conditions of the final component.

Rules and knowledge based systems have proven the capabilities in other domains like logistics, medical, accountancy, fisheries etc. Even many researchers have worked on such useful tool in manufacturing sectors like welding, casting, machining and forming. Era and Dias [2] developed the expert system for die casting dies. It decides most appropriate casting process parameters. Also in 2002, Lee and Luo [3] generated a system for self learning and reasoning for die casting of metals. An automated nesting and piloting system was developed by Ghatrehnaby and Arezoo [4]. They developed CAD system for automated nesting and piloting of geometry for progressive die. Chul Kim and Chul Woo Park [5] used DEFORM and ANSYS to develop expert system for axi-symmetric cold forging components. Metal stamping die set tool selection system was given by Kumar and Singh [6]. To predict the deep drawing behavior of tailor welded blanks, Veera Babu et al. [7] produced an expert system using PAM-STAMP FE code. Naranje and Kumar [8] discussed about the AI applications in sheet metal stamping die. Horikoshi et al. [9] concentrates on expert system of deep drawing process using high velocity water jet. The fluid dynamics concept was incorporated along with Reynold's equation. However some local issues of the micro, small and medium scale industries are still not addressed. Hence, in present study, tooling design is concentrated as it is a very important in small metal forming industries.

Generally micro, small and medium scale industries produce such tooling based on the personnel experience and trial & error method. Here, an attempt is made to eliminate such practice which reduces lead time and cost. It also enhances quality and productivity. In connection to this, force approximation system have been developed using VB and AUTOCAD based on the rules (IF; THEN) and knowledge. The rules and knowledge is framed using design data book. The module can automatically predict the punch force, blank size, punch diameter, die diameter, number of draws etc. The inputs from the user are geometry and material properties only. It will help industry to react instantly with fluctuating market demand. It also provide an advantage that a semi skilled labor can also easily operate this system.

2 Development of System

2.1 Technical Detail

In the present work, rules and knowledge have been framed using the industrial data and design data book of deep drawing process. A sample of framed rules is given in Appendix I. The system architecture is given in Fig. 2.

All the rules are executed in the given fashion to arrive at the desired results. Here, Fig. 2 depicts that, architecture of the methodology to be performed during the execution of the system. The system data includes following checks.

(1) Input Parameters

Input parameters are the basic parameters which are required to provide for the system execution. The input parameters include diameter of the cup/component, height of cup and final thickness of the component which is desired. Figure 3a shows the main input parameter dialogue box. Further, different material is having different properties and it is required to be incorporated to execute rules. So that, material property dialogue box is given as per Fig. 3b. Here user need to feed material type, yield stress and ultimate stress of the particular material along with draw ratio.

(2) Blank design

Blank is nothing but the initial work piece which is converted into the finished component. The blank diameter can be calculated using Eq. (1) [10].

$$D = (d^2 - 4dh)^{0.5} \tag{1}$$

where, h = cup height (mm), d = cup diameter (mm) and D = blank diameter (mm)

(3) Determination of Clearance

Clearance is the gap between the work material and tool. It may be between work material and punch and work material and die. Normally in industrial practice it is taken as 15–20 times of blank thickness based on the workpiece material. It will produce air gap between tooling and workpiece which prevent sticking during operation.

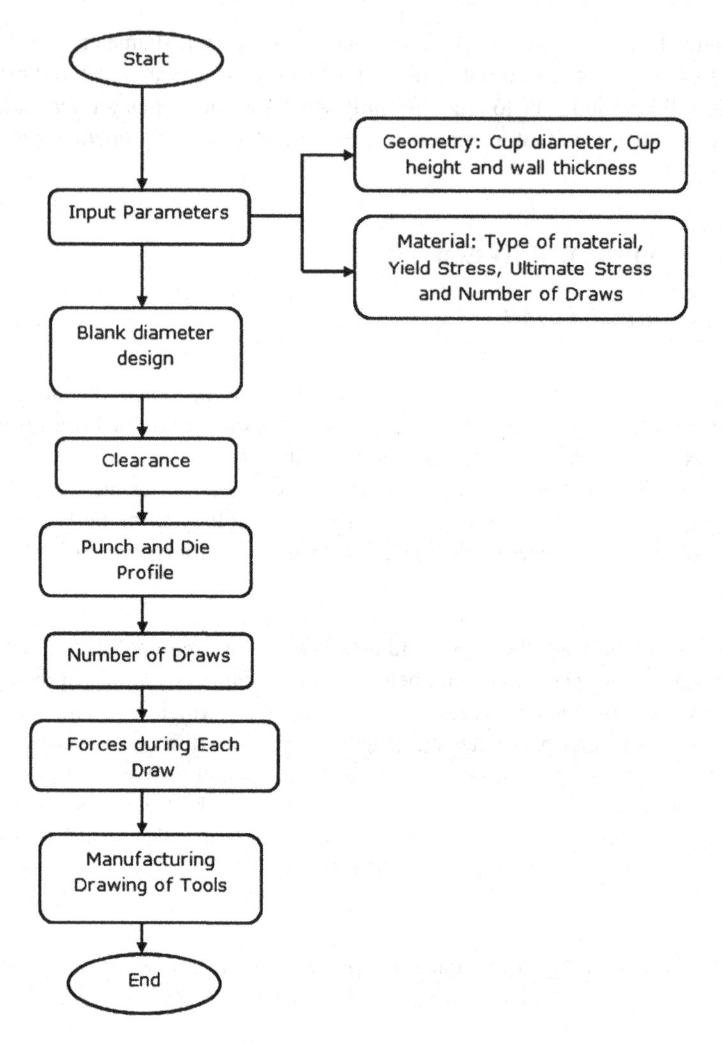

Fig. 2 System architecture for execution

(4) Determination of Punch and Die profile

The punch profile is very important during the operation because it is going to replicate the profile of the finished product. Moreover, the die profile radius is also vital as it is a negative impression of the punch profile. Such die profile radius is required in closed die operation. The punch and die profile radius ranges from 4-10 times of the material thickness. Figure 4 shows the clearance, punch and die profile and blank diameter dialogue box.

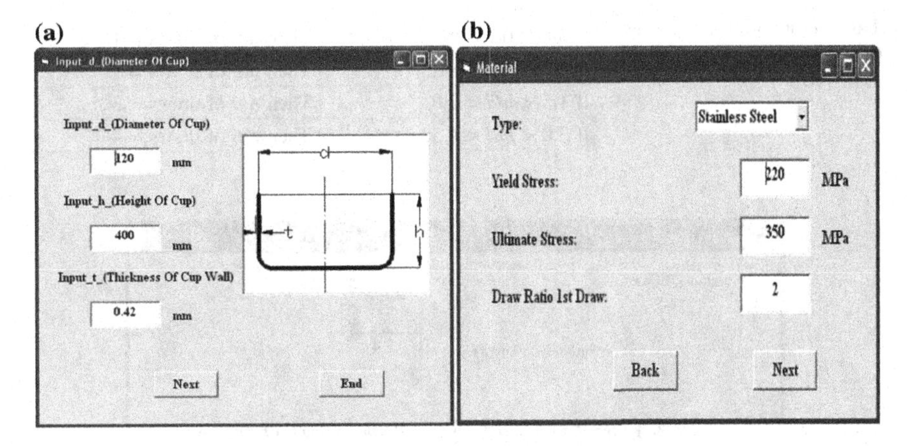

Fig. 3 Input parameter dialogue box **a** geometry, **b** material property

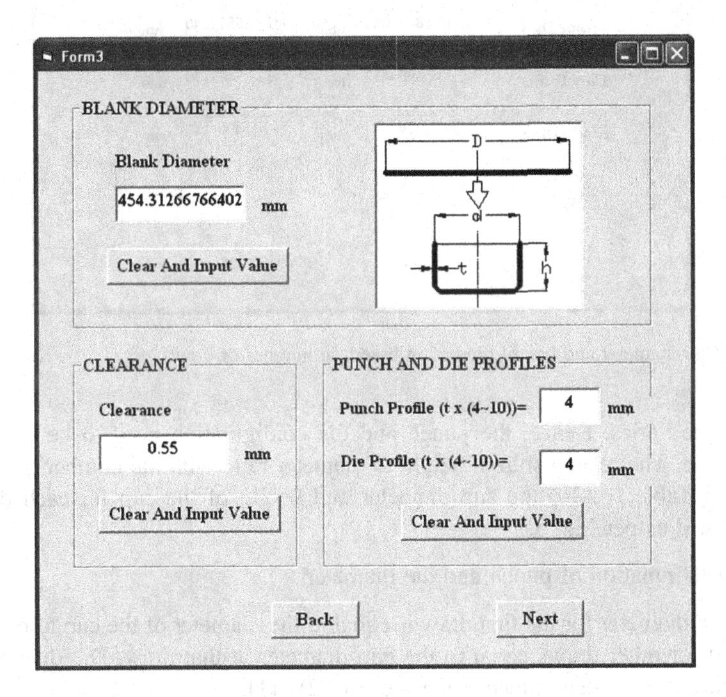

Fig. 4 Clearance, punch and die profile and blank diameter dialogue box

(5) Determination of number of Draws

Many times the component is required to be work out many times. It is known as number of draws. It is required when the height to diameter ratio [draw ratio (*h/d*)] is high for component. As the number of draw increases corresponding punch and

Table 1 Relationship of the (h/d) ratio and number of draw [10]

If $h/d < 0.75$	Then no. of draws = 1
If $0.75 < h/d < 1.5$	Then no. of draws = 2
If $1.5 < h/d < 3.0$	Then no. of draws = 3
If $3.0 < h/d < 4.5$	Then no. of draws = 4

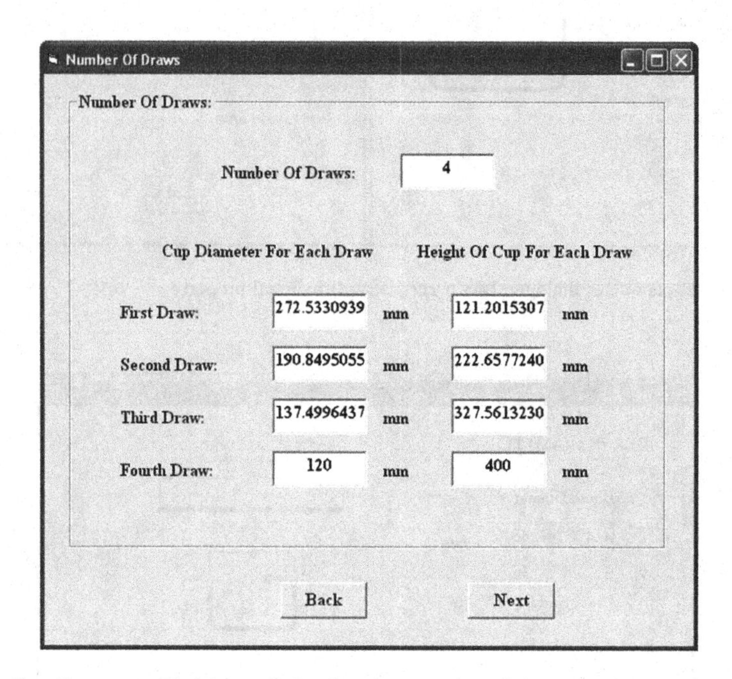

Fig. 5 Cup diameter and height prediction based on number of draws

die profile varies. Hence, the punch and die configuration need to be calculated every time. The relationship of height to diameter ratio with the number of draw is given in Table 1. Also the cup diameter and height of the cup for each draw is determined as per Fig. 5.

(6) Determination of punch and die diameter

Punch diameter for the first draw is equal to the diameter of the cup to be drawn (d), and for other draws equal to the cup diameter at that draw. Die diameter for each draw can be determined using the Eq. (2) [11].

$$\text{Die diameter} = \text{punch diameter} + (2 * \text{clearance}) \tag{2}$$

(7) Determination of punch and blank holder Force

The punch is normally operated by hydraulic piston and cylinder assembly for precise and controlled motion. The force encountered during the operation can be calculated by the Eq. (3) [11]. Further, blank holder is required to hold the blank/workpiece during the operation. So that, it cannot be displace during working. Hence blank holder force is provided and it play vital role. The reason is that, if the blank holder force is too high then material does not go into die and higher thinning occurs, leads to failure of component. On the other hand, if blankholder force is too less then blank lifts up from its position and wrinkles produced on its flange area. Hence accurate blank holding force is very much important to predict. The empirical formula to predict blank holding force is given by Eq. (4).

The total force is calculated based on the summation of maximum punch force and blank holder force. Figure 6 shows the dialogue box which indicates punch and die force and diameter after each draw.

$$F_{\max} = \pi \cdot d \cdot t \cdot \sigma_{ult}((D/d) - 0.7) \tag{3}$$

where, $F_{\max}$ = maximum drawing force (N), t = initial blank thickness (mm), σ_{ult} = ultimate tensile strength (MPa), D = blank diameter (mm) and d = punch diameter (mm)

$$\text{Blank holder force} = 1/3 * \text{maximum punch force} \tag{4}$$

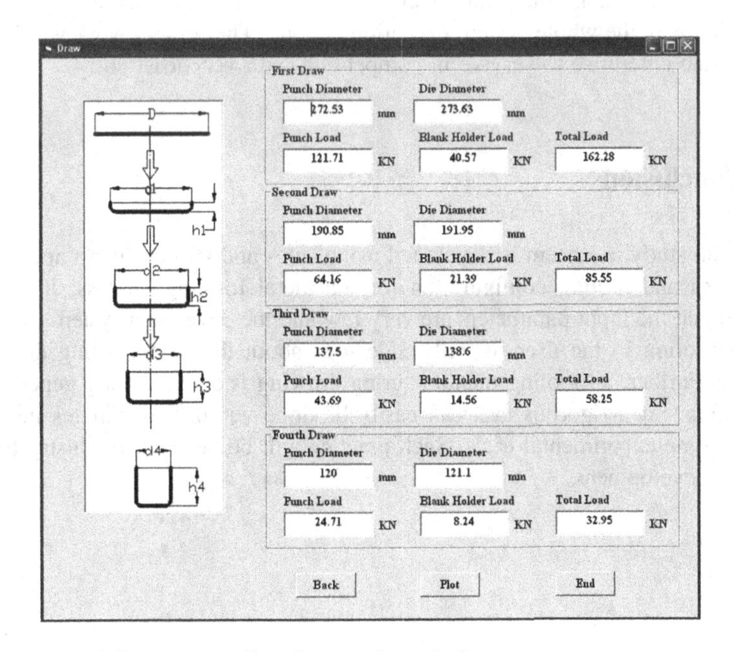

Fig. 6 Punch and die force as well as diameter for each draw

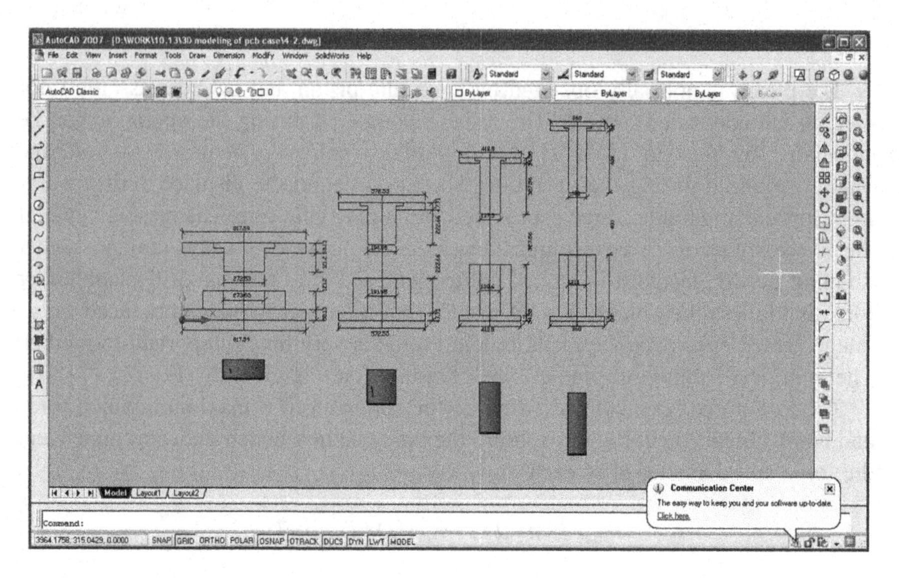

Fig. 7 Punch and die force as well as diameter for each draw

(8) Development of Manufacturing drawing of tools

The system is synchronized with the AUTOCAD (draughting tool) in such way that, after execution of each check, it will generate the manufacturing drawing of the punch, die, blank holder and blank after each draw as shown in Fig. 7.

Such a way the whole system execution is done. The present work will help the small scale enterprise to survive in competition with very least cost.

3 Conclusion

In current study, a system is developed using rules and knowledge to approximate the forces and tooling configuration for the metal forming process. It has been noticed that the input parameters are very few and the developed system gives a lot of information to the user. It will guide user about the forces acting during the process. Further, all tooling manufacturing drawing is automatically generated for fabrication. Such rigorous task can easily be done within few minutes instead of cumbersome experimental trials. Such practice will be helpful to industry for new product development.

Appendix I

```
Private Sub next_b_Click()
    If Val(diameter_t.Text) = 0 Then
    MsgBox "Enter Value of Diameter.(Diameter of Cup)",
    vbOKOnly, "Warnning"
        Exit Sub
    End If
    If Val(height_t.Text) = 0 Then
    MsgBox "Enter Value of Height.(Height of Cup)",
    vbOKOnly, "Warnning"
        Exit Sub
    End If
    If Val(thickness_t.Text) = 0 Then
    MsgBox "Enter Value of Thickness.(Thickness of
    Cup)", vbOKOnly, "Warnning"
        Exit Sub
    End If

    Form1.Hide
    Form2.Show
End Sub
```

References

1. ASM Handbook, forming and forging, **14**, (2012)
2. Era, A., Diasb, R.: A rule—based expert system approach to process selection for cast components. Knowl. Based Syst. **13**(4), 225–234 (2000)
3. Lee, K.S., Luo, C.: Application of case-based reasoning in die-casting die design. Int. J. Adv. Manuf. Technol. **20**(4), 284–295 (2002)
4. Ghatrehnaby, M.: Barezoo.: a fully automated nesting and piloting system for progressive dies. J. Mater. Process. Technol. **209**(1), 525–535 (2009)
5. Chul, K., Chul, W.P.: Development of an expert system for cold forging of axisymmetric product. Int. J. Adv. Manuf. Technol. **29**(5), 459–474 (2006)
6. Kumar, S., Singh, R.: A low cost knowledge base system framework for progressive die design. J. Mater. Process. Technol. **1**, 958–964 (2004)
7. VeeraBabu, K., Ganesh Narayanan, R., Saravana Kumar, G.: An expert system for predicting the deep drawing behavior of tailor welded blanks. Expert Syst. Appl., **37**, 7802–7812 (2010)
8. Naranje, V., Kumar, S.: AI applications to metal stamping die design—a review. World Acad. Sci., Eng. Technol. **4**, 08–22 (2010)
9. Horikoshi, Y., Kuboki, T., Murata, M., Matsui, K., Tsubokura, M.: Die design for deep drawing with high-pressured water jet utilizing computer fluid dynamics based on Reynolds equation. J. Mater. Process. Technol. **218**, 99–106 (2015)
10. Hosford, W.F., Caddell, R.M.: Metal Forming: Mechanics and Metallurgy. Cambridge University Press (2011)
11. Ghosh, A., Malik A.K.: Manufacturing Science. East West Press Pvt. Ltd. (2009)

Matchmaking of Web Services Using Finite State Automata

Sujata Swain and Rajdeep Niyogi

Abstract Recent works in web services have employed finite state machines for solving different problems, like matchmaking of web services, modelling of web service composition and verification of web service composition. Annotated Deterministic Finite State Automata (ADFSA) is used for matchmaking of web services. ADFSA is the combination of deterministic finite state automata (DFA) with logical annotation of transitions in state. BPEL4WS is a high level programming language to express the execution behavior of web services but this language is Turing-complete. For matchmaking of web services, only a fragment of BPEL suffice which is equivalent to regular language. A complex web service is obtained from simpler web services where each simpler web service is modeled as Communicating Automata (CA). A CA is a Non-deterministic Finite State Automata (NFA). In this paper, we show how CA can easily be used for matchmaking of services. For this purpose, we give translations of CA to ADFSA. Thus matchmaking of services can be carried out even when the services are modeled using CA.

Keywords Annotated deterministic finite state automata · Deterministic finite state automata · Communicating automata · Web service matchmaking

1 Introduction

Web services play an important role in E-business, E-commerce, and E-learning. Web service providers provide their services through Universal Description Discovery and Integration (UDDI). When a requester sends his request to the UDDI

S. Swain (✉) · R. Niyogi
Department of Computer Science and Engineering, Indian Institute of Technology Roorkee, Haridwar 247667, Uttarakhand, India
e-mail: sujataswain019@gmail.com

R. Niyogi
e-mail: rajdpfec@iitr.ernet.in

S.C. Satapathy and S. Das (eds.), *Proceedings of First International Conference on Information and Communication Technology for Intelligent Systems: Volume 1*, Smart Innovation, Systems and Technologies 50, DOI 10.1007/978-3-319-30933-0_58

(broker), it selects a web service URL (Uniform Resource Locator) by string comparison with description of service available at UDDI and send to the requester. But UDDI does not know that the selected web service is compatible or not to the requester. To know the compatibility of two web services, we need to carry out matchmaking process. In this process, the receiver must have the capability to receive all types of messages from the sender. If this is not the case, then both the sender and receiver will go into an indeterminate state.

Recent works in web services have employed finite state machines for solving different problems, like matchmaking of web services [1] and web services composition [2]. In [1], an ADFSA is proposed for matchmaking of web services. An ADFSA is essentially a DFA together with annotations on states. Such annotations are propositional logic formulae that capture the structure of the automata. A web service is defined by BPEL4WS (Business Process Execution Language for Web Services), a high level programming language especially designed for specifying the execution behavior of web services. BPEL4WS is very expressive that is Turing-complete. For matchmaking purpose, only a regular fragment of BPEL4WS suffice, which turns out to be as expressive as the regular languages. Thus now the matchmaking problem reduces to the intersection of DFAs [3]. In [1], a similar construction for ADFSAs has been suggested. In [2], a complex web service is obtained from simpler web services, where each web service is modeled as a CA which is a planning operator. The goal state is obtained from initial state by applying CA as action. A web service is specified using Promela [4] and the corresponding finite state machine is modeled as a CA which is a finite state machine (NFA) with send/receive messages comprising the alphabet of the automata.

BPEL4WS language does not provide matchmaking process in service discovery. But, at times of matchmaking, we only require a fragment of BPEL code. That fragment of BPEL code is a regular subset of BPEL, which can be represented by a DFA. So, to solve the matchmaking problem, DFA is a suitable starting point to model the business process. Matchmaking process is the non-empty intersection of DFA. We can understand the matchmaking process by giving a simple example of compatible and incompatible business process.

Let there be two business process, i.e., customer process and vendor process. Customer process starts with the order request message, it wants delivery before payment. Vendor process expects to receive order request followed by a payment, then it sends the delivery.

These two business processes match at the level of individual message sequence. But the intersection of two process is empty due to the different order of delivery and payment. So, they are incompatible and they cannot interact successfully. We need to take into account the message sequence rather than the individual messages to avoid the incompatible matching of business process [5].

There is also a case where the intersection is not empty, but the process cannot be matched. The process starts with the order message followed by a delivery message, and either a credit card payment or invoice payment message. If the ordered product is not available, then the vendor rejects the order and sends no stock available message. The vendor involves two types of messages: Mandatory message and Optional message. When it receives order request, it insists on the availability of both that is delivery and no stock. So, these two messages are mandatory message. On the other hand, it receives two alternative payment options, called optional messages.

Let the customer process has the transition to receive the delivery message, but there is no stock transition. Here, intersection is not empty. But, when vendor sends no stock message, then customer process cannot handle it and hence cannot reach an end state. So, this is called false match.

To provide matchmaking process, ADFSA is used, in [1, 6] where mandatory messages are represented as annotation of logical AND (mandatory message) and optional messages do not required any annotation because that is equivalent to standard automata.

An ADFSA A is represented as a tuple,

$$A = \langle Q_D, \Sigma_D, \delta_D, q_D^0, F_D, QA \rangle$$

where, Q_D is a finite set of states, Σ_D is a finite set of messages, $\delta_D : Q_D \times \Sigma_D \rightarrow Q_D$ represents a transition, q_D^0 is a start state with $q_D^0 \in Q_D$, $F_D \subseteq Q_D$ is the final states, $Q_A : Q_D \times$ E is a finite relation of states and logic terms within the set E of propositional logic terms. The terms in E are standard Boolean formulae.

Thus web services can be modeled using different forms of FA, namely; ADFSA and CA. Although, these automata are different we can show how CA can easily be used for matchmaking of services using the framework in [1]. For this we first give translations of CA to ADSFA and then use the construction suggested in Sect. 3 to obtain the matchmaking of services. Thus matchmaking of services can be carried out even when the services are modeled using CA.

USER CASE DOMAIN: ***Running Information of Trip Process***

A car maker would like to provide top model cars with a software. Through the software user can access all the remote services. Now a days, top model cars contain Global Positioning System (GPS) tracking system in which location of the car can be tracked. The software helps the user by observing the schedule and get updated information about the travel status to the user. The user uploads his plan schedule of travel onto the software. According to the plan and navigation system, the service knows the location of the car. The software automatically check the plan to make sure that he is on time or not. The software provides the running information of the trip. If any conflict occurs then the user will take proper action to change the schedule. This process consists of three processes, namely; user process, software process and navigation system, as shown in Fig. 1. It checks the user agenda with the above information. Then it sends conflict/no conflict information to the user.

To solve the matchmaking problem, there are several solutions provided by different authors. In [7–10], a graph based approach is used for matchmaking. Service matching is reduced to graph matching. The matchmaking process formulated as a search for graph or subgraph isomorphism. In [11], a multi-layer

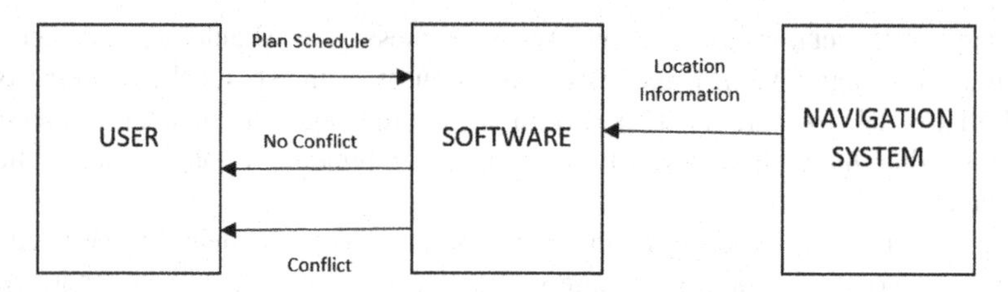

Fig. 1 Running information of trip process

UDDI architecture is proposed, so that we can choose partner, according to both enterprise commercial data and web services technical information. UDDI provides keyword based search facilities. This keyword-browsing method is limited, so it is insufficient. To overcome this limitation, in [12, 13] a similarity based matchmaking is used. There is not many formal methods to solve matchmaking process in service discovery.

Section 2 discusses the CA is used for modeling the web services. In Sect. 3, we give an algorithm for translation of CA to ADFSA for matchmaking. Section 4 discusses the conclusion.

2 Communicating Automata

In [2], a complex web service is obtained from simpler web services, where each web service is modeled as a CA which is a planning operator. The goal state is obtained from initial state by applying CA as action. A web service is specified using Promela [4] and the corresponding finite state machine is modeled as a CA. A CA is a finite state machine with send/receive messages comprising the alphabet of the automata. Promela language is input for SPIN [14–16] which is used for verification of the complex web service.

A communicating automata [2, 17] A is represented as a tuple,

$$A = \langle Q_C, \Sigma_C, \delta_C, q_c^0, F_C \rangle$$

where, Q_C is a finite set of states of the automata, Σ_C is the set of messages that can be sent to or received from a communication channel by A.

"$\langle$*channelname*$\rangle$! *message-name*" has a guard that returns true only when the channel is non full. Its effect changes the value of channel by appending the message name.

"$\langle$*channelname*$\rangle$? *message-name*" has a guard that returns true only when the channel holds a message and its effect part deletes that message from channel.

$\Sigma_C = \{!, ?\}\Sigma$. $\delta : \mathrm{Q} \times \Sigma_C \rightarrow 2^Q$ is the transition function. $q_c^0 \in Q_C$ is the initial state of A. $F_C \subseteq Q_C$ is the set of end states of A.

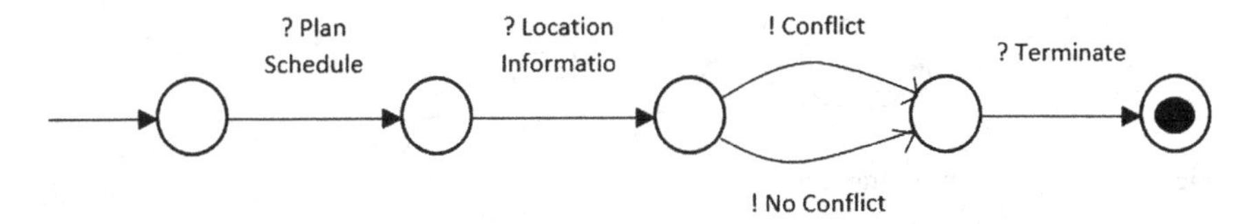

Fig. 2 Software process in communicating automata

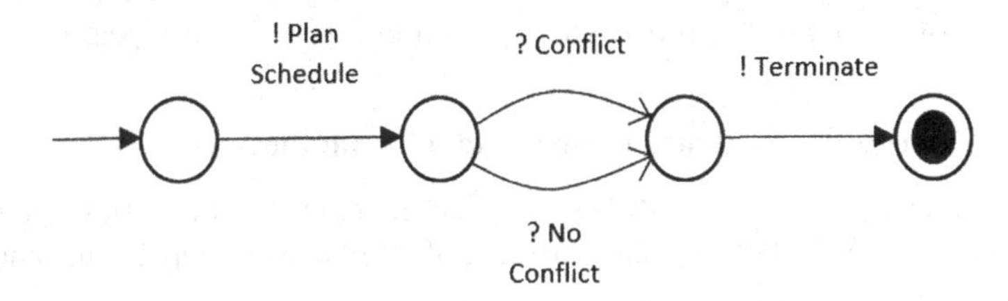

Fig. 3 User process in communicating automata

The communicating automata for software and user processes is shown in Figs. 2 and 3.

3 Matchmaking Through Communicating Automata

The web services can be modelled using CA for the verification of web service composition. Although, the web services are presented in CA, we cannot do matchmaking of the web services. In this section, we give an algorithm where we can do matchmaking of the web services by translating the CA to ADFSA.

There be two communicating automata A_1 and A_2.

$$A_1 = (Q_1, \Sigma_1, \delta_1, q_{10}, F_1, QA_1) \quad \text{and} \quad A_2 = (Q_2, \Sigma_2, \delta_2, q_{20}, F_2, QA_2)$$

Take $\Sigma = \Sigma_1 \cap \Sigma_2$ if $\Sigma \neq \phi$ then there may be a possibility for matchmaking, otherwise there is no possibility for matchmaking between automata.

3.1 Algorithm-1: Matchmaking of Web Services Through Communicating Automata (Translation of CA to ADFSA)

Step-1 **Remove the irrelevant transitions from A_1 and A_2.**

For all transitions in $A_1(\forall e_1 \in A_1)$, let $\delta_1(q_1, e_1) = q_2$, means $e_1 \in \Sigma_1$, $e_1 \in \Sigma$ or $e_1 \notin \Sigma$. If $e_1 \notin \Sigma$ then remove e_1 by following the procedure mentioned under.

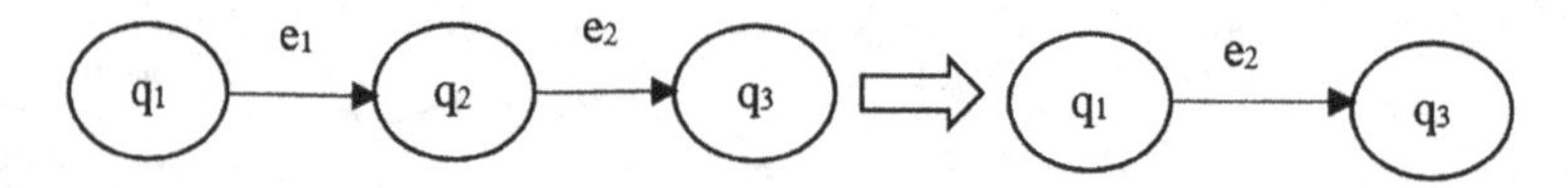

Fig. 4 Remove irrelevant transitions

Let $\delta_1(q_2, e_2) = q_3$ then make transition $\delta_1(q_1, e_2) = q_3$. Now remove q_2 and also remove the incoming transitions associated with q_2 and add the outgoing transitions of q_2 to q_1. This procedure is shown in Fig. 4. Now, repeat the process for A_2.

Step-2 **Remove the send and receive symbol from message.**

For all transitions in $A_1 (\forall e_1 \in A_1)$, $e_1 =$ "!*Message*-name "or " ?*Message-name*" then remove ! or ? symbol from all transitions after remaining Step-1. Then the same process is for A_2.

Step-3 **Now, construct DFA from NFA by using subset construction** [3].

Subset Construction

Any NFA $N = \langle Q_N, \Sigma, \delta_N, q_0, F_N \rangle$ is equivalent to a DFA $D = \langle Q_D, \Sigma, \delta_D, q_0, F_D \rangle$ where D can be defined in terms of N as follows.

Start state of D is the set containing only the start state of N, Q_D is the set of subsets of Q_N that is Q_D is the power set of Q_N. If Q_N has n states then Q_D has 2^n states. Many states are not accessible. Such inaccessible states can be eliminated. So, number of states of D can be much smaller than 2^n. F_D is the set of subsets S of Q_N such that $S \cap F_N \neq \phi$. For each set $S \subseteq Q_N$ and for each input symbol a in Σ, $\delta_D(S, a)$ is

$$\delta_D(S, a) = \bigcup_{p \in S} \delta_N(p, a)$$

Step-4 **Put the logical annotation of transitions wherever it is required within state for knowing the mandatory messages.**

For all states in $A_1 (\forall Q \in A_1)$, put the logical AND ($\wedge$) in between transitions for the mandatory transitions in states and stores in QA_1. Repeat the process for A_2.

Step-5 **Perform the intersection of A_1 and A_2.**

The intersection of $A_1 \cap A_2$ is $A = (Q, \Sigma, \delta, q_0, F, QA)$, with $Q = Q_1 \times Q_2$, $\Sigma = \Sigma_1 \cap \Sigma_2$, $\delta((q_{11}, q_{21}), \alpha) = (q_{12}, q_{22})$ with $\delta_1(q_{11}, \alpha) = q_{12} \wedge \delta_2(q_{21}, \alpha) = q_{22}$, $q_0 = (q_{10}, q_{20})$, $F = F_1 \times F_2$, and the transition function of intersection is defined as

$$\forall((n, n'), l, (m, m')) \in T : (l \in \Sigma \cap (n, l, m)) \cap (l \in \Sigma' \cap (n', l, m'))$$

Equation of QA part in intersection of two automata is

$$QA = \bigcup_{q_1 \in Q_1, q_2 \in Q_2} \begin{cases} ((q_1, q_2), e_1 \Lambda e_2) & if (q_1, e_1) \in QA_1, (q_2, e_2) \in QA_2 \\ ((q_1, q_2), e_1) & if (q_1, e_1) \in QA_1, q_2 \in Q_2, \nexists e'(q_2, e') \in QA_2 \\ ((q_1, q_2), e_2) & if (q_2, e_2) \in QA_2, q_1 \in Q_1, \nexists e'(q_1, e') \in QA_1 \\ \phi & otherwise \end{cases}$$

Step-6 **If $A_1 \cap A_2 \neq \phi$ then return yes, otherwise return no.**

If $L(A_1) \cap L(A_2) = \phi$ then there is no matchmaking between web services. If $L(A_1) \cap L(A_2) \neq \phi$ then there is matchmaking in between two web services.

3.2 *Illustration*

Software process has irrelevant transitions because it also communicates with navigation system. So in the matchmaking process of software and user process, these transitions are removed. Retain only the transitions in between software and user process, others are removed as shown in Fig. 5. After removing irrelevant transitions, the next step is removing the send and receive symbol from the software and user process as shown in Fig. 6. Next, construct DFA from NFA. But the software and user automata are already in DFA. So there is no need of translation of NFA to DFA in this case. Then next step is putting the logical annotation into transitions in states of software process as shown in Fig. 7. The intersection of user automata and software automata is similar as shown in Fig. 6. The intersection is non empty, so the user process is matched with software process.

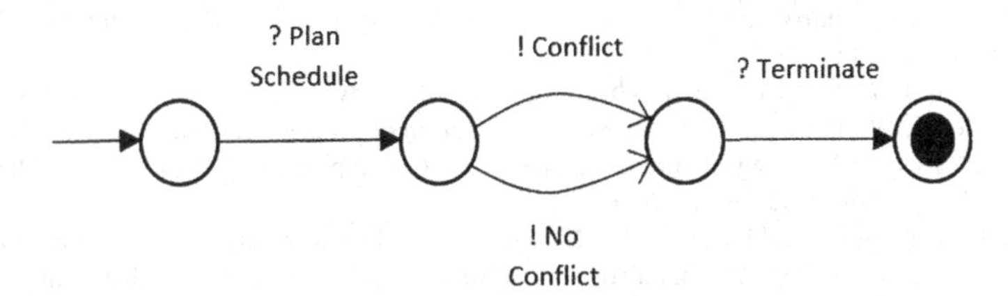

Fig. 5 Removal of irrelevant transitions from software process

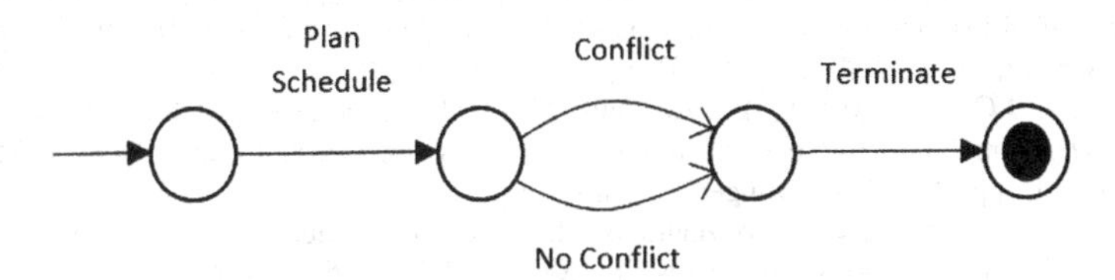

Fig. 6 Removing send and receive symbol from software process

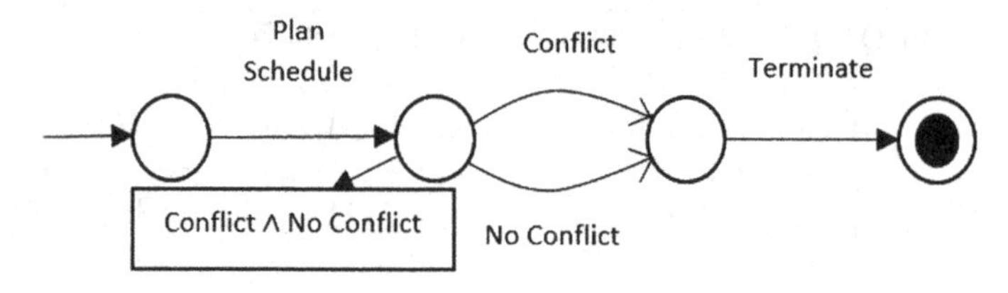

Fig. 7 Put the logical annotation in state

4 Conclusion

We assert that BPEL4WS is insufficient for matchmaking. For matchmaking process, a regular fragment of BPEL4WS is required, which can be represented by DFA. For matchmaking, ADFSA is used. A complex web service is obtained from simpler web services. The simpler web service is in form of CA. But through CA, we cannot do matchmaking of web services. To solve this problem, this paper suggests an algorithm that translates CA to ADFSA. Therefore, if the web service is in form of CA, we can do matchmaking by applying the propose algorithm.

References

1. Wombacher, A., Fankhauser, P., Mahleko, B., Neuhold, E.: Matchmaking for business processes based on choreographies. In: 2004 IEEE International Conference on e-Technology, e-Commerce and e-Service, EEE'04, pp. 359–368. IEEE (2004)
2. Koehler, J., Srivastava, B.: Planning with communicating automata. Technical Report, RI08006 (2008)
3. Hopcroft, J.E.: Introduction to automata theory, languages, and computation. Pearson Education, India (1979)
4. Holzmann, G.J.: Software model checking with spin. Adv. Comput. **65**, 77–108 (2005)
5. Rinderle, S., Wombacher, A., Reichert, M.: Evolution of process choreographies in dychor. In: On the Move to Meaningful Internet Systems 2006: CoopIS, DOA, GADA, and ODBASE, pp. 273–290. Springer, Berlin (2006)
6. Wombacher, A., Fankhauser, P., Neuhold, E.: Transforming BPEL into annotated deterministic finite state automata for service discovery. In: Proceedings. IEEE International Conference on Web Services, 2004, pp. 316–323. IEEE (2004)
7. Grigori, D., Corrales, J.C., Bouzeghoub, M.: Behavioral matchmaking for service retrieval. In: International Conference on Web Services, ICWS'06, pp. 145–152. IEEE (2006)
8. Cong, Z., Fernández, A.: Behavioral matchmaking of semantic web services. In: Proceedings of the 4th International Joint Workshop on Service Matchmaking and Resource Retrieval in the Semantic Web (SMR2), vol. 667, pp. 131–140 (2010)
9. Corrales, J.C., Grigori, D., Bouzeghoub, M.: BPEL processes matchmaking for service discovery. In: On the Move to Meaningful Internet Systems 2006: CoopIS, DOA, GADA, and ODBASE, pp. 237–254. Springer, Berlin (2006)
10. Grigori, D., Corrales, J.C., Bouzeghoub, M.: Behavioral matchmaking for service retrieval: application to conversation protocols. Inf. Syst. **33**(7), 681–698 (2008)

11. Young, C., Yu, S., Le, J.: Research on partner-choosing and web services composition for B2B E-commerce in virtual enterprises. In: Advanced Web Technologies and Applications, pp. 814–823. Springer, Berlin (2004)
12. Wu, J., Wu, Z.: Similarity-based web service matchmaking. In: IEEE International Conference on Services Computing, vol. 1, pp. 287–294. IEEE (2005)
13. Peng, H., Niu, W., Huang, R.: Similarity based semantic web service match. In: Web Information Systems and Mining, pp. 252–260. Springer, Berlin (2009)
14. Ben-Ari, M.: Principles of the Spin Model Checker. Springer Science & Business Media, Berlin (2008)
15. Holzmann, G.J.: The model checker spin. IEEE Trans. Softw. Eng. **23**(5), 279–295 (1997)
16. Holzmann, G.J.: The SPIN Model Checker: Primer and Reference Manual. Addison-Wesley Reading, Reading (2004)
17. Muscholl, A.: Analysis of communicating automata. In: Language and Automata Theory and Applications, pp. 50–57. Springer, Berlin (2010)

Partial Satisfaction of User Requests in Context Aware Settings

Sujata Swain and Rajdeep Niyogi

Abstract Context-aware web service composition is a challenging research topic. In context aware application, a context changes over a period of time. In order to satisfy a user request, a set of services will be updated according to the context. Thus, services may be added and/or removed from the set of existing services. A context-aware application may not satisfy a user request because of unavailability of services. In this paper, we address these type of scenarios and suggest a method for web services composition.

Keywords Approximate solution · Context-aware · Partial solution · Web service composition

1 Introduction

A web service is a functional unit transforming input data into output data. Web services are obtained from web to satisfy the needs of the users. Sometimes a user's request is quiet complex. A single web service cannot satisfy a complex request. A composition of multiple web services are required to satisfy such requests. This process is called web service composition [1].

A pervasive computing environment has computing and communication capability, yet so gracefully integrated with users that it becomes a "technology that disappears". Context aware web service composition (CA-WSC) is needed to perform a user's task in a pervasive environment. CA-WSC is the process of composing a web service from other web services on the basis of actual and relevant context information. It is a challenge for web service composition to

S. Swain (✉) · R. Niyogi
Department of Computer Science and Engineering, Indian Institute of Technology Roorkee, Roorkee 247667, India
e-mail: sujataswain019@gmail.com

R. Niyogi
e-mail: rajdpfec@iitr.ernet.in

S.C. Satapathy and S. Das (eds.), *Proceedings of First International Conference on Information and Communication Technology for Intelligent Systems: Volume 1*, Smart Innovation, Systems and Technologies 50, DOI 10.1007/978-3-319-30933-0_59

perceive dynamic changes of external environment and adapt to the changes of business process rapidly.

Context includes information about the user and his/her environment. This is a nonfunctional attribute which is used to describe the user, location, software, internet connection and available devices. Context values change over a period of time. An application is context aware if it uses context value to provide relevant services to the user. This relevancy depends on the user's request [2]. The Context aware Application (CA) aims at automatic updation of a set of services by collecting the context values of user and environment. Thus, CA is more sensitive to the context.

Problem Statement:

Context aware applications are created for satisfying a user's goals. When a context changes, the set of services which are required for satisfying the request are also updated. Sometimes, the required services may or may not be available. Due to the unavailability of required services, context aware application may not satisfy a user's request. In this paper, we suggest methods to find the approximate and partial solution.

The rest of the paper is organized as follows. Section 2 describes literature survey. Section 3 provides the proposed approach. Section 4 gives the implementation and results. Finally, Sec. 5 provides conclusions and future work.

2 Literature Survey

Different aspects of context aware applications that have been mostly studied include, i.e., service discovery, searching and matching algorithm, composition methods, middleware building. We discuss some of these techniques used in CA-WSC.

Context values are collected from different heterogeneous systems. They have their own local clock and are not synchronized globally. To solve the synchronization problem, a logical clock is used in a heterogeneous system [3].

There are various middlewares proposed for services integration in pervasive environment. MySIM is a middleware [4], to provide spontaneous service integration in pervasive environments. PERSE is a semantic middleware [5], to provide service registration, discovery and composition. An extension of pi-calculus is used to formalize the dynamic web service composition [6].

Vukovic and Robinson [7] developed a framework, i.e., GoalMorpho, for providing partial satisfaction of user request when some services are not available. Hussein et al. [8] suggests a context aware adaptive service framework. There are very few who have focused on service unavailability. We proposed two methods for this.

3 Proposed Approach

Context aware applications are developed for satisfying a user's goal with minimal or no human intervention. This can only be done by having the current context information about it's user and his/her environment. The CA automatically runs the program, an according to context and it may require a set of web services to achieve the user goal. When context changes, the CA adapts to changes made in context. According to the current context, the CA fulfills user requirement. For this reason, the new services may be added to or existing services may be removed from the set of services. However, sometimes new services may not be available because of the server being down or due to some other failure.

Due to unavailable services, CA fails to fulfill the user request. To handle these types of problems, we provide different solutions in the current works.

3.1 *Case Study-1: Travel Agent Assistant*

The travel agent assistant is an application which takes care of the travel plan of a user. It takes user agenda as input, which contains the meeting details of the user. In this application, the context is user agenda. When a user has a new meeting schedule then, the application adds into the user agenda. The user agenda is updated from time to time and so, context continuously changes. According to new context value, the assistant tries to find the services which will satisfy the user agenda.

Consider an employee who needs to attend an urgent meeting in Chennai. For this he has to travel by Air-India. Travelling by Air-India is necessary for reimbursement purpose. Travel agent assistant is shown in Fig. 1.

However there may be two cases:

1. Air India flight is not available for the travel location.
2. Ticket is not available for Air India flight.

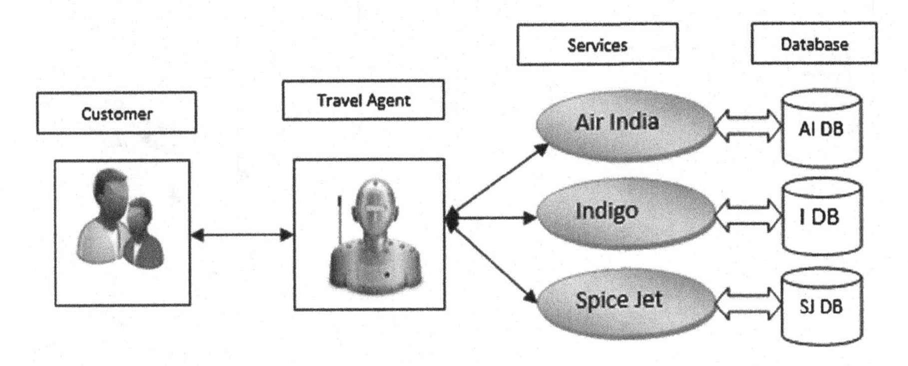

Fig. 1 Travel agent assistant

So in the above cases, the user request cannot be satisfied as it is. Most of the CA available in the literature will not fulfill the request of users due to unavailable services. In this case we try to satisfy this request with some alternatives such as by providing some other carriers (Spice Jet, Indigo).

3.2 Case Study-2: Medicine Monitoring System

The Medicine Monitoring System (MMS) is a context aware application which monitors medicines dosage for a user. Dosage is the size or frequency of a dose of a medicine to be taken.

In this application, context is user activity and time. The application selects medicine according to the context values. When context value changes, MMS checks the medicine time and user activity with time and MMS gives an alarm for taking the medicine based on output.

Suppose that a patient is being treated at home. She is taking medicines as prescribed by the doctor. There may be a situation where, she has to take medicine after dinner; but it is not available. At that time the near by medical stores may be closed and so she cannot purchase the medicine. To handle the situation, she seeks advice from MMS which retrieves information about available medicine from medicine box. MMS advises the patient to take a set of medicines which is functionally equivalent to the required medicine. In this way, the condition of the patient would be under control. Medicine monitoring system is shown in Fig. 2.

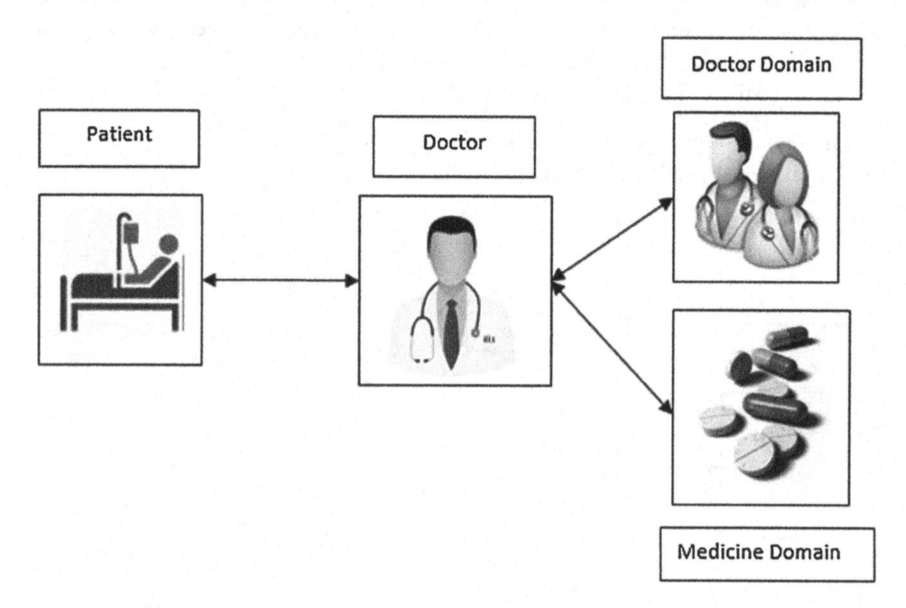

Fig. 2 Medicine monitoring system

3.3 Proposed Methods

Let there be a complex request which can be satisfied by a composition of services. Suppose that one or more web services are unavailable. Our aim is to satisfy those requests which can not be satisfied due to unavailability of services.

3.3.1 Approximate Solution

Request R1 is satisfied in context C1, but when context change to C2, R1 is not satisfied due to unavailability of some web services. In this solution, our main aim is to the provide unavailable web service any-how. We will find out the set of web services which is semantically equivalent to unavailable web service. Once we find it, then the request R1 can be satisfied without any failure. Semantic equivalence implies that the functionality of web services is same. In this solution, the safety property is maintained. Algorithm-1 shows the steps of approximation solution.

Algorithm-1: (Approximation Solution)

Input: Request (R), required services to compose (S).

Output: Possible or not Possible

Step-1: If composition is successful then provide the composition of required services to satisfy user request.

Step-2: If composition fails, then find a set (s) which contains unavailable services. ($s \subseteq S$)

Step-3: Find out a set (S') which contains all available services in registry.

Step-4: For each service (s') $\in s$ //find out replacement of s from S'.

Step-5: For each combination of services (s'') from S'.

Step-6: If (functionality (s') = = functionality (s'')) then replace s' in s with the services present in s''.

Step-7: If there is not any combination which will functionally equivalent to s', then the application cannot satisfy the request.

3.3.2 Partial Solution Request R1 is satisfied in context C1, but when the context change to C2, R1 is not satisfied due to unavailability of some web services. We now find alternate web services which will satisfy the main subgoals [7] of the request.

Algorithm-2: (Partial Solution)

Input: Request (R), required services to compose (S)

Output: Possible or not Possible

Step-1: If composition is successful, then provide the composition of required services to satisfy user request.

Step-2: If composition fails, then find the main subgoal of request.

Step-3: Find out a set (S') which contain all available services in registry.

Step-4: Find a set (S'') from S' which will satisfy the main subgoal.

Step-5: If S'' possible, then provide the composition of services in S'' to satisfy the base core goal of request.

Input of MMS Software:

<soapenv:Envelope xsi:schemaLocation= “http://schemas.xmlsoap.
org/soap/envelope/ http://schemas.xmlsoap.org/soap/envelope/”>
<soapenv:Body>
<<fs2:localServiceOperation>>
<RequiredMedicine> Paracetamol </RequiredMedicine>t
<AvailableMedicine> sinarest, Aspirin </AvailableMedicine>
</fs2:localServiceOperation> </soapenv:Body>
</soapenv:Envelope>

4 Implementation

We have implemented the above examples in OpenEsb toolkit [9]. To satisfy the user request, an application is developed. For this purpose, the application needs services which are to be composed and the final service will be sent to the user. Sometimes, the web service may be unavailable and composition get fails. First, we will find out the reason of service composition failure and then we will find out the solution for satisfying the goal.

Output of MMS Software:

<?xml version=“1.0” encoding=“UTF-8” standalone=“no”?>
<SOAP-ENV:Envelope xmlns:SOAP-ENV=“http://schemas.xmlsoap.
org/soap/envelope/”xmlns:xsi=“http://www.w3.org/2001/XMLSchema-
instance” xsi:schemaLocation=“http://schemas.xmlsoap.org/soap/
envelope/ http://schemas.xmlsoap.org/soap/envelope/”>
<SOAP-ENV:Body>
<m:localServiceOperationResponse xmlns:m=“FS2-CA” xmlns:msgns=
“http://j2ee.netbeans.org/wsdl/HS2-BPEL/src/newWSDL”>
<Result xmlns:msgns=“http://me/” xmlns:ns0=“http://j2ee.net
beans.org/wsdl/HS2-BPEL/src/newWSDL”>sinarest, Aspirin</Result>
</m:localServiceOperationResponse>
</SOAP-ENV:Body>
</SOAP-ENV:Envelope>

Implementation of Medicine Monitoring System:

Medicine monitoring system described in Sect. 3.2. Here context is current time and user activity, i.e., time is late night and activity is after food.

The system will take two inputs from the user namely, name of required medicine and the availability medicine in medicine box. First, we are trying to find out if required medicines functionality is equal to the functionality of any combination of the available medicine that can control the diseases for next few hours.

The experiment carried out for this purpose. We have provided two inputs, i.e., required medicine and available medicines. Paracetamol is considered as the required medicine and sinarest, asprin and namcold are considered as the medicines available in the medicine box. The system will find out combination of available medicine, which can replace the required medicine.

The output of MMS Software is shown in above. It shows paracetamol works same as the combination of sinarest and asprin. The system will return the result sinarest and asprin as output.

Implementation of Travel Agent Assistant (TAA):

In travel agent assistant scenario, the CA may not fulfill the user request due to unavailability of tickets in Air-India flights.

Output of Travel Agent Assistant:

<?xml version="1.0" encoding="UTF-8" standalone="no" ?>
<SOAP-ENV:Envelope xmlns:SOAP-ENV="http://schemas.xmlsoap.
org/soap/envelope/"xmlns:xsi="http://www.w3.org/2001/XMLSchema-
instance" xsi:schemaLocation="http://schemas.xmlsoap.org/soap/
envelope/ http://schemas.xmlsoap.org/soap/envelope/">
<SOAP-ENV:Body>
<m:localServiceOperationResponse xmlns:m="FS-CA" xmlns:msgns=
"http://j2ee.netbeans.org/wsdl/HS2-BPEL/src/newWSDL">
<result>Seats is available </result>
<flightno xmlns:msgns="http://is/"xmlns:ns0="http://j2ee.net
beans.org/wsdl/FS-BPEL/src/newWSDL">A101</flightno>
<flightnamexmlns:msgns="http://is/"xmlns:ns0="http://j2ee.net
beans.org/wsdl/FS-BPEL/src/newWSDL">AirIndia</flightname>
<cost xmlns:msgns="http://is/"xmlns:ns0="http://j2ee.net
beans.org/wsdl/FS-BPEL/src/newWSDL">12000</cost>
<available-seats xmlns:msgns="http://is/" xmlns:ns0="http://
j2ee.netbeans.org/wsdl/FS-BPEL/src/newWSDL">8</available-seats>
</m:localServiceOperationResponse>
</SOAP-ENV:Body>
</SOAP-ENV:Envelope>

So, in this case the composition get failed. This is the case of partial solution. Here the base core goal is need a ticket from Delhi to Chennai. Therefore, we will search other air-line or train services which will provide a ticket from Delhi to Chennai at the best price.

For this application, we are taking a small number of web services to illustrate Algorithm-2. In this application, the web services are Air-India, Indigo and Spice Jet. All the web services are connected to their own databases. OpenESB tool [9] has been used to develop the services for this application. These services are orchestrated in BPEL (Business Process Execution Language) process using OpenESB tool.

The experiment carried out for this purpose. We have provided inputs, i.e., source, destination and preference of Air Line. First TAA checks the Air India ticket from Delhi to Chennai. If the ticket is not present or the web service is unavailable then TAA will search for other Air Lines ticket from Delhi to Chennai at the best price.

The output of Travel Agent Assistant is shown in above. It shows flight number is I101, flight name Indigo, price 11,000 and available seats as 12.

5 Conclusion and Future Work

In this work, we have provided two solutions to the problems that occurred in the composition of services due to unavailability of services. The solutions are approximation solution and partial solution. The approximate solution will find out the other set of services which provide the functional closure of unavailable service. The partial solution will provide the service from the available web service to satisfy the base core goal. These two methods are semi-automatic in nature. In the future, we intend to develop a fully-automatic approach for handling the above problems.

References

1. Abowd, G.D., Dey, A.K., Brown, P.J., Davies, N., Smith, M., Steggles, P.: Towards a better understanding of context and context-awareness. In: Handheld and Ubiquitous Computing, pp. 304–307. Springer, Berlin (1999)
2. Satyanarayanan, M.: Pervasive computing: vision and challenges. Pers. Commun. **8**(4), 10–17 (2001)
3. Yang, Y., Huang, Y., Cao, J., Ma, X., Lu, J.: Formal specification and runtime detection of dynamic properties in asynchronous pervasive computing environments. IEEE Trans. Parallel Distrib. Syst. **24**(8), 1546–1555 (2013)
4. Ibrahim, N., Le Mouël, F., Frénot, S.: Mysim: a spontaneous service integration middleware for pervasive environments. In: Proceedings of the 2009 International Conference on Pervasive Services, pp. 1–10, ACM (2009)
5. Mokhtar, S.B.: Semantic middleware for service-oriented pervasive computing. Ph.D. dissertation, Université Pierre et Marie Curie-Paris VI (2007)
6. Swain, S., Niyogi, R.: Modeling and verification of dynamic web service composition. In: 7th International Conference on Data Mining and Warehousing (ICDMW 2013), Bangalore, India, August 9-11, pp. 1–8 (2013)
7. Vukovic, M., Robinson, P.: Context aware service composition. University of Cambridge, Technical Report UCAM-CL-TR-700 (2007)
8. Hussein, M., Han, J., Yu, J., Colman, A.: Scenario-based validation of requirements for context-aware adaptive services. In: 2013 IEEE 20th International Conference on Web Services (ICWS), pp. 348–355 (2013)
9. Openesb toolkit. http://www.open-esb.net/

Printed in the United States
By Bookmasters